High-T$_c$ Superconductors and Related Materials

Material Science, Fundamental Properties, and Some Future Electronic Applications

NATO Science Series

A Series presenting the results of activities sponsored by the NATO Science Committee. The Series is published by IOS Press and Kluwer Academic Publishers, in conjunction with the NATO Scientific Affairs Division.

A. Life Sciences	IOS Press
B. Physics	Kluwer Academic Publishers
C. Mathematical and Physical Sciences	Kluwer Academic Publishers
D. Behavioural and Social Sciences	Kluwer Academic Publishers
E. Applied Sciences	Kluwer Academic Publishers
F. Computer and Systems Sciences	IOS Press

1. Disarmament Technologies	Kluwer Academic Publishers
2. Environmental Security	Kluwer Academic Publishers
3. High Technology	Kluwer Academic Publishers
4. Science and Technology Policy	IOS Press
5. Computer Networking	IOS Press

NATO-PCO-DATA BASE

The NATO Science Series continues the series of books published formerly in the NATO ASI Series. An electronic index to the NATO ASI Series provides full bibliographical references (with keywords and/or abstracts) to more than 50000 contributions from international scientists published in all sections of the NATO ASI Series.
Access to the NATO-PCO-DATA BASE is possible via CD-ROM "NATO-PCO-DATA BASE" with user-friendly retrieval software in English, French and German (WTV GmbH and DATAWARE Technologies Inc. 1989).

The CD-ROM of the NATO ASI Series can be ordered from: PCO, Overijse, Belgium

Series 3. High Technology – Vol. 86

High-T$_c$ Superconductors and Related Materials

Material Science, Fundamental Properties, and Some Future Electronic Applications

edited by

S.-L. Drechsler

Institute for Solid State Research and Materials Science,
IFW Dresden e.V.,
Department of Solid State Theory,
Dresden, Germany

and

T. Mishonov

University of Sofia,
Department of Theoretical Physics, Bulgaria and
Katholieke Universiteit Leuven,
Laboratorium voor Vaste-Stoffysica en Magnetisme,
Leuven, Belgium

Kluwer Academic Publishers

Dordrecht / Boston / London

Published in cooperation with NATO Scientific Affairs Division

Proceedings of the NATO Advanced Study Institute on
High-T$_c$ Superconductors and Related Materials: Material Science, Fundamental
Properties, and Some Future Electronic Applications
Albena, Bulgaria
13–26 September 1998

A C.I.P. Catalogue record for this book is available from the Library of Congress.

ISBN 0-7923-6872-X (HB)
ISBN 0-7923-6873-8 (PB)

Published by Kluwer Academic Publishers,
P.O. Box 17, 3300 AA Dordrecht, The Netherlands.

Sold and distributed in North, Central and South America
by Kluwer Academic Publishers,
101 Philip Drive, Norwell, MA 02061, U.S.A.

In all other countries, sold and distributed
by Kluwer Academic Publishers,
P.O. Box 322, 3300 AH Dordrecht, The Netherlands.

Printed on acid-free paper

Printed in the Netherlands.

LIST OF CONTENTS

FOREWORD AND PREFACE

In these Proceedings selected problems of high-temperature superconductors and related compounds are reviewed, which range from fundamental questions of basic research till future electronic applications. At first
glance it seems to be an unrealistic aim for a single school since it covers
a quite broad field. On the other hand there is an increasing number of
fundamental properties of these compounds which are relevant to future
applications, opening new possibilities. Some of them, such as the high
transition temperature and the upper critical field, are apparently helpful,
whereas other properties, such as the short anisotropic coherence length
and the d-wave symmetry of the superconducting order parameter, which
affect the possible critical currents, yield strong restrictions for potential
applications.

It is noteworthy that the achieved high technological level in producing
thin films and nanostructures allows to perform sophisticated experiments
intended to study in detail such phenomena as subgap-plasmonic excitations, which have never been performed or even been considered theoretically. Last but not least several conceptions and ideas created in the field
of classical superconductivity, such as the spatial structure of the superconducting order parameter in the vicinity of interfaces and/or twinning
planes, have a significant impact on other condensed matter problems such
as wetting phenomena, and this impact is reciprocal.

In general, the following issues frequently considered during the whole
ASI might be pointed out: (i) Fermi liquid properties and determination of
the Fermi surface, (ii) different vortex states in superconductors and related
phenomena, (iii) fluctuations of the superconducting order parameter, (iv)
order parameter symmetry and anisotropy in relation to the underlying
pairing mechanism and general consequences for experimentally accessible
properties, (v) the influence of real structure effects such as grain boundaries, nonlocality effects and various kinds of local disorder. For more detailed information on the considered scientific topics the reader is referred
to the Content part.

In our opinion young people should be involved as much as possible in learning and eventually developing coherent views on these broad areas. In this context we consider the stimulating atmosphere of an ASI, having a long tradition in supporting science and in particular the physics of superconductivity and the materials science of superconductors, ideally appropriate to achieve this purpose. In the course of the ASI in addition to the main lectures, most of them presented in this book, several miniworkshops available upon request and devoted to plasmon excitations, borocarbides, problems of Ginzburg-Landau theory, and the Bernoulli effect in superconductors were organized for the young scientists.

We would like to note that several lecturers, working hard on their manuscripts, tried to update their lectures to provide the reader with comprehensive and tutorial introductory reviews of their topics. In our opinion those valuable contributions compensate the retardation we had in the editing the manuscript of the present Proceedings. We are indebted to KLUWER Publishing Company and especially to Annelies Kersbergen for their patience and support.

The nice Bulgarian Black Sea coast is a traditional scientific meeting point, in particular for scientists coming from Western and Eastern Europe. It is noteworthy to remember that organizing an ASI gives also some support to young scientists from Eastern European and other NATO cooperative partner countries, which is difficult to overestimate in the present complicated transition period for science in that region. Therefore, the large percentage of participants from those countries can be regarded as a success for Albena ASI. Furthermore we are glad to mention that several of our ASI-students I. Askerzade, A. Kordyuk, A. Mikitin, E. Ozkan, E. Penev, D. Nikolova, A. Posazhennikova, N. Zahariev, and others succeeded in the meantime to establish a collaboration with ASI lecturers.

The welfare of an ASI depends not only on the purely scientific aspects but also very much on the venue and its meeting facilities. Thus the Albena resort, which has good conference centers, accommodation, meeting facilities and sport centers, has been selected. The Cultural Center in Albena where the ASI took place, from 13 to 26 September 1998, with its neighbouring hotels, offered excellent conditions for the participants. Situated only 3 min. from the Sea coast, we succeeded to minimize not only the costs (in the after-season) but also any perturbation from tourist activities. Thus not only during the lecture sessions but also during the free time intensive scientific discussions could take place among the participants. Especially for the young scientists the discussions with the lecturers were a unique experience and gave them together with the oral presentations and posters (some of them included in the present book too) the opportunity to present and critically discuss their projects and create new common proposals.

This ASI was made possible due to the generous support by the NATO Scientific and Environmental Affairs Division, Brussels, Belgium. The directors wish also to express their gratitude for additional support by the Institute of Solid State and Materials Research Dresden (IFW-Dresden) and the University of Sofia. We are grateful to the conference secretary, Valya Mishonova, for her exquisite organization and continuous kind attention towards the participants of the NATO-ASI and for preparing and producing the booklet. We are deeply indebted to Joseph Indekeu for his indispensable help in the finishing of these proceedings enhancing the cooperation between us and the last delayed lecturers. Andrea Runow and Peter Joehnk (IFW Dresden) are acknowledged for their assistance in coordinating invaluable financial and technical support for the ASI.

It is a pleasure to thank K. Christova, E. Kutsilova, G. Toteva, R. Aleksieva, R. Mitanova, D. Andonova, M. Daneva, and K. Rachev from the Albena A.D. for their kind attention and well-organized collaboration. We are thankful to V. and V. Penchevi from Academic Travel for their help at the initial stage of organizing the school. Finally, we are very much indebted to Evgeni Penev for preparing the web-page with the advertisement of our ASI, and for the help with other computer-related issues.

S.-L. Drechsler T. Mishonov

Co-Directors of the ASI

THE ELECTRONIC STRUCTURE OF HIGH-T_c SUPERCONDUCTORS: INTRODUCTION AND RECENT STUDIES OF MODEL COMPOUNDS

J. FINK, M. S. GOLDEN, M. KNUPFER, TH. BÖSKE, S. HAFFNER, R. NEUDERT, S. ATZKERN, C. DÜRR, Z. HU, S. LEGNER, T. PICHLER, H. ROSNER, S.-L. DRECHSLER, R. HAYN, J. MÁLEK, H. ESCHRIG, K. RUCK, G. KRABBES

Institut für Festkörper- und Werkstofforschung Dresden, Postfach 270016, D-01171, Dresden, Germany

Abstract

An introduction into the electronic structure of high-T_c superconductors is given. This introduction is illustrated by recent studies of model compounds containing Cu-O planes, ladders and chains, performed using high-energy spectroscopies such as photoemission, x-ray absorption and electron energy-loss spectroscopy.

1. Introduction

High-T_c superconductivity was discovered 1986 by Bednorz and Müller [1] in a complex ceramic containing two-dimensional CuO_2 planes. Thirteen years later many new high-T_c cuprate superconductors (HTSC) have been synthesized with critical temperatures up to 133 K at ambient pressure. The HTSCs were a real surprise after a long, unsuccessful search for higher T_c's in conventional superconductors. In these materials the transition temperature, T_c, had only increased by a few degrees in decades. As a result of this, and as their high T_c's could lead to many applications of the cuprate superconductors, an impressive amount of work has been devoted to HTSCs. We now know many details of HTSCs. Some ingredients of the Bardeen-Cooper-Schrieffer (BCS) theory are known to be applicable. Many properties of HTSCs can be explained by phenomenological models but at present there is no generally accepted microscopic theory for superconductivity in the cuprates.

One reason for this is that even the normal state properties of these materials are not understood. For example, the resistivity of optimally doped HTSC material (i.e. that with highest T_c) shows a linear temperature dependence,

1

S.-L. Drechsler and T. Mishonov (eds.), High-T$_C$ Superconductors and Related Materials, 1–38.
© *2001 Kluwer Academic Publishers. Printed in the Netherlands.*

extending in some systems from 10 to 1000 K and extrapolating to zero resistance at zero degrees. This singular behaviour is to be contrasted with the behaviour of conventional metals where the resistivity is linear above some fraction of the Debye temperature, ($\Theta_D \sim 200$ K for cuprates), cuts the temperature axis at a finite temperature ($\sim \Theta_D$) and saturates at high temperatures. There are also other anomalous normal state transport properties of these materials, e.g. the temperature dependence of the Hall effect or the logarithmic divergence of the conductivity in the normal state at low temperatures (achieved by high magnetic fields) in underdoped systems. For all these strange normal state properties there is also no generally accepted microscopic model. As all of these normal state transport properties are closely related the low-energy electronic structure the latter represents a key question in the understanding of cuprates.

The HTSCs all have one structural element in common, namely the CuO_2 planes. They determine the electronic structure close to the Fermi level. Band structure calculations in the local density approximation (LDA) fail in describing the low-energy electronic structure of CuO_2 planes since in this approximation correlation and exchange effects are treated as in a free-electron metal. As in other transition metal compounds, however, the Coulomb repulsion between holes on the transition metal sites, i.e. correlation effects, play an important role in these compounds. The understanding of the electronic structure is further complicated because correlation effects and hybridization between the metal ions and the ligand are of the same order.

In this situation all kinds of experiments giving information on the microscopic electronic structure are valuable. In this article we will review more recent studies of the electronic structure of model compounds of HTSCs by high-energy spectroscopies such as photoemission (PES), electron energy-loss spectroscopy (EELS), and x-ray absorption spectroscopy (XAS). Not only compounds containing CuO_2 planes are treated but also low-dimensional Cu-O ladder and chain compounds in which correlation effects are even more important will be discussed. We start this article with an introduction into the electronic structure of cuprates. Although many review articles on the electronic structure of these systems exist [2-5], this is reasonable since many basic elements are still not widely disseminated and there are still many discussions on fundamental points not only among young scientists entering the field but also among experts working in the field since many years.

2. The electronic structure of cuprates - an introduction

One of the prototypes of the HTSCs is La_2CuO_4 in which the trivalent La ions are partially replaced by divalent Ba or Sr ions. This compound was the first

HTSC discovered. By substitution of La^{3+} by Sr^{2+}, La_2CuO_4 is p-type doped. A schematic phase diagram of $La_{2-x}Sr_xCuO_4$ is shown in Fig. 1. The undoped system is an insulating antiferromagnet.

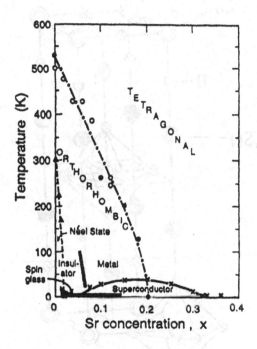

Figure 1. Schematic phase diagram of $La_{2-x}Sr_xCuO_4$ [6]

As the system is doped with holes the antiferromagnetic order is supressed and the system eventually becomes metallic and superconducting. For $x \sim 0.15$, T_c reaches its optimum. For this doping concentration (known generally as optimal doping) one observes the strange normal state properties mentioned in the Introduction. For higher x values, T_c decreases and more conventional normal state transport properties are observed. A similar but inequivalent phase diagram is also observed for the n-type cuprate superconductor $Nd_{2-x}Ce_xCuO_4$ in which trivalent Nd ions are replaced by tetravalent Ce ions.

The structure of La_2CuO_4 is shown in Fig. 2. In this compound there are CuO_2 planes separated by two rock-salt-like LaO planes. The doped La(Sr)O planes form the block layers which capture electrons from the CuO_2 planes. In an ionic picture, the positive La ions are trivalent, the negative O ions are divalent and in order to conserve charge neutrality of the unit cell, the Cu ions must be in a Cu^{2+} state. This is achieved by its losing one 4s electron and one 3d electron. This creates a hole in the 3d shell and thus Cu^{2+} has a net spin of ½ in the crystal. These spins interact with each other via other ions and form the antiferromagnetic order in the undoped case.

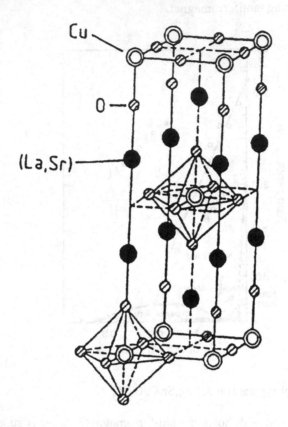

Figure 2. Crystal structure of La_2CuO_4

In a slightly different view of the structure of La_2CuO_4, the Cu ions are surrounded by oxygens making octahedra (also shown in Fig. 2). This perovskite structure is often emphasized by crystallographers and would indicate that the so-called apical O ions (situated above and below Cu ions) are as important as the O ions in the CuO_2 planes. However, the octahedron is strongly distorted. The Cu-O distance in plane is ~ 1.9 Å while the distance between Cu ions and the apical O is about 2.4 Å. Therefore, the in-plane Cu-O bond is much stronger than the out-of-plane Cu-O bond and the importance of the apical oxygens is somewhat questionable. As already mentioned, in an ionic picture the Cu^{2+} ions are in a $3d^9$ configuration, i.e. there is one hole per Cu site, while the O ions have a $2p^6$ configuration, i.e. the O2p shell is completely filled. This is illustrated on the left side and the right side of Fig. 3.

A regular octahedron of O^{2-} ions produces a crystal field which splits the Cu3d levels into e_g and t_{2g} states. The crystal field of the distorted O octahedron further splits the e_g levels into $3d_{x^2-y^2}$ and $3d_{3z^2-r^2}$ states. A similar splitting occurs with the t_{2g} levels. Also the O2p states are split into σ levels, the orbitals of which point to the Cu sites and into π levels, the orbitals of which are perpendicular to the Cu-O bonds. Hybridization of the Cu3d states with the O2p states leads to the formation of bands, illustrated in the middle of Fig. 3. The hybridization is strongest for the σ bonds between in-plane $Cu3d_{x^2-y^2}$ and in-plane $O2p_{x,y}$ orbitals. This leads to bonding σ and antibonding σ^* $Cu3d_{x^2-y^2}$-$O2p_{x,y}$ bands.

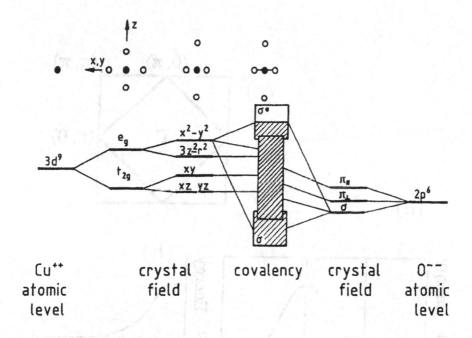

Figure 3. The formation of the electronic structure of the CuO_6 octahedra in La_2CuO_4. Left and right part: atomic levels of Cu^{2+} and O^{2-} ions. By introduction of the crystal fields due to a regular and a distorted octahedron of O ions, the 3d levels are split. Upon hybridization of Cu3d and O2p orbitals, bands are formed. The antibonding dpσ^* band is half-filled.

In between these bands there are other bands formed from hybrids which have a smaller splitting between bonding and antibonding states. Filling all these bands with the appropriate number of electrons leads to a half-filled antibonding σ^* band.

6

In this model where the energy of the electrons are assumed to be independent from each other, La_2CuO_4 should be a metal - in contradiction with the observation of an antiferromagnetic insulator with a gap of ~ 2 eV. The states at the Fermi level should have $Cu3d_{x^2-y^2}$ and $O2p_{x,y}$ character. If we want to describe the low-energy properties of La_2CuO_4, we can therefore possibly neglect 19 orbitals from the 22 orbitals which play a role in the chemical bonding (5Cu3d, 1Cu4s, 3O2p, 7La4f, 5La5d, and 1La6s) and we are left with 3 orbitals, namely the $Cu3d_{x^2-y^2}$, $O2p_x$, and $O2p_y$ orbitals. The antibonding σ^* band formed from $Cu3d_{x^2-y^2}$ and $O2p_{x,y}$ orbitals is schematically shown in Fig. 4a.

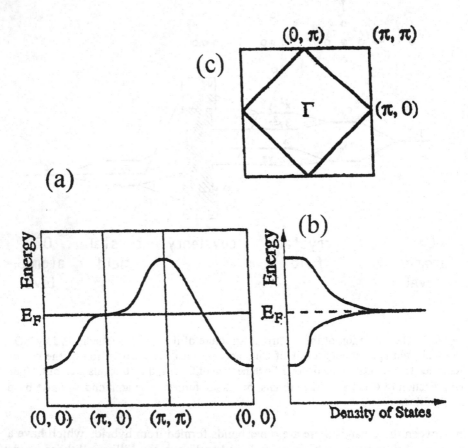

Figure 4. (a) Band structure of the antibonding $Cu3d_{x^2-y^2}$-$O2p_{x,y}$ band; (b) density of states of this band; (c) Fermi surface of this half-filled band.

In a simple single band tight-binding scheme one obtains a band structure which describes the antibonding σ^* band:

$$E(k) = E_0 - 2t (\cos k_x a + \cos k_y a).$$

E_0 is the energy corresponding to the centre of the σ^* band and t is the effective transfer integral. This is related to the hopping integral between the Cu and the O sites, t_{pd}, by the relation $t \sim t^2_{pd}/\Delta$, whereby Δ is the energy difference between the O and Cu levels. The density of states (DOS) of this band is shown in Fig. 4b. The DOS has a singularity at E_0 which is called a van-Hove-singularity. The Fermi level for the half-filled band is right in the van-Hove-singularity. Several theoretical models for high-Tc superconductivity in cuprates are related to this van-Hove singularity. The Fermi surface of this half-filled band is a square as shown in Fig. 4c. The corners of the square are at the saddle points of the 2D band structure at $(0,\pi)$ which is an abbreviation for $(0,\pi/a)$ where a is the in-plane lattice constant. E(k) has a minimum at the Γ point and maxima at (π,π) $(\pi,-\pi)$ $(-\pi,\pi)$, and $(-\pi,-\pi)$. A similar half-filled antibonding σ^* band is also derived from LDA band structure calculations [7]. This means that these calculations also predict a metallic state for La_2CuO_4. In LDA band structure calculations, correlation and exchange effects are taken into account within the approximation of a homogeneous electron liquid. This is a good model for simple metals like Na or Al, in which the conduction electrons are nearly-free and are therefore homogeneously distributed over the lattice. On the other hand, in transition metal (TM) compounds such as La_2CuO_4 or NiO, the TM 3d electrons are strongly localized at the TM sites and hop from time to time from one site to the next. In this case, the Coulomb interaction, U, between two holes (or two electrons) on the same site will be important and may be the reason for the insulating behaviour of La_2CuO_4. If the Coulomb interaction between two holes on a Cu site is much larger than the bandwidth, and knowing that each Cu site has one hole, it is evident that it is difficult to move holes from one site to the next site and therefore an insulating character is expected.

The situation is illustrated in Fig. 5 by a chain of hydrogen atoms, each having one electron per atomic level. When moving one electron to the neighbouring site one has to overcome the Coulomb repulsion, U, between two electrons. In the localized limit, where the width of the levels is much smaller than U, an insulating behaviour results. When removing an electron (e.g. by photoemission) it costs no additional energy while adding an electron (e.g. by inverse photoemission) costs the Coulomb repulsion, U, between two electrons. This process can be performed on all H atoms. Therefore a so-called upper Hubbard band (UHB) is formed. The sum of the half-filled levels is called the lower Hubbard band (LHB).

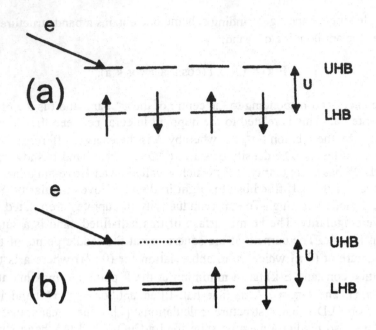

Figure 5. Energy levels of a chain of hydrogen atoms. (a) half filled; (b) p-type doped

Replacing the electrons and the H atoms by holes and Cu sites, respectively, one would arrive at the level scheme shown in Fig. 6a, and could explain the insulating character of La_2CuO_4. However, these considerations would result in a large energy gap of about 10 eV for cuprates, as PES measurements [8], Auger data [9] and constraint LDA band structure calculations [3,10,11] predict values for U of about 10 eV. On the other hand the optical gap for La_2CuO_4 is only about 2 eV. Already before the discovery of the HTSCs it was realized that the gap in the late transition metal compounds is too small to be explained within a normal Mott-Hubbard picture [12]. The smaller gap can be explained by a O2p valence band between the UHB and the LHB, which is separated from the UHB by the charge-transfer energy Δ, which is of the order of 4 eV. Then the gap is determined by Δ and the width of the O2p and Cu3d bands. This is the model for the so-called charge-transfer insulator - a first approximation for undoped late transition metal compounds. The level scheme for this model is shown in Fig. 6b. In this model the result of p-type doping by the $(La,Sr)_2O_2$ block layers would be that the Fermi level would move into the O2p band. The charge carriers would then have exclusively O2p character and correlation effects would not be important.

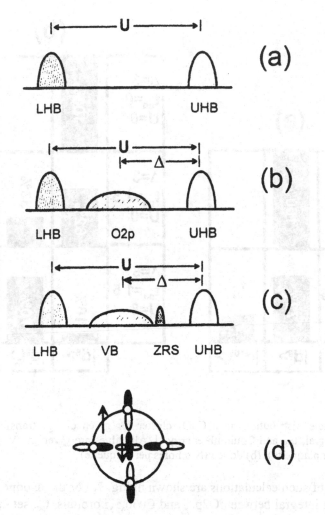

Figure 6. Level schemes for models under discussion for the electronic structure of CuO₂ planes. (a) Mott-Hubbard model; (b) charge-transfer insulator without hybridization; (c) charge-transfer insulator with hybridization; (d) Zhang-Rice singlet.

The situation changes when one takes into account the strong hybridization between $O2p_{x,y}$ and $Cu3d_{x^2-y^2}$ states. This hybridization leads to a strong mixing of the O2p and Cu3d states. This is best illustrated in simple calculations for a square planar CuO_4 cluster, which is the basic unit of the CuO₂ planes.

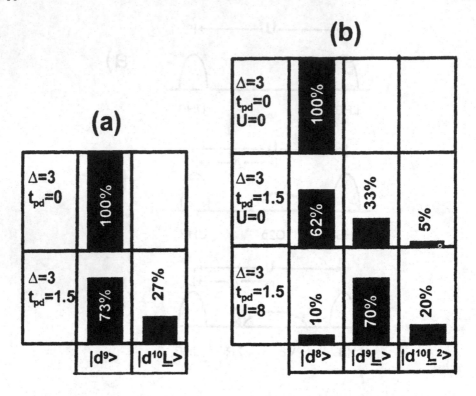

Figure 7. Hole distribution on a CuO₄ cluster for given charge transfer energy, Δ, hopping integral, t_{pd}, and Coulomb energy U. All values are given in eV. (a) Undoped (one hole per plaquette); (b) doped (two holes per plaquette)

The results of such calculations are shown in Fig. 7. For the undoped case, with the hopping integral between O2$p_{x,y}$ and Cu3$d_{x^2-y^2}$ orbitals, t_{pd}, set equal to zero (no hybridization), and a finite Δ there would be one hole on the Cu site and no hole density on the O site. The wave function would thus be $|\psi\rangle = \alpha|d^9\rangle + \beta|d^{10}\underline{L}\rangle$ with $\alpha = 1$ and $\beta = 0$, where $\underline{L}$ denotes a hole on the ligands, i.e. on the O sites. Switching on the hybridization and taking Δ = 3 eV and t_{pd} = 1.5 eV (typical values for these parameters) we arrive at $\alpha^2 = 0.73$ and $\beta^2 = 0.27$, which means that the hole density on the Cu sites is reduced and there will be hole density on the O sites. The situation is more complicated for the doped case (1 additional hole per CuO₄ cluster). The wave function is then given by $|\psi\rangle = \alpha|d^8\rangle + \beta|d^9\underline{L}\rangle + \gamma|d^{10}\underline{L}^2\rangle$. For a finite Δ and $t_{pd} = U = 0$, only α would be different from zero and therefore the wave function would be $|\psi\rangle = |d^8\rangle$. For Δ = 3 eV and t_{pd} = 1.5 eV, most of the holes are still on the Cu sites but due to hybridization a large percentage is also on the O sites. The probability that two holes are on the O sites is small (see Fig. 7b). Finally, when switching on correlation effects by using a finite U = 8 eV, the probability of two holes on Cu sites is strongly reduced. The probability for an equal distribution of the two holes on the Cu site and on the O sites as well as the probability for two holes on O sites has increased. This clearly indicates that the hole distribution in the doped cluster is strongly dependent on the hybridization (t_{pd}) and the Coulomb repulsion between two holes on the Cu site (U).

The strong hybridization also changes the level scheme for the charge transfer insulator as shown in Fig. 6. The UHB no longer has pure Cu3d character but has a considerable O2p admixture. As shown above there is a high probability that in a doped system there is one hole on the Cu site (as in the undoped ionic model) and one on the O sites. Both holes have a spin and there is a strong exchange interaction leading to singlets and triplets. Calculations [13] predict that those states having the lowest ionization energy are the singlets which often are called Zhang-Rice singlet states [14]. When doing PES experiments, one hole is created in the photoemission process, i.e. p-type doping with extremely low doping concentration is achieved, and those states having the lowest binding energy should be these Zhang-Rice singlet states (see Fig. 6d). Moreover, for p-type doping by chemistry, i.e. via the block layers, at low doping concentration these states will also have Zhang-Rice singlet character. The level scheme for the charge transfer insulator with hybridization is shown in Fig. 6c. On the low-energy scale the model then transforms to an effective Mott-Hubbard model with an effective U of the order of Δ.

A further simplification can be introduced as is done in the t-J model [14] in which a double occupancy of the Cu sites with two holes is forbidden (U = ∞). In this effective Mott-Hubbard model, with increasing dopant concentration, more and more states of the LHB become unoccupied (see Fig. 5b) and therefore electrons can move and a normal metallic state may form. This is actually seen in the overdoped cuprates where typical properties of normal metals are observed. With increasing dopant concentration, at those sites where doped holes reside, there is no Coulomb repulsion for adding an electron. Therefore, at these sites, the UHB loses spectral weight by an amount proportional to the dopant concentration x and the LHB gains spectral weight in electron addition spectroscopies (inverse PES or XAS) proportional to 2x. It is important to note that this so-called spectral weight transfer from the high-energy scale to the low-energy scale only occurs in doped correlated systems but not in doped semiconductors [15]. Upon p-type doping of cuprates a similar spectral-weight-transfer from the UHB to the ZRS-band should be observed in electron addition spectroscopies, provided an effective Mott-Hubbard model is applicable.

3. High-energy spectroscopies

High-energy spectroscopies form an important tool for the investigation of the electronic structure of HTSCs.

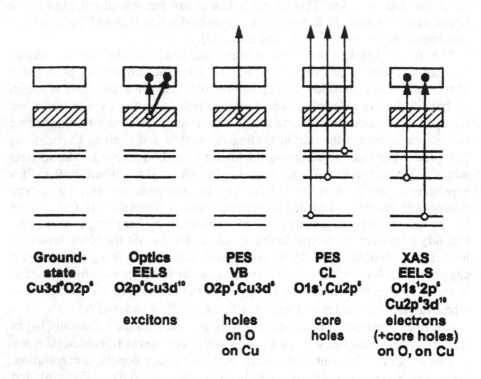

Ground-state	Optics	PES	PES	XAS
$Cu3d^9O2p^6$	EELS	VB	CL	EELS
	$O2p^5Cu3d^{10}$	$O2p^5, Cu3d^8$	$O1s^1, Cu2p^5$	$O1s^12p^6$
				$Cu2p^53d^{10}$
	excitons	holes	core	electrons
		on O	holes	(+core holes)
		on Cu		on O, on Cu

Figure 8. High-energy spectroscopies with excitations of a CuO_2 plane. Ground state and excited final states are given in an ionic approximation.

In Fig. 8 we show typical excitations of CuO_2 planes, which are performed in high-energy spectroscopies. In the ionic limit the ground state is $Cu3d^9O2p^6$. Excitations from the valence to the conduction bands can be performed by optics and EELS. The final state is then an $O2p^5Cu3d^{10}$ configuration. Using EELS, momentum dependent excitations can also be performed, yielding information on the dynamics of electron-hole pairs. Using PES, electrons can be removed from O or Cu sites. Information on final states reached by electron removal can be obtained.

Using angular resolved PES (ARPES), information on the dynamics of holes can be derived. In normal metals ARPES yields information on the dispersion of quasi-particles, i.e. one can perform band mapping. In correlated systems much of the spectral weight measured by ARPES may not stem from coherent quasi-particles states but may form satellites - so-called incoherent spetral weight [16]. PES from core levels results in core holes at O or Cu sites. These holes may be screened by charge-transfer excitations between the valence and the conduction band. Such measurements yield information on the response of the electronic system of a CuO_2 plane to a positive charge, e.g. on the Cu sites.

Finally, excitations from core levels to the unoccupied states yield information on the density of unoccupied states, as long as the interaction with the core hole is not too strong. Since the wave functions of a core level are quite localized, finite transition probabilities can only be expected for conduction band wave functions which overlap with the core wave function. So starting from an O1s core level and taking into account dipole selection rules, only O2p states can be reached. On the other hand, starting from the Cu2p levels, essentially Cu3d states can be reached. Thus, the important unoccupied states of CuO_2 planes close to the Fermi level can be studied site-selectively using such methods (EELS and XAS). Moreover, performing momentum dependent EELS or polarization dependent XAS measurements, information on the symmetry of the unoccupied states could be obtained [17]. In correlated systems these methods yield information on final states reached by the addition of electrons to CuO_2 planes (when neglecting the interaction of these electrons with the core holes).

When discussing these methods one should emphasize that PES measurements are extremely surface sensitive. Since the discovery of HTSCs the preparation of surfaces which represent bulk properties has presented a real challenge. On the other hand, EELS in transmission and XAS in the fluorescence mode are not surface sensitive and reliable results can be obtained more easily.

Before going on one should mention that these high-energy spectroscopies themselves have undergone an enormous technical development since the discovery and by the discovery of the HTSCs. For example, at present, band mapping and mapping of Fermi surfaces can be performed with energy resolutions as high as 5 meV while in the 1980's typical energy resolutions where about 100 meV. This development and the large superconducting gaps in HTSCs led to the first observation of such superconducting energy gaps by PES.

4. Studies of model compounds

4.1 MODEL COMPOUNDS

Parallel to the study of the highly complex HTSCs themselves, much can be learned regarding the electronic structure of the doped cuprates by high-energy spectroscopic investigation of undoped model compounds. As shown in Section 2 the important units in HTSCs are the CuO_2 planes as depicted in Fig. 9e. Completely undoped CuO_2 planes are realized e.g. in $Sr_2CuO_2Cl_2$ which has the same structure as La_2CuO_4. The LaO block layers are replaced in this case by SrCl layers. Compared to La_2CuO_4, $Sr_2CuO_2Cl_2$ has the great advantage that good surfaces for PES measurements can be prepared easily by cleaving. In $Sr_2CuO_2Cl_2$ there is also no problem of a small dopant concentration, encountered in $La_2CuO_{4+\delta}$ due to excess oxygen. Therefore, $Sr_2CuO_2Cl_2$ has become the ideal model compound for the undoped parent compounds of HTSCs.

Using the CuO_4 building blocks many other cuprates can be formed [18] (see Fig. 9), many of which, with respect to the electronic structure, show low-dimensional behaviour. The edge-sharing plaquettes in Li_2CuO_2 form a linear chain (see Fig. 9a). This chain, however, is not a quasi-one-dimensional electronic system since there is only a weak coupling between neighbouring CuO_4 plaquettes due to a 90° Cu-O-Cu interaction. Therefore, Li_2CuO_2 is a model compound for isolated CuO_2 plaquettes, i.e. a quasi zero-dimensional system. On the other hand, the corner-sharing plaquettes in Sr_2CuO_3 form a quasi-one-dimensional (1D) system because there is a strong 180° Cu-O-Cu interaction along the chain backbone as in the CuO_2 planes of the HTSCs. This interaction, in conjunction with interplane or interchain exchange, is responsible for the antiferromagnetic ordering observed in the undoped cuprates.

Connecting two corner-sharing chains one obtaines the zig-zag chains of $SrCuO_2$ (see Fig. 9c). Connecting zig-zag chains, one obtains, for instance, the 2-leg ladder systems (see Fig. 9d) of $Sr_{14}Cu_{24}O_{41}$, which in addition contains edge-sharing chains analogous to those in Li_2CuO_2. These ladder systems represent Cu-O compounds in which the dimensionality can be varied between one and two. Indeed, superconductivity was predicted [19] and experimentally a $T_c=12$ K has been observed in $(Sr,Ca)_{14}Cu_{24}O_{41}$ under high pressure (3 GPa) [20].

Finally, we mention the compound $Ba_2Cu_3O_4Cl_2$ which contains Cu_3O_4 planes (see Fig. 9f) which are composed of two sub-systems: a Cu_AO_2 plane analogous to that of the HTSCs and an additional Cu_B site situated in every second open plaquette of the Cu_AO_2 plane. The two subsystems can be envisaged as a pair of interpenetrating 2D Heisenberg antiferromagnetic systems, with quite different Néel temperatures: T_N^A=330 K, T_N^B=31 K. The coupling between the systems A and B is weak and mainly due to quantum spin fluctuations since it is based upon a 90° Cu_A-O-Cu_B coupling and since the coupling of Cu_A to Cu_B is frustrated.

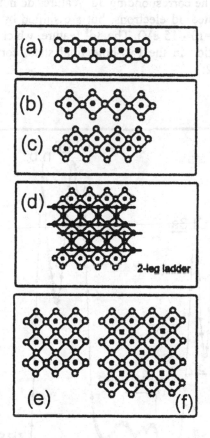

Figure 9. Schematic representation of cuprate networks. (a) The chain of edge-sharing plaquettes of Li_2CuO_2; (b) the chain of corner-sharing plaquettes of Sr_2CuO_3; (c) the zig-zag chain of $SrCuO_2$; (d) the Cu-O 2-leg ladder of $Sr_{12}Cu_{24}O_{41}$ (e) the CuO_2 plane of the HTSCs; (f) the Cu_3O_4 plane of $Ba_2Cu_3O_4Cl_2$

4.2 PHOTOEMISSION OF THE VALENCE BAND OF $Sr_2CuO_2Cl_2$

In Fig. 10 we show an XPS (photon energy hv = 1487 eV) valence-band spectrum of $Sr_2CuO_2Cl_2$ [21]. In the photoemission process, electrons can be

16

removed from different sites and different orbitals. In the ground state, the Cu atoms are mainly in the $3d^9$ configuration (one hole in the 3d shell). Removing one electron in the photoemission process from the Cu3d orbitals, the final state (whose energy is the quantity actually measured) is in a Cu$3d^8$ configuration. The Coulomb interaction, U ~ 10 eV, between the two holes in the 3d shell leads to the fact that the corresponding $3d^8$ features do not appear at the binding energy of two separated 3d electrons, but are shifted by U ~ 10 eV to higher binding energies (E_B=10 - 15 eV). The $3d^8$ feature, which is further split due to the multiplet interaction in the two-hole final state, corresponds to the LHB discussed in Section 2.

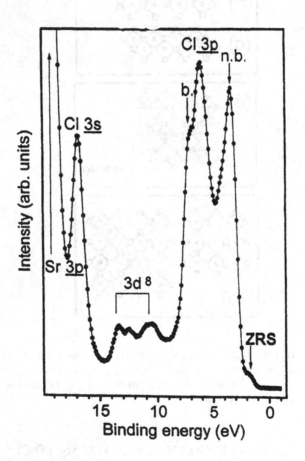

Figure 10. Valence-band photoelectron spectrum of $Sr_2CuO_2Cl_2$

Due to hybridization in the ground state some of the Cu atoms are also in the $3d^{10}$ configuration. In this case, correlation effects are not important (the final state is $3d^9$) and one measures the normal density of states in the energy range 3-8 eV, together with Cl states and O2p states. Between 3 and 5 eV the non-bonding states show a pronounced peak. According to the discussion in Section 2, the small shoulder at ~ 2 eV should be assigned to the Zhang-Rice singlet states, in which the intrinsic hole on the Cu site interacts with the hole (on the O sites) formed in the photoemission process.

Due to the relevance to high temperature superconductivity, the dispersion relation of the Zhang-Rice singlet states - which describes the dynamics of a single hole in an undoped antiferromagnetic CuO_2 plane - has attracted both a lot of experimental and theoretical attention. Here we present ARPES data for $Sr_2CuO_2Cl_2$, recorded along the main directions in the Brillouin zone for different orientations of the polarization vector of the synchrotron radiation with respect to the emission plane [22]. The measured photocurrent is determined by

$$I \sim \sum_{i,f} |M|^2 S(k,E) f(E)$$

where $S(k,E)$ is the spectral function, $f(E)$ is the Fermi function, and M is the matrix element, which in the dipole approximation is given by

$$M \sim <f|Ar|i>,$$

whereby A is the vector potential of the photon field. Putting the detector in the emission plane means $<f|$ must be even relative to this plane in order to obtain a finite intensity in the detector. For A perpendicular to the emission plane Ar is odd and in order to get a finite matrix element, $|i>$ must be odd. In the same way, for A parallel to the emission plane, Ar is even and therefore $|i>$ must be also even. Thus by performing such polarization dependent measurements information on the symmetry of the initial state can be obtained.

In Fig. 11 we show ARPES data for $Sr_2CuO_2Cl_2$ recorded along the Γ-(π,π) direction for two different orientations of the polarization vector of the synchrotron radiation with respect to the emission plane. In the left panel the polarization vector is perpendicular to the emission plane, and a clear dispersive feature is seen having minimal binding energy (highest kinetic energy) and maximal intensity at $(\pi/2,\pi/2)$. Since in this geometry odd initial states are sampled and since the Zhang-Rice singlet states are odd with respect to the chosen emission plane, the results support the assignment of these states to Zhang-Rice singlet states. On the other hand, for the polarization parallel to the emission plane, where even states are sampled (Fig. 11, right panel), almost no intensity is observed. This is compatible both with the symmetry of the Zhang-

18

Rice singlet states and the picture in which these states are the first electron removal states.

Also the intensities for measurements along the Γ-$(\pi,0)$ direction (not shown) are in agreament with the expectations based upon the symmetry of the Zhang-Rice singlet states.

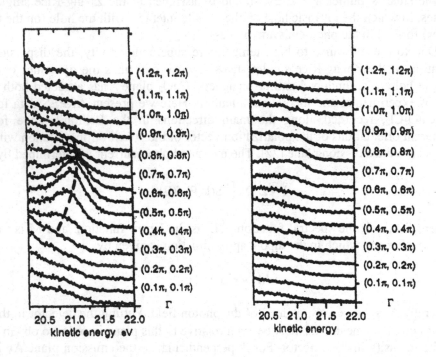

Figure 11. ARPES spectra of $Sr_2CuO_2Cl_2$ recorded along the Γ-(π,π) direction. Left panel: Polarization of the photons perpendicular to the emission plane. Right panel: polarization of the photons parallel to the emission plane.

In Fig. 12 we plot the dispersion relation of the Zhang-Rice singlet states in the two directions in **k** space studied in these experiments, together with analogous data from other groups [23, 24]. Comparing these results with the band structure plotted in Fig. 4, one could speculate that in the undoped case a gap has opened close to the Fermi level at $(\pi,0)$ and $(\pi/2,\pi/2)$. On the other hand, one realizes that the total bandwidth is only ~ 350 meV which is much smaller than that estimated from band structure calculations. This can be explained in the following way: when moving a hole through the antiferromagnetic CuO_2 plane one has to reverse the direction of Cu spins. The further the hole has been moved the more misaligned spins are created. This means that in a Néel state, the holes are localized. Only by spin fluctuations, determined by the exchange integral J, the frustrated spins can flip back to an antiferromagnetic configuration. Therefore the dynamics of the holes are not determined by the hopping integral t but by the exchange integral J, which is of the order of 120 meV. This is in agreement with theoretical studies using a t-J model [25-26]. However, the experimental dispersion does not agree with calculations within the simple t-J model (see Fig. 12). There the energy at $(\pi/2,\pi/2)$ and $(\pi,0)$ should be almost degenerate, while in the experiment the energy at $(\pi,0)$ is almost at the bottom of the band. Better agreement between experiment and theory can be achieved within an extended t-J model, where hopping to the next-nearest and the next-next-nearest neighbours (t-t'-t''-J model) is taken into account [27]. From the comparison of those calculations with experimental data, information on the hopping integrals t, t' and t'' can be achieved.

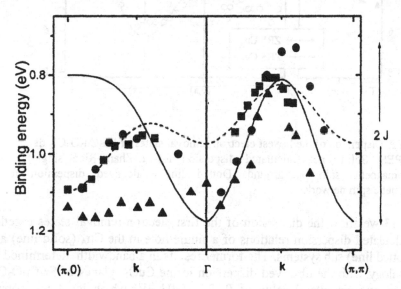

Figure 12. Dispersion of the Zhang-Rice singlet states in $Sr_2CuO_2Cl_2$. ● Dürr et al. [22]; ▲ Wells et al. [23]; ■ La Rosa et al. [24]. Solid line: t-J model, dashed line extended t-J model

Finally, we present ARPES measurements of the hole dynamics for the more complicated Cu_3O_4 plane in $Ba_2Cu_3O_4Cl_2$ (see Fig. 13). As mentioned in Section 4.1, the two subsystems into which this plane can be split have radically different T_N's, thus offering the unique possibility of investigating the dispersion of a single hole in both an antiferromagnetic and paramagnetic spin background simultaneously [28, 29]. Performing ARPES measurements at or slightly above room temperature, the hole can either be injected into the essentially antiferromagnetically ordered Cu_A sublattice or the paramagnetic subsystem of the Cu_B spins.

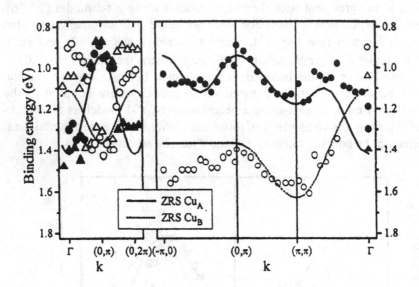

Figure 13. Dispersion of the lowest electron removal states of $Ba_2Cu_3O_4Cl_2$ determined by ARPES. Solid line: Calculated dispersion for a Zhang-Rice singlet in an antiferromagnetic spin background. Dotted line: Calculated dispersion for a paramagnetic spin network

In Fig. 13 we show the dispersion of the first electron removal states together with calculated dispersion relations of a single hole in the Cu_A (solid line) and Cu_B (dotted line) sub systems. The former results in a bandwidth determined by J, in analogy with the observed dispersion in the CuO_2 planes of $Sr_2CuO_2Cl_2$, giving an experimental J-value of 0.22 ± 0.03 eV, which is of the correct magnitude. Note that the k-vector $(\pi,0)$ where the first electron removal states have the lowest binding energy is the equivalent point to $(\pi/2,\pi/2)$ in $Sr_2CuO_2Cl_2$. The hole in the paramagnetic spin background (Cu_B subsystem), however, follows a tight-binding-like dispersion relation with a hopping integral $t_B = -0.13 \pm 0.05$ eV. The value of t_B agrees roughly with the estimated hopping integral between the Cu_B sites.

In summary, as far as the spin background is concerned, the ARPES data from the Cu_3O_4 planes of $Ba_2Cu_3O_4Cl_2$ represent simultaneously the low and very high doping limits in 2D cuprate materials.

4.3 UNOCCUPIED STATES PROBED BY EELS AND XAS

As discussed in Section 3, transitions from the core levels into the conduction bands yield information on the site-selective unoccupied density of states, when core-excitonic effects are not too strong. For correlated systems these excitations represent an electron addition spectroscopy.

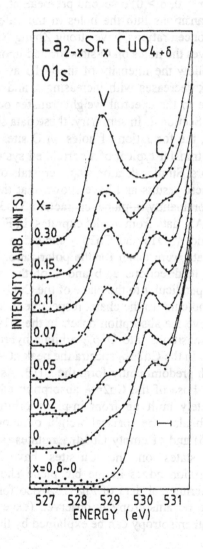

Figure 14. O1s absorption edges of $La_{2-x}Sr_xCuO_{4+\delta}$ measured using electron energy-loss spectroscopy

Before we discuss such XAS studies on model compounds this method is illustrated by EELS measurements of $La_{2-x}Sr_xCuO_{4+\delta}$ [30]. In Fig. 14 we show O1s absorption edges for various dopant concentrations. For the undoped system ($x = 0$, $\delta \sim 0$) a pre-edge (labelled with C) at about 530.2 eV is observed. It is assigned to a transition from the O1s core level into $O2p_{x,y}$ states admixed due to hybridization to the UHB [($3d^{10}2p^5$) $\Rightarrow$ O1s $3d^{10}2p^6$ transition)]. At finite dopant concentrations ($x > 0$, $\delta > 0$) a second pre-peak at 528.7 eV is observed which is assigned to transitions into the holes in the valence band, which, at least for small dopant concentrations, have strong Zhang-Rice singlet character. Thus one directly observes the holes on O-sites formed upon p-type doping.

It is interesting to follow the intensity of the UHB as a function of dopant concentration. The peak decreases with increasing x and is almost not visible for $x = 0.3$. This is due to the spectral weight transfer occuring in correlated systems as discussed in Section 2. In summary, these data show an admixture of O2p states to the UHB, the formation of holes on O-sites upon p-type doping and the spectral weight transfer typical of a correlated system and not observed in normal semiconductors. Similar data on single crystals of $La_{2-x}Sr_xCuO_4$ using XAS have supported these results and have shown that the hole states on the CuO_2 planes have predominantly in-plane character [31, 32].

We now turn to XAS data from model cuprates. In Fig. 15 and Fig. 16 we show polarization-dependent O1s and $Cu2p_{3/2}$ excitation spectra, respectively, of Li_2CuO_2 [33]. The data were taken for the polarization vector E parallel to the three crystallographic directions a, b and c of the orthorhombic crystal structure, where a is perpendicular to the plane of the CuO_4 plaquettes and b and c are parallel and perpendicular to the chain, respectively. We will mainly focus on the peak directly above the absorption onset. In the O1s spectra this peak at ~530 eV is related to transitions into $O2p_{y,z}$ orbitals hybridized with $Cu3d_{y2-z2}$ states forming the UHB. In the $Cu2p_{3/2}$ spectra the peak at ~931.8 eV is ascribed to $Cu3d_{y2-z2}$ states which predominantly form the UHB. As in all other cuprates [17], the O1s data and those of the $Cu2p_{3/2}$ absorption edges indicate that the UHB is nearly completely built up from in-plane orbitals: in this case the $Cu3d_{y2-z2}$ and $O2p_{y,z}$ orbitals. The spectral weight of unoccupied out-of-plane $O2p_x$ states (see Fig. 15) and of empty $Cu3d_{3x2-r2}$ states (see Fig. 16) is small. Since the in-plane hole states on the Cu sites have $3d_{y2-z2}$ character, no anisotropy of the absorption edges in the b,c plane should be observed, in agreement with the experimental data of Fig. 16. Also for the O1s absorption edge in the chain plane no anisotropy is observed (except a small 150 meV shift). This absence of an anisotropy can be explained by the fact that all O sites are equivalent.

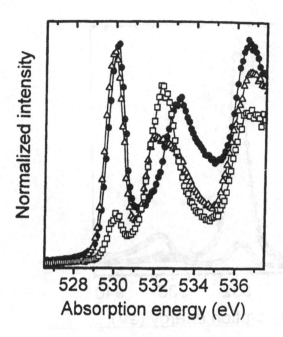

Figure 15. Polarization dependent O1s absorption edges of Li$_2$CuO$_2$ measured with x-ray absorption spectroscopy for the electric field vector **E** parallel to the three crystallographic axes. •: **E**||**b**; △: **E**||**c**; □: **E**||**a**.

The situation is completely different in the corner-shared chain Sr$_2$CuO$_3$, in which there is a strong coupling between the plaquettes. In Fig. 17 we show polarization dependent O1s XAS spectra of Sr$_2$CuO$_3$ [34]. The a and b axes lie in the plane of the chain and the b axis lies parallel to the chain direction. Again we focus on features below 530.5 eV, which have their origin in O2p$_{x,y}$ orbitals hybridized with the (predominantly) Cu3d$_{x^2-y^2}$ derived UHB. The two different peaks (for **E**||**a** at 529.6 eV and for **E**||**b** at 529.1 eV) follow naturally from the two inequivalent O sites, whereby the energetically lower peak is due to transitions into unoccupied O2p$_x$ states of the two peripheral O(2) sites and the higher lying peak corresponds to unoccupied O2p$_y$ states of the central O(1) sites connecting the plaquettes. The difference in energy between the two peaks may be explained by different Madelung potentials acting on the O1s levels and/or the UHB states of the two O sites. Again, the intensity corresponding to unoccupied out-of-plane O2p$_z$ states is small (see. Fig. 17, **E**||**c**). The different intensities of the two in-plane peaks can be explained in the following way. The

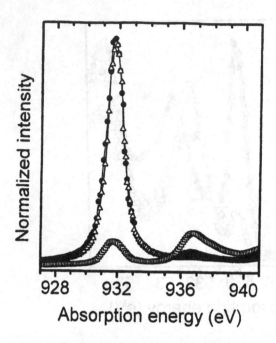

Figure 16. Polarization dependent $Cu2p_{3/2}$ absorption edges of Li_2CuO_2 measured by X-ray absorption spectroscopy for the electric field vector **E** parallel to the three crystallographic axes. ●: **E**||b; Δ: **E**||c; □: **E**||a.

O(1) site has two Cu neighbours compared to a single Cu neighbour for each of the O(2) sites. Correspondingly, as this O1s XAS intensity is an expression of the hybridization with the Cu orbitals forming the UHB, in the simplest picture one would expect twice the hole occupation for O(1) compared with O(2). Upon analysis of the peak areas, however, one finds a ratio for the hole occupation numbers, $n[O(1)]/ n[O(2)] \sim 0.6$ instead of 0.5. Thus, holes are pushed out from the central O(1) sites to the peripheral O(2) sites. This can be explained by a non-zero intersite Coulomb interaction V_{pd} (the central O(1) atom has 2 Cu neighbours and therefore experiences this interaction more strongly) or by a difference of the charge-transfer energy $\Delta_{pp} = \Delta_{O(2)} - \Delta_{O(1)}$ for the two O sites, resulting from different Madelung potentials at the two O sites. A more quantitive analysis gives $V_{pd} \sim 2\text{-}2.5$ eV and $\Delta_{pp} \sim 0.5\text{-}1$ eV. This example shows that from such measurements valuable information regarding the magnitude of important model parameters can be obtained.

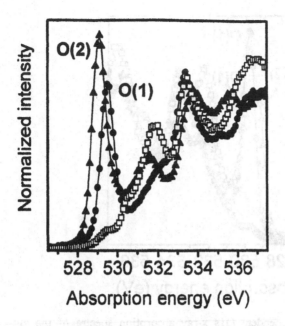

Figure 17. Polarization dependent O1s x-ray absorption spectra of Sr_2CuO_3 for the electric field vector **E** parallel to the crystallographic axis. ▲: **E**||a; ●: **E**||b; □: **E**||c.

Similar measurements have been performed on the zig-zag chain $SrCuO_2$ [35]. In this case the central O sites have not only O2p orbitals hybridized with Cu sites along the chain direction, but also perpendicular to the chain direction. Therefore, for **E** perpendicular to the chain axis and in the plane of the plaquettes (**E**||b), contributions from the central (at E = 529.8 eV) and the peripheral (at E = 529.3 eV) O sites are observed (see Fig. 18) while for **E**||c (parallel to the chain direction) only a peak from the central O sites is observed at E ~ 529.8 eV. The reason that two different energies for the transitions into the UHB is observed could arise from different binding energies of the O1s level in the two O sites. Also on this more complicated system, detailed information on the distribution of holes in different orbitals at different O sites can be obtained which may give, by comparison with calculations, information on the parameters describing the electronic structure of these materials.

More recently, we have performed similar measurements [36] on the even more complicated compound $(La,Sr,Ca)_{14}Cu_{24}O_{41+\delta}$ which contains edge-sharing chains and two-leg ladders formed from zig-zag chains (see Section 4.2). For high concentrations of trivalent La, the system is only slightly doped.

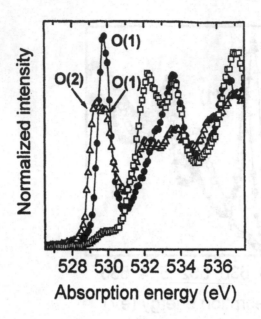

Fig. 18: Polarization dependent O1s x-ray absorption spectra of the zig-zag chain system $SrCuO_2$ for the electric field vector **E** parallel to the crystallographic axes. □: **E**∥a; △: **E**∥b; ●: **E**∥c.

In this case polarization dependent O1s x-ray absorption edges in the energy range of Zhang-Rice singlet states and UHB states show no anisotropy when **E** is in the plane of the Cu-O systems, similar to Li_2CuO_2. Therefore, one can conclude that at these dopant concentrations all holes are located on the edge-sharing chains.

For the compound $Sr_{14}Cu_{24}O_{41}$, two Zhang-Rice singlet features are observed, one showing an anisotropy. This anisotropic line is ascribed to holes formed in the two-leg ladder system. When going to the compound which shows superconductivity at high pressure (most of the Sr is in this case replaced by isoelectric Ca), the number of holes in the two-leg ladder system increases slightly. Thus, even in this complex compound, where holes can be formed in five different planar O2p orbitals (two in the chain and 3 in the ladder system), information on the hole distribution in the two sub-systems can be obtained. The interpretation is supported by a comparison with measurements which only contain one sub-system, e.g. Li_2CuO_2.

4.4 SCREENING OF CORE HOLES

We now turn again to the question on the dynamics of holes which can also be gained from investigating the screening response to the creation of a Cu2p core-hole in high resolution PES. Fig. 19 shows the Cu2p$_{3/2}$ photoelectron spectra (hv = 1487 eV) of Sr$_2$CuO$_2$Cl$_2$, Li$_2$CuO$_2$ and Sr$_2$CuO$_3$. All spectra [38] show,

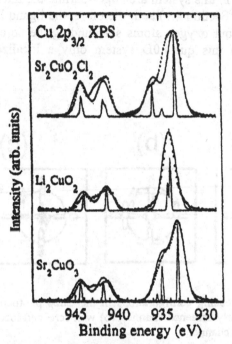

Figure 19. Cu2p$_{3/2}$ core-level photoemission spectra of cleaved single crystals of Sr$_2$CuO$_2$Cl$_2$, Li$_2$CuO$_2$ and Sr$_2$CuO$_3$. Also shown as solid lines are calculated spectra from a three-band Hubbard model [37], the results of which have either been broadened with a Gaussian of either 0.2 (to show the individual final states) or 1.8 eV (for direct comparison with experiment).

besides the so-called main line at around 933 eV, a multiplet-split satellite feature between 940 eV and 946 eV. The satellite is due to a final state in which the core hole is poorly screened, i.e. a Cu2p3d^9 configuration, where 2p denotes a core hole in the 2p shell. The main line then corresponds to a well-screened final state, generally denoted as Cu2pd^{10}L, in which the core hole is screened by an excitation from the ligands (L = 'ligand hole') to the Cu site containing the core hole [39]. It has been pointed out [40] that there may be different screening channels for the main line, for instance a more localized screening, in which the hole is close to the excited Cu site and a more delocalized screening in which the hole has moved away from the excited Cu site. Thus, by these measurements, information on the dynamics of holes in this screening process can be obtained.

Starting with the spectrum of Li_2CuO_2, we see that the main line is composed of a single component. In this case, the nature of the $Cu2pd^{10}\underline{L}$ final state is clear. As the plaquettes in this system are edge-sharing, the interaction between plaquettes is small (see Section 4.1) and, therefore, the ligand hole is localized predominantly on the four oxygen atoms surrounding the core-ionized copper site (see Fig. 20b). In this quasi-0D system only a localized screening is possible.

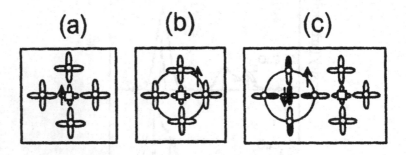

(a)   **(b)**   **(c)**

Figure 20. Core hole screening configurations in the Cu2p photoemission process in divalent cuprates. (a) Poorly-screened channel; (b) well screened localized channel; (c) well screened delocalized channel

Also shown in Fig. 19 are calculated $Cu2p_{3/2}$ spectra based upon a three-band Hubbard model using a cumulant projection technique [37]. The advantage of this method is the treatment of infinite systems, thus circumventing some of the finite size problems which limit exact diagonalization studies. As can be seen from Fig. 19, the spectrum of Li_2CuO_2 can be well reproduced by calculations based on an isolated CuO_4 plaquette, supporting the idea discussed above that only local screening is relevant for the main line in this compound. The model parameters required are $\Delta=3.5$ eV, $t_{pd}=1.3$ eV, $t_{pp}=0.65$ eV, $U_{dc}=8.5$ eV and $U=8.8$ eV, where t_{pp} is the hopping integral between O sites and U_{dc} is the Coulomb interaction between the core-hole and a d electron.

In Sr_2CuO_3, the main line is clearly composed of two features, with the main component located at lower binding energies than the locally screened final state in Li_2CuO_2. In the corner-sharing chain, Sr_2CuO_3, the interaction between the plaquettes is strong and, therefore, there is a high probability for a screening channel in which the ligand hole delocalizes away from the core-ionized Cu site. The calculated spectrum, in which this delocalized screening process is dominant, agrees well with experiment (using $\Delta = 2.7$ eV, with the other parameters as above). Examination of the theoretical hole distribution in the final states shows that the dominant screening channel is one in which the ligand hole has moved to the neighbouring plaquette forming a Zhang-Rice singlet (see Fig. 20c). In this screening process, the formation energy of a Zhang-Rice singlet is gained and, therefore, the spectral weight for this process appears at the lowest binding energy. The next feature at higher binding energy corresponds to the locally screened final state shown in Fig. 20b.

Finally, we mention that for the 2D system $Sr_2CuO_2Cl_2$ the comparison between the calculated and the experimental spectrum of the main line is less convincing (see Fig. 19). Possibly, besides the localized and the Zhang-Rice singlet screening processes there are other screening channels which do not appear in the calculations.

Recently, the experimental data presented here and other experimental data [38] on compounds discussed in Section 4.1 were compared with calculations using an Anderson impurity model [41]. For most of the measured compounds (but not the CuO_2 plane system $Sr_2CuO_2Cl_2$) good agreement between experiment and calculation was obtained.

4.5 PARTICLE-HOLE EXCITATIONS

As pointed out in Section 3, EELS provides us with information on momentum dependent excitations between the valence and the conduction bands, i.e. on particle-hole excitations [42]. Actually, the momentum and energy dependent loss function Im $(-1/\varepsilon(q,\omega))$ is measured, which describes collective excitations. On the other hand, using a Kramers-Kronig analysis the imaginary part of the dielectric function, ε, and the optical conductivity can be derived. In insulating systems such as the present model systems, the low-energy loss function near the fundamental gap is determined by interband plasmons which appear at slightly higher energy than the corresponding interband transitions.

In Fig. 21 we compare the loss functions for small momentum transfer ($q=0.08\text{Å}^{-1}$) of the corner-sharing chain Sr_2CuO_3 and the edge-sharing chain Li_2CuO_2 [43,44]. The corner-sharing chain, with its strongly interacting plaquettes, shows a strong excitation at about 2 eV as do all undoped 2D cuprates. In Li_2CuO_2, in which the plaquettes are electronically isolated, the spectral weight in this energy range is weak and the first prominent feature appears at 4.6 eV. This indicates that the 2 eV excitation is not related to a localized excitation on a single plaquette, but stems from an excitation in which electron and hole are delocalized.

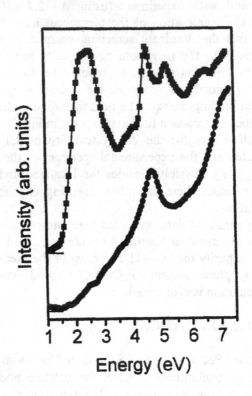

Figure 21. Loss function of Sr_2CuO_3 () and Li_2CuO_2 (•) for q=0.08 Å^{-1} parallel to the chain directions.

Calculations on the possible excitations in a single CuO_4 plaquette indicate that the 4.6 eV feature corresponds to a dipole allowed transition between non-bonding O2p states into the unoccupied $Cu3d_{x^2-y^2}$ states [44,45]. The final state has then one hole on O sites and a completely filled Cu3d shell. The 2 eV transition in Sr_2CuO_3 can then be explained by a final state in which the hole has moved to the neighbouring plaquette forming there a Zhang-Rice singlet. In this process the system gains the formation energy for the Zhang-Rice singlet and therefore the first excitation in Sr_2CuO_3, in which the delocalization is possible, appears at lower energy (~ 2 eV). The two final states are illustrated in Fig. 22. When looking at the momentum dependence of the loss function for q parallel to the chain direction, the 4.6 eV transition in Li_2CuO_2 shows no dispersion as expected since the interaction between the plaquettes in this system is weak.

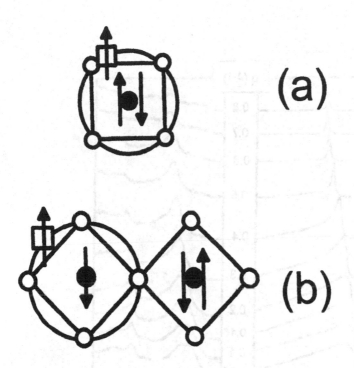

Figure 22. Particle-hole excitations in Cu-O systems illustrated in an ionic picture. (a) localized charge-transfer excitation in an isolated plaquette forming a hole on the non-bonding O2p states (□) and an additional electron in the $Cu3d_{x^2-y^2}$ orbital; (b) delocalized excitation in which the hole has moved to antibonding O2p states of the neighbouring plaquette, forming a Zhang-Rice singlet.

On the other hand, in Sr_2CuO_3 the 2 eV feature shows a dispersion of almost 1 eV when going to the zone boundary at $q=0.8$ Å^{-1} (see Fig. 23), indicating that the dispersion of a particle-hole excitation is much larger than the dispersion of a single hole in a CuO_2 plane (see Section 4.2). The dynamics of holes and electrons after a particle-hole excitation in a 1D chain is illustrated in Fig. 24. The first line corresponds to the excitation shown in Fig. 22 b. When moving the hole (holon) or the doubly occupied site (doublon) no spin defects are created, which is different from moving a hole in a 2D antiferromagnetic lattice as was discussed in Section 4.2. Therefore, the dynamics of the holon and the doublon are not determined by J, but by the hopping integral t. In this model, the holon and doublon 'bands' are given by $E_h(k) = -2t_h \cos ka$ and $E_d(k) = 2t_d \cos ka + U$.

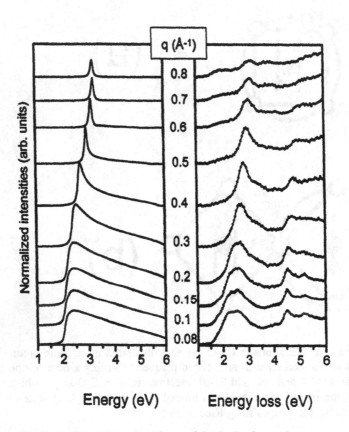

Figure 23. Right panel: loss functions of Sr_2CuO_3 for various momentum transfers **q** parallel to the chain direction. Left panel: calculated loss functions within an extended one-band Hubbard model

The holon and doublon bands are plotted in Fig. 25a together with vertical (q=0) and non-vertical (q≠0) transitions between them. The continuum of possible excitations between U-4t and U+4t is shown for the first Brillouin zone by the hatched area in Fig. 25b. The inclusion of a nearest-neighbour Coulomb interaction, V, is illustrated in Fig. 26. In the localized excitation, the energy of the system is reduced by 2V around the hole and increased by V at the doublon.

Thus the energy of this localized excitation, corresponding to excitations at the zone boundary, is reduced by V and the total energy is U-V. For a delocalized excitation (holon and doublon are separated), which corresponds to excitations near the zone center, the system gains 2V at the holon site and loses 2V at the doublon site. Therefore, at small q the excitations are not influenced by V.

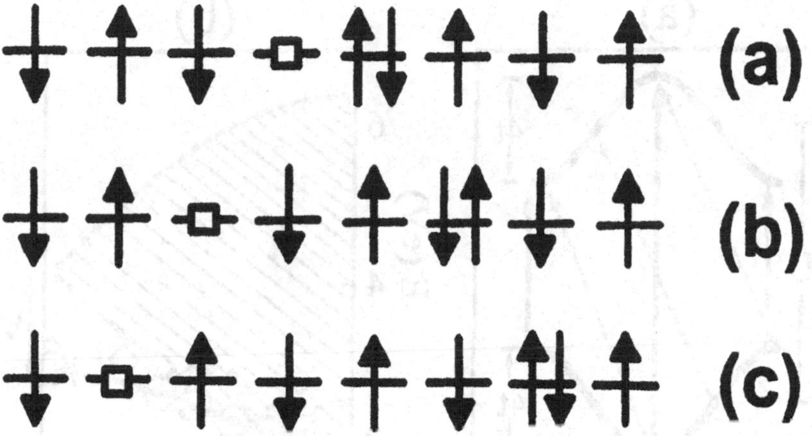

Figure 24. Dynamics of holes and electrons after a particle-hole excitation in a 1D chain. A □ stands for a hole and ↑↓ represents a doubly occupied site. As the holon and doublon separate from each other, it can be seen that no misaligned spins are created.

In summary, the nearest-neighbour Coulomb interaction, V, leads to excitonic excitations below the continuum, separated from the continuum by V at the zone boundary. This is illustrated in Fig 25b by the solid line. Calculated spectra are shown in Fig. 23. They show a continuum of excitations at small momentum transfer and a narrow excitonic line at the zone boundary. This is in qualitative agreement with the experimental data of Fig. 23, where a broad feature is seen at small q and a narrow line is observed for $q=0.8 \text{Å}^{-1}$. The dispersion of the 2 eV feature in the loss function is shown in Fig. 25b by diamonds. The corresponding maximum of the optical conductivity is given by circles. The parameters used in this calculation are $t^{(1B)}=0.55$ eV, $U^{(1B)}=4.2$ eV, and $V^{(1B)}=0.8$ eV. Both the dispersion and the lineshape provide strong constraints on the parameters. The relative large value of $V^{(1B)}$ compared to values derived for 2D compounds may be explained by a reduced screening in the 1D systems and/or by a larger extension of the atomic wave functions due to the 1D confinement.

Finally, we mention that similar particle hole excitations were also measured for the 2D system $Sr_2CuO_2Cl_2$ [46, 47]. A large anisotropic dispersion of the same magnitude as that of Sr_2CuO_3 is observed.

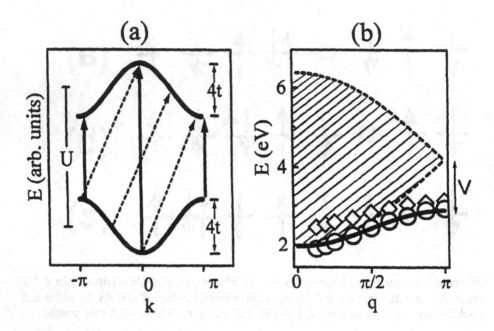

Figure 25. (a) Schematic picture of the 'band structure' of a holon and a doublon in a one-dimensional Mott-Hubbard insulator together with vertical and non-vertical excitations. (b) Excitations for various energies, E, and momentum transfers, q. Continuum of possible excitations: hatched area. Excitonic excitations: solid line; ◊: maxima of the loss function; ○: maxima of the optical conductivity.

Therefore, the theoretical explanation of the dispersion of the particle-hole excitations is not based on the dynamics of separated particles and holes (which would be determined in the 2D system by J) but by the dynamics of a localized exciton which is not determined by frustrated spins [48]. However, the more recent high resolution data [47] contain features which cannot be explained in the theoretical work [48] mentioned above.

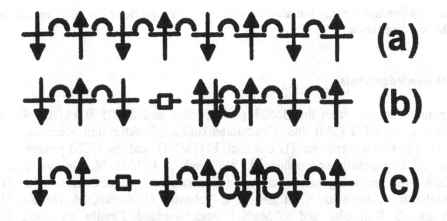

Figure 26. The effect of the nearest-neighbour Coulomb interaction, V, on the dynamics of holons and doublons in a one-dimensional chain (single band model). V is illustrated by ⌒. (a) Ground state; (b) localized excitation; (c) delocalized excitation

5. Conclusions

The aim of the first part of this paper was to give a simple introduction into the electronic structure of high-T_c superconductors and related compounds. In the second part we illustrated the wealth of information that can be obtained regarding the character, the location, the orbital symmetry and the dynamics of charge carriers in one- and two-dimensional cuprates, when using high-energy spectroscopies such as photoemission, x-ray absorption spectroscopy and electron energy-loss spectroscopy. Through such studies we gain an understanding of the effects of electron correlation, the influence of the spin background and the dimensionality on the electronic structure and the electronic properties of cuprates. Moreover, those measurements can serve of constraints for the parameter-space available to model calculations developed to describe not only the high-energy but also the low-energy properties of HTSCs and their related compounds.

Acknowledgements

Financial support from the BMBF (13N 6599/9 and 05605 BDA/ 05 SB 8 BDA6), the DFG (SFB 463, Graduiertenkolleg "Struktur und Korrelations-effekte in Festkörpern" der TU Dresden, FI439/7-1), and the HCM program of the EU is gratefully acknowledged. We thank S. Uchida, N. Motoyama, H. Eisaki for providing single crystals. Experimental collaboration with G. Reichardt, C. Janowitz, R. Müller, R. L. Johnson, G. Kaindl, M. Domke, N. Nücker, S. Schuppler, and K. Maiti is acknowledged. Finally we thank W. Stephan, K. Penc, C. Waidacher, K.W. Becker, K. Okada, A. Kotani, V.Y. Yushankhai for theoretical support.

References

1. J.G. Bednorz and K.A. Müller, *Z. Phys. B* **64**, 189 (1986)
2. K.C. Hass, *Solid State Physics* **42**, 213 (1989)
3. W.E. Pickett, *Rev. Mod. Phys.* **61**, 433 (1989)
4. E. Dagotto, *Rev. Mod. Phys.* **66**, 763 (1994)
5. W. Brenig, *Phys. Reports* **251**, 154 (1995)
6. B. Keimer, N. Belk, R.J. Birganeau, A. Cassanho, C.Y. Chen, M. Greven, M.A. Kastner, A. Aharony, Y. Endoh, R.W. Erwin, and G. Shirane, Phys. Rev. B **46**, 14034 (1992)
7. L.J. Mattheiss, *Phys. Rev. Lett.* **58**, 1028 (1987)
8. H. Eskes, L.H. Tieng, and G.A. Sawatzky, Phys. Rev. B **41**, 288 (1990)
9. J.C. Fuggle, P.J. Weijs, R. Schoorl, G.A. Sawatzky, J. Fink, N. Nücker, P.J. Durham, and W.M. Temmerman, Phys. Rev. B **37**, 123 (1988)
10. A.K. Mc Mahan, J.F. Annett, and R.M. Martin, Phys. Rev. B **42**, 6268 (1990)
11. M.S. Hybertsen, E.B. Stechel, M. Schlüter, and D.R. Jennison, Phys. Rev. B **41**, 11068 (1990)
12. J. Zaanen, G.A. Sawatzky, and J.W. Allen, Phys. Rev. Lett. **55**, 418 (1985)
13. H. Eskes and G.A. Sawatzky, Phys. Rev. Lett. **61**, 1415 (1988)
14. F.C. Zhang and T.M. Rice, Phys. Rev. B **37**, 3759 (1987)
15. H. Eskes, M.B.J. Meinders, and G.A. Sawatzky, Phys. Rev. Lett. **67**, 1035 (1991)
16. L. Hedin and S. Lundquist, Solid State Phys. **23**, 1 (1969)
17. J. Fink, N. Nücker, E. Pellegrin, H. Romberg, M. Alexander, and M. Knupfer, J. Electron Spectr. Rel. Phenom. **66**, 395 (1994)

18. H. Müller-Buschbaum, Angew. Chemie **89**, 704 (1977)

19. E. Dagotto, J. Riera, and D. Scalapino, Phys. Rev. B **45**, 5744 (1992). E. Dagotto and T.M. Rice, Science **271**, 618 (1996)

20. M. Uehara, T. Nagata, J. Agimitsu, H. Takahashi, N. Môri, and K. Kinoshita, J. Phys. Soc. Jpn. **65**, 2764 (1996)

21. T. Böske, O. Knauff, R. Neudert, M. Kielwein, M. Knupfer, M.S. Golden, J. Fink, H. Eisaki, S. Uchida, K. Okada, and A. Kotani, Phys. Rev. B **56**, 3438 (1997)

22. C. Dürr, S. Legner, Z. Hu, R. Hayn, M. Knupfer, M.S. Golden, J. Fink, S. Haffner, H. Eisaki, S. Uchida, G. Reichardt, C. Janowitz, R. Müller, and R.L. Johnson, unpublished

23. B.O. Wells, Z.-X. Shen, A. Matsuura, D.M. King, M.A. Kastner, M. Greven, and R. Birgeneau, Phys. Rev. Lett. **74**, 964 (1995)

24. S. La Rosa, I. Vobornik, F. Zwick, H. Berger, M. Grioni, G. Margaritondo, R.J. Kelley, M. Onellion, and A. Chubukov, Phys. Rev. B **56**, 525 (1997)

25. E. Dagotto, Rev. Mod. Phys. **66**, 763 (1994)

26. A. Nazarenko, K. Vos, S. Haas, E. Dagotto, and R.J. Gooding, Phys. Rev. B **51**, 8676 (1995)

27. B. Kyung and R.A. Ferrell, Phys. Rev. B **54**, 10125 (1996)

28. M.S. Golden, H.C. Schmelz, M. Knupfer, S. Haffner, G. Krabbes, J. Fink, V.Y. Yushankhai, H. Rosner, R. Hayn, A. Müller, and G. Reichardt, Phys. Rev. Lett. **78**, 4107 (1997)

29. S. Haffner, M. Knupfer, S.R. Krishnakumar, R. Ruck, G. Krabbes, M.S. Golden, J. Fink and R.L. Johnson, Eur. Phys. J. B, submitted,

30. H. Romberg, M. Alexander, N. Nücker, P. Adelmann, and J. Fink, Phys. Rev. B **42**, 8768 (1990)

31. C.T. Chen, F. Sette, Y. Ma, M.S. Hybertsen, E.B. Stechel, W.M.C. Foulkes, M. Schlüter, S.-W. Cheong, A.S. Cooper, L.W. Rupp, Jr., B. Batlogg, Y.L. Soo, Z.H. Ming, A. Krol, and Y.H. Kao, Phys. Rev. Lett. **66**, 104 (1991)

32. E. Pellegrin, N. Nücker, J. Fink, S.L. Molodtsov, A. Gutiérrez, E. Naras, O. Strebl, Z. Hu, M. Domke, G. Kaindl, S. Uchida, Y. Nakamura, J. Markl, M. Klauda, G. Seemann-Ischenkoo, A. Krol, J.L. Peng, Z.Y. Li, and R.L. Greene, Phys. Rev. B **47**, 3354 (1993)

33. R. Neudert, H. Rosner, S.-L. Drechsler, M. Kielwein, M. Sing, M. Knupfer, M.S. Golden, J. Fink, N. Nücker, S. Schuppler, N. Motoyama, H. Eisaki, S. Uchida, Z. Hu, M. Domke, and G. Kaindl, Phys. Rev. B, in press

34. R. Neudert, S.-L. Drechsler, J. Málek, H. Rosner, M. Kielwein, M. Knupfer, M.S. Golden, J. Fink, N. Nücker, S. Schuppler, N. Motoyama, H. Eisaki, S. Uchida, Z. Hu, M. Domke and G. Kaindl, unpublished

35. M. Knupfer, R. Neudert, H. Kielwein, S. Haffner, M.S. Golden, J. Fink, C. Kim, Z.-X. Shen, M. Merz, N. Nücker, S. Schuppler, N. Motoyama, H.

38

Eisaki, S. Uchida, Z. Hu, M. Domke and G. Kaindl, Phys. Rev. B **55**, R 7291 (1997)

36. N. Nücker, H. Merz, C. Kuntscher, S. Schuppler, R Neudert, M.S. Golden, J. Fink, D. Schild, V. Chakarian, J. Freeland, Y.U. Idzerola, V. Conder, M. Uehara, T. Nagata, J. Goto, J. Akimitsu, S. Uchida, U. Ammerahl, and A. Revcolevschi, to be published

37. C. Waidacher, J. Richter, and K.W. Becker, Europhys. Lett. **47**, 77 (1999)

38. T. Böske, K. Maiti, O. Knauff, K. Ruck, M.S. Golden, G. Krabbes, J. Fink, T. Osafune, N. Motoyama, H. Eisaki, and S. Uchida, Phys. Rev. B **57**, 138 (1998)

39. S. Larsson, Chem. Phys. Lett. **32**, 401 (1975)

40. M. van Veenendaal and G.A. Sawatzky, Phys. Rev. Lett. **70**, 2459 (1993)

41. K. Karlsson, O. Gunnarsson and O. Jepsen, Phys. Rev. Lett. **82**, 3529 (1999)

42. J. Fink, Adv. Electron. Electron Phys. **75**, 121 (1989)

43. R. Neudert, M. Knupfer, M.S. Golden, J. Fink, W. Stephan, K. Penc, N. Motoyama, H. Eisaki, and S. Uchida, Phys. Rev. Lett. **81**, 657 (1998)

44. S. Atzkern, R. Neudert, M. Knupfer, M.S. Golden, J. Fink, C. Waidacher, J. Richter, K.W. Becker, H. Eisaki, and S. Uchida, to be published

45. see also J. Mizuno, T. Tohyama, S. Maekawa, T. Osafune, N. Motoyama, H. Eisaki and S. Uchida, Phys. Rev. B **57**, 5326 (1998)

46. J.Y. Wang, F.C. Zhang, V.P. David, K.K. Ng, M.V. Klein, S.E. Schnatterly, and L.L. Miller, Phys. Rev. Lett. **77**, 1809 (1996)

47. R. Neudert, PhD Thesis, TU Dresden (1999)

48. F.C. Zhang and K.K. Ng, Phys. Rev. B **58**, 13520 (1998)

FERMI SURFACE MAPPING BY ANGLE-SCANNED PHOTOEMISSION

P. AEBI

Institut de Physique Université de Fribourg,
Pérolles, CH-1700 Fribourg, Switzerland

1. Introduction

Photoemission has a long-standing tradition in surface analysis. In the ultraviolet (UV) regime, using UV photoelectron spectroscopy, typically with a He discharge lamp, the acquisition of energy distribution curves (EDC's) gives information on the electronic structure at and near the surface. On single crystals, angle-resolved photoemission is able to study the dispersion of bands. Traditionally angle-resolved experiments have been performed for rather few photoelectron emission angles preferentially along high symmetry directions where complete EDC's were collected. Here we report on experiments using extensive angle-scanning, covering much of the hemisphere above the sample surface as illustrated in Fig. 1.

Photoemission is also considered to be a key experiment to elucidate the electronic structure of high temperature superconductors. But despite many photoemission experiments a clear-cut picture has not been reached for many issues. In this lecture therefore, we try to gain insight using angle-resolved photoemission in a different mode being complementary to the traditional one. Instead of collecting complete EDC's at few $\vec{k}$ space locations (photoelectron emission angles), a full $\vec{k}$ space mapping of photoelectron intensities within a small energy window is performed.[1] This procedure allows to map $\vec{k}$ space locations where direct transitions move through the Fermi level (E_F) very accurately. As a matter of fact, previously unobserved features, so called "shadow bands" (SB), have been observed on the Fermi surface (FS) of $Bi_2Sr_2CaCu_2O_{8+\delta}$ (Bi2212).[2] Alas, a complementary view brings not only new light into the subject but also additional questions about the discovered features.

39

S.-L. Drechsler and T. Mishonov (eds.), High-Tc Superconductors and Related Materials, 39–50.
© 2001 *Kluwer Academic Publishers. Printed in the Netherlands.*

Angle-scanned photoemission

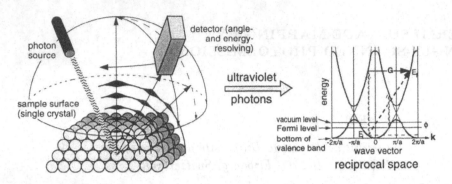

Figure 1. General principle of angle-scanned photoemission: Photoelectron intensities are collected over much of the hemisphere above the sample surface. Using UV photons reciprocal-space information can be gained. (see text)

2. Angle-Scanned Photoemission

2.1. EXPERIMENTAL

Figure 1 presents the general idea of the experiment. The UV photon source is fixed in space and consists, in the present case, of a He lamp (HeI ($h\nu$=21.2 eV)/HeII ($h\nu$=40.8 eV)). The electrostatic, hemispherical, angle and energy resolving analyzer is also fixed in space, and angle-scanning is performed via motorized sequential sample rotation.[3] Typically 4000 to 6000 angular settings are scanned homogeneously distributed over the hemisphere above the sample.

Experiments presented here were performed in a VG ESCALAB Mk II spectrometer. Photoelectrons excited with He I radiation were analyzed with a 150 mm radius hemispherical analyzer operating with an angular resolution of 1° full cone and an energy resolution of about 30-40 meV. The base pressure was $2 \cdot 10^{-11}$ mbar and the He partial pressure during operation of the differentially pumped He discharge lamp reached $3 \cdot 10^{-9}$ mbar. Clean surfaces of the Bi-cuprates were prepared by adhesive tape cleaving at room temperature. The typical measuring time was 12 hours for the experiments on high T_c superconductors. Depending on the spectral weight and cross sections experiments can be completed in 15-20 minutes (e.g. for Ni). The data presented in this work has not been symmetry averaged, i.e., for every polar angle measured, photoelectron intensities have been recorded over the whole range of 360° of azimuthal angles.

For the transmission electron microscopy (TEM) experiments a Hitachi HF-2000 cold-field emission microscope was used. Very thin platelets were removed from the single crystals and fixed on standard TEM copper grids whereby the crystal c-axis was oriented approximately parallel to the electron beam direction. The areas selected to obtain the selected area electron diffraction (SAED) patterns were always unstrained parts and typically 1 mm in diameter.

2.2. FERMI SURFACE MAPPING

Intrinsically angle-resolved UV photoelectron spectroscopy (ARUPS) is suited best for two-dimensional (2D) electron systems, i.e., where the perpendicular component of the wave vector $\vec{k}_\perp$ does not play any role. This is the case because in ARUPS the parallel component of the wave vector $\vec{k}_\parallel$ of the photoelectron is conserved but not $\vec{k}_\perp$. Furthermore, attractive measuring modes made it possible to map the FS of 2D systems directly.[4, 2] Nevertheless one wishes to gain information on 3D FSs as well. Very promising results have been obtained for Cu [1] and Ni [5]. The dilemma of visualizing bulk states in general is indicated in Fig. 2: For every particular $\vec{k}$ vector many different energies may be a solution to Schrödinger's equation. Therefore, either one chooses only few $\vec{k}$ vectors, preferentially along high symmetry directions within the Brillouin zone (BZ) as in Fig. 2(a) [6] to display all energies, or one chooses one single energy, e.g., E_F, and displays all corresponding $\vec{k}$ vectors ending up with a constant energy surface, e.g., the FS, as in Fig. 2(b) [7]. The analogy between the two ways of looking ($E(\vec{k})$ versus $\vec{k}(E)$) is seen by following a path from the L-point to the W-point in $\vec{k}$ space. Very shortly after L the FS is crossed. This is also evident from the $E(\vec{k})$ in Fig. 2(a). From the experimental point of view the situation is very much the same in that either energy spectra can be measured for a few angles (Fig. 2(c)) or for, e.g., E_F all the angles. This is displayed in Fig. 2(d) where intensities from E_F are plotted using a linear gray scale with high intensities in black. The data are plotted linearly in $\vec{k}_\parallel$ (via $\vec{k}_\parallel = |\vec{k}_0| \sin\theta$, $|\vec{k}_0|$ the absolute value of the final state $\vec{k}$ vector in vacuum and θ the photoelectron emission angle with respect to the surface normal) to compare directly with theory in Fig. 2(e). Normal emission occurs at the center of the image and grazing emission towards the outside. Intensities collected at E_F along the azimuthal scans taken at a polar angle of 66° off normal in Fig. 2(c) are indicated with an arrow in the Fermi scan of Fig. 2(d). Assuming a free electron model for the final state we get high intensity in our energy window centered at E_F when the free electron final state (FEFS) sphere intersects the FS in the extended zone scheme, i.e., conditions for energy and momentum conservation are fulfilled. This is illustrated in Fig. 2(f) for a plane containing the (001), (111) and (110) directions. Final state spheres for different photon ener-

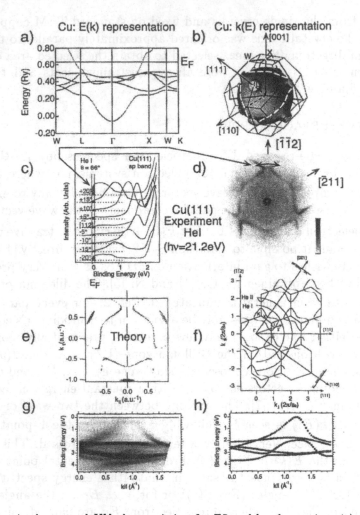

Figure 2. Angle-scanned UV photoemission for FS and band mapping: (a) $E(\vec{k})$ representation of the bandstructure [6]; (b) $\vec{k}(E_F)$), FS of Cu [7]; (c) He I excited EDC's from Cu(111) at a polar angle of 66°; azimuthal angles around the [$\bar{1}\bar{1}2$] direction are indicated; vertical lines indicate the energy window used in the angle scans of (d); (d) E_F scan on Cu(111), i.e., the total intensity of He I excited photoelectrons collected in the energy window sketched in (c), measured as a function of $\vec{k}_\parallel$; normal emission is located in the center; the gray-scale is linear with high intensities in black; (e) cuts through the bulk FS using a FEFS (see text); (f) high symmetry plane perpendicular to the [$\bar{1}10$] direction; the solid line polygons correspond to the cut through the bulk BZ; partly dotted, a calculation of the FS; the typical necks at the L-point and bones around X are evident; the half circles around the Γ point represent the FEFS wave vector for the HeI and II photon energies; (g) "carpet", i.e., EDC's continuously taken as a function of polar angle on Cu(001) and interpolated in $\vec{k}_\parallel$, for band mapping, linear gray-scale representation with high intensities in black; (h) corresponding calculation (see text).

gies (HeI, HeII) appear as circles with different radii. As a matter of fact, using tunable synchrotron radiation it should be possible to slice the FS piece by piece. In Fig. 2(e) the drawing plane of Fig. 2(f) is represented by the dashed line entering the plot from the upper side through the "neck" and ending up in the "bone".[1] The dotted lines mark all the intersecting points of the HeI FEFS sphere with the calculated FS in the extended zone scheme, analogous to what is done for a single plane in Fig. 2(f). Theory in Fig. 2(e) fits very nicely with experiment in Fig. 2(d) except for the small ring in the center representing the well known surface state of the Cu(111) surface and which, of course, is not reproduced in a bulk calculation. The white dashed line in Fig. 2(b) indicates the measured spherical cut across the bulk FS. Following this technique, other groups have measured E_F scans on Cu as well. [8]

Yet another mode of angle-scanned UV photoemission is indicated in Fig. 2(g) and Fig. 2(h): A large set of complete EDC's is taken along a high symmetry azimuth and assembled into a so-called "carpet". The data are taken as a function of angle and then mapped onto a regular $\vec{k}_\parallel$ grid. Here, $\vec{k}_\parallel=0$ corresponds to the Cu(001) surface normal emission direction and increases towards the (111) direction, i.e., the measurements follow the HeI FEFS circle in Fig. 2(f). At approximately $\vec{k}_\parallel =1$ Å^{-1} (Fig. 2(g))the sp band is seen to cross the Fermi level towards empty states and to come down again. This corresponds to the intersection of the neck and the bulk BZ in Fig. 2(f). For the calculation shown in Fig. 2(h), again, a FEFS has been assumed. For different binding energies the radii of the circles representing the FEFS have been chosen accordingly. A point is then plotted in the case of energy and momentum conservation.[9] We find all experimental bands in the calculation. However, agreement of the relative positions might be better. A possible explanation is that the FEFS is not a good enough approximation. Depending on the slope of the initial state constant energy surfaces a slight deviation from a sphere in the final state results already in a considerable $\vec{k}$ displacement of the crossings and therefore in shifts in Fig. 2(h). Nevertheless in such a manner extended band mapping is possible.

As a second example, showing temperature (T) dependent experiments, we may discuss Ni, a magnetic material and prototypical itinerant ferromagnet. Analogous to the case of Cu the intensity of photoelectrons from E_F has been mapped over much of the hemisphere.[5] Figure 3 displays E_F maps as a function of T for Ts below and above the Curie temperature T_c together with a spin resolved calculation where spin up and down regions are marked in light and dark gray, respectively. On the right side a series of measurements covering 1/4 of the azimuth is presented collected with increasing T, visualizing the development while going from $T < T_c$ to $T > T_c$. Because of the good agreement of theory and room temperature

44

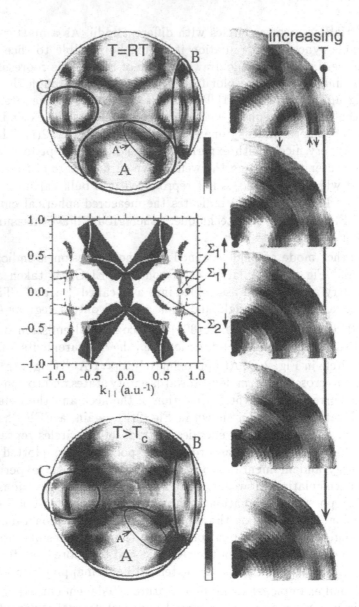

Figure 3. Experimental E_F scans on Ni(110) as a function of T together with a spin resolved calculation [5] (left side, center). A linear gray-scale is applied with high intensities in black. For the experiments with increasing T only 1/4 of the azimuthal range is measured and a maximal contrast has been chosen for every polar angle. The complete measurements at RT and above T_c have been normalized to a smooth polar angle dependent background for better visualization. The scaling is linear in $\vec{k}_\parallel$ and normal emission is situated in the center (corner) of the circle (1/4 circle) which represents grazing emission of photoelectrons. A,B,C,A' mark regions of interest.

(RT) experiment up and down spins can readily be identified without explicit spin detection and sample magnetization. Regions marked A, B, C all show moving bands as T is increased. Some spectral weight around A', however, seems to remain in place as opposed to what is expected for a Stoner behavior of magnetism. The region of A' consists of overlapping up and down bands as can be seen from the calculation. Recent measurements and calculations analogous to the ones presented above identified the character of the moving and remaining bands.[10] It turns out that the bands staying in place are up and down sp-bands, so steeply dispersing in energy that they are not resolved in angle in the experiment. Because of the large slope, moving up respectively down in energy of the sp-bands while collapsing will not result in any observable movement in $\vec{k}$ space. Therefore, the observation of these bands staying in place as we go across T_c does not oppose a Stoner-like behavior. For a detailed discussion of the experiments on Ni and implications on models of the magnetism see Ref. [5, 10]

3. Results on $Bi_2Sr_2CaCu_2O_{8+\delta}$

Figure 4(a) shows the Bi2212 $\vec{k}_{\parallel}$ mapping of the intensity of He I excited photoelectrons collected within an energy window of about 30 meV centered at E_F.[2] A sketch of the measurement is shown in Fig. 4(b). In Fig. 4(c) the FS mapping experiment on a lead-doped modulation-free sample is presented[11]. These crystals have a Pb to Bi ratio of 0.42:1.73 and a critical temperature T_c of 83K. The idea of having modulation- free crystals arises from structural complications (discussed below) created by the lattice modulation present in Bi2212 [12]. Whereas modulation related diffraction spots as observed by TEM and low energy electron diffraction (LEED) disappear in the Pb doped samples a weak set of spots related to a superstructure appears, as will be discussed below.

We observe several different features in Fig. 4(a). First, there is a strong set of lines drawn as strong solid lines in Fig. 4(b), where also high symmetry points are indicated. For a truly 2D system the high symmetry points Γ and Z would be equivalent. Points X and Y correspond to $(\pi/a, \pi/a)$ with respect to the Cu-O planes and are not equivalent because of a slightly different lattice constant of the a and b-axis and because of the lattice modulation along the b-axis. Note that no pronounced nesting is observed among these lines and there is a rather good agreement of the overall FS shape and Fermi wave vectors with the large FS obtained in Fermi liquid theory.[2]

Second, there is a weaker set of lines corresponding to SB, appearing as dashed lines in Fig. 4(b). If we take a copy of the stronger set of lines centered at Γ and center it at X or Y, it covers the weaker set of lines (see

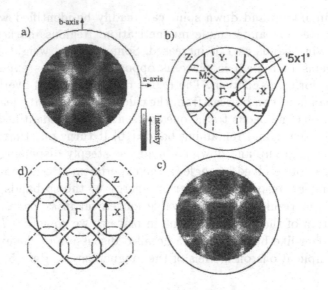

Figure 4. (a) Bi2212 $\vec{k}_{\|}$ mapping of the HeI (21.2 eV) excited photoelectrons collected within an energy window of about 30 meV centered at E_F. A logarithmic intensity scale is used to enhance weaker features. (b) Outline of (a) indicating high symmetry points and different sets of lines. (c) Same as (a) but for a modulation-free lead-doped sample (see text). (d) Taking the stronger set of lines and superposing it with a shift (arrow) reproduces the pattern of (b) except for the features labeled '5x1' in (b). The circle indicates the range mapped by the experiment.

Fig. 4(b)). The weaker set thus seems to behave around Y or X just as the main set of lines around Γ. This behavior is further confirmed with the dispersion observed in constant energy maps for different binding energies, where the strong lines close-in towards Γ and the weaker lines towards X or Y.[2, 13] X and Y points acting like Γ points, in a simplified picture, corresponds to a reduction of the BZ and in real space to a larger unit cell as it is illustrated in Fig. 5. The transformation of a real space lattice into reciprocal space for different structural arrangements is shown. Real space is represented by a primitive square lattice of points in Fig. 5(a), (b) and (c). In our case these points can be the metal atoms of a particular layer if a is given a value of ~3.8 Å. For photoemission from states at E_F we are considering the Cu-O planes. Points in reciprocal space can be looked at as Γ points or as diffraction spots. In Fig. 5(a) the transformation of the primitive lattice is shown, in (b) a $\sqrt{2}$ times larger and 45° rotated, so called $(\sqrt{2} \times \sqrt{2})$R45° or c(2x2) unit cell results in a smaller BZ. This is the case if the Cu lattice is occupied with antiferromagnetically (AF) correlated spins. The fine lines observed on the FS also coincide with SB for AF correlated metals predicted by Kampf and Schrieffer[14]. On the other hand a reconstruction of the atomic structure with the same change of the

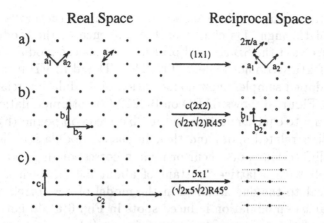

Real Space   Reciprocal Space

Figure 5. Transformation of a 2D real space lattice into reciprocal space for (a) a primitive (b) a c(2x2) or $(\sqrt{2} \times \sqrt{2})R45°$ and (c) a "5x1" or $(\sqrt{2} \times 5\sqrt{2})R45°$ superstructure. The real space lattice constant is represented by a. Vectors mark the unit cells. (see text)

unit cell has the same consequences on the FS. Note that our experiment does not specifically detect the magnetism and therefore cannot distinguish between the two effects in a direct way.

The third feature in Fig. 4(a), plotted as pieces of fine solid lines in Fig. 4(b) and marked as '5x1', occurs near the Γ and Z points with a banana-like shape. These features have been identified[13] as due to the incommensurate "5x1" lattice modulation along the crystal b-axis. This is explained in Fig. 5(c) where the behavior of a unit cell corresponding to a "5x1" or $(\sqrt{2} \times 5\sqrt{2})R45°$ superstructure, approximately present in Bi2212, is simulated. The different size of points in reciprocal space indicates that in a diffraction experiment the spots can have different intensities. It also means that contributions of the electronic structure are added up with different intensities around the different Γ points. Then, the banana-like features can be understood as the appearance of such replica of the main FS.

Note that at locations of the additional spots in Fig. 5(b), i.e. at $(\pi/a, \pi/a)$, there are also spots in Fig. 5(c). In order to estimate intensities of contributions to different Γ points from first principles, detailed calculations on a *known* structure are necessary. An experimental estimate might be given by LEED. There, spot intensities are modulated [15] and decrease when going away from the main Γ points suggesting only a small contribution from the "5x1" arrangement. Also from studying the FS map itself in Fig. 4(a), additional features along ΓY would have to be visible in the case of significant contributions from the "5x1" induced Γ points away from the main Γ points.

Figure 4(c) displays the FS mapping experiment on the lead- doped, modulation-free sample. The idea was to obtain a structurally simplified situation and through this clarity on the influence of the modulation on the SB. Notice that the SB are still present whereas the modulation related banana-like features do not appear. TEM, SAED and LEED measurements on these Pb doped samples show no indication of modulation related super-lattice spots. Figure 6 shows SAED on Bi2212 (a) with modulation induced spots compared to equivalent data on the Pb doped compound (b). Whereas the modulation related spots and their intensity modulation disappear, a set of faint diffraction spots occur or remain corresponding to a real space layer unit cell with a lattice constant of about 5.4 Å. Such a layer unit cell is identical to the one assumed in the model of AF correlations. Note however, that the modulation induced spots in Fig. 6(a) do not exhibit an accentuation of the superstructure appearing in (b).

To understand the diffraction pattern in Fig. 6(b) we observe the top and left most four diffraction spots composing a small square with one strong and three weak spots. This square corresponds to a layer real space unit cell of about 5.4 Åby 5.4 Å(see Fig. 5(b)). For the $\sqrt{2}$ times smaller and 45° rotated primitive 2D unit cell (Fig. 5(a)) with a lattice constant of 3.8 Åwe would expect diffraction spots arranged in a $\sqrt{2}$ times larger and 45° rotated square pattern instead of the little one we are observing. A dynamical TEM SAED computer simulation however shows that for the 3D unit cell extinction occurs such that only the strong spots would appear.

Thus, the Pb doping provides the intended elimination of the modu-lation but it also introduces, somewhere, a layer unit cell larger than the primitive, Cu-O plane related one. Therefore, the situation with regard to the AF correlations is not clear-cut. If all layers in the crystal have the larger unit cell, main bands and SB on the FS should exhibit almost iden-tical intensities and strong interaction should take place between the two sets. This is clearly not the case. More structural information is needed to determine the atomic arrangement in this Pb-doped unmodulated high T_c superconductor. From photoelectron diffraction, a structural method deter-mining the local environment of a selected chemical species, we know that Pb occupies Bi sites.[11] Also, computer simulations of TEM diffraction patterns show that a partial, ordered substitution of Bi with Pb produces the weak additional spots. The question is to which extent only the Bi-O layer is concerned, if Pb is inducing the larger unit cell or if it is an intrinsic property of the Bi-O layers. Furthermore, scanning tunneling microscopy should be able to detect a larger unit cell in a Bi-O surface layer. Truly magnetic measurements are necessary or experiments on compounds with well known structures without complications. Recent experiments, measur-ing the SB to main band ratio as a function of doping indicate a decrease with underdoping.[16] This is in contrast to what is expected for AF cor-

a)   b)

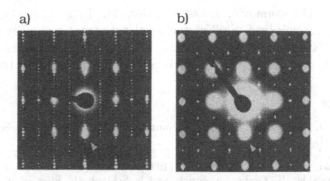

Figure 6. Fig. 5: TEM SAED pattern of (a) modulated Bi2212 and (b) unmodulated Pb doped Bi2212.

relations when approaching the AF insulator. However, there should also be T dependent measurements confirming such a result independently on a single sample in a well-defined state.

4. Conclusions

Angle-scanned photoemission represents an attractive and accurate tool to map FS transitions with a selectable uniform point density.[1] From such experiments on Bi2212 [2] we find rather good agreement of the overall FS shape and Fermi wave vectors with the large FS obtained in Fermi liquid theory. No pronounced nesting is observed. A weak superstructure on the FS is found. It coincides with SB predicted by Kampf and Schrieffer [14] based on the presence of AF spin fluctuations in the metallic state. A second superstructure present on the FS appears to be due to the incommensurate lattice modulation present in the Bi-cuprates[13].

On Pb-doped, modulation free crystals SB also occur. However, weak diffraction spots are observed in TEM SAED and LEED, corresponding to a structural layer unit cell identical to the one for a model with AF correlated spins. Therefore, even though the modulation disappears upon Pb doping the situation with respect to the origin of the SB is not clarified. More structural information, truly magnetic and T dependent measurements and/or FS maps on samples not presenting structural ambiguities are required.

Acknowledgment A special thank goes to P. Schwaller for all concerning the high T_c superconductors. Samples were generously provided by T. Mochiku, K. Kadowaki, Japan and by H. Berger at the EPFL. TEM experiments were done by C. Beeli at the EPFL. Fruitful discussions and collaboration with E. Boschung, M. Bovet, R. Fasel, T. Greber, J. Hayoz,

T.J. Kreutz, D. Naumović, J. Osterwalder, Th. Pillo, L. Schlapbach, are gratefully acknowledged. Skillful technical assistance was provided by E. Mooser, O. Raetzo, Ch. Neururer and F. Bourqui. This Project has been supported by the Fonds National Suisse de la Recherche Scientifique.

References

1. P. Aebi, J. Osterwalder, R. Fasel, D. Naumović, L. Schlapbach, Surf. Sci. **307-309**, 917 (1994).
2. P. Aebi, J. Osterwalder, P. Schwaller, L. Schlapbach, M. Shimoda, T. Mochiku, K. Kadowaki, Phys. Rev. Lett. **72**, 2757 (1994).
3. J. Osterwalder, T. Greber, A. Stuck, and L. Schlapbach, Phys. Rev. B**44**, 13764 (1991); D. Naumović, A. Stuck, T. Greber, J. Osterwalder, and L. Schlapbach, Phys. Rev. B **47**, 7462 (1993).
4. A. Santoni, L.J. Terminello, F.J. Himpsel, and T. Takahashi, Appl. Phys. A **52**, 229 (1991).
5. P. Aebi, T.J. Kreutz, J. Osterwalder, R. Fasel, P. Schwaller and L. Schlapbach, Phys. Rev. Lett. **76**, 1150 (1996).
6. V.L Moruzzi, J.F. Janak, A.R. Williams, *Calculated Electronic Properties of Metals* (Pergamon, New York, 1978).
7. N.W. Ashcroft, N.D. Mermin, *Solid State Physics* (Saunders College, Philadelphia) p. 290.
8. A.P.J. Stampfl, J.A. Con Foo, R.C.G. Leckey, J.D. Leckey, J.D. Riley, R. Denecke, L. Ley, Surf. Sci. **331-333**, 1272 (1995); J. Avila, C. Casado, M.C. Asensio, J.L. Perez, M.C. Muoz, F. Soria, J. Vac. Sci. Technol. A **13**, 1501 (1995); Z. Qu, A. Goonewardene, K. Subramanian, J. Karunamuni, N. Mainkar, L. Ye, R.L. Stockbauer, R.L. Kurtz, Surf. Sci. **324**, 133 (1995).
9. M. Bovet, Diploma Thesis, Université de Fribourg, 1997.
10. T.J Kreutz, PhD Thesis, Universität Zürich, 1997; T. Greber, T.J. Kreutz, J. Osterwalder, Phys. Rev. Lett. **79**, 4465 (1997).
11. P. Schwaller, PhD Thesis, Universität Zürich, 1997.
12. A. Yamamoto, M. Onoda, E. Takayama-Muromachi and F. Izumi, Phys. Rev. B **42**, 165 (1989).
13. J. Osterwalder, P. Aebi, P. Schwaller, L. Schlapbach, M. Shimoda, T. Mochiku, K. Kadowaki, Appl. Phys. A **60**, 247 (1995).
14. A.P. Kampf, J.R. Schrieffer, Phys. Rev. B **42**, 7967 (1990); Phys. Rev. B **41**, 6399 (1990).
15. P.A.P. Lindberg, Z.-X. Shen, B.O. Wells, D.B. Mitzi, I. Lindau, W.E. Spicer, A. Kapitulnik, Appl. Phys. Lett. **53**, 2563 (1988).
16. P. Schwaller, T. Greber, J.M Singer, J. Osterwalder, P. Aebi, H. Berger, L. Forró, Phys. Rev. B, submitted.

Magnetic Properties of Low-Dimensional Cuprates

K.-H. MÜLLER

IFW Dresden
POB 270016
D-01171 Dresden, Germany

ABSTRACT. The interest in quasi low-dimensional cuprates originated from the discovery of high-Tc superconductors typically consisting of intermediate valence ("doped") copper oxide planes with strongly correlated d-electrons. For understanding the mechanism of superconductivity in these materials their magnetic properties, even in the non-doped state, have to be considered. The magnetism of the cuprates mainly originates from the d-electrons of copper in the oxidation states Cu^I, Cu^{II} or Cu^{III}. In an ionic approximation the Cu species are described by charge states, e.g. Cu^{II} by Cu^{2+}. According to this approximation the ground state of Cu^I compounds has no magnetic moment and they exhibit diamagnetism or *van Vleck* paramagnetism. Cu^{2+} has an odd number of d-electrons resulting in so called *Kramers* degeneracy. Its paramagnetic moment is well approximated by the spin-only value of 1.73 Bohr magnetons. Depending on its anionic surrounding the paramagnetic moment of Cu^{3+} corresponds to the high-spin value of 2.83 Bohr magnetons (as in K_3CuF_6) or it may be zero (low spin) as usually in cuprates. In the ionic approximation a certain overlap of wave functions results in exchange interactions of the magnetic moments which may lead to antiferromagnetic or ferromagnetic long-range order. In the case of cuprates the dominating type of interaction is superexchange via oxygen anions. In a more realistic description covalence or, more generally, overlapping electron-wave functions combined with electron correlation have to be taken into account in order to explain effects as (i) the existence of well localized magnetic moments (in spite of delocalized wave functions), (ii) the real electric charge of the species usually being much smaller than following from the formal valency, (iii) the metallic or insulating behavior of particular compounds as e.g. $LaCuO_3$ or La_2CuO_4, respectively, and (iv) the special crystallographic structures which in the case of Cu^I compounds usually contain O-Cu-O dumbbells. In most cases the Cu^{II} cuprates contain quasi two-dimensional or quasi one-dimensional networks of CuO_4 plaquettes and they mostly behave like quantum spin-1/2 antiferromagnets of low dimensionality (d = 1 or d = 2). Usually a crossover to three-dimensional behavior occurs in these compounds at sufficiently low temperatures because, actually, they are three-dimensional solids.

51

S.-L. Drechsler and T. Mishonov (eds.), High-T$_C$ Superconductors and Related Materials, 51–80.
© *2001 Kluwer Academic Publishers. Printed in the Netherlands.*

1. Introduction

The discovery of high-T_c superconductivity in the La-Ba-Cu-O system by Bednorz an Müller [1] led to a world wide effort to explore compounds containing copper oxide layers or chains and related oxychlorides because they may become superconductors when doped. Fig. 1 shows the crystal structure of two typical high-T_c cuprates characterized by a network of corner-sharing CuO_4 plaquettes. Regarding the large spacing between the CuO_2 planes in these cuprates and their electronic structure they can be considered as consisting of weakly interacting two-dimensional subsystems [3, 4]. Experimental data on the structural, electronic and magnetic properties of the compounds $La_{2-x}Sr_xCuO_4$ (i.e. the Bednorz/Müller system with Sr instead of Ba) are summarized in the phase diagram of Fig. 2. Since La is expected to be always in the oxidation state (+3) and O in (-2) the electronic configuration of Cu in the undoped compound La_2CuO_4 is $3d^9$ with one hole per copper site. In a naive point of view such a material is expected to be a metal, but in reality it is an antiferromagnetic insulator because electron correlations prevent the charge fluctuations required for metallic conduction. As suggested by Mott and Hubbard, the electron correlations split a half-filled conduction band into two, with the lower band being filled and the upper band remaining empty. Antiferromagnetic insulators of this type are called Mott-Hubbard insulators [3]. It is generally accepted that strong electron correlation has to be taken into account to understand phase diagrams as that of Fig. 2. And it has been even suggested that the correlations are responsible for the electron attractions which are required to obtain the high superconducting transition temperatures observed in the cuprate superconductors [5]. Fig. 2 shows that, in particular at zero temperature, with increasing doping rate x various magnetic and electronic states occur which are separated by critical points: Néel-type antiferromagnetism, spin glass, superconductivity, normal-state metallic behavior etc. Additionally a quantum critical point is expected to lie hidden within the superconducting range of x, separating a non-Landau-Fermi liquid from a Landau-Fermi liquid [6]. Experimental and theoretical investigations have shown that even the undoped compound La_2CuO_4 has an intermediate valence ground state i.e. the Cu configurations fluctuate between $3d^9$ and $3d^{10}$ [3]. In La_2CuO_4 doped with Sr or Ba the added holes go predominantly into oxygen 2p dominated states because otherwise the Cu sites would be doubly occupied by holes [3, 7]. A similar phase diagram as that in Fig. 2, with antiferromagnetic and superconducting ground states, has been realized in $Nd_{2-x}Ce_xCuO_4$ i.e. by adding electrons instead of holes. The proximity of antiferromagnetism to superconductivity in both electron and hole doped cuprates has been interpreted to suggest that superconducting electron pairing in the cuprates may be mediated by antiferromagnetic spin fluctuations i.e. superconductivity is based on a magnetic pairing mechanism [8]. In doped cuprates above the superconducting transition temperature the added holes tend to segregate into charge stripes which act like domain walls between antiferromagnetic regions. Those

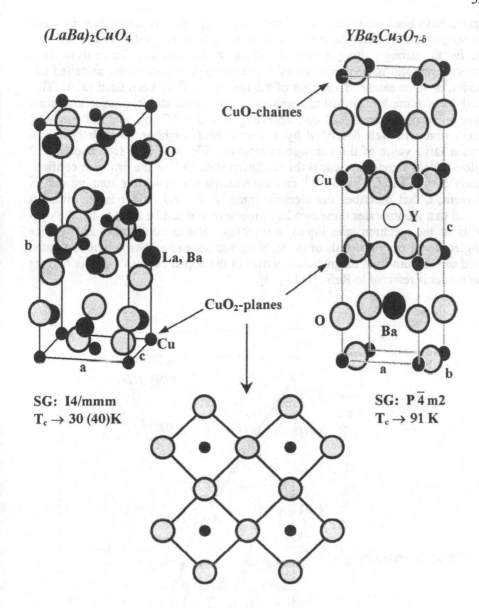

(LaBa)₂CuO₄

SG: I4/mmm
T_c → 30 (40)K

CuO-chaines

Cu

Y

CuO₂-planes

O

Ba

YBa₂Cu₃O₇₋δ

SG: P4̄m2
T_c → 91 K

O

La, Ba

Cu

Figure 1. Crystal structure of the typical high-T_c cuprate superconductors (La,Ba)₂CuO₄ [1] and YBa₂Cu₃O₇₋δ [2] and the nearly planar CuO₂ network consisting of corner-sharing CuO₄-plaquettes. SG and T_c are the space group and the achieved superconducting transition temperatures, respectively.

54

stripes have been observed by neutron scattering [9]. The interaction between the stripes and antiferromagnetic domains have been suggested to be responsible for the strong pairing interaction found in the cuprates. To analyse these observations and other properties of cuprate superconductors the so called t-J-model, or some extended versions of it have successfully been used [3, 6]. This model has a simple chemical interpretation. The parent stoichiometric insulators have an oxidation state Cu^{II} corresponding to a $S = 1/2$ ion, and the magnetic interactions are well described by a simple next-neighbor Heisenberg model with a large value of the exhange interaction, $J/k_B \approx 1500K$ (see Section 2.2 below). Hole doping introduces the oxidation state Cu^{III} in the low spin configuration $S = 0$. The Cu^{II} and Cu^{III} can interchange places with a hopping matrix element, t, that describes the electron transfer. The t-J model is the simplest model that incorporates these two key processes and can be used to handle concepts as the quantum spin liquid, resonating valence bond state, Zhang-Rice singlet etc [5, 6]. For details of the very interesting and extensively investigated field of electronic and magnetic properties of the doped cuprate superconductors the reader is referred to Refs. [1] to [13].

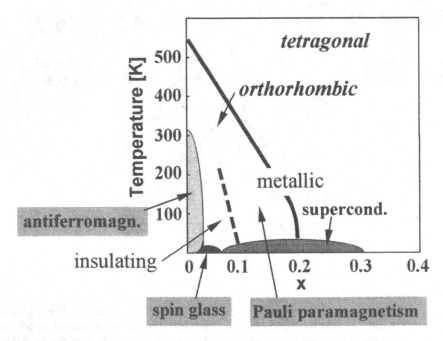

Figure 2. Magnetic / electronic / structural composition-temperature phase diagram for $La_{2-x}Sr_xCuO_4$ (after [4]).

In this study we will restrict ourselves to magnetic properties of some nominally integer valence cuprates and oxychlorides. Even under this restriction a large variety of phenomena of copper magnetism has been reported and many problems are still unsolved. The schematic presentation of Fig. 3 shows some aspects of the magnetism in cuprates. To start with the consideration of single Cu ions is suggested by the mentioned above fact that, due to strong electron correlation, a description by simple bands of independent electrons will definitely fail. The copper ions experience the influence of their anionic (or more or less covalent) anisotropic environment in the solid. Hence, even in this most simple approximation the magnetic properties will be anisotropic. In a more careful analysis the interaction between the magnetic ions in the solid have to be considered. Different types of magnetic anisotropy have to be taken into account and to be distinguished from the anisotropy connected with the (quasi) low dimensionality. The next step is to find a relevant spin Hamiltonian in order to describe and to predict the magnetic properties of the considered material. In the approximations needed for this purpose, the (quasi) dimensionality of the system has carefully to be treated because it is known to have crucial influence on the magnetic properties.

In the following Section few examples of magnetic properties of quasi low-dimensional compounds will be taken from literature and will be discussed in the simple ionic approximation. In Sections 3 and 4 some recent results obtained for two quasi low-dimensional copper oxychlorides will be presented.

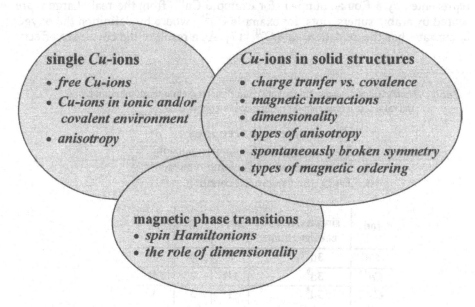

Figure 3. Aspects of magnetism in low-dimensional *Cu*-compounds.

2. Ionic compounds as a simplified picture

As already discussed in the previous section, Mott-Hubbard insulators are expected to be well described by electrons being localized on atoms or ions [14]. We will first consider the electronic and magnetic properties of free ions. If a magnetic field H is applied to an ion its quantum mechanical states will be modified resulting in small negative or positive contributions to its magnetic moment, known as diamagnetism or van Vleck paramagnetism, respectively. Additionally, if the atomic shells of the ions are only partially filled by electrons a magnetic moment may occur which does not vanish for $H \to 0$. This magnetic moment is closely related to the angular momentum of the ion which is a good quantum number because the ion is a rotationally symmetric object. Strictly speaking there are three types of angular momentum quantum numbers: total spin, S, total orbital momentum, L, and total momentum, J, the values of which are well determined by three Hund's rules [14, 15]. As indicated in Table 1 the first two Hund's rules that determine S and L are a result of the intraatomic electron correlation (Coulomb repulsion) whereas the third rule is based on spin-orbit interaction. The Hund's rule values of S, L, J for copper ions are presented in Table 1. In the solid cuprate structure the copper ions experience an interaction with their anionic environment characterized by (i) electrostatic interactions (crystal fields) and (ii) overlap of wave functions resulting in (indirect) exchange interaction with neighbor Cu ions and covalence effects based on hybridization of various electronic configurations [16]. Covalency manifests itself, e.g., by remarkable deviations of the oxidation state (or valence) here represented by a Roman number, for example Cu^{II}, from the real charge represented by arabic superscripts, for example Cu^{2+}, where by definition the oxygen in cuprates has the oxidation state O^{-II} [17]. As a result of the covalence effects

Table 1. Hunds' rules determining the total spin S, orbital angular momentum L and the total angular momentum J, applied to copper ions.

Hunds rules for free ions :

1. $S \to$ maximum (Coulomb repulsion)
2. $L \to$ maximum (Coulomb repulsion)
3. $J = L \pm S$ (spin-orbit coupling)

ion	single electron configuration	S	L	J
Cu^+	$3d^{10}$	0	0	0
Cu^{2+}	$3d^9$	1/2	2	5/2
Cu^{3+}	$3d^8$	1	3	4

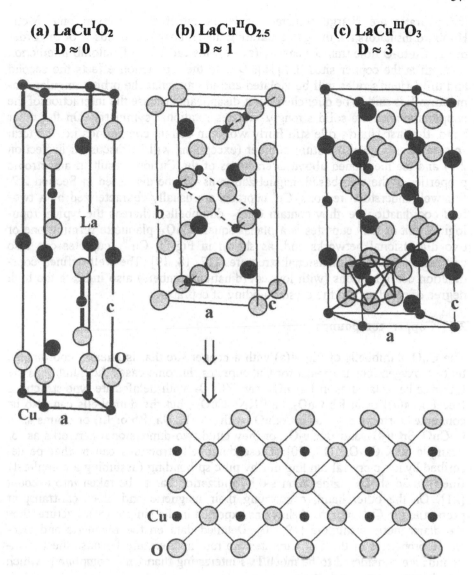

(a) LaCu^I O_2
D ≈ 0

(b) LaCu^II O_2.5
D ≈ 1

(c) LaCu^III O_3
D ≈ 3

Figure 4. Change of valence and crystal structure by oxidation. D is the (quasi) dimensionality of the lattice structure.
(a) LaCuO$_2$ (SG: R$\overline{3}$m) - isolated O-Cu^I-O dumbbells in the delafossite structure.
(b) La$_2$Cu$_2$O$_5$ (SG: Pbam) - two-leg ladder with Cu^{II}. The ladder is formed by corner-sharing CuO$_4$ plaquettes (lower part) being only nearly planar with the Cu-O-Cu bonding angle along the ladder leg of about 177°.
(c) LaCuO$_3$ (SG: R$\overline{3}$c) is a distorted perovskite. The CuO$_6$ octahedra are corner sharing and form a three-dimensional network.

the cuprates are charge transfer insulators rather than conventional Mott-Hubbard insulators i.e. the gap separating the Hubbard subbands is of the order of the Cu-to-O hole transfer energy (ε_p - ε_d) instead of the Coulomb interaction strength at the copper sites, U_d [16]. Due to the interaction effects the second and third Hund's rules will be violated and in particular the orbital angular momentum L is said to be quenched i.e. it disappears because the interaction of the copper ion with the solid strongly violates rotational symmetry. On the other hand, the first Hund's rule still fairly works in certain compounds i.e. the total spin remains a good quantum number (exceptions will be discussed in Section 2.3) and the mentioned above interactions of the Cu ions result in anisotropic properties of the spin based magnetization as will be discussed in Section 2.2. For well understood reasons, Cu^I cuprates are usually characterized by a two-fold coordination i.e. they contain CuO_2 dumbbells, whereas the typical topological unit of Cu^{II} cuprates is a planar square CuO_4 plaquette forming one or two dimensional networks and, as shown in Fig. 4, Cu^{III} cuprates can also crystallize in three-dimensional structures [17, 18, 19]. The well defined coordination configurations (with low coordination numbers) also indicate the high degree of covalency in the crystal binding of cuprates.

2.1 Copper^I compounds

The CuO_2 dumbbells of Fig. 4(a) with a copper site that is linearly coordinated by two oxygen ions is typical for Cu^I cuprates. In some cases the dumbbells are found to be isolated as in $LaCuO_2$ and $YCuO_2$ with delafossite type structures (see Fig. 4(a)) or in Rb_3CuO_2 and KNa_2CuO_2, but the dumbbells can also be connected forming rings as in ACuO (with A = K, Na, Rb or Li) or chains as in CsCuO, $SrCu_2O_2$ and $BaCu_2O_2$ or they build two-dimensional networks as for example in $K_3Cu_5O_4$ [19, 20]. These dumbbell structures can neither be described by ionic crystal binding nor by pure sp bonding (assuming a completely filled Cu 3d shell). Rather than s-d hybridization has to be taken into account [21]. On the other hand, concerning their magnetic and electrical-transport properties the Cu^I cuprates behave as expected in the simple ionic picture: they are diamagnetic insulators [19, 20]. Detailed data on the magnetic and electronic properties of the Cu^I cuprates are rare in literature because these compounds are considered to be much less interesting than Cu^{II} compounds which may become superconducting if doped.

As can be seen in Fig. 4 the formal oxidation state of $LaCuO_2$ can be increased from I to II or III by increasing the oxygen stoichiometry from 2 to 2.5 or 3. A similar "doping" procedure has been done for $YCuO_2$ where the valency of Cu could also be increased by partial substitution of Y by Ca. For the latter case contributions of Curie paramagnetism as well as Pauli paramagnetism to the magnetic susceptibility have been reported and suggested to indicate the presence of localized magnetic moments as well as quasi free electrons [20].

2.2 Copper^{II} compounds

The magnetic properties of cuprates with two-valent copper, Cu^{II}, have been extensively investigated because among them are the undoped parent compounds of high-T_c superconductors containing two-dimensional networks of corner-shared CuO_4 plaquettes as shown in Fig. 1. Most of these compounds are insulators and the ionic picture describing Cu^{II} as a $S = 1/2$ ion (see Table 1) works rather well. The contribution of such an ion to the magnetization (measured, e.g., at very low temperature and high magnetic field) is

$$\mu = gS\mu_B \approx 1\mu_B \qquad (1)$$

where $g \approx 2$ is the g factor or Landé factor of an ion with vanishing (i.e. totally quenched) orbital momentum L. The contribution of the same ion to Curie paramagnetism is

$$\mu_p = g\sqrt{S(S+1)}\,\mu_B \approx 1.73\,\mu_B \qquad (2)$$

The difference between μ and μ_p is due to quantum effects and it has its maximally achievable value for $S = 1/2$ [15]. In the cuprates the Cu^{II} magnetic moments dominantly interact with their next Cu neighbors via indirect exchange interaction through oxygen. Therefore these materials are often considered as ideal model systems for spin 1/2 quantum magnetism that can well be described by Heisenberg type Hamiltonians as

$$H = J\sum_{(i,j)} S_i S_j \qquad (3)$$

where (i,j) is over nearest-neighbor bonds and $S_i S_j$ is the scalar product of the quantum operators of spin 1/2 vectors. In the typical Cu-O-Cu straight bonds of corner-shared plaquettes (as in Figs. 1 and 4) the coupling is strongly antiferromagnetic ($J/k_B \approx 1500$ K) and results from hybridization of the oxygen $2p_\sigma$ orbitals with the neighboring 3d Cu orbitals (superexchange). In the large family of Cu^{II} cuprates there are also cases where the indirect exchange through

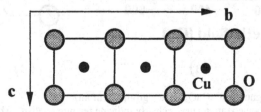

Figure 5. Edge-sharing CuO_4 plaquettes forming chains in Li_2CuO_2 [22, 23].

60

oxygen is not along straight lines [19]. Examples are compounds with edge-sharing CuO_4 plaquettes as in Li_2CuO_2 (Fig. 5) or compounds with more or less isolated plaquettes as in La_2BaCuO_5 (Fig. 6). In these cases the strength of the exchange interaction is reduced and J may even be negative (ferromagnetic coupling) [25-28]. Thus, for edge-sharing plaquettes the $2p_\sigma$ orbitals are almost orthogonal to the 3d orbital on the nearest Cu ion and the exchange via a single O ion is blocked. Then the effective coupling is given by higher-order perturbation terms and has a value of typically $J/k_B = -460$ K [25].

Magnetic order and ground states in quasi low-dimensional Cu^{II} cuprates

All materials including cuprates are three-dimensional magnetic systems. However, as discussed above, the strength of exchange interaction between the magnetic moments in Cu^{II} cuprates strongly depends on the local topology and the angle of Cu-O-Cu bonds. Therefore Cu^{II} cuprates usually consist of relatively weakly exchange-coupled subsystems with a dimensionality D lower than 3 where the exchange coupling within the subsystem is relatively strong. To some approximation such materials can be considered as being of (quasi low) dimensionality D [29]. In this sense the compounds of Figs. 1 are of D ≈ 2, those of Figs. 5 and 7 of D ≈ 1 and that of Fig 6 of D ≈ 0. Many rigorous theoretical results on magnetic properties of low-dimensional Cu^{II} cuprates have been de-

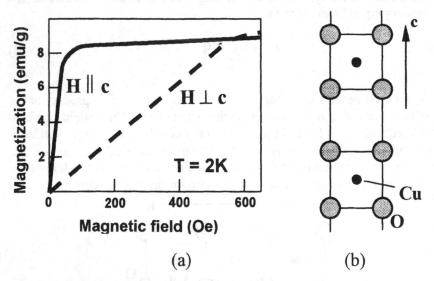

(a) (b)

Figure 6. The ferromagnet La_2BaCuO_5 (space group P4/mbm) .
    (a) The demagnetization curves clearly indicate ferromagnetism, where the tetragonal c-axis is the magnetically easy direction. Curie temperature T_c ≈ 6.5K [24].
    (b) CuO_4 plaquettes (parallel to the tetragonal c-axis) as typical for Cu^{II} cuprates.

rived by analysing the spin 1/2 Hamiltonian (3). For $J < 0$ and arbitrary dimensionality $D = 0, 1, 2$ or 3 the ground state can easily be shown to be ferromagnetic with totally aligned individual moments, each with its maximum value of 1 Bohr magneton. The direction of the collective magnetization can be fixed by a very small external field or a weak magnetic anisotropy. With increasing temperature, the magnetization decreases and vanishes at the Curie temperature T_c. However, a finite T_c (in the order of $|J|/k_B$) only exists for $D = 3$, whereas T_c is zero for $D < 3$ i.e. at finite temperatures ferromagnetic order cannot exist in finite clusters, linear chains or square lattices [31]. In the case of antiferromagnetic coupling ($J > 0$) in Equ. (3) the situation is more complicated. Now, even at $T = 0$ finite clusters as well as linear chains are in a singlet state i.e. the Cu^{II} sites do not exhibit any individual magnetic moments. The difference between the two systems is that in finite clusters ($D = 0$) the magnetic excitations are separated from the singlet ground state by a finite gap [32] called spin gap whereas for $D = 1$ there is no spin gap [33]. On the other hand for $D = 3$ [34] as well as the square lattice ($D = 2$) [35] a Néel type antiferromagnetic order with a staggered magnetization on the individual $copper^{II}$ sites has been theoretically shown to exist in the quantum mechanical ground state. Due to quantum fluctuations the individual staggered moment per Cu^{II} site in the square lattice is about $0.6\mu_B$ instead of the full value of Equ. (1) [36]. On the square lattice ($D = 2$) the magnetic order cannot survive at finite temperatures i.e. the Néel temperature T_N is zero [31]. Because the Néel type order violates the rotational symmetry of the Hamiltonian (3) there must exist zero-energy Goldstone modes (spin waves) and consequently there is no spin gap in that case [37]. For the

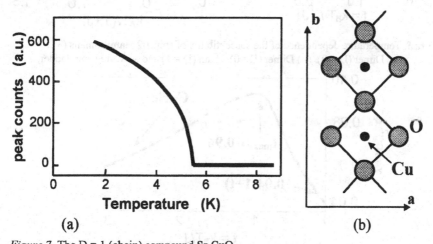

Figure 7. The $D = 1$ (chain) compound Sr_2CuO_3.

    (a) Temperature dependence of the intensity of the $(0,1/2,1/2)$ neutron diffraction peak, which is proportional to the staggered magnetization [30].

    (b) Corner-sharing CuO_4 plaquettes along the b axis (SG: Immm).

simplest $D = 0$ cluster, namely a dimer of two interacting spin 1/2 moments, the spin gap and the susceptibility χ can easily be calculated [38]. Upon cooling the antiferromagnetically coupled dimers from a high temperature, χ will increase similar as in a Curie paramagnet. On the other hand, the spin gap causes $\chi = 0$ at $T = 0$. Consequently, the temperature dependence of χ has a maximum as shown in Fig. 8. The maximum in $\chi(T)$ is a typical feature of low-dimensional Heisenberg antiferromagnets independent of the existence or non-existence of a spin gap (see Figs. 8 and 9). Two-leg spin ladders (as that in Fig. 4(b)), or more generally even-number leg ladders, show a spin gap and an exponential decay of spin-spin correlations. Their coherent singlet ground state is considered to be a realization of the quantum spin liquid or resonating valence bond state proposed by Anderson [5]. On the other hand, odd-number leg ladders (including uncoupled chains as that of Fig. 7) have a slow decay of spin-spin correlations and no spin gap [39].

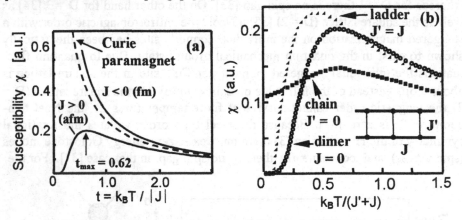

Figure 8. Temperature dependence of the susceptibility of spin 1/2 Hamiltonians (3). (a) Dimer ($D = 0$). (b) Dimer ($D = 0$), chain ($D = 1$) and a two-leg spin ladder.

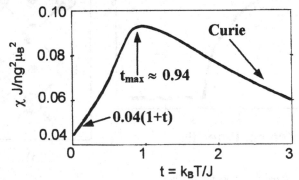

Figure 9. Susceptibility of the square-lattice spin 1/2 Heisenberg antiferromagnet, where n is the volume density of the spin 1/2 moments.

Dimensionality crossover and quantum critical points in Cu^{II} cuprates

Contrary to the discussed above theoretical predictions concerning the magnetic properties of low-dimensional magnetic systems, many quasi-low-dimensional Cu^{II} cuprates show long-range magnetic order and magnetic phase transitions at finite temperatures (as e.g. shown in Figs. 2, 6 and 7). It is generally accepted that the reason for these observations is the (relatively small) exchange interaction between Cu^{II} magnetic moments of low-dimensional subsystems. For quasi 2D Cu^{II} cuprates (as the undoped parent compounds of typical high-T_c superconductors (see Figs. 1 and 2)) Soukoulis et al. derived the formula

$$k_B T_N = \frac{4\pi J}{3\ln(32 J/J_\perp)} \tag{4}$$

using low-temperature spin wave theory [40]. In Equ. (4) T_N is the Néel temperature, J is the main antiferromagnetic exchange interaction (within the plaquettes of Fig. 1) and $J_\perp$ is the interplanar antiferromagnetic exchange interaction. This approximation works rather well for the rhombohedral La_2CuO_4 where the typical data are $T_N \approx 300$ K, $k_B J \approx 1500$K and $J_\perp/J \approx 10^{-5}$ [4]. Thus at high enough temperature the influence of $J_\perp$ is negligible and the quasi-low-dimensional materials behave as really low-dimensional ones. For example, the layered Cu^{II} cuprates will show a temperature dependence of susceptibility χ as shown in Fig. 9 which however is difficult to be observed because the maximum of χ and the Curie-type behavior will occur at temperatures as high as or even higher than 1500 K. At sufficiently low temperatures the influence of $J_\perp$ results in 3D magnetic ordering as presented in Fig. 2 and described by Equ. (4). Also a crossover from decreasing χ with decreasing temperature (as expected from Fig. 9 for $t < t_{max}$) to a special behavior of $\chi(T)$ connected with 3D ordering has been observed for La_2CuO_4 [41]. Such a dimensionality crossover upon cooling from lower to higher dimensionality has often been observed in quasi-low-dimensional materials [29]. In the case of La_2CuO_4 details of this crossover are influenced by anisotropic (Dzyaloshinsky-Moriya) exchange interaction. For tetragonal layered compounds (as e.g. $Sr_2CuO_2Cl_2$) $J_\perp/J$ is orders of magnitude smaller compared to La_2CuO_4 but T_N and J have similar values and therefore Equ. (4) cannot be valid for the magnetic phase transition in tetragonal cuprates. It is suggested that in these materials T_N is strongly influenced by anisotropic exchange interaction [42,43] (see next subsection).

A variation of the exchange coupling $J_\perp$ between low-dimensional subsystems can result in further phenomena that have to be distinguished from the dimensionality crossover discussed above. In the following this will be discussed for the two-leg ladder material La$_2$Cu$_2$O$_5$ (see Fig. 4(b)) which was first synthesized by Hiroi and Takano [44]. As a good approximation the coupling along the rungs of the ladder (J' in Fig. 8(b)) can be considered to be equal to the coupling J along the legs [45]. Although the single ladder clearly has dimensionality D = 1 its behavior remarkably deviates from that of a single chain i.e. it is a spin liquid with a spin gap resulting in zero susceptibility at zero temperature (see Fig. 8(b)). If now an interaction $J_\perp$ (of the same strength as J) is switched on the system will be a 3D antiferromagnet with staggered magnetic moments at the Cu^{II} sites and a gapless spin-wave spectrum. Thus an interesting question is how the system transforms between the magnetically ordered state and the spin-liquid state if $J_\perp$ is varied between 0 and J. It was shown by quantum Monte Carlo simulations that the spin gap opens below a critical interladder coupling $J_\perp \approx 0.11$ J [45]. This value can be considered as a quantum critical point (a term introduced by Chakravarty et al. [35]) characterizing the quantum phase transition between the ordered state and the spin liquid at zero temperature. Using and generalizing the bond-operator mean-field theory the Néel temperature of the system could be calculated for the whole range $0 \leq J_\perp \leq J$ (see Fig. 10) and a critical value $J_\perp /J \approx 0.12$ was found [46]. Taking into account these theoretical approaches the behavior of La$_2$Cu$_2$O$_5$ could be well understood [45, 46]. The value T$_N$ = 117 K (or $k_B T_N \approx 0.1$ J) indicates that the system is above but rather close to the quantum critical point. At intermediate temperatures the system crosses over to the 1D decoupled-ladders regime.

Systems of weakly coupled even-number-leg ladders have been predicted to become superconducting if doped with holes [39]. In La$_2$Cu$_2$O$_5$ a transition to metallic behavior occurs upon doping with Sr but no superconductivity has been found so far. However superconductivity has been reported for the two-leg-ladder compound Sr$_x$Ca$_{14-x}$Cu$_{24}$O$_{41}$ which has weak interladder couplings [47].

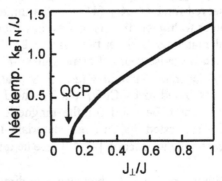

Figure 10. Néel temperature T$_N$ of interacting two-leg ladders. $J_\perp$ is the strength of the coupling between the ladders, J the coupling strength along the legs and along the rungs of the ladders. QCP marks the quantum critical point at $J_\perp \approx 0.12$ J [46].

Magnetic anisotropy in Cu^{II} cuprates

In condensed matter a Cu^{II} site has always a discrete environment of neighbors, i.e. anions or cations etc., which represent a local anisotropy. In crystalline solids the local anisotropy adds up to the macroscopic crystalline anisotropy which is manifested by anisotropic physical properties of the considered material. In the previous subsection we discussed the directional dependence of the strength of exchange interactions, resulting in quasi-low dimensionality which is a special type of anisotropy. Note that low dimensionality has to be distinguished from real magnetic anisotropy which is the subject of this subsection. For example, the Heisenberg Hamiltonian (3) is isotropic i.e. rotationally symmetric also for chains (D = 1) or square lattices (D = 2). Magnetic anisotropy is the dependence of magnetic properties on the direction (with respect to the crystal axes). A typical example is the directional dependence of demagnetization curves shown in Fig. 6. Further examples will be presented in Sections 3 and 4.

As discussed above the magnetic moment on a Cu^{II} site is based on the copper spin and therefore the anisotropy of the solid can be transmitted to the magnetization only by spin-orbit (LS) interaction or by magnetic dipolar interaction which however can be neglected in most cases. Thus in any approach to magnetic anisotropy in Cu^{II} cuprates LS coupling has to be explicitly taken into account. (In the case of rare-earth magnetism the LS coupling is responsible for the third Hund's rule of Table 1 and therefore it does not need to be considered explicitly). Lets first consider the magnetic anisotropy on a single Cu^{II} site. Usually crystalline-electric-field (CEF) splitting in cuprates is large compared to both $k_B T$ (temperature) and LS coupling. Therefore the orbitally non-degenerated CEF ground state with quenched orbital angular momentum and twofold spin degeneracy is a good zero order approximation [14, 38]. The LS coupling (of strength λ) and the Zeman energy can be considered as a perturbation

$$H_1 = \lambda LS + \mu_B H (2S + L) \tag{5}$$

Collecting the terms proportional to the field H from the result of first and second order perturbation theory results in an effective Zeman energy

$$H_{eff} = \sum_{\mu\nu} 2\mu_B H_\mu (\delta_{\mu\nu} - \lambda\Lambda_{\mu\nu}) S_\nu \tag{6}$$

with

$$\Lambda_{\mu\nu} = \sum_n \frac{\langle 0|L_\mu|n\rangle\langle n|L_\nu|0\rangle}{E_n - E_0} \tag{7}$$

where E_n and $|n\rangle$ are the CEF energy levels and eigenfunctions, respectively, and L_μ are the components of the orbital angular momentum operator. Taking into account Equ. (6) the g factor (1) has to be replaced by the g tensor

$$g_{\mu\nu} = 2(\delta_{\mu\nu} - \lambda\Lambda_{\mu\nu}) \tag{8}$$

For Cu^{II} in elongated tetragonal surroundings with C_{4v} symmetry the second term in Equ. (8) is positive and typical values of $g_{\mu\nu}$ are $g_{zz} = 2.20$ and $g_{xx} = g_{yy} = 2.08$ [38]. The increase of g (> 2) is due to an induced orbital momentum which arises from mixing with higher-energy orbital states due to the LS coupling. The g tensor (8) represents the magnetic anisotropy of a single Cu^{II} moment. The perturbation term proportional to λ^2 contributes to magnetic anisotropy only for $S > 1/2$. Therefore it is not effective for Cu^{II} moments.

A further type of magnetic anisotropy in Cu^{II} cuprates is superexchange modified by LS coupling, which is usually much more efficient than the anisotropy of g. In Hamiltonians as that in Equ. (3) the spin is present only in order to fulfill the Pauli principle. Therefore Hamiltonians describing pure exchange interactions are invariant under (physically meaningless) rotations in spin space [32]. Taking the LS coupling on two neighboring Cu^{II} sites, together with the isotropic superexchange between them, as a perturbation to the same nonperturbed states as considered above for the calculation of the g tensor, perturbation theory results in the following spin Hamiltonian that has to be added to the isotropic Hamiltonian (3)

$$H_a = \alpha_{DM} J \sum_{(i,j)} d_{ij}(S_i \times S_j) + \alpha_{xy} J \sum_{(i,j)} S_i^z S_j^z \tag{9}$$

where $\alpha_{DM} \sim \lambda$ and $\alpha_{xy} \sim \lambda^2$ and d_{ij} are vectors depending on the bond (i,j) and the symmetry of its environment [13, 14, 48, 49]. The first term in Equ. (9), the Dzyaloshinsky-Moriya type antisymmetric exchange interaction, was introduced by Dzyaloshinsky [50] by symmetry analysis on a macroscopic scale in order to explain the phenomenon of weak ferromagnetism which occurs also in Cu^{II} cuprates if the crystalline anisotropy is sufficiently low [49, 4, 13] (see also Section 3). The second term in Equ. (9) may result in a Kosterlitz-Thouless phase transition even if the (i,j) in Equs. (3) and (9) run over a 2D square lattice [51]. The 2D low-temperature Kosterlitz-Thouless phase does not carry a staggered magnetization. However it is generally accepted that the xy term in Equ. (9) together with the interlayer coupling $\alpha_\perp = J_\perp /J$ (see previous subsection) results in a 3D xy transition which is assumed to assist the development of Néel type magnetic order in tetragonal layered Cu^{II} cuprates as $Sr_2CuO_2Cl_2$ where $\alpha_\perp \ll \alpha_{xy}$ [42, 43]. This approach results in the following expression for the Néel temperature of layered cuprates

$$T_N = \frac{0.3\,\pi k_B\, J}{ln(4\alpha_{eff} /[0.3\pi^2 ln(4\alpha_{eff}/\pi)])} \tag{10}$$

with $\alpha_{eff} = 4\alpha_{xy} + 2\alpha_\perp$ [43,13].

2.3 Copper^{III} compounds

As shown in Fig. 11 the ionic compound Cs_3CuF_6 contains isolated $Cu^{III}F_6$ octahedra and it is an ideal Curie paramagnet with a paramagnetic moment of 3 Bohr magnetons per copper site [52]. This is in good agreement with the ionic approximation which predicts $S = 1$ (see Table 1) and $\mu_p = 2.8$ μ_B (see Equ. (2)) for Cu^{III} ionic compounds.

A net oxidation state Cu^{III} can also be realized in cuprates. But, so far, these compounds are not intensely investigated because in the high-T_c superconductor scene they are considered as overdoped Cu^{II} cuprates which do not show superconductivity. A common feature of the Cu^{III} cuprates is that they do not have a magnetic moment corresponding to $S = 1$. In Fig. 12 various mechanisms are outlined which, in principle, could be responsible for the Cu^{III} magnetic moment to disappear. The first mechanism is zero-field splitting (Fig. 12(a)) resulting in the $S_z = 0$ ground state which can occur e.g. in tetragonal crystalline electric fields. For temperatures small compared to the energy difference of the $S = 0$ and the $S_z = \pm 1$ level the system will behave like a $S = 0$ (singlet) system [38]. On the other hand, low-spin states (Fig. 12(b)) occur if the crystal-field splitting of the single-electron levels is strong compared to the intraatomic interaction normally ensuring the 1st Hund's rule. A subset of lower crystal-field levels will be filled violating 1st Hund' rule and, consequently, the total spin

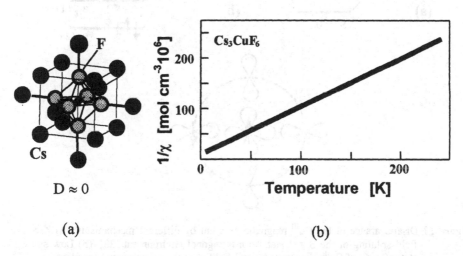

D ≈ 0

(a)   (b)

Figure 11. (a) Local environment of Cu^{III} in Cs_3CuF_6 : The Cu^{3+} ion is surrounded by an ideal octahedron of 6 F⁻ ions. These octahedrons are isolated in the crystal structure, i.e. Cs_3CuF_6 can be considered as quasi zero dimensional.
(b) The inverse-susceptibility-vs.-temperature curve shows that Cs_3CuF_6 is an ideal *Curie* paramagnet. Its paramagnetic moment is $3\mu_B$ [52].

vanishes [53]. In recent years a further concept has been generally accepted for doped cuprates which explicitly takes into account the observed presence of a hole near the oxygen sites and the high degree of covalency in these compounds. According to this model a hole on the copper site forms a singlet, called Zhang-Rice singlet [54], with a second hole being located on (and shared by) the surrounding oxygen ions. Among the Cu^{III} cuprates diamagnetic insulators, as e.g. $ACuO_2$ (A = Li, Na, K, Rb, Cs) [55] or La_4LiCuO_8 [56], as well as metals with Pauli paramagnetism as e.g. $LaCuO_3$ [57] have been found. The crystal structure of $LaCuO_3$ is shown in Fig. 4(c).

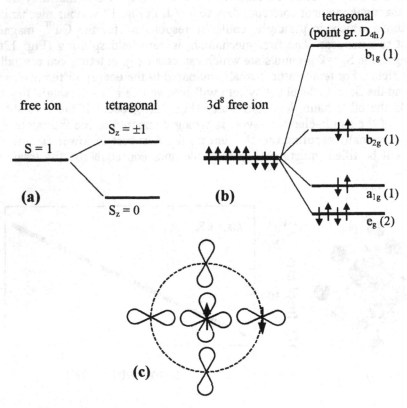

Figure 12. Disappearance of the Cu^{III} magnetic moment by different mechanisms: (a) Zero field splitting of the S = 1 state in a tetragonal environment [38]. (b) Low spin state (S = 0) of Cu^{III} in a strong crystal field caused by a tetragonal (or planar) environment: splitting of the 3d level of the free ion. The orbital degeneracy is given in brackets. The arrows indicate the occupation of the levels according to a modified Hund's rule [53]. According to first Hund's rule the free ion is in the high-spin state S = 1. (c) Two holes, one on a copper and the other on an oxygen site, forming a Zhang-Rice singlet [54, 3].

3. Weak ferromagnetism in $Ba_2Cu_3O_4Cl_2$

Examples for well investigated copper oxychlorides are $Ba_2Cu_3O_4Cl_2$ and $Sr_2Cu_3O_4Cl_2$ (space group I4/mmm, [58]). Their lattice structure contains two Cu sites, Cu_A and Cu_B (see Fig. 13(a)). The Cu_A ions form the typical, for Cu^{II} compounds, planar Cu_AO_4 plaquettes which build a planar corner-sharing network (Fig. 13(b)). The Cu_B ions fill half of the remaining empty oxygen squares. Therefore, the magnetic behavior of these compounds is expected to be that of 2D quantum Heisenberg antiferromagnets. The two compounds differ only little in their ordering temperatures and we focus here on $Ba_2Cu_3O_4Cl_2$. For all Cu_A sites the bond angle Cu_A-O-Cu_A is 180°, whereas the angle for the Cu_A-O-Cu_B bonds is 90° and the Cu_B-Cu_B interaction is expected to be very small. Two antiferromagnetic ordering temperatures are observed: At $T_{NA} \approx 337K$ the Cu_A ions order antiferromagentically, whereas the Cu_B ions order at a quite lower temperature $T_{NB} \approx 33K$ ($Ba_2Cu_3O_4Cl_2$) [59-61]. As the magnetization-versus-field curves for applied fields parallel to the [001] direction pass the origin (see Fig. 14(a)), the easy direction of the staggered magnetization is in the (a,b) plane. The finite magnetization, M_0, measured for H → 0 in both directions in the basal plane, [100] and [110], indicates the presence of a small spontaneous magnetization [60-62]. A comparison of the M(H) curves in Fig. 14(a) shows that the easy direction for the staggered magnetization is [110].

space group: I4/mmm

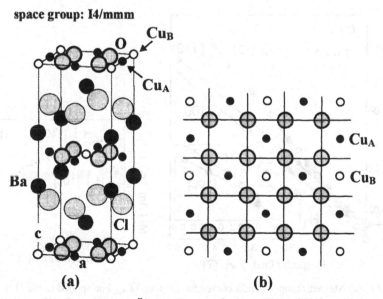

(a) **(b)**

Figure 13. Crystal structure of the Cu^{II} compound $Ba_2Cu_3O_4Cl_2$ [58]. (a) unit cell. (b) planar network of corner-sharing Cu_AO_4 plaquettes and isolated Cu_BO_4 plaquettes. The Cu_BO_4 plaquettes share their edges with those of the Cu_AO_4 plaquettes.

70

Furthermore, some metamagnetic behavior i.e. a small kink of the M(H) curve at a threshold field H_t is found for H || [100]. At higher fields the demagnetization curve measured for H || [100] (cf. Fig. (14a)) approaches a straight line passing through the origin. Thus, the "kink" is suggested to be some kind of spin-flop transition of the staggered magnetization into the direction perpendicular to the field direction. The temperature dependence of the residual magnetization M_0, measured for H || [100] and H || [110], is given in Fig. 14(b). The value of M_0 is not much different for both directions and it vanishes at $T \approx 337$ K which is close to the reported [60, 61] upper Néel temperature T_{NA}. Significantly smaller values of M_0 are observed for H || [100], only below the lower of the two Néel temperatures, $T_{NB} \approx 33$ K, which can be determined from the cusps in the temperature dependence of the inverse magnetic susceptibility, $1/\chi$, presented in Fig. 15.

Contrary to the results of Ref. 60 where ferromagnetic hysteresis was reported for H || [110] but not for H || [100], in later investigations [61] similar types of hysteresis curves were found for both directions. This is not surprising in view of the square symmetry in the tetragonal basal plane. Typical hysteresis loops measured for H || [100] are presented in Fig. 16. Above the lower Néel temperature T_{NB} the coercive field H_c is more than one order of magnitude smaller than below. Between T_{NB} and T_{NA} $\mu_0 H_c$ is only about 0.5 mT, for both directions [110] and [100], and does not much depend on temperature [61].

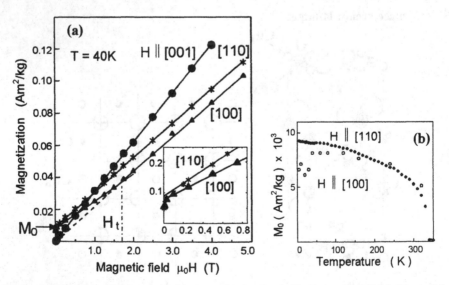

Figure 14. (a) Magnetization-vs.-field curves for $Ba_2Cu_3O_4Cl_2$. For applied fields H parallel to [110] or [100] a small residual magnetization M_0 is observed. For H parallel to [100] a kink is found at a threshold field H_t. Above H_t the magnetization curves for [110] and [100] are parallel to each other. (b) Temperature dependence of M_0. (Ref. 61)

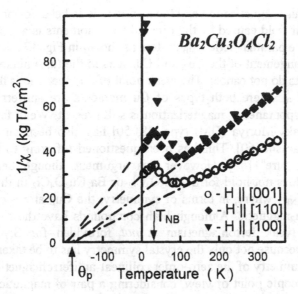

Figure 15. Temperature dependence of the inverse susceptibility $1/\chi$ of $Ba_2Cu_3O_4Cl_2$ [61].
For applied fields parallel to [110] or [100] a cusp at the ordering temperature T_{NB}
of the Cu_B moments is observed. θ_p is the paramagnetic Curie temperature.

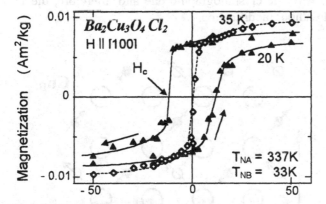

Figure 16. Above T_{NB} a residual net magnetization but (nearly) no hysteresis is observed.
Below T_{NB} where the Cu_B moments are also antiferromagnetically ordered, a
hysteresis is found as typical for ferromagnets. (Ref. 61)

The microscopic mechanism for the magnetic properties of $Ba_2Cu_3O_4Cl_2$, as
presented in the Figures and reported in Refs. 60, 61 and 62 remains to be clarified. An anisotropic pseudodipolar interaction between the antiferromagnetically ordered Cu_A and the paramagnetic Cu_B moments has been successfully
proposed [62] to explain, for the temperature range $T_{NB} \leq T \leq T_{NA}$, the presence

of a spontaneous magnetization and the metamagnetic behavior of $Sr_2Cu_3O_4Cl_2$.
A quasi-dipolar field caused by the ordered Cu_A moments is assumed to induce
a moment in the paramagnetic Cu_B system as shown in Fig. 17. Note, that due to
the special arrangement of the Cu_A and Cu_B ions in the (a,b) planes, these fields
and their effects do not cancel. The weak point of this scenario is that it does not
work below T_{NB} where both types of Cu moments are antiferromagnetically
ordered but a spontaneous magnetization is still present. Weak ferromagnetism
of Dzyaloshinsky-Moriya (DM) type [48, 50] has also been quoted to be the
relevant mechanism [60]. This has been questioned referring to the "perfectly
tetragonal structure" [62]. However this argument, though being valid for
$Sr_2CuO_2Cl_2$, does not hold for $Sr_2Cu_3O_4Cl_2$ or $Ba_2Cu_3O_4Cl_2$ in the temperature
range above T_{NB}, where, in terms of symmetry, the Cu_B atoms can be considered to be non-magnetic. Although both compounds have the same tetragonal
space group, weak ferromagnetism is *not* forbidden for $Sr_2Cu_3O_4Cl_2$ and
$Ba_2Cu_3O_4Cl_2$ because not only the crystal symmetry has to be taken into account
but also the symmetry of the zero-order collinear antiferromagnetic order. From
a more microscopic point of view, considering a pair of magnetic ions one has
to take into account the symmetry of the whole environment (see Section 2.2,
[48]) which here is also determined by the Cu_B ions. Below T_{NB} two different
types of Cu^{++} ions are antiferromagnetically ordered and the magnetic unit cell
is not identical with the crystallographic one and, therefore, the DM scenario
cannot be used without modification.

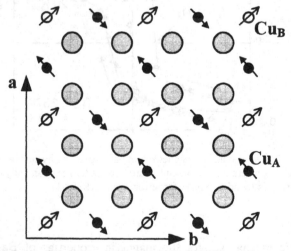

Figure 17. Suggested magnetic structure of $Ba_2Cu_3O_4Cl_2$ in the temperature range $T_{NB} < T < T_{NA}$. The local direction of the moments is indicated by arrows. The Cu_A moments show a Néel-type antiferromagnetic order and they induce a small net magnetization in the Cu_B sublattice (after Ref. 62).

4. Spin flop transition in $Ba_3Cu_2O_4Cl_2$

A further example for compounds containing planar corner-sharing CuO_4 plaquettes is $Ca_3Cu_2O_4Cl_2$ (see Fig. 18) [63]. It is an insulator that becomes superconducting if doped with Na [64]. Néel-type antiferromagnetic ordering has not yet been reported. The complete substitution of Ca by the chemically very similar Ba, which however has a much larger ionic radius, results in drastic changes: The structure of the corresponding compound $Ba_3Cu_2O_4Cl_2$ is orthorhombic (space group Pmma). The two Cu sites, Cu_A and Cu_B, within the unit cell and the surrounding oxygen ions form nearly regular squares. However, these plaquettes are edge-sharing and build folded chains (see Fig 18, [65]). The axes of the chains are parallel to the orthorhombic a-axis. Thus, the two struc-

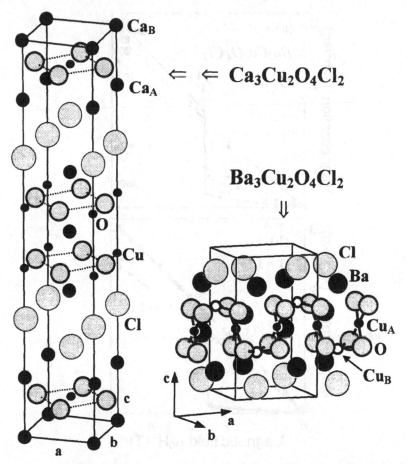

Figure 18. Crystal structures of the D ≈ 2 compound $Ca_3Cu_2O_4Cl_2$ (SG I4/mmm) with an ideally planar CuO_2 network of CuO_4 squares [63] and the folded-chain compound $Ba_3Cu_2O_4Cl_2$ (SG Pmma) with two types of Cu^{II}-sites, Cu_A and Cu_B [65].

74

tures are not related to each other by a small distortion only. The field dependence of the magnetization, M, measured along the a-axis is shown in Fig. 19(a).
Below $T_N \approx 20$ K these M(H) curves show a metamagnetic transition [67] i.e. a
strong upward curvature in a limited range of the magnetic field H. At a threshold field, H_t of about 2.6T the M-vs.-H curves measured at low temperatures
jump between two straight lines. Above the jump, the slope is considerable
larger than below. Whereas M, and thereby the susceptibility χ, increases nearly
linearly with increasing temperature at fields below H_t and below T_N, above H_t
only a weak dependence on temperature is observed. No such jump of the magnetization has been found for $H \parallel b$ or c (cf. Fig. 19(b)). These observations
indicate a spin-flop transition [68] in the case $H \parallel a$ i.e. a transition from a collinear antiferromagnetic structure with localized moments aligned along the

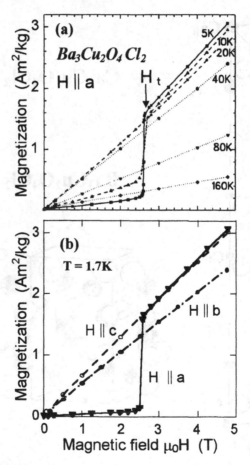

Figure19. (a) For fields H applied parallel to the a-axis, below the Néel temperature $T_N \approx 20$K:
spin flop transition characterized by a threshold field H_t. Above T_N : paramagnetic
behavior. (b) For H parallel to the b- or c-axis: no spin-flop. [61, 66]

easy a-axis to a configuration perpendicular to the field. From $K_1 = H_t^2(\chi_\perp - \chi_\parallel)/2$ [14] with the susceptibilities measured in fields parallel ($\chi_\parallel$) and perpendicular ($\chi_\perp$) to the easy a-axis the corresponding anisotropy constant is estimated to be $K_1 \approx 1.1 \cdot 10^4$ VAs/m^3 (T = 1.7K). Above the spin-flop transition the moments progressively rotate towards the field direction. In that field range the increase of M is mainly determined by exchange interaction and therefore only weakly depends on temperature. The temperature dependence of the inverse susceptibility $1/\chi$, shown in Fig. 20, is consistent with the concept of spin flopping. For $T < T_N$ and $H < H_t$, χ remains nearly constant for $H \parallel$ b or c, as expected for a classical antiferromagnet, but it increases with decreasing temperature for $H \parallel$ a. For sufficiently high temperatures all the χ(T)-curves are well described by the Curie-Weiss law, $\chi = C/(T - \theta)$, with the two fitting parameters C (Curie constant) and θ (paramagnetic Curie temperature). The positive value of $\theta \approx 21$K (being within the estimated error the same for all three directions) manifests that the Cu^{II} moments experience a ferromagnetic exchange interaction in addition to the dominating antiferromagnetic exchange interaction. The Curie constants are different for different directions which is compatible with an anisotropic g-factor. As the Cu^{II} sites are surrounded by four O^{2-} ions forming squares, the g-factors, i.e. the components of g averaged with respect to the two Cu sites should be larger than 2 (cf. Equ. (8), [38, 69]), which is in agreement with the experimental data g$_a$ = 2.37, g$_b$ = 2.19 and g$_c$ = 2.28. In Fig. 19(b) the M(H)-curves for applied fields parallel to the a-, b-, and c-axis are compared for T = 1.7K. Whereas the M-vs.-H curve for $H \parallel$ c has nearly the same slope as that measured for $H \parallel$ a above the threshold field, the slope of M-vs.-H for $H \parallel$ b is considerably lower. This can qualitatively be understood considering the orientation of the field H with respect to the CuO$_4$ squares in the folded chain. For $H \parallel$ a as well as for $H \parallel$ c for one half of the squares H is in the square plane, for the other half nearly perpendicular to this plane. On the other hand, for $H \parallel$ b for every square H is in the plane of the CuO$_4$ squares, which gives rise to the different M(H) dependence. The spin structure of Ba$_3$Cu$_2$O$_4$Cl$_2$ has not yet been determined experimentally. From symmetry arguments it follows that (i) the moments of crystallographically equivalent Cu moments are parallel, (ii) the moments of the non-equivalent sites are antiparallel, and (ii) spin canting of the Cu$_A$ moments leading to weak antiferromagnetism cannot be excluded (Fig. 21) [70].

Thus, despite the occurrence of one-dimensional CuO$_2$-chains Ba$_3$Cu$_2$O$_4$Cl$_2$ is a classical 3D anisotropic antiferromagnet. It is suggested that this is closely related to results of LDA band structure calculations which give a relatively strong dispersion of the energy bands perpendicular to the chain axis i.e. the next nearest neighbor transfer integrals are of the same order as some inter-chain hopping integrals [71,72]. Therefore Ba$_3$Cu$_2$O$_4$Cl$_2$ has to be considered as being, at least, two-dimensional. This example shows that the physical dimensionality may be quite different from that supposed by a first glance at the lattice structure.

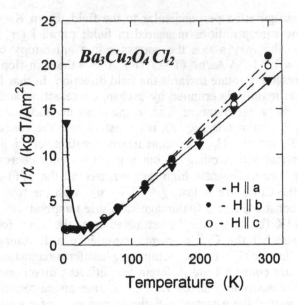

Figure 20. The susceptibility shows the typical behavior of a classical antiferromagnet: for applied fields H parallel to one direction, here the a-axis, $1/\chi$ diverges for vanishing temperature. For the other two directions it goes to a finite and constant value. [61]

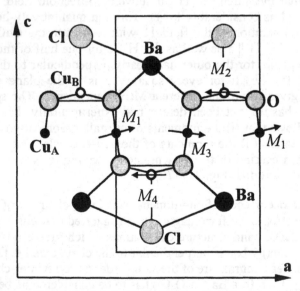

Figure 21. Folded CuO_2 chain in $Ba_3Cu_2O_4Cl_2$ and weak antiferromagnetic structure following from symmetry arguments. [70]

5. Summary and Conclusions

This study was restricted to copper magnetism in integer-valence cuprates (and copper oxyhalogenides). A natural and important extension would be to consider also rare-earth containing cuprates (where La is substituted by 4f elements) and mixed-valence compounds containing strongly correlated itinerant d electrons. Even under the mentioned above limitation the cuprates show a large variety of magnetic phenomena originating from the d-electrons of copper in the oxidation states Cu^{I}, Cu^{II} or Cu^{III}. The ionic approximation for the Cu species, describing oxidation states by charge states works fairly well. According to this approximation the ground state of Cu^{+} compounds has no magnetic moment and they exhibit diamagnetism or van Vleck paramagnetism. The orbital angular momentum of Cu^{II} and Cu^{III} is quenched due to crystalline-electric-field effects and covalency. Cu^{2+} has an odd number of d-electrons i.e. it is a so called *Kramers* ion with a doubly degenerated ground state. Its paramagnetic moment is well approximated by the spin-only value of 1.73 Bohr magnetons. The magnetic moment of Cu^{3+} in cuprates is usually zero (low spin) due to a higher degree of covalency. The Cu^{III} cuprates, in particular their magnetic properties, are not much investigated so far. Concerning the linkage of the copper sites by overlapping wave-functions (in the localized-electron picture) or by dispersion of energy bands (in the description by itinerant electrons) most of the cuprates can be considered as quasi-two or one or zero dimensional systems. The majority of published experimental and theoretical results on copper magnetism concern the quasi-two dimensional Cu^{II} cuprates because they contain most of the undoped parent compounds for high-T_c superconductors. In the ionic approximation the overlap of wave functions results in exchange interactions of the magnetic moments which may lead to antiferromagnetic or ferromagnetic long-range order. In the case of cuprates the dominating type of interaction is antiferromagnetic superexchange via oxygen anions. Therefore the mentioned parent compounds can be considered, in a good approximation, as 2D square lattice Heisenberg antiferromagnets. However, for understanding the Néel type magnetic ordering of these materials at finite temperatures the crossover to dimensionality three, mediated by interlayer exchange coupling of strength $\alpha_{\perp}$, and to Kosterlitz-Thouless type ordering due to symmetric anisotropic exchange (strength α_{xy}), have to be taken into account. These crossover phenomena are not yet well understood. Progress in this field will depend on whether a successful combination of analytical approaches with computational modeling can be developed. One of the main aims of this research will be to establish complete T-$\alpha_{\perp}$-α_{xy}-phase diagrams. A further problem is how Dzyaloshinsky-Moriya type interaction modifies these phase diagrams or causes spin canting that may result in weak ferromagnetism or weak antiferromagnetism. A promising modification of the layered cuprates is to break-up the CuO_2 planes into weakly coupled ladders. Such spin ladders can be used to study quantum phase transitions and crossover phenomena between dimensionality 0, 1, 2 and 3.

6. References

1. Bednorz, J.G., and Müller, K.A. (1986) Possible High T_c Superconductivity in the Ba-La-Cu-O system, *Z. Phys. B* **64**, 189-193.
2. Wu, M.K., Ashborn, J.R., Torng, C.J., Hor, P.H., Meng, R.L., Gao, L., Huang, Z.J., Wang, Y.Q., and Chu, C.W. (1987) Superconductivity at 93K in a New Mixed-Phase Y-Ba-Cu-O Compound System at Ambient Pressure, *Phys. Rev. Lett.* **58**, 908-910.
3. Fulde, P. (1995) *Electron Correlations in Molecules and Solids*, Springer, Berlin.
4. Johnston, D.C. (1997) Normal-State Magnetic Properties of Single-Layer Cuprate High-Temperature Superconductors and Related Materials, in K.H.J. Buschow (ed.), *Handbook of Magnetic Materials*, Elsevier, Amsterdam, vol. 10, p.1.
5. Anderson, P.W. (1987) The Resonating Valence Bond State in La_2CuO_4 and Superconductivity, *Science* **235**, 1196-1198.
6. Rice, T.M. (1997) Reviews, Prospects, and Concluding Remarks. High-T_c Superconductivity – Where Next ?, *Physica C* **282-287**, xix-xxiii .
7. Nücker, N., Fink, J., Renker, B., Ewert, D., Politis, C., Weijs, P.J.W., and Fuggle J.C. (1987) Experimental Electronic Structure Studies of $La_{2-x}Sr_xCuO_4$, *Z. Phys. B* **67**, 9-14.
8. Maple, M.P. (1998) High Temp. Superconductivity, *J. Magn. Magn. Mater.* **177-181**, 18-30.
9. Tranquada, J.M., Sternlieb, B.J., Axe, J.D., Nakamura, Y., Uchida, S. (1995) Evidence for Stripe Correlations of Spins and Holes in Copper Oxide Superconductors, *Nature* **375**, 561-563.
10. Dagotto, E. (1994) Correlated Electrons in High-Temperature Superconductors, *Rev. Mod. Phys.* **66**, 763-840.
11. González, J., Martín-Delgado, M.A., Sierra, G., and Vozmediano, A.H. (1995) Quantum Electron Liquids and High-T_c Superconductivity, Springer, Berlin.
12. Anderson, P.W. (1997) *The Theory of Superconductivity in the High-T_c Cuprates*, Princeton University Press, Princeton NJ.
13. Kastner, M.A., Birgeneau, R.J., Shirane, G., and Endoh, Y. (1998) Magnetic, Transport, and Optical Properties of Monolayer Copper Oxides, *Rev. Mod. Phys.* **70**, 897-928.
14. Yosida, K. (1996) *Theory of Magnetism*, Springer Series in Solid State Sciences 122, Springer, Berlin.
15. Craik, D. (1995) *Magnetism*, John Wiley & Sons, Chichester.
16. Zaanen, J., and Sawatzky, G.A. (1990) Systematics in Band Gaps and Optical Spectra of 3D Transition Metal Compounds, *J. Sol. State Chem.* **88**, 8-27.
17. Sleight, A.W. (1988) Chemistry of High-Temp. Superconductors, *Science* **242**, 1519-1527.
18. Müller-Buschbaum, Hk. (1989) Zur Kristallchemie der oxidischen Hochtemperatur-Supraleiter und deren kristallchemischen Verwandten, *Angew. Chemie* **101**, 1503-1524.
19. Müller-Buschbaum, Hk. (1991) Zur Kristallchemie von Kupferoxometallaten, *Angew. Chemie* **103**, 741-761.
20. Isawa, K., Yaegashi, Y., Komatsu, M., Nagano, M., Sudo, S., Karppinen, M., and Yamauchi, H. (1997) Synthesis of Delafossite-Derived Phases, $RCuO_{2+\delta}$ with R = Y, La, Pr, Nd, Sm, and Eu, and Observation of Spin-Gap-like Behavior, *Phys. Rev. B* **56**, 3457-3464.
21. Sleight, A.W. (1989) Crystal Chemistry of Oxide Superconductors, *Physica C* **162-164**, 3-7.
22. Losert, W., Hoppe, R. (1984) Zur Kenntnis von $Li_2[CuO_2]$, *Z. anorg. allg. Chem.* **515**, 95-100.
23. Mizuno, Y., Tohyama, T., Maekawa, S., Osafune, T., Motoyama, N., Eisaki, H., and Uchida, S. (1998) Electronic States and Magnetic Properties of Edge-Sharing Cu-O Chains, *Phys. Rev. B* **57**, 5326-5335.
24. Mizuno, F., Masuda, H., Hirabayashi, I., Tanaka, S., Hasegawa, M., and Mizutani, U. (1990) Low-Temperature Ferromagnetism in $La_4Ba_2Cu_2O_{10}$, *Nature* **345**, 788-789.
25. Tornow, S., Entin-Wohlman, O., and Aharony, Amnon (1999) Anisotropic Superexchange for Nearest and Next-Nearest Coppers in Chain, Ladder and Lamellar Cuprates, *Phys. Rev. B* **60**, 10206-10215.

26. Anderson, P.W. (1950) Antiferromagnetism. Theory of Superexchange Interaction, *Phys. Rev.* **79**, 350-356.

27. Goodenough, J.B. (1955) Theory of the Role of Covalence in Perovskite-Type Manganites [La,M(II)]MnO$_3$, *Phys. Rev.* **100**, 564- 573.

28. Kanamori, J. (1959) Superexchange Interaction and Symmetry Properties of Electron Orbitals, *J. Phys. Chem. Sol.* **10**, 87-98.

29. de Jongh, L.J. (1990) *Magnetic Properties of Layered Transition Metal Compounds*, Kluwer Academic Publishers, Dordrecht.

30. Kojima, K.M., Fudamoto, Y., Larkin, M., Luke, G.M., Merrin, J., Nachumi, B., Uemura, Y.J., Motoyama, N., Eisaki, H., Uchida, S., Yamada, K., Endoh, Y., Hosoya, S., Sternlieb, B.J., and Shirane, G. (1997) Reduction of Ordered Moment and Néel Temperature of Quasi-One-Dimensional Antiferromagnets Sr$_2$CuO$_2$ and Ca$_2$CuO$_3$, *Phys. Rev. Lett.* **78**, 1787-1790.

31. Mermin, N.D., and Wagner, H. (1966) Absence of Ferromagnetism or Antiferromagnetism in One- or Two-Dimensional Isotropic Heisenberg Models, *Phys. Rev. Lett.* **17**, 1133-1136.

32. Mattis, D.C. (1965) *The Theory of Magnetism*, Harper & Row, New York.

33. Bethe, H. (1931) Zur Theorie der Metalle, *Z. Physik* **71**, 205-226.

34. Kennedy, T., Lieb, E.H., and Shastry, B.S. (1988) Existence of Néel Order in some Spin-1/2 Heisenberg Antiferromagnets, *J. Stat. Phys.* **53**, 1019-1030.

35. Chakravarty, S., Halperin, B.I., and Nelson D.R. (1989) Two-Dimensional Quantum Heisenberg Antiferromagnet at Low Temperatures, *Phys. Rev. B* **39**, 2344-2371.

36. Manousakis, E. (1991) The Spin-1/2 Heisenberg Antiferromagnet on a Square Lattice and its Application to the Cuprous Oxides, *Rev. Mod. Phys.* **63**, 1-62.

37. Wagner, H. (1966) Long-Wavelength Excitations and the Goldstone Theorem in Many-Particle Systems with "Broken Symmetries", *Z. Physik* **195**, 273-299.

38. Kahn, O. (1993) *Molecular Magnetism*, VCH Publishers, New York.

39. Dagotto, E. and Rice, T.M. (1996) Surprises on the Way from One- to Two-Dimensional Quantum Magnets: The Ladder Materials, *Science* **271**, 618-623.

40. Soukoulis, C.M., Datta, S., and Lee, Y.H. (1991) Spin-Wave Theory for Anisotropic Heisenberg Antiferromagnets, *Phys. Rev. B* **44**, 446-449.

41. Johnston, D.C., Matsumoto, T., Yamaguchi, Y., Hidaka, Y., and Murakami, T. (1992) Magnetic Susceptibility Anisotropy in High T_c Cuprates, in T. Oguchi, K. Kadowaki, and T. Sasaki (eds.), *Electronic Properties and Mechanisms of High T_c Superconductors*, Elsevier, Amsterdam, pp. 301-306.

42. Matsuda, M., Yamada, K., Kakurai, K., Kadowaki, H., Thurston, T.R., Endoh, Y., Hidaka, Y., Birgeneau, R.J., Kastner, M.A., Gehring, P.M., Moudden, A.H., and Shirane, G. (1990) Three-Dimensional Magnetic Structures and Rare-Earth Magnetic Ordering in Nd$_2$CuO$_4$ and Pr$_2$CuO$_4$, *Phys. Rev. B* **42**, 10098-10107.

43. Keimer, B., Aharony, A., Auerbach, A., Birgeneau, R.J., Cassanho, A., Endoh, Y., Erwin, R.W., Kastner, M.A., and Shirane, G. (1992) Néel Transition and Sublattice Magnetization of Pure and Doped La$_2$CuO$_4$, *Phys. Rev. B* **45**, 7430-7435.

44. Hiroi, Z., and Takano, M. (1995) Absence of Superconductivity in the Doped Antiferromagnetic Spin-Ladder Compound (La,Sr)CuO$_{2.5}$, *Nature* **377**, 41-43.

45. Troyer, M., Zhitomirsky, M.E., and Ueda, K. (1997) Nearly Critical Ground State of LaCuO$_{2.5}$, *Phys. Rev. B* **55**, R6117-R6120.

46. Normand, B., and Rice, T.M. (1997) Dynamical Properties of an Antiferromagnet near the Quantum Critical Point : Application to LaCuO$_{2.5}$, *Phys. Rev. B* **56**, 8760-8773.

47. Uehara, M., Nagata, T., Akimitsu, J., Takahashi, H., Mori, N., and Kinoshita, K. (1996) Supercond. in the Ladder Material Sr$_{0.4}$Ca$_{13.6}$Cu$_{24}$O$_{41.84}$, *J. Phys. Soc. Jpn.* **65**, 2764-2767.

48. Moriya, T. (1960) Anisotropic Superexchange Interaction and Weak Ferromagnetism, *Phys. Rev* **120**, 91-98.

49. Coffey, D., Rice, T.M., and Zhang, F.C. (1991) Dzyaloshinskii-Moriya Interaction in the Cuprates, *Phys. Rev. B* **44**, 10112-10116.

80

50. Dzyaloshinsky, I.J. (1958) A Thermodynamic Theory of "Weak" Ferromagnetism of Antiferromagnets, *J. Phys. Chem. Sol.* **4**, 241-255.
51. Kosterlitz, J.M., and Thouless, D.J. (1973) Ordering, Metastability and Phase Transitions in Two-Dimensional Systems, *J. Phys. C: Sol. State Phys.* **6**, 1181- 1203.
52. Hoppe, R., and Wingefeld, G. (1984) Zur Kenntnis der Hexafluorocuprate (III), *Z. anorg. allg. Chem.* **519**, 195-203.
53. Vulfson, S.G. (1998) *Molecular Magnetochemistry*, Gordon and Breach Science Publishers, Australia.
54. Zhang, F.C., and Rice, T.M. (1988) Effective Hamiltonian for the Superconducting Cu Oxides, *Phys. Rev.* B **37**, 3759-3761.
55. Mizokawa, T., Fujimori, A., Namatame, H., Akeyama, K., and Kosugi, N. (1994) Electronic Structure of the Local-Singlet Insulator $NaCuO_2$, *Phys. Rev.* B **49**, 7193-7204.
56. Yu, Z.G., Bishop, A.R., and Gammel, J.T. (1998) Low-Energy Magnetic Excitations in $La_2Cu_{0.5}Li_{0.5}O_4$, *J. Phys.: Condens. Matter* **10**, L437-L443.
57. Zhou, J.-S., Archibald, W., and Goodenough, J.B. (1998) Pressure Dependence of Thermoelectric Power in $La_{1-x}Nd_xCuO_3$, *Phys. Rev* B **57**, 2017-2020.
58. Kipka, R., and Müller-Buschbaum, Hk. (1976) Zur Kenntnis von $Ba_2Cu_3O_4Cl_2$, *Z. anorg. allg. Chem.* **419**, 58-62.
59. Yamada, K., Suzuki, N., and Akimitsu, Y. (1995) Magnetic Properties of $(Sr,Ba)_2Cu_3O_4Cl_2$: Two-Dimensional Antiferromagnetic Cuprates Containing Two Types of Cu Sites, *Physica* B **213&214**, 191-193.
60. Ito, T., Yamaguchi, H., and Oka, K. (1997) Anisotropic Characterisic of Cu_3O_4 planes in $Ba_2Cu_3O_4Cl_2$, *Phys. Rev.* B 55, R684-R687.
61. Eckert, D., Ruck, K., Wolf, M., Krabbes, G., and K.-H. Müller, (1999) Magnetic Behavior of the Low-Dim. Compounds $Ba_2Cu_3O_4Cl_2$ and $Ba_3Cu_2O_4Cl_2$, *J. Appl. Phys.* **83**, 7240-7242.
62. Chou, F.C., Aharony, Amnon, Birgeneau, R.J., Entin-Wohlman, O., Greven, M., Harris, A.B., Kastner, M. A., Kim, Y.J., Kleinberg, D.S., Lee, Y.S., and Zhu, Q. (1997) *Phys. Rev. Lett.* **78**, 535-538.
63. Sowa, T., Hiratani, M., and Miyauchi, K. (1990) A New Chlorooxocuprate, $Ca_3Cu_2O_4Cl_2$, with an Oxygen Defect Intergrowth, *J. Sol. State Chem.* **84**, 178-181.
64. Zenitani, Y., Inari, K., Sahoda, S., Uehara, M., Akimitsu, J., Kubota, N., and Ayabe, M. (1995) Superconductivity in $(Cu,Na)_2CaCu_2O_4Cl_2$: The New Simplest Double-Layer Cuprate with Apical Chlorine, *Physica* C **248**, 167-170.
65. Kipka, R., and Müller-Buschbaum, Hk. (1976) Zur Kenntnis von $Ba_3Cu_2O_4Cl_2$, *Z. anorg. allg. Chem.* **422**, 231-236.
66. Wolf, M., Ruck, K., Eckert, D., Krabbes, G., and Müller, K.-H. (1999) Spin-Flop Transition in the Low-Dimensional Compound $Ba_3Cu_2O_4Cl_2$, *J. Magn. Magn. Mater.* **196-197**, 569-570.
67. Gignoux, D. and Schmitt, D. (1995) Metamagnetism and Complex Magnetic Phase Diagrams of Rare Earth Intermetallics, *J. Alloys & Comp.* **225**, 423-431.
68. Néel, L. (1936) Propriétés Magnétiques de l'État Métallique et Énergie d'Interaction entre Atomes Magnétiques, *Ann. Phys.* (Paris) **5**, 232-279.
69. Barbara B., Gignoux D., and Vettier, C. (1988) *Lectures on Modern Magnetism*, Springer, Berlin.
70. Müller, K.-H. and Wolf, M. Weak Ferromagnetism in Low-Dimensional Cuprates $Ba_2Cu_3O_4Cl_2$ and $Ba_3Cu_2O_4Cl_2$, J. Appl. Phys. in press
71. Yushankai, V., Wolf, M., Müller, K.-H., Hayn, R., and Rosner, H. Weak ferromagnetism due to Dzyaloshinskii-Moriya Interaction in $Ba_3Cu_2O_4Cl_2$, to be published.
72. Drechsler, S.L., Rosner, H., Málek, J., and Eschrig, H. (1998) Electronic and Magnetic Properties of Cuprate Chains and Related Materials, *These Proceedings* .

ELECTRONIC AND MAGNETIC PROPERTIES OF CUPRATE CHAINS AND RELATED MATERIALS

From bandstructure to aspects of many-body physics in real materials

S.-L. DRECHSLER[a], H. ROSNER[a,b], J. MÁLEK[a], H. ESCHRIG[a]

[a] *Institut für Festkörper- und Werkstofforschung Dresden*
D-01171 Dresden, Postfach 270116, Germany
[b] *Dept. of Physics, University of California, Davis, USA*

1. Introduction

2. Bonding and architecture of cuprates

3. Electronic structure from LDA-bands

4. Model Hamiltonians

5. Magnetic Properties: fluctuations, anisotropy, and frustrations

6. Effects of strong electron-electron interactions

7. Interplay of el-el with el-ph interactions

8. 1D-Aspects of Superconductivity in 2D and 3D Cuprates

9. Summary and Outlook

Abstract. Band structure calculations for CuO_3 and CuO_2 chain compounds as well as for zigzag Cu_2O_4 double chain compounds are reported. These cuprate chains form the basic elements of a rich and beautiful variety of cuprate structures. From the dispersion of the antibonding band crossing the Fermi energy the exchange interaction is estimated and qualitative trends in the antiferromagnetic ordering at low temperatures are well explained. From the comparison of exact diagonalization studies of corresponding periodic chains with frequently used approximations a restricted validity of the latter with respect to the hole doping amount is found. The possibility of CDW-BOW states is discussed on the basis of self-consistent adiabatic studies. Chain aspects for superconductivity and the possible relationship to the stripe scenario are briefly discussed.

Some figures in this lecture appear in colour in the preliminary electronic version; xxx./cond-mat 0012... ; see also [1]

S.-L. Drechsler and T. Mishonov (eds.), High-T$_c$ Superconductors and Related Materials, 81–134.

1. Introduction - Why is it worth to study chain cuprates and their derivatives ?

Compounds containing chains as structural elements comprise an important class of the rich and beautiful cuprate family[2, 3]. While the planar CuO_2 network is so far mandatory for the appearance of high-T_c superconductivity among those compounds having only cuprate chains superconductivity is rather the exception. Only one or two directly related composite ladder chain superconductors (under pressure with a relative low T_c value of 12 K) such as $(Ca,Sr)_{14}Cu_{24}O_{41}$ [4, 5] have been found. The whole class contains many quite interesting physical objects worth to be studied on their own right. An extensive investigation of their physical properties started only recently. In particular, several typical representatives attracted general attention in connection with the following hot topics in solid state physics:

(i) The undoped and insulating cuprates provide the long sought nearly ideal spin-1/2 antiferromagnetic Heisenberg model systems, in one and two dimensions and with all kinds of anisotropies and frustration.

(ii) The best realization of the spin-1/2 antiferromagnetic Heisenberg chain (AHC) known at present are Sr_2CuO_3 and the isostructural Ca_2CuO_3, respectively [6].

The AHC is the simplest highly nontrivial many-body model studied for several decades by a large number of theorists (see e.g. [7]).

(iii) There is experimental evidence for spin-charge separation in one dimension: $SrCuO_2$ [8] and Sr_2CuO_3 [9],

(iv) Interplay of spin-Peierls transition, dilution induced antiferromagnetism [10], and frustration: $CuGeO_3$.

(v) Coexistence of spin gap behaviour with dilution and/or doping induced antiferromagnetism: $SrCu_2O_3$ and $(Sr,Ca)_{14}Cu_{24}O_{41}$[11]

(vi) Antiferromagnetism at large hole doping: $Ca_{0.83}CuO_2$ and $Sr_{0.73}CuO_2$ [12].

In addition, with respect to complex theoretical problems, notions and approximations developed originally for their "big brother"- the celebrated high-T_c layered cuprates, which are the central object of the present book, can be tested. Further general questions such as the role of intervening cations, charge transfer etc. may be studied also for these chain-like compounds, both theoretically and experimentally.

On the other hand the well-known superconductors $RBa_2Cu_3O_{7-\delta}$ and $RBa_2Cu_4O_8$ (R=Y,rare earth) being composite compounds, contain Cu-O chains which control the hole count in the cuprate planes and participate also in the plane induced superconductivity. Thus their properties are often related to or reflect plane properties. In describing some chain related properties one has often to adopt parameter sets for a corresponding model

Hamiltonian. Because there are usually too many parameters to be derived from one experiment only, it is tempting to use plane derived parameters eventually scaled in an appropriate way to account for changed character-istic distances. However, as we shall see this is an unreliable procedure in some cases. In contrast the more anisotropic properties of chains allow in principle the extraction of several parameters from the experimental data easier than for the layered case.

Last but not least and quite interestingly, even the seemingly at first glance completely different 2D cuprate planes exhibit after hole doping within the so called stripe scenario essential 1D-like properties. All these topics require much more detailed studies. Hopefully, there will be a positive feedback for our understanding of the cuprate superconductivity in general.

It is the aim of the present lecture to introduce the reader, especially young scientists, into this fastly growing field, to emphasize the common generic features as well as the large diversity moving from one compound to the other. Furthermore, the lecture is intended to illustrate how various model descriptions occur at different levels, to show how these models are related to each other. Last but not least, this lecture should promote further investigation of the topics arisen here in the future.

2. Bonding and architecture of cuprates

2.1. THE SINGLE CuO$_4$ PLAQUETTE AND RELATED COMPOUNDS

Since the basic common structural element of all cuprate compounds is the planar CuO$_4$ plaquette (COP)[13], we shall start from a consideration of covalency, doping and correlation for this unit (see Fig. 1).

As discussed in the lecture of J. Fink [14] for the prototypical parent compound La$_2$CuO$_4$ the orthorhombic crystal field splits the levels of the Cu $3d$ states and the O $2p$ states in such a way that Cu $3d_{x^2-y^2}$ and the O $2p_x$, $2p_y$ states near the Fermi level should be taken into account to first approximation in order to describe the low-energy physics of the cuprate compounds under consideration. The shell occupation of the Cu^{+2} ion is $3d^9$ and that of the O^{-2} ion is $2p^6$ and the nominal anionic redox number (total charge) of the undoped COP complex [CuO$_4$]$^{-6}$ is very high: -6. The COP obeys two main prerequisites for strong covalency: (i) there is nearly a resonance of orbital energies, i.e. the energetical distance $\Delta_{pd}=2.5$ to 4 eV of the relevant Cu $3d$ orbital and the O $2p$ one is smaller than the covalent splitting $\sim 2zt_{pd}$, where $+(-)t_{pd}$ is the transfer integral for an electron(hole) hopping from the Cu to the O site and z is the number of nearest neighbour oxygens; (ii) the resulting antibonding molecular state is partially filled.

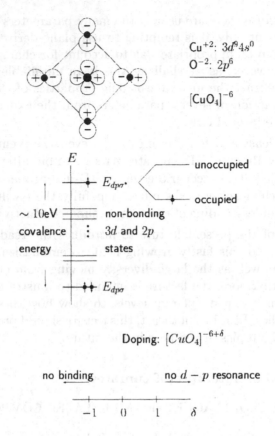

Figure 1. The CuO$_4$ plaquette (COP) and its relevant orbitals.

"Isolated" COPs appear in Bi$_2$CuO$_4$ (ideal squares with a typical Cu-O distance of 1.936 Å [15]), in R$_2$BaCuO$_5$ (R=La, Pr, Nd); in Ta$_2$CuO$_6$, CaCu$_3$Ge$_4$O$_{12}$ [2], as well as in CaCuSi$_4$O$_{10}$ and in several other (at least 4) much less investigated natural minerals [16]. To ensure charge neutrality of the whole crystal the COPs are surrounded by (large) cations, see Fig. 2.

The face to face oriented COPs in Bi$_2$CuO$_4$ form staggered chains along the c-axis with a twist angle ϕ being 33.3°. Therefore despite the relatively short COP-COP distance d_{Cu-Cu}=2.907 Å, the interstack interactions are weak but larger than the direct intrastack ones. This is a first instructive example when considering a "chainlike" structure without knowledge of the relevant orbitals and their interaction one might readily come to erroneous conclusions (see also below). Another example is given by the mentioned above La$_2$BaCuO$_5$ [2] which exhibits four planar, i.e. unfolded chains of separated COPs.

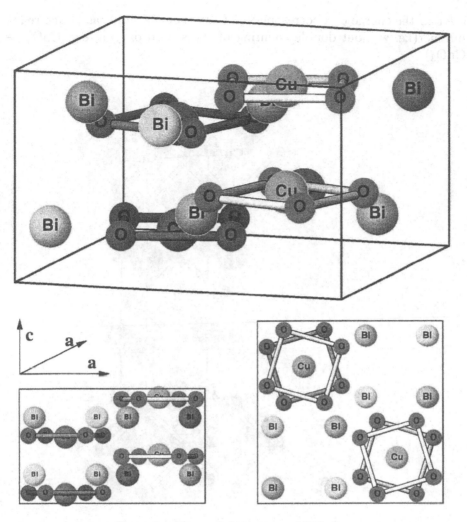

Figure 2. The crystal structure of BiCuO₄.

2.2. THE COPPER OXYGEN NETWORKS

The idea to build up planar networks by linking COPs which share in oxygen ions is due to Müller-Buschbaum [19]. He performed first a systematic study of cuprate structures long before the HTSC's were discovered. In order to reduce the electrostatic energy without loss of covalency, two linked COPs sharing in one " corner" O(1) become $[CuO_3]^{-4}$ periodic units after continuation. They are realized in the infinite chain limit in the so-called straight or corner-sharing CuO_3 chains in the isostructural compounds Ca_2CuO_3 and Sr_2CuO_3 (see Fig. 3). In the following we shall refer to the charge of a corresponding COP-monomer unit as to the redox number(i.e. without double counting of the shared oxygen; e.g. $CuO_4 \rightarrow CuO_3$).

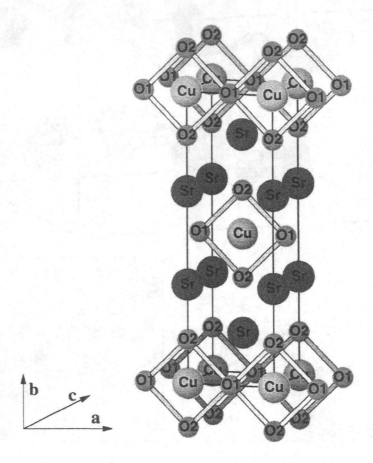

Figure 3. Crystal structure of the linear chain corner-shared compounds Sr_2CuO_3 (and the isostructural Ca_2CuO_3). The CuO_3 chains run along the *a*-direction.

Again charge neutrality is attained by the two earth alkaline ions, one above and one below the side oxygens O(2). Next we may link two CuO_3 chains at one row of these side oxygens (or equivalently building ribbons from corner-shared double COPs, i.e. from Cu_2O_5 monomers). Then one arrives at the so called "isolated" two-leg ladder realized in $LaCuO_{2.5}$ where the nominal redox number is now reduced to -3.

If one of the side oxygens O(2) is involved in an additional longitudinal 'inkage and the chain oxygen O(1) also in a transverse linkage, we arrive at the double or zigzag chain realized in $SrCuO_2$ (see Fig. 4) where the redox number is now -2.

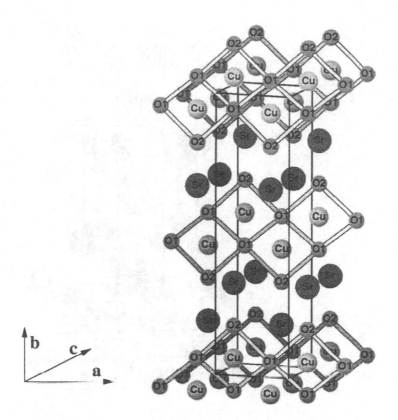

Figure 4. Crystal structure of the zigzag (linear double chain) Cu_2O_4 compound $SrCuO_2$. The zigzag chains, best seen in the center of the unit cell, run along the $4a$-direction.

The low redox number -2 can be also attained connecting two oxygen ions on each side of a COP. Then we arrive at the so called edge-shared chains realized for instance in Li_2CuO_2 (see Fig. 5) or in the celebrated first inorganic spin-Peierls compound $CuGeO_3$ where rectangular edge-shared COP chains alternate with zigzag GeO chains (see Fig. 6).

In the former case the singly charged cationic Li^+ ions reside above and below the external oxygen-oxygen *bonds*. Here the crystal field is relatively weak. While in $CuGeO_3$ and in the isostructural $CuSiO_3$ the quaternary charged cations reside in front of the oxygen *sites*. Here the crystal field is strong and as a consequence a sizable splitting of the onsite energies between oxygen $2p$ orbitals oriented along and perpendicular to the CuO_2 chain axis occurs (compare Sec. 4.3).

In CuO the sharing is at maximum and the redox number is zero. Here each oxygen belongs at the same time to four COPs. This is achieved by

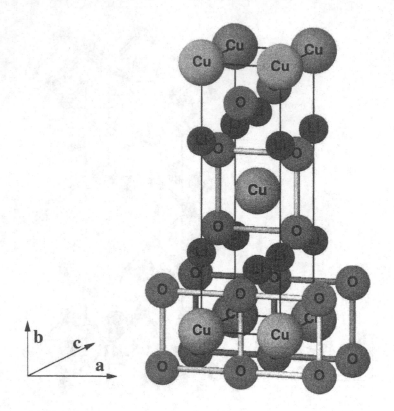

Figure 5. Crystal structure of the linear edge-shared CuO_2 chain compound Li_2CuO_2.

stacking alternating layers of differently oriented parallel CuO_2 chains (see Fig. 7).

In each layer the CuO_2 chains run parallel whereby there projections on the boarder plane cross with an angle of 75°. This compound is interesting among other things for the absence of a cationic counter ion framework.

In all chains mentioned above the COP, remaining planar, is slightly deformed becoming a rhombus, an asymmetric rhombus or a rectangle, respectively. Since the magnetic properties of the chain compounds depend via the nearest neighbor exchange integral $J_\parallel$ very sensitively on the Cu-O-Cu bond along the chain direction, we recall that this angle amounts 180° in the case of the corner-sharing CuO_3 chains and in the two-leg ladder and slightly exceeds 90° only in the case of edge-shared CuO_2 chains. In those compounds the situation is similar to the mentioned above situation in the spin-Peierls compounds.

In addition there are also tubelike chains in $Ba_3Sc_4Cu_3O_6$ [20] and zigzag distorted single chains with intermediate bond angles of about 124°

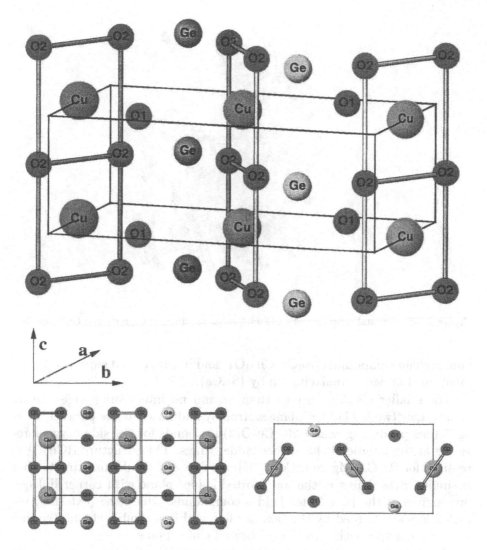

Figure 6. Crystal structures of $CuGeO_3$ and analogously for the isostructural $CuSiO_3$.

in $BaCu_2Si_2O_7$ and of about $135°$ in $BaCu_2Ge_2O_7$ [21, 22]. The latter compounds consist of folded and twisted corner-shared COPs supported by a second Ge(Si)-O network. A similar situation is realized in the popular $CuGeO_3$ compound where rectangular edge-shared COP chains alternate with zigzag GeO chains.

Finally, an infinite planar assembling of Cu_2O_4 double (zigzag) chains mentioned above leads to the Cu_2O_3 two-leg ladder compounds, where the redox number is -3. It is realized in $SrCu_2O_3$ and in the composite

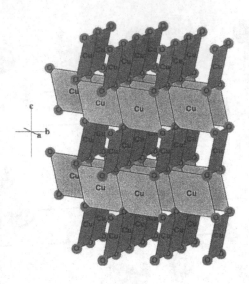

Figure 7. The crystal structure of CuO with cross-touching of edge-shared CuO_2 chains.

ladder chain compounds $La_6Ca_8Cu_{24}O_{41}$ and in the related superconductor mentioned in Sec. 1 replacing La by (Sr,Ca).

For smaller counter cations than Sr and no intervening edge-sharing chains, the "two-leg ladder" plane is strongly undulated at the parallel "linkage" lines producing nearly 90° Cu-O(2)-Cu bonds for the side oxygens residing at the middle of the former ladder rungs. This structure RCu_2O_3 is realized for R=Ca, Mg, and Co [2, 35]; see Fig. 8). While inserting further n single CuO_3 chains in the undistorted ladder-plane with corner linkage, one arrives at the $(n + 2)$-leg ladder compounds, where the (external) so-called legs are formed by the boarder line within a double chain. Tending $n \to \infty$, one approaches to the celebrated CuO_2 plane.

2.3. 3D-ARRANGEMENTS OF NETWORKS

The 3D-arrangements of the low dimensional networks is of great interest in the context of phase transitions and ordering at low temperature (Mermin-Wagner theorem; see Sec. 5.1.) as well as for some details related to the mutual doping of different cuprate structures. Thus, for the high-

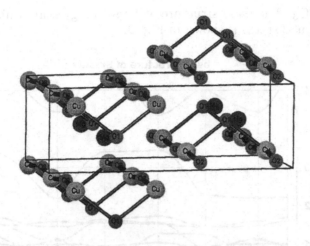

Figure 8. The crystal structure of $MgCu_2O_3$ with folded ladders.

T_c superconductors $RBa_2Cu_3O_7$ and $RBa_2Cu_4O_8$ a slab of parallel CuO_3 corner-sharing chains and Cu_2O_4 (double (zigzag) chains, respectively, located in between the bilayers of cuprate planes is the reason for self-doping in these compounds, i.e. chains dope the planes and vice versa. Then in both subsystems the redox number of the corresponding COPs is further reduced: $-2+\delta$ for the planes, $-3-2\delta$ for the CuO_3 chains and $-1.5-\delta$ for the edge-shared linear double (zigzag) chains.

For the magnetic properties at low temperature the 3D arrangement of the 1D networks is of great importance since any ordering depends crucially on the interaction between these subunits. Beside the interchain distance also the presence or absence of frustrations can be quite important.

3. Electronic structure from LDA-bands

In order to understand which orbitals are important for a realistic description we performed local density approximation (LDA) LCAO band structure calculations for several prototypical compounds, namely Sr_2CuO_3, Ca_2CuO_3, $SrCuO_2$, $CuGeO_3$, $Ba_3Cu_2O_4Cl_2$, CuO, and related layered cuprates such as $CaCuO_2$, La_2CuO_4, and $Sr_2CuO_2Cl_2$ [1, 24, 27]. The inspection of all calculated bandstructures (see e.g. Figs. 9–11) exhibits a half-filled band at the Fermi energy, i.e. pointing to a metallic ground state in sharp contrast to the experimentally found insulating behaviour. This failure can be removed adding Coulomb interactions to the TB-Hamiltonian (the corresponding matrix elements are taken from the literature or from experimental electron spectroscopy data). Let us start with the simplest case of Sr_2CuO_3. The band structure at large energy scale which contains all $O2p$ and $Cu3d$ states is shown in Fig. 9.

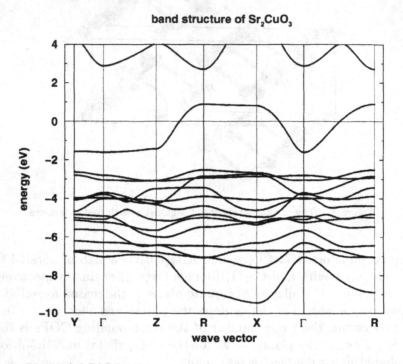

Figure 9. The LDA-LCAO bandstructure of Sr_2CuO_3 calculated within the LDA-LCAO scheme near the Fermi Energy (E=0) including all CuO pd-derived states.

In Fig. 10 the same is depicted for the total density of states (DOS) and the atomic partial DOS.

For the topmost band crossing E_F one observes a strong dispersion $\sim$2eV in chain direction and a much weaker dispersion $\sim$ 100 meV in the z-direction perpendicular to the COPs and an even smaller dispersion $\sim$ 40 meV, but yet visible in the third perpendicular direction.

4. Model Hamiltonians

On the basis of our LDA-calculations and available experimental data we shall derive/parameterize several extended Hubbard-type Hamiltonians

$$H = \sum_i \varepsilon_i \hat{n}_i + \sum_{<i,j>,s} t_{ij}(c_{i,s}^\dagger c_{j,s} + \text{H.C.}) + \sum_i U_i \hat{n}_{i,\uparrow} \hat{n}_{i,\downarrow}$$
$$+ \sum_{<i,j>} V_{ij} \hat{n}_i \hat{n}_j,$$

$$(1)$$

density of states of Sr₂CuO₃

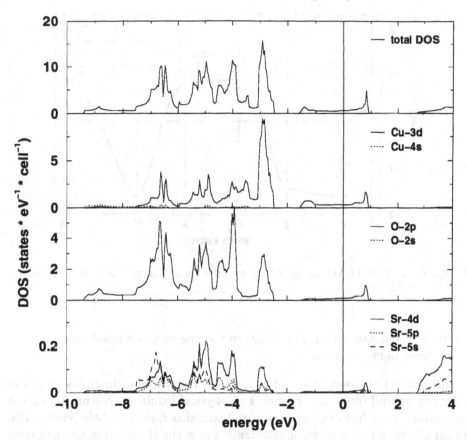

Figure 10. The LDA-LCAO result for the total and the atomic partial density of states (DOS).

and/or spin-1/2 Heisenberg type Hamiltonians

$$H_{spin} = \sum_{<ij>} J_{ij} \vec{S}_i \vec{S}_j, \qquad (2)$$

in dependence on which property we wish to describe. In Eq. (1) $c_{i,s}^{\dagger}$ describes the creation of a hole with spin projection $\pm 1/2$ at site i, $\hat{n}_{i,s} = c_{is}^{\dagger} c_{is}$ denotes the number operator, and $\hat{n}_i = \sum_s c_{is}^{\dagger} c_{is}$. The symbol $< i, j >$ stands for the summation over interacting pairs (e.g. nearest neighbour (nn) sites). Here the site index i includes also the orbital index. In the few-band case it means the site of a radical containing several atomic sites and

94

antibonding band of Sr₂CuO₃

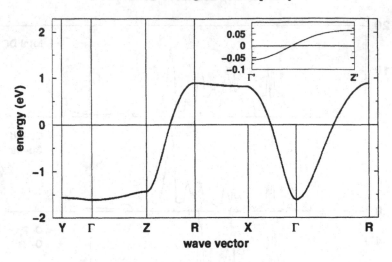

Figure 11. LDA-LCAO Energy bands near the Fermi Energy (E=0) for Sr₂CuO₃.

orbitals, for instance CuO_3 or CuO_2 in the one and two-band models to be considered below in detail.

In general the derivation of any model Hamiltonian includes elements of mapping and often also elements of physical intuition. Naturally, such a procedure is far from being unique and contains various subtle points. The main aim of deriving or parameterizing a new Hamiltonian is to work with an extended or a reduced manifold of states in order to cover more physical phenomena or to improve the computational conditions.

Let us consider for instance the cluster mapping of a given extended Hubbard type Hamiltonian (Eq. (1)) onto the spin-Hamiltonian given by Eq. 2. Apparently, the number of monomer units, say e.g. of CuO_3 units in the case of corner-shared chains, and that of the target units i.e. the number of effective sites (spins) has to be conserved. For an even membered ring (with one hole per monomer unit) the ground state is a spin singlet. We shall consider besides the ground state also the energies of the lowest lying excited states with total spin $S = 1$ (triplet), $S = 2$ (quintuplet), etc. The corresponding energy differences of the Hubbard model on a finite ring, say $N = 12$ sites, can be compared with those of the spin-1/2 antiferromagnetic Heisenberg model. Then for each transition $S = 0 \rightarrow S=1,2,3,...$, we can derive an effective exchange integral J_S. Their dependence on U (both measured in units of the n.n. transfer integral t) are shown in Fig. 12. In addition, the effective exchange integral

$$J_{eff} = \frac{2v_s\hbar}{a\pi} = \frac{4tI_1(z)}{\pi I_0(z)}, \tag{3}$$

defined via the spinon velocity v_s (taken from the analytically available Bethe-ansatz Lieb-Wu-solution for the Hubbard model) is shown, too. In Eq. (3) the notation $z = 2t/\pi U$ has been introduced, I_0, I_1 denote modified Bessel functions, and a is the in chain lattice constant. Here we adopted for the sake of simplicity $V = 0$.

One realizes that in the strongly correlated limit $U \gg 4t$ all curves approach to each other and tend also from below to the well-known expression given by the second order perturbation theory

$$J = 4t^2/U. \tag{4}$$

Note that the latter expression diverges in the uncorrelated limit $U \to 0$ whereas v_s in Eq. (3) tends to the Fermi velocity $\hbar v_F = 2ta$ as it should be. In this weakly correlated limit various spin excitations cannot be described by a single parameter J and/or higher order terms in the spin operators as well as non n.n. exchange processes must be taken into account. In other words, the mapping onto the "simple" n.n. bilinear Heisenberg model becomes problematic. Anyhow, in the case of undoped corner-shared linear chain cuprates $U/t > 5$ holds (see Sec. 4.1) and the mapping on the Heisenberg model is justified. Using Eq. (4) one may slightly overestimate by $\sim$ 10 to 20 % the value of J.

We may proceed in the same way for a two-band model (see Fig. 13 and Sec. 4.2) or any other multi-band model. Again the frequently used lowest order (here fourth order) perturbation theory result $J \propto t_{AB}^4$ is approached only in the strongly correlated limit. However, in the realistic parameter regime for the same compound the two-band perturbational approach may overestimate the exchange integral $J \sim 250$ meV by a factor of two. Hence, in such a situation the empirical exchange integral can be analyzed self-consistently numerically (cluster mapping) or by more complicated higher order expressions, only.

These procedures make sense, if the excitations of the spin degrees of freedom are well separated from the charge excitations. In the case of one hole (electron) per unit cell this is ensured by rather different energy scales $J \leq 0.1$ to 0.2 eV for the former and U or $\Delta \leq 2.5$ to 5 eV for the latter. Then also in the general multi-band case one may compare the difference of the lowest eigenvalues for systems with the same number of effective sites which equals the number of involved COPs, similarly as demonstrated above.

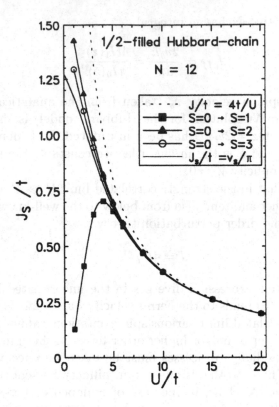

Figure 12. The cluster mapping of the one-band Hubbard model at half-filling and for a $N = 12$ site periodic ring onto an antiferromagnetic spin-1/2 Heisenberg ring. The undoped antiferromagnetic spin-1/2 corner-shared A_2CuO_3 chain compounds with $A=$Sr, Ca correspond to $J_s/t \sim 0.45$ and $U/t \sim 7.6$ values which are reasonably well described by the asymptotic expression (Eq. (4)).

In other cases one calculates an analogous physical quantity for the two Hamiltonians or the two approaches which should be employed. Usually it is assumed that there is a hierarchy of models which is valid for different energy scales.

The simplest mapping is the transition from the LDA-band structure via an extended tight-binding model to a single band extended Hubbard-model and further to a Heisenberg model for the description of the magnetic properties (see Sec. 5). The parameterization of various multi-band model

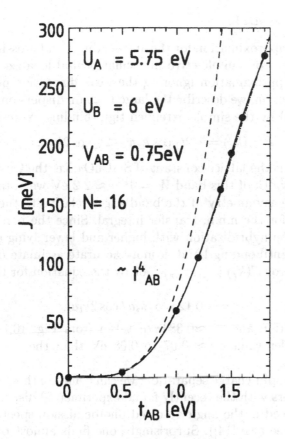

Figure 13. The effective exchange integral derived from the singlet-triplet separation for a two-band extended Hubbard model and 3/4-filling which corresponds to the 1/2-filling in the single-band case (see Fig. 12). The depicted parameter set with Δ_{AB}=1.23 eV and $t_{AB} \approx 1.5$ eV represents an approximate description of the undoped antiferromagnetic spin-1/2 corner-shared A_2CuO_3 chain compounds, A=Sr, Ca; (see Sec. 4.2.). Broken line: fourth order perturbation theory, full line: cluster mapping.

Hamiltonians such as the widely used σ-pd model is a little bit more elaborate. To illustrate this point let us consider the prototypical single linear chain compound Sr_2CuO_3 [24, 73] adopting various approximations.

4.1. ONE-BAND MODELS

In the one-band approximation for the antibonding band crossing the Fermi energy one starts from an effective molecular orbital localized on a CuO_3 unit. In a first approximation ignoring the weak dispersion perpendicular to the chain direction, we describe the large in chain dispersion of this band (shown in Fig. 9) by the simple extended tight-binding expression

$$E_{1D}(k_x) = -2t_1 \cos k_x a - 2t_2 \cos 2k_x a, \tag{5}$$

where $a = 3.9$ Å is the lattice constant of Sr_2CuO_3, i.e. the Cu-Cu distance. From the total width of this band $W = 4t^{(0)} \approx 2.2$ eV we obtain $t_1 \approx 0.55$ eV and from the asymmetry of the band edges relative to the Fermi level $t_2 \approx 0.1$ eV[24] for the n.n.n. transfer integral. Since the band edges may be affected by the hybridization with higher and lower lying orbitals, less involved in the antibonding band, a more accurate estimate of t_2 is given by the curvature of $E(k_x)|_{k_x=k_F=\pi/2a}$ or by the equation for the inflection point $k_i \neq k_F$.

$$t_2 = -0.125 \cos k_i a / \cos 2k_t a. \tag{6}$$

With this definition and $k_t \approx 0.384\pi/a$ taken from Fig. 10, we arrive at a little bit smaller value $t_2 \approx 0.07$ to 0.08 eV than the value of 0.1 eV estimated before.

Owing to the spin-charge separation, we can compare these theoretically predicted numbers with the recently found experimental dispersion seen by the holon produced in the angle resolved photoemission spectroscopy process((ARPES), see also [14]). Surprisingly, one finds almost perfect agreement: $2t_1^{ARPES} + 4t_2 \approx 1.5$ eV [9], where the experimental curve (see Fig. 3 of [9]) has been linearly extrapolated to $k_x = 0$ and our LDA t_2-value has been adopted. Notice that the n.n. in chain transfer integral is somewhat larger than the corresponding value for planar cuprates $t_{2D} \sim 0.4$ to 0.45 eV (see Tab. 1).

From Eq. (5) one would expect to see asymmetric square root van Hove-singularities for the calculated density of states $N(E)$ at the band edges $E_\pm = \pm 2t_1 - 4t_2$. However, the inspection of Fig. 10 tells us that there are only more or less sharp asymmetric peaks which can be considered as singularity remnants due to the finite interchain interaction ignored so far. The corresponding weak dispersion of ~ 0.1 eV in c-direction. i.e. perpendicular to the COPs is clearly visible in Fig. 9. Refining Eq. (5), we write the quasi-one-dimensional dispersion law as

$$
\begin{aligned}
E_{3D}(\vec{k}) = \ & E_{1D}(k_x a) - 2t_y \cos k_y b - 2t_z \cos k_z c - 4t_{xz} \cos k_x a \cos k_z c + \\
& -8t_{xyz} \cos \frac{k_x a}{2} \cos \frac{k_y b}{2} \cos \frac{k_z c}{2},
\end{aligned}
\tag{7}
$$

where the new small transfer integrals are self-explaining. Considering the cosine-like dispersion at $k_x = \pi/2a$ along k_z (see the inset in Fig. 11) one obtains t_z=25 meV. The asymmetry of the analogous dispersions starting in the same direction from (0,0,0) or (1,0,0), respectively, is caused by the presence of the interchain "diagonal" transfer term t_{xz} acting in the basal plane and its spatial counter part $t_{xyz} \approx 11$ meV. Finally, the smallest transfer integral t_y=1.8meV is obtained considering the tiny deviation from the symmetric cosine shape along Γ–$(0,2\pi/b,0)$. Of course, all these small terms can be neglected considering any electron spectroscopy problem but they must be retained in estimating the origin of the magnetic order occurring at low temperatures (see Sec. 5).

From the large-gap-insulating state of Sr_2CuO_3 *not* reproduced by our LDA calculation one concludes that strong el-el correlation must be present. In the one-band model under consideration this is described by the Hubbard on-site U interaction, i.e. the Coulomb repulsion which feel two holes on the same CuO_3 unit. From the ARPES data [9] and the relation suggested by the Bethe-Ansatz solution for the charge gap one estimates the gap as $E_g \sim U - 4t_1 \approx 1.7$ eV , or taking the transfer integral t_1= 0.55 to 0.6 eV determined above, one arrives at $U \sim$4eV. This is in accord with a recent extended one-band Hubbard model analysis of the EELS-data (electron energy loss spectroscopy) [26, 14] which yields U=4.2 eV. In addition, there is also a substantial n.n. intersite Coulomb interaction $V \sim$0.8 eV detected from the evidence of strong excitonic effects seen at the Brillouin-zone boundary $q_x = \pi/a$.

We close this subsection comparing the obtained main parameters for 1D and 2D corner-sharing compounds (see Tab. 1). The Coulomb interaction interaction parameters for La_2CuO_4 are taken from [38] and the n.n.n. neighbour integral for $Sr_2CuO_2Cl_2$ (denoted usually as t'' in the t-J model literature) has been taken from [39]. The inspection of the shape of the calculated density of states $N(E)$ for at first glance chain-like compounds Bi_2CuO_4, $Ba_3Cu_2O_4Cl_2$, and Li_2CuO_2 (see Figs. 14-16, respectively) included in Tab. 1, too allows a quick classification with respect to the electronic dimensionality. One realizes in all cases symmetric sharp peaks for the total DOS in the middle of the antibonding band(s) crossing the Fermi energy. Hence, the effective electronic dimension of these compounds is not near 1D but instead near 2D or 3D with important consequences for the magnetic properties at low temperature (see Sec. 5).

This situation is most clear for Li_2CuO_2 where one of the cleanest one-band descriptions in the cuprates at all is possible[28]. It reads

$$E_{3D}(\vec{k}) = E_{1D}(k_x a) - 2t_y \cos k_y b - 2t_z \cos k_z c - 4t_{xz} \cos k_x a \cos k_z c +$$
$$-8t_{xyz} \cos \frac{k_x a}{2} \cos \frac{k_y b}{2} \cos \frac{k_z c}{2} - 8t'_{xyz} \cos \frac{3k_x a}{2} \cos \frac{k_y b}{2} \cos \frac{k_z c}{2}.$$

$$(8)$$

(see Fig. 16 or Fig. 15 for $Ba_3Cu_2O_4Cl_2$)

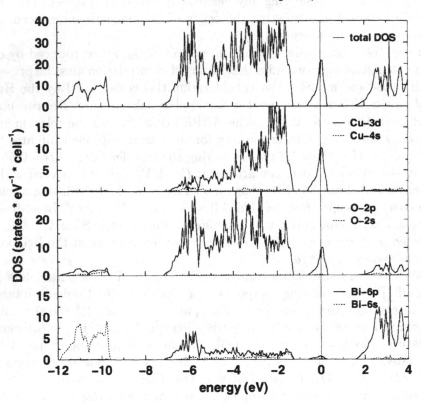

Figure 14. Total and partial DOS of Bi_2CuO_4.

In fact, the fit of the single band near E_F reveals $t_1 = -52$ meV (-63 meV), $t_2 = -80$ meV (-94 meV), for the in chain transfer integrals and $t_{xyz} = -12$ meV (-16 meV), $t'_{xyz} = -44$ meV (-44 meV), for the n.n. and n.n.n. interchain transfer integrals, respectively. The numbers in parentheses are the result obtained by Weht and Pickett [28] within the linearized augmented plane wave (LAPW) method. Note that the n.n. in chain transfer integral $| t_1 |$ is about ten times smaller than the corresponding value of the corner-sharing chain compound Sr_2CuO_3, reflecting the hindered hopping along the CuO_2 chain due to the 93° Cu-O-Cu bond angle, typical for edge-shared chains. In addition, the n.n. interchain transfer integral is somewhat enhanced and of the same order of magnitude as t_1 giving rise to an essential 3D-like behaviour. In this context it is interesting to compare

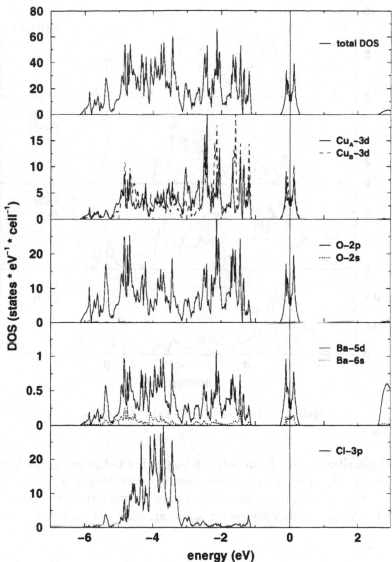

Figure 15. Total and partial DOS of $Ba_3Cu_2O_4Cl_2$.

the DOS of this nonmagnetic band structure with an itinerant magnetic solution derived from spin-polarized calculations by Weht and Picket [28]. There, the antiferromagnetic alignement of ferromagnetic chains together with the large moment suppresses the interchain interaction and 1D-van

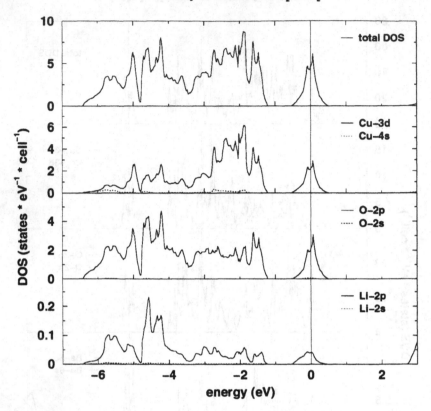

Figure 16. Total and partial DOS of Li_2CuO_2.

Hove singularities occur. However, the insulating behaviour of Li_2CuO_2 is of course not caused by the magnetic ordering occurring at $T = 9$ K as this result might suggest but caused by the strong correlation. Hence the gap is again of charge transfer character as for all undoped cuprates.

The inspection of Tab. 1 shows that the obtained parameters of the 1D system Sr_2CuO_3 differs somewhat from those for the layered cuprates. All quantities are enhanced at least by 20% . The investigation of the multi-band models provides some insight for these values since the formulation in atomic site specific orbitals is more appropriate for a detailed microscopic analysis than the molecular (radical) absorbing many degrees of freedom into few parameters.

TABLE 1. Comparison of typical isolated COP- , corner- and edge-sharing cuprates within the extended one-band Hubbard. All units in eV

Compound	COP sharing	effective dimension	t_1	t_2	U	V
Bi_2CuO_4	isolated COP	3D	0.1	0.1	4	
La_2BaCuO_5	isolated COP	3D	0.08			
Li_2CuO_2	edge	3D	0.1	0.1	4	?
$CuGeO_3$	edge	3D	0.1	0.1	4	?
Sr_2CuO_3	corner	1D	0.55 to 0.6	-0.1	4.2	0.8
La_2CuO_4	corner	2D	0.4 to 0.43	0.1	3.5	0.15
$Sr_2CuO_2Cl_2$	corner	2D	0.5	0.077	3.5	0.15
$Ba_3Cu_2O_4Cl_2$	folded edge	2D	0.1	0.077	3.5	0.15

4.2. TWO-BAND MODELS

The one-band model considered in Sec. 4.1. is thought to be valid at the low-energy scale provided the effective radical orbital remains robust with respect to internal changes. Obviously, this is violated when local deformation, caused for instance by atomic displacements, must be considered since they change the transfer integrals and on-site energies etc. affecting this way the internal admixture of the involved orbitals. The simplest generalization is given by an effective two-band or AB-model, where A stands for the antibonding state of a rigid CuO_2 unit, i.e. including the antisymmetric combination of the two side oxygen $2p_y$ states and the Cu $3d_{x^2-y^2}$ orbitals. The second subunit B is given by the chain oxygen $2p_x$ orbital. The half-filled situation in 4.1 corresponds now to a 3/4 (1/4) filled two-band complex.

$$E_{2B}(k_x) = \bar{\epsilon} - 2\bar{\tau} \cos k_x a \pm \sqrt{\epsilon_0^2 + (\Delta \tau)^2 \cos^2 k_x a + 4t_{AB}^2 \sin^2 \frac{k_x a}{2}}, \quad (9)$$

where the averaged quantities

$$\bar{\epsilon} = 0.5(\varepsilon_A + \varepsilon_B), \qquad \bar{\tau} = 0.5(t_{AA} + t_{BB}), \quad (10)$$

and the differences

$$\varepsilon = \varepsilon_A - \varepsilon_B = 2\epsilon_0, \qquad \Delta \tau = t_{AA} - t_{BB}, \quad (11)$$

of the onsite energies and of the second neighbour transfer integrals, respectively, have been introduced. These differences are responsible for the gap in the middle of the band complex and for deviation from the $\cos k_x a$

like shape of the upper (antibonding) and the lower (bonding) bands. The resulting asymmetry relative to E_F has been modeled in the one-band description of Sec. 4.1. by the t_2 transfer integral. In general it is clear that any dispersion involving a square root dependence can be only approximately simulated by few Fourier-components $\cos nk_x a$ as in the one-band description. Thus, the price we would have to pay for an exact mapping of a two-band model onto a one-band would be the occurrence of many small higher order Fourier coefficients, i.e. artificial long-range transfers in the language of the one-band model. If $\epsilon_0 \gg t_{AB}$, one gets from Eq. (9) up to a constant

$$E_{2B} \approx 2(t_{AB}^2/\epsilon_0)\sin^2 k_x a/2 = -(t_{AB}^2/\epsilon_0)\cos k_x a,$$

i.e. the leading term of the one-band model. In the general case introducing for instance a common Fermi velocity $\hbar v_F = \partial E(k_x)/\partial k_x \,|_{k_x=\pi/2a}$, we can connect both models by

$$\hbar v_F/2a = t_1 = \bar{\tau} + 0.5t_{AB}^2/\sqrt{\epsilon_0^2 + 2t_{AB}^2}, \tag{12}$$

and similarly for t_2 considering the curvature of $E(k)$ at $k = k_F$:

$$8t_2 = t_{AB}^4 \left(\epsilon_0^2 + 2t_{AB}^2\right)^{-3/2}. \tag{13}$$

From Eqs. (12,13) upper bounds follow: $t_1 < \bar{\tau} + 2^{-1.5}t_{AB} \approx \bar{\tau} + 0.354t_{AB}$ and $t_2 < 2^{-4.5}t_{AB} \approx 0.041t_{AB}$. Thus, the n.n. transfer integral of the single-band model is practically always smaller than the n.n. two-band value.

Considering the lowermost band in Fig. 18, starting near -7 eV as the bonding counter part to the antibonding band under consideration, we estimate $\epsilon_0 \approx 2.5$ eV. In our previous papers devoted to the study of the interplay of el-el and el-ph interaction we used t_{AB}=1.4 eV and $2\epsilon_0 = 1.25$ eV derived from bandstructure calculations for the chain related band in $YBa_2Cu_3O_7$.

We conclude this section with a brief estimate of the expected magnitude of the Coulomb interactions. By definition, we have $U_B \equiv U_p$. Hence, values of $U_p \approx 4$ to 6 eV can be recommended. For U_A one would expect a value intermediate between U_p and $U_d \approx 8$ to 10.5 eV. However, for strong inter-site Coulomb repulsion acting between the states on the A-radical (Cu-O_2) this value can be significantly enhanced resulting in $U_A \sim U_d$. Concerning V_{AB}, we would recommend to use values ~ 0.5 eV (see Sec. 5).

4.3. FOUR- AND HIGHER MULTI-BAND MODELS

The deepest insight in the difference and common properties of 2D and 1D cuprates can be gained comparing them at the atomic pd level. The analysis of the partial/total DOS (see Fig. 10) shows that mainly three orbitals are involved in the antibonding band. Cu $3d_{x^2-y^2}$, O(1) $2p_x$, and O(2) $2p_y$. The relevant on-site energies and transfer integrals are shown in Fig. 17.

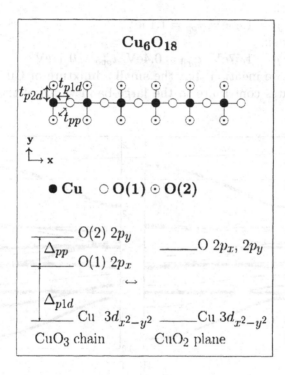

Figure 17. Typical $(CuO_3)_N$, $N = 4, 6$ cluster with periodic boundary conditions used in the exact diagonalization reported in Sec. 6.

Thus, in addition to the Coulomb interactions U_d, U_p , and at least one (nearest neighbor) intersite interaction V_{pd}, three transfer integrals t_{pd1}, t_{pd2}, t_{pp}, and two onsite energies Δ_{pd1}, $\Delta_{pd2} \equiv \Delta_{pd1} + \Delta_{pp}$ have to be determined. The antibonding band behaves too smoothly to determine so many parameters in an unique fashion from a correct fit. Therefore some of the bonding and/or non-bonding bands must be included in the analysis. However, a non-negligible hybridization with further oxygen orbitals occurs (non σ overlap which is of less interest for many physical properties at the low-energy scale). The inspection of the analysis of the corresponding orbital weights of the states mentioned above (see Fig. 18, left panel) reveals a discontinuity of the O(2) $2p_y$ character of the region of the non-bonding O derived states near -4 to -5 eV, if we restrict ourselves to the four-band model, only. Therefore another fit was performed within the extension to a seven band model, where the O(1) $2p_y$ and the two O(2) $2p_x$ orbitals were taken into account. To illustrate the improvement achieved by the seven-band fit with respect to reasonably small n.n. transfer integral t_{pp}, we list the obtained transfer integrals:

four-band model:

$t_{pd2} = 1.8\text{eV}$, $t_{pd1} = 1.45\text{eV}$, $t_{pp} = 1.15\text{eV}$,

seven-band model:

$t_{pd2} = 1.8\text{eV}$, $t_{pd1} = 1.57\text{eV}$, $t_{pp\parallel} = 0.4\text{eV}$, $t_{ppo} = 0.41\text{eV}$.

In this context we mention that the small admixture of Cu $4s$ states and of the O $2s$ states contribute to the large band width of the antibonding band.

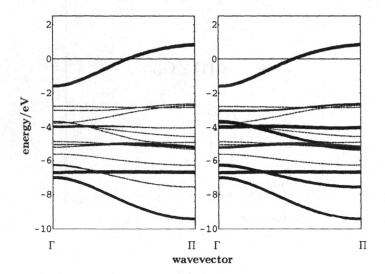

Figure 18. LCAO-LDA band structure of Sr_2CuO_3 along the chain direction. The thickness of the lines is scaled with the orbital weight of the four-band (left) and of the seven-band model (right).

Such a procedure – to determine the relevant states on a larger energy scale including much more states than are needed directly for the many-body problems which must be kept still solvable numerically at least for small clusters – seems to us prefavourable compared with other techniques used in the literature where these additional states enter implicitly the down-mapped parameters in an unphysical way.

5. Magnetic Properties: fluctuations, anisotropy, and frustrations

In the following we shall assume that to a first approximation the magnetic properties of undoped cuprate chains can be described by the spatially anisotropic spin-1/2 Heisenberg model

$$H_{spin} = \sum_{<ij>} J_{ij} \vec{S}_i \vec{S}_j = \sum_i (J_1 S_i S_{i+1} + J_2 S_i S_{i+2}), \qquad (14)$$

where J_{ij} is the exchange integral between nearest neighbor (n.n.) sites i and j. We restrict ourselves to these terms since within the strongly correlated limit of the extended single band Hubbard model the antiferromagnetic superexchange process is related to the square of the effective transfer integrals

$$J_{ij} = \frac{4t_{ij}^2}{U_i - V_{ij}}. \qquad (15)$$

The effective transfer integral is a rapidly vanishing function of the intersite distance. Only in some special cases where the n.n. interaction is suppressed for special geometrical (bonding) reasons such as in the case of edge-sharing CuO_2 chains, the corresponding next nearest neighbour terms J_2 will be taken into account. If $J_1 \cdot J_2 > 0$, i.e. both interactions have the same sign, such an interaction introduces qualitative new features by the "frustrational" physics. In particular, even a relatively weak frustration tends to produce in 1D a spin gap at $J_2 \geq 0.241 J_1$. In general it competes this way with the Néel ordering. The magnitude of J_2 is mainly governed by t_2 discussed in detail above. For the corner-sharing CuO_3 chains we obtain $t_1 \sim (6 \text{ to } 9)t_2$ and therefore $J_2 \ll J_1$ in contrast to some CuO_2 chains such as $CuGeO_3$ where the critical ratio 0.241 is presumably exceeded. As a result, the spin-Peierls state is strongly supported (we recall that the lattice dimerization also produces a spin gap for finite spin-phonon coupling, see also Sec. 7).

Consequences of further generalizations such as weak spin anisotropy and/or additional terms are discussed in the lecture of Müller [29]. For additional information covering also cuprate chain and ladder compounds the interested reader is referred to comprehensive and excellent review articles of Johnston[7] and Dagotto[11].

5.1. HOW STRONG IS THE INTERCHAIN INTERACTION?

In low-dimensional systems all thermodynamic quantities are strongly affected by thermal and quantum fluctuations governed by the interchain interactions. Their influence is also seen in the temperature dependence of the magnetic susceptibility $\chi(T)$. For a general spin-1/2 Heisenberg antiferromagnet $\chi(T)$ exhibits for $T \lesssim J$, (J is the dominant exchange integral) a maximum at some characteristic temperature $T_{\chi_{max}}$ (affected also by the coordination number and possible frustration etc.)[7]. The ratio $r_J = T_N/T_{\chi_{max}}$ can be used to classify approximately the dimensionality of an antiferromagnetic system[34]. For $r_J > 0.9$ the system can be regarded as three dimensional, while $0.25 < r_J < 0.5$, stands for quasi-2D AFM, and $r_J < 0.1$ for quasi-1D systems. According to this classification Bi_2CuO_4 (see Table 1) is clearly a 3D-magnetic system albeit from the point of view of the electron spectroscopy it can be regarded as zero-dimensional [14].

Here we shall concentrate on the Néel temperature T_N and the ordered magnetic moment $\langle\mu\rangle$ at very low temperature. An overview is given in Table 2. The first inspection of this Table reveals that the magnitude of $\langle\mu\rangle$ correlates with the anisotropy ratio $r_J = J_\perp/J_\parallel$.

TABLE 2. Undoped chain cuprates and related layered/ladder compounds

Compound	Type	d_{Cu-Cu} [Å]	$\langle\mu\rangle$ [μ_B]	T_N [K]	Anisotr.	$J_\parallel$ K
Bi_2CuO_4	single plaquette	2.907	0.63	44	$4 \cdot 10^{-1}$	
Sr_2CuO_3	corner 1D	3.494	0.06	5.5	$3 \cdot 10^{-3}$	2200
Ca_2CuO_3	corner 1D	3.263	0.09	8	$7 \cdot 10^{-3}$	2200
$BaCu_2Si_2O_7$	zigzag corn. 1D	3.481	0.16	8	$9 \cdot 10^{-3}$	280
$BaCu_2Ge_2O_7$	zigzag corn. 1D	3.481	0.16	8	$8 \cdot 10^{-3}$	540
Li_2CuO_2	edge 1D	3.263	0.92	9	$7 \cdot 10^{-3}$	
$CuGeO_3$	edge 1D	4.81	0.24	9	$-1 \cdot 10^{-2}$	
$Ca_2Y_2Cu_5O_{10}$	edge 1D	3.263	0.92	30	$7 \cdot 10^{-3}$	
$SrCuO_2$	zigzag double 1D	3.563	0.09	5	$1 \cdot 10^{-3}$	
$Ca_{0.86}Sr_{0.15}CuO_2$	corner 2D	3.193	0.51	540	$6 \cdot 10^{-2}$	
$YBa_2Cu_3O_{6.15}$	corner 2D	3.3	0.48	400	$8 \cdot 10^{-2}$	
$SrCu_2O_3$	2-leg	3.46	0.06	8	$5 \cdot 10^{-2}$	
$Sr_2Cu_3O_5$	3-leg	3.495	0.06	55	$8 \cdot 10^{-2}$	
$LaCuO_{2.5}$	2-leg	~ 3.4	0.5 ± 0.2	125	$3 \cdot 10^{-3}$	
CuO	edge	~ 3.4	0.68	230	$3 \cdot 10^{-3}$	
$MgCu_2O_3$	fold. 2-leg	~ 3.2	0.32	95	$6 \cdot 10^{-2}$	
$Ba_3Cu_2O_4Cl_2$	fold.-edge	3.263	0.92	10	$7 \cdot 10^{-3}$	

In the highly anisotropic case $J_y \ll J_\perp \ll J_\parallel$ the magnetic moment at $T = 0$ depends only on r_J whereas the Néel temperature T_N is affected also by the weakest coupling in accord with the Mermin-Wagner theorem which requires a finite interaction in all three directions for a 3D-ordering at finite temperature $T \neq 0$. In our analysis we shall adopt the recent results of modern quantum spin chain theory: the work of Schulz [65] and the multi-chain mean-field theory by Sandvik[64]

$$\langle \mu \rangle = 0.53 g_L \sqrt{r_J} (1 + 0.1 r_J) \ln^\gamma (1.3/r_J), \qquad \gamma = 1/3, \qquad (16)$$

where $g_L \approx 2.13$ is the material dependent Landé factor. A similar result without logarithmic correction has been obtained by Schulz [65]. The important message of Eq. (16) is that increasing the interaction in the inchain direction being the strongest one, the 3D-Néel-type ordering is *suppressed*. Thus the main influence on the ordering is not given by the interaction strengths but by the anisotropy. This explains at least qualitatively why materials with weak interactions may exhibit higher values of the ordered moments and of T_N as well.

Combining Eq. (15) with Eq. (16) implies that $\langle \mu \rangle$ yields a direct measure of the strongest interchain interaction, provided the main interaction remains nearly constant. The ordered moment scales then roughly with t_z in the notation of Sec. 4.2. Comparing our results for Ca_2CuO_3, Sr_2CuO_3, and $SrCuO_2$ one indeed observes a nearly linear dependence on t_z. With respect to the inherent frustrations of the zigzag (double) Cu_2O_4 chain (i.e. the interaction of the two subchains) the question arises[70],[66] what is the origin of the further reduced magnetic moment compared with Sr_2CuO_3[69] and Ca_2CuO_3[71]? In this context the findings for $Mg_2Cu_2O_3$ [35] exhibiting similar zigzag chains but with significant inter(double)chain interaction are of interest. According to the authors, T_N is as large as 95 K and the ordered moment is $0.32 \pm 0.02 \mu_B$. Thus, all characteristics of the Néel state are enhanced by an order of magnitude and we may conclude that the peculiar behaviour of $SrCuO_2$ must be attributed to the specific interchain interactions. This conclusion is supported also by our band structure based analysis similarly to that given here for Sr_2CuO_3.

In contrast, the Néel temperature T_N is an increasing function of all coupling strengths although there may be a quasi-plateau near cross-over temperatures [72] where the dependence on certain interaction constants is very weak. Anyhow, in the present case the situation is more complex due to alternating ordering of chain planes in y direction (see Fig. 3) which leads in the classical approach to decoupling in nearest neighbour approximation since the interaction of the basal planes with the spins in the middle plane is *frustrated*. According to our LDA-calculation there is a weak transfer $t_y \sim$ 2 meV in b-direction to the n.n.n. plane which gives rise to a very weak antiferromagnetic coupling $\sim 5 \cdot 10^{-3}$ meV. Then antiferromagnetic ordering in this direction might be expected. However, the neutron data[69] show *integer* Bragg reflections in magnetic scattering in b-direction and half odd integers in the $a-$ and c-directions, which must be interpreted as ferromagnetic and antiferromagnetic ordering, respectively. Thus, we are forced to

110

assume, that either quantum fluctuations affect the frustrated spins at the intermediate chain plane leading finally to an unexpected significant ferromagnetic long-range coupling or the ferromagnetic correlation seen in neutron scattering is caused by the interference of inequivalent exchange paths similarly as it was suggested for the layered cuprate with two inequivalent Cu sites $Ba_2Cu_3O_4Cl_2$[27], K_2CuF_4, and the "isolated" COP compound in the framework of the standard pd-model [36]. The former scenario is on the lines of the so-called "order from (quantum) disorder" concept for frustrated magnetic systems with a classically degenerate ground state introduced by Shender [37]. Thus, it is quite amazing that counterintuitively quantum fluctuation may suppress and/or induce order in the same system.

We conclude this section with a short consideration of the effective dimensionality of $Ba_3Cu_2O_4Cl_2$ (its interesting field dependent physical

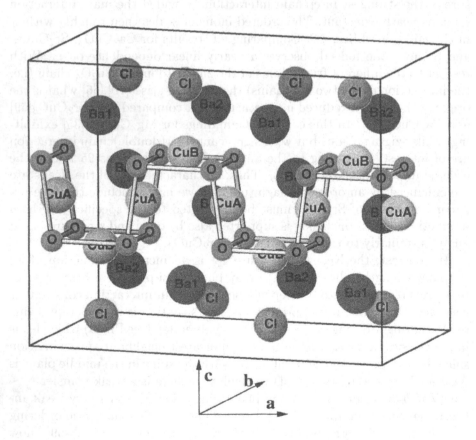

Figure 19. The pseudo chain structure of the quasi-two-dimensional $Ba_3Cu_2O_4Cl_2$ compound. Due to the two inequivalent Cu sites "A" and "B" at least a two-band description is required.

properties are discussed by Müller [29]). At first glance one might suppose again a quasi-one-dimensional character. However, the LDA bandstructure calculations show stronger dispersion perpendicular to the folded chain direction resulting from an effective interaction of topmost COPs. Thus, this compound should be considered as an effective anisotropic 2D or 3D material.

Another example is given by $CaCuGe_2O_6$ [17] which exhibits zigzag Cu chains but magnetically it must be described as consisting of weakly interacting dimers[18].

5.2. THE LARGE J-PROBLEM AND OTHER PECULIARITIES OF CORNER-SHARED CHAINS

Comparing the experimental value of $J_{\|}$ estimated for instance from the above mentioned ARPES-data [9] with the Néel temperature $T_N \approx 5K$, one finds from their ratio, the so-called Ramirez ratio, $r = J_{\|}/k_B T_N \sim 300$ to 600, i.e. a record value so far observed for any spin-1/2 antiferromagnet, irrespective to the precise value of $J_{\|}$. This suggests that at high temperature a pure 1D-description should be an excellent approximation. Traditionally the magnetic susceptibility data $\chi(T)$ of many quasi-1D-magnetic spin-1/2 systems have been analyzed for more than two decades up to now by many experimentalists to extract the exchange integral J from an approximate closed-form expression suggested by Estes *et al.* [46]

$$J_{\|}\chi_{BF}(T) = \frac{1}{4}\frac{\sum\limits_{m=0}^{n-1} a_m x^m}{\sum\limits_{m=0}^{n} b_m x^m}, \qquad x = \frac{k_B T}{J_{\|}}, \qquad n = 3, \qquad (17)$$

where

$$\begin{aligned} a_0 &= 0.3009400 & a_1 &= 0.299900, & a_2 &= b_3 = 1.0 , & (18)\\ b_0 &= 0.757825, & b_1 &= 0.172135, & b_2 &= 0.993100 . \end{aligned}$$

This expression is based on the pioneering numerical exact diagonalization results obtained by Bonner and Fischer (BF) [31] in the 60/70-ties for very small clusters ($N \leq 11$ sites only, for which such calculations could be performed at that time, see Fig. 20) However, in a recent paper by Eggert *et al.* [40] the occurrence of *logarithmic* terms $\propto 1/\ln(x_0/x)$ with $x_0 \approx 7.7$, for the low-temperature region has been stressed, thus giving for $x < 0.1$

$$J_{ch}\chi(T) \approx \tilde{\chi}(0)\left(1 + \frac{1}{2\ln(x_0/x)} + O(\ln^3(x_0/x))\right), \qquad x \ll 1, \qquad (19)$$

where $\tilde{\chi}(0) = 1/\pi^2$. To our knowledge there is up to now no approximate closed-formula for the whole T-region. In Ref. [45] we proposed to use instead of Eq. (17), valid only for $x \geq 0.3$, the following rational polynomial expression

$$
\begin{aligned}
\tilde{\chi} &= R(x)/Q(x), & (20)\\
R(x) &= 1 + 0.5(1 + 0.74x + 0.4x^2)\tanh(y) + 1.12x^2 + 1.16x^3 + 1.57x^4 + \\
&\quad + 2.85x^5 + 2.7x^6 + \pi^2 x^8, \\
Q(x) &= \pi^2(1 + 0.37x^3 + 4.23x^4 - 0.17x^5 + 1.45x^6 + 2.5x^7 + 2x^8 + 4x^9), \\
y(x) &= (1 - x/x_0)/\ln/(x_0/x).
\end{aligned}
$$

The graph of $\tilde{\chi}(T)$ approximated by Eqs. (20) is shown together with that of the BF-curve (Eq. 17) in Fig. 20. Although, the general behaviour around and above the maximum at $x = 0.6408$ is quite similar, non-negligible deviations are clearly seen. The BF-curve (Eqs. (17,18)) underestimates slightly the reduced susceptibility but *overestimates* its slope until a factor of ≈ 1.57 at the lower inflection point $x_i = 0.087$. For $x_i < x < 0.25$, this artifact of the BF–expression may pretend an enhanced exchange integral J_{ch} as discussed below. Contrarily, at very low T, i.e. far below x_i, the reduced temperature derivative is strongly underestimated.

Therefore the first statements of unusually large J-values ≈ 245 meV based on analysis of Eq. (17) by Ami *et al.* [41] have been shortly after questioned [42],[45]. In fact, the determination of J from the fit to the numerical available Bethe-ansatz solution gave a little bit smaller J-value ≈ 190 meV [6], exceeding nevertheless significantly the typical layered cuprate values by a 50 to 60 % . The analysis of Eggert which gave $J_{\parallel} \approx 147$ meV has been criticized by Johnston [7] with respect to the use g_L as a second fit parameter in addition to $J_{\parallel}$ resulting in a too small value of the Landé factor $g_L < 2$ entering the data which must be fitted. Shortly after, Suzuura al. [48] came up with midinfrared optical data exhibiting a sharp cusp or singularity remnant like feature near 0.48 eV for Sr_2CuO_3 and 0.47 eV for Ca_2CuO_3. This feature has been interpreted by the authors and later on similarly in Ref. [49] as evidence for a phonon assisted spinon absorption (recall that the elementary excitations of the 1/2- AFM Heisenberg chain are spinons, thereby the triplet excitation being in usual antiferromagnetic systems a magnon which here becomes unstable and decays into two spinons, more precisely, into an antispinon and a spinon for topological reasons). The

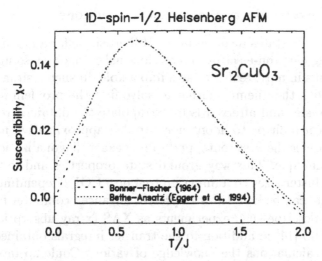

Figure 20. Reduced magnetic susceptibility $J_\| \chi$ vs. reduced temperature $k_B T / J_\|$ for the spin-1/2 antiferromagnetic chain. Our proposed approximation given by Eq. which simulates on one hand the Bethe-ansatz solution of Eggert *et al.* at low and intermediate $T/J_\| \leq 2$ as well as the Curie-Weiss behaviour at arbitrary high T. The simulation of the Bonner-Fisher curve (dashed curve) used for the fit of the experimental data by Ami *et al.* [41].

sharp peak associated with the logarithmic singularity at

$$\hbar\omega_{ir} = \hbar\omega_{ph} + (2/\pi)J_\|, \tag{21}$$

where $\hbar\omega_{ph} \approx 70$ meV being a typical phonon energy, allows in principle a very accurate determination of $J_\|$ from Eq. (21). In the analysis of Suzuura as well as in that of Lorenzana and Eder it gave again a very large number of 260 to 246 meV. Irrespective to the actual value, the resolution of the clear discrepancy (~ 70 meV) between these infrared and the susceptibility data remains a challenging problem. We briefly mention several factors which might contribute to this unexpected difference:

(i) the Heisenberg Hamiltonian might be generalized with respect to additional terms resulting for instance from finite correlation corrections yielding quartic spin-operators in the generalized Hamiltonian. Similarly long-range interaction or anisotropy effects might affect the final results

(ii) the Landé factor entering the fit of the susceptibility data exceeds the empirical Cu $^{2+}$ value $g_L = 2.13$ used in Ref. [6].

(iii) since thermal lattice expansion affects the magnitude of $J_\|$, Takigawa *et al.* [50] consider them as a likely reason which would explain also the discrepancy between the susceptibility data and the infrared data. In addition excitations of phonons might also be important.

6. Effects of strong electron-electron interactions

The huge number of states involved in the Hilbert space for a model Hamiltonian including electron-electron interaction makes rigorous solutions for extended systems in most realistic cases impossible. In such a situation one is confronted with the dilemma either to solve first the problem for a relatively small cluster and afterwards to extrapolate the obtained results to the infinite chain limit or to adopt uncontrolled approximative methods. Here we shall prefer the first route, performing exact diagonalizations with the Lanczos-technique. This way ground state properties and low-energy excitations in cluster sizes containing up to 4 or 6 COPs depending on the hole doping state can be handled with present advanced computer facilities.

In order to describe such experiments as XAS (x-ray absorption spectroscopy or EELS [14] in addition to the transfer integrals obtained above from the LDA-calculations the knowledge of various Coulomb interaction parameters is required. In this respect the analysis of various spectroscopic experiments is very helpful and challenging. One would like to learn to which interactions a particular spectroscopy in combination with the symmetry properties of a given material is most sensitive.

6.1. EVOLUTION OF THE ORBITAL DISTRIBUTION OF HOLES WITH DOPING - COMPARISON OF THEORETICAL APPROACHES

Before turning to one instructive example, namely the XAS O $1s$ spectroscopy on doped and undoped CuO_3 chains, let us consider how accurate is the el-el interaction handled by various popular theoretical methods. For this purpose we shall consider the related problem of orbital hole distribution [43]. In order to compare with other results we adopted for the time being a plane-derived parameter set proposed by Oleś and Grzelka[44] who performed detailed calculations applying the Hartree-Fock (HFA) and the local-ansatz (LA) methods. In the latter method with a special variational ansatz for the ground state wave function some local correlations are approximately taken into account. In other methods being very popular in the cuprate field a reduction of the Hilbert-space of states is achieved neglecting special configurations such as double occupancies (NDO) of two holes on the same Cu site since the onsite repulsion $U_d \approx 8$ to 10 eV is by far the largest energy in the pd problem. This NDO-approximation is the basic assumption of the t, t', t''-J models. We may check these approaches numerically by adopting unusual large U-values, say of ~ 100 eV and extrapolating the obtained U-dependences in the limit $U_d \to \infty$.

Now we consider the orbital distribution of the holes present in differently charged $(CuO_3)_N$ clusters, i.e. the occupation numbers $\langle G \mid n_d \mid G \rangle$ and analogously for the two nonequivalent oxygen sites. The result is shown in Fig. 21. Although the general tendency in all approaches is similar there is nevertheless significant different behaviour with respect to the absolute values. For small doping ratio, i.e. near one hole per COP, the mean-field HFA is wrong by about 0.2 holes per Cu shifting them to the neighboring O sites, i.e. the covalency is strongly overestimated. The situation is slightly improved within the more sophisticated LA. Neverthelss a 20% effect remains. As expected here an excellent description is achieved by the $U_d = \infty$ approximation and also by the spinless fermion approximation putting additionally also $U_p \to \infty$. However, for large doping ratios ≥ 0.15 sizable deviations start to develop. At 0.5 doping ratio all approximations underestimate n_d by about 15% . Approaching the closed shell situation with one additionally doped hole the HFA and the LA yield a reasonable description. The NDO fails most strongly by more than 25% . This means that finite Coulomb repulsion U_d is essential at high doping level such as in the 1/4-filled CuO_3 chains in $YBa_2Cu_3O_7$ or in the charged stripes (see Sec.

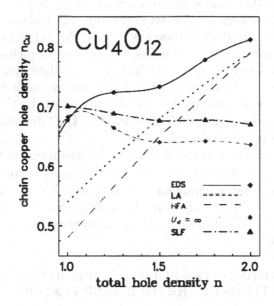

Figure 21. Cu $3d_{x^2-y^2}$ hole occupation number n_d vs. doping rate for various treatments of the el-el interaction: exact diagonalization (EDS), local ansatz (LA), Hartree-Fock-approximation (HFA), no double occupancy at Cu (NDO, i.e. $U_d = \infty$), and spinless fermion (SLF, i.e. $U_d = U_p = \infty$).

8.2). How can one understand the failure of the NDO? Indeed, for realistic U_d-values according to our exact diagonalizations the double occupancy is really small. But it enters various physical quantities as a product with U and the doping ratio δ yielding an uncertainty like $0 \times \infty$ and this way a significant finite effect. At the same time this explains also the success of the NDO at low doping $\delta \ll 1$.

6.2. EXTRACTION OF HAMILTONIAN PARAMETERS FROM COMPARISON WITH XAS AND LDA

Now we consider the O $1s$ polarization dependent XAS measurements for Sr_2CuO_3 and $YBa_2Cu_3O_7$ where the ground state hole ratio

$$R = 2\langle G \mid n_{p2} \mid G\rangle/\langle G \mid n_{p1} \mid G\rangle, \tag{22}$$

of both nonequivalent oxygen sites $O(2)$ and $O(1)$ in the CuO_3 chain can be directly measured [14].

In order to find reliable estimates for the mentioned above transfer integrals, we analyze at first the orbital character of the bandstructure of Sr_2CuO_3 near the Fermi energy E_F [24]. At first glance only the half-filled antibonding band near E_F needs to be considered. However, the extended tight-binding fit for this band is numerically unstable, and the three different values of transfer integrals required in the four-band pd model cannot be determined from this band alone (See Fig. 17 and Eq. (1)) for our notation of axes, sites, and parameters.). Thus, we were forced to include also additional energetically lower lying bands which have bonding and nonbonding character. But due to the non-negligible hybridization with further O $2p$ orbitals, within the four band model, a discontinuity of the $O(2)$ $2p_y$ character is observed in the region of the nonbonding $O(2)$ derived bands near -4 and -5 eV (see Fig. 18 , left panel).

In Sec. 6.3 we will arrive at significantly enhanced intersite Coulomb interaction V_{pd} in the prototypical single corner-sharing chain cuprate Sr_2CuO_3. One might look for other consequences of this fact. For this purpose and to illustrate the interplay of various model description, we consider in Sec. 6.3 a pd-model analysis of the empirical one-band parameters derived from the EELS data [26].

6.3. INTERPRETATION OF THE EMPIRICAL ONE-BAND COULOMB INTERACTION IN TERMS OF THE FOUR-BAND pd-MODEL

As it was already mentioned [14] in the EELS data of Sr_2CuO_3 with the transferred momentum parallel to the chain directions at the Brillouin zone boundary a relatively sharp peak near 3.2 eV has been observed. The quantitative analysis of this phenomenon in terms of the one-band extended

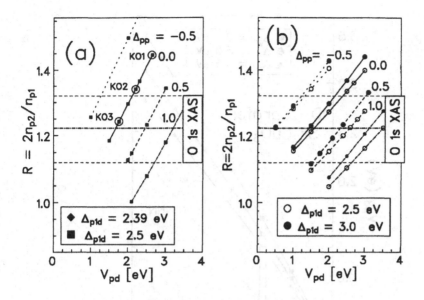

Figure 22. Oxygen hole ratio vs. intersite Coulomb interaction V_{pd} for an undoped CuO_3 chain described in the four-band pd model for various differences of oxygen onsite energies Δ_{pp} (panel (a): using parameter sets proposed in Ref. [74], panel (b) using our LDA-LCAO derived transfer integrals (see Fig. 18).

Hubbard model [26] revealed in addition to the agreement with our predicted LDA value [24] for the n.n transfer integral $t_1 = 0.55 \pm 0.05$ eV and the later ARPES-data [9] mentioned above also somewhat unexpected values for the onsite and intersite Coulomb interactions $U_{1B}=4.2$ eV and especially $V_{1B}=0.8$ eV, respectively. The first value is surprising since in typical layered cuprates one has usually $U_{1B} \approx \Delta_{pd} \approx 3.5$ to 4 eV but in the present case significantly *lowered* values of about 2.5 eV have been deduced from the analysis of Cu $2p$ core-level x-ray photoemission spectra (XPS) [75].

Comparing the expectation value of the onsite Coulomb interaction contribution to the ground state energy in a *completely* filled one-band chain, i.e. with *two* holes per COP $\equiv U_{1B}$ per definition with the direct contributions of Coulomb terms within a CuO_3 unit, we consider the quantity $\tilde{u}$ defined as

$$\tilde{u} = D_{d,2}U_d + (2D_{p2,2} + D_{p1,2})U_p + [\langle n_d n_{p1}\rangle_2 + 2\langle n_d n_{p2}\rangle_2]V_{pd}$$
$$+ 2\langle n_{p1}n_{p2}\rangle_2 V_{pp} + \langle n_{p2a}n_{p2b}\rangle_2 \tilde{V}_{p2p2} + [n_{p1,2} - n_{p1,2}^{(0)}]\Delta_{p1d} +$$
$$+ 2[n_{p2,2} - n_{p2,2}^{(0)}][\Delta_{pp} + \Delta_{p1d}], \qquad (23)$$

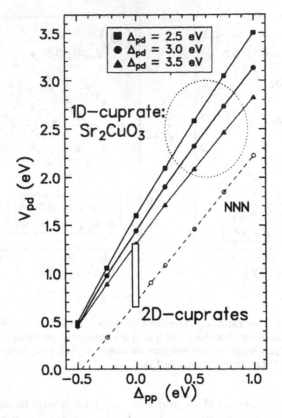

Figure 23. Intersite Coulomb repulsion V_{pd} vs. oxygen onsite energy difference Δ_{pp}

which serves as an useful interpolation formula $\tilde{u} \approx U^{(1B)}$ at arbitrary parameters. Here the double occupancies $D_{i,2}$ ($i = d, p_1, p_2$), the correlation functions $\langle...\rangle_2$, and the occupation number s $n_{i,2}$ entering Eq. (23) have been calculated for the case of two holes per unit cell, and $\langle n_{p2a} n_{p2b} \rangle$ denotes the correlation function between the two equivalent peripheral oxygens "O(2a) " and "O(2b)" residing on the same chain unit. The occupation numbers with the superscript "(0)" are calculated setting the Coulomb interactions to zero, i.e. in the one-particle limit.

In the weak correlation limit $u \to U_0^{(1B)}$ given by

$$
\begin{aligned}
U_0^{(1B)} = {} & \frac{1}{4} \left[n_d^2(2) U_d + \left(2n_{p2}^2(2) + n_{p1}^2(2) \right) U_p \right] + \\
& + n_d(2) \left[n_{p1}(2) + 2n_{p1}(2) \right] V_{pd} + \\
& + 2n_{p1}(2) n_{p2}(2) V_{pp} + n_{p2}^2(2) \tilde{V}_{p2p2},
\end{aligned} \tag{24}
$$

where $n_i(2)$ $i = d, p1, p2$ denotes the occupation numbers for the case of two holes (one doped hole) per CuO_3 (COP) unit. The result is shown in Fig. 24. One realizes that starting at $V_{pd}=0$ $U_{1B} < \Delta_{p1d}$ (presumably due to the kinetic energy related to the motion inside the COP). At "normal" V_{pd}-values of about 1eV, we have like in the standard 2D case $U_{1B} \approx \Delta_{p1d}$. But for the enlarged values estimated from the O $1s$-XAS data discussed in Sec. 6.1 we arrive at about 4 eV as in the EELS.

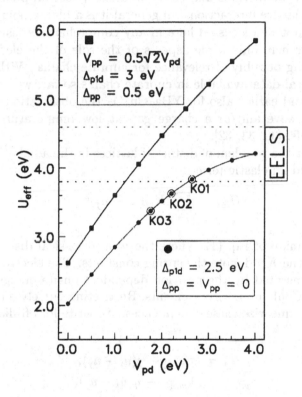

Figure 24. Effective onsite Coulomb repulsion $U_{eff} \equiv U_{1B}$ of the effective single-band model vs. intersite Coulomb interaction V_{pd} of the four-band pd-model.

An analysis for the intersite Coulomb interaction V reveals that our enhanced V_{pd} values cover only about 50% of the empirically required 0.8 eV. This discrepancy can be removed if an additional small Cu-Cu Coulomb interaction of about 0.7 to 0.8 eV is added to the pd-Hamiltonian. Strictly speaking within a more sophisticated approach all parameters of the Hamiltonian are renormalized, if new terms are added.

The necessity to add V_{dd} seems to be quite natural, since the inelastic scattering of electrons is governed by the long-range Coulomb interaction and V_{dd} is just a link between short and long range.

7. Interplay of el-el with el-ph interactions

It is well-known that quasi-one dimensional systems exhibit a pronounced instability to various ground states. In the case of commensurate band filling the spin-Peierls state, various bond or charge density wave states are likely candidates. Which of these ground states will be realized at low temperatures depends among many other factors essentially on the interplay of in chain el-el interactions and electron-lattice (electron-phonon (el-ph)) as well as spin-lattice interactions. In general it is a highly nontrivial problem which cannot be discussed here in any comprehensive fashion. Let us briefly consider here only some aspects of the role of the electron-lattice interaction being possibly of relevance for cuprate chains. With respect to few experimental data available in cuprate chains so far, we note that for $PrBa_2Cu_3O_7$ and earlier also for $YBa_2Cu_3O_7$ the observation of a charge (spin) density wave and/or a charge gap at low temperatures has been reported in Refs. [30, 33, 32].

We consider the problem of distorted lattices in the adiabatic limit. For this aim we add an elastic term

$$H_{lat} = \sum_{<i,j>} K_{ij}(u_i - u_j)^2, \qquad (25)$$

to the Hamiltonian of Eq. (1), where the u_i are classical displacements of the site i and the K_{ij} denote the spring constants. The electron-lattice interaction is generated by the position dependent onsite energies, transfer integrals, and Coulomb matrix elements. Restricting ourselves to small displacements, we linearize these dependences. We write for off-diagonal el-ph interaction

$$t_{i,j} = t_{ij;(0)} - \gamma(u_i - u_j), \qquad (26)$$
$$V_{pd;i,j} = V_{pd;(0)} - \eta(u_i - u_j), \qquad (27)$$

and analogously for diagonal interactions

$$\varepsilon_i = \varepsilon_{(0);i} - \alpha u_i, \qquad (28)$$
$$U_i = U_{(0);i} - \beta u_i. \qquad (29)$$

For instance, the off-diagonal electron-phonon coupling parameter γ can be determined either from adopting empirical exponential or power-law distant dependences of the transfer integrals in the case of pd-models or by considering deformations in the band structure due to elongations of selected atomic sites from their equilibrium position [55]. The spring constants K_{ij} were determined fitting experimental values of frequencies of Raman and infrared active phonon modes [57, 56].

In the noncorrelated state and for a single interaction channel the strength of the el-ph interaction is measured by the competition of the lattice stiffness (K), the band filling ratio, the band width $(W \propto t_1)$, and the introduced above linear expansion parameters which enters the in the one-band half-filled case the dimensionless coupling constants g or more traditionally λ_{el-ph},

$$g = \gamma/\sqrt{Kt_0} < 1, \quad \lambda_{el-ph} = 2g^2/\pi. \tag{30}$$

For dimerized systems it is convenient to express the dimerization (bond) amplitude $2u_0$ in dimensionless units d

$$d = 2u_0\sqrt{K/t_0} < 1. \tag{31}$$

The necessity of el-ph interactions of the type described by Eq. (29) occurs in effective one- and two-band models. In the strongly correlated limit $W = 4t_1 \le U$ when the one-band extended Hubbard model can be mapped on the antiferromagnetic spin-1/2 Heisenberg model, the combination of Eqs. (27,29) results in a distant dependent n.n. exchange integral J

$$J_{i,i+1} = J_0 - \gamma_{sp}(u_{i+1} - u_i). \tag{32}$$

In all cases the lattice distortions have been found from the minimum of the total energy of $H_{el}+H_{lat}$ with respect to $\{u_i\}$. For small clusters the new equilibrium positions can be found directly considering the corresponding energy functional. But for large clusters an equivalent procedure, namely the iterative solution of the self-consistency equations, is more convenient to obtain the set u_i characterizing the new distorted ground state. In the case of the extended Hubbard model these equations read [52]

$$Kv_i \equiv K(u_{i+1} - u_i) = \Lambda/N - \gamma P_{i,i+1} + \eta D_{i,i+1}, \tag{33}$$

where v_i is the bondlength change and the Lagrange-multiplier

$$\Lambda = \sum_i (2\gamma P_{i,i+1} - \eta D_{i,i+1}),$$

expresses the convenient fixed length constraint $\sum_i v_i = 0$, $P_{i,i+1}$ denotes the bond order and $D_{i,i+1}$ being the density-density correlator in the ground state $|G\rangle$

$$P_{i,i+1} = 0.5 \sum_s \langle G \mid c^\dagger_{i,s} c_{i+1,s} \mid G \rangle \tag{34}$$

$$D_{i,i+1} = \langle G \mid n_i n_{i+1} \mid G \rangle. \tag{35}$$

In many cases of practical interest the actual dimerization amplitude is very small if *even* rings (chains with periodic boundary conditions) are considered and no self-consistent solution $d \neq 0$ is found. However, this difficulty can be circumvented for reasonable strength of el-ph interactions and arbitrary strength of el-el correlations if *odd* rings are considered. Here we exploit the generic property of such odd (neutral, i.e. undoped) rings that their ground state is always given by a spin soliton. The lattice distortions for such a ring are shown in Fig. 25. The bond in front of the soliton cen-

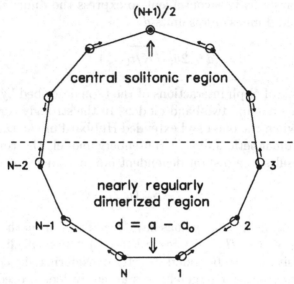

Figure 25. Schematical view of the lattice distortions in 1/2-filled odd rings. Undistorted sites are denoted by $\circ(\bullet)$.

ter $(u_{(N+1)/2}=0)$, i.e. between the sites N and 1 approaches for large N to the nearly regularly dimerized state. Hence, varying the model parameters, insight into behaviour of d_∞ can be gained already at finite N from the study of $d(N)$. Let us now consider how the onsite interaction U does affect the dimerization d (see Fig. 26). We remind that some special features of the interplay of on-site correlation U and off-diagonal *el-ph* interaction have been pointed out first in Refs. [58], [59]. Employing for the Peierls Hubbard model the Gutzwiller approximation (GA) (strictly valid only for $4t \gg U$), it was shown that for weak and intermediate *el-ph* interaction strength the dimerization d with increasing U is first enhanced reaching a maximum near $U/t_0 \approx 2.5 - 3$ after which it is suddenly suppressed [59]. However, the region just above this maximum cannot be described quantitatively by the GA since in the exponential ansatz for the ground state wave function usually an expansion up to second order in the double occupancy is made. The geminal approximation of Kuprievich [60] shows a smooth decrease of d with increasing U in the opposite strongly correlated limit. Exact cluster calculations for $4n$- and $4n+2$-membered open and periodic chains (rings) by Dixit and Mazumdar [61] as well as by Hayden and Soos [62] demonstrate that the GA predicted d-enhancement due to U is somewhat overestimated. The dimerization d can be estimated from above using $4n$-membered open chains and from below by $4n\pm2$- membered rings.

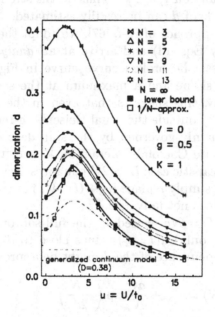

Figure 26. Dimerization in the one-band Peierls-Hubbard model vs. onsite interaction U (1/2-filled band case) for el-ph coupling constant $g = 0.5$

Unfortunately, for the parameter region of interest for real systems of interest the numerically accessible rings show $d = 0$, only. Thus one is forced to extrapolate the dimerization amplitude of infinite rings only from above using periodic $4n$-membered and antiperiodic $4n+2$-membered rings or using phase randomization, i.e. an averaging over different boundary conditions [63]. Quite remarkably, the self-consistent treatment of odd-membered rings yields a finite d in the above discussed sense for *any* coupling strength. Our approach reproduces the GA and the geminal approximation predictions qualitatively. Quantitatively, we obtained slightly smaller values of d for the same small cluster sizes where exact diagonalizations can be performed as well as for the extrapolated large ones. In the large U-limit our solutions tend from below to the approximate solution of the spin-Peierls problem obtained by Inagaki and Fukuyama [67] rewritten in our notation as

$$d = 3\lambda_{el-ph} 8t/3U^{3/2}, \tag{36}$$

where the well-known asymptotic expression of the exchange integral $J = 4t^2/U$ has been used. According to Eq. (36) d depends linearly on the el-ph interaction constant in sharp contrast to the weakly correlated regime where it depends exponentially on $1/\lambda_{el-ph}$. Thus, in the strongly correlated limit the order of magnitude of d can be readily estimated.

Generalizing the approach of Ref. [67], we adopt the effective exchange integral J defined by Eqs. (3) and arrive at an *analytical* expression for intermediate U-values (the broken curve curve in Fig. 26). Surprisingly it exhibits a similar shape and a maximum at the same position as the ED-curves. Physically, this suggests that even in the case of conducting polymers, being clearly outside the usual Heisenberg-regime, the dimerization mechanism is mainly governed by the spin degrees of freedom. The presence of the intersite Coulomb interaction V and its derivative (η, see Eq. 34 in the usual realistic case $U > 2V$ enhances d as expected. Thereby for $U \gg t, V$ one can simply replace $U \to U - V$. For smaller U-values and $V = t_0$, the behaviour is not universal.

For ring lengths $L = Na$ exceeding the full soliton width 2ξ (feasible for Heisenberg rings, only) $d(L_{max})$-values close to d_∞ can be expected. Therefore the size dependence of the d might be approximated by

$$d(N) = d_\infty + \frac{a_1}{N} \exp\left(\frac{-N}{2\xi}\right) + \cdots . \tag{37}$$

Ignoring higher order terms in Eq. (36) we estimate for the upper curves in Fig. 27 described by the parameter set $g = 0.6$, $U = 13$, $V = t_0 = K = 1$; $d_\infty = 0.0765$, $a_1 \approx 0.5$ and $\xi/a = 5.26$ in accord with the numbers derived from the selfconsistent shape of the soliton $v_n \times (-1)^n \approx d_\infty \tanh(na/\xi)$. The even ring curve reveals a slightly larger value $d_\infty = 0.078$. But the prefactor and the exponent differ significantly $a_1 = 11.8$ and $\xi/a = 1.85$. Taking the DMRG-values for $N \sim 100$ as granted, we conclude that the accuracy of the solitonic estimate of d_∞ is $\sim \pm 0.02$ to 0.03. From the other hands, fitting the decreasing curvature of the solitonic curve at large N by a simple parabola, the slightly smaller solitonic value obtained above from Eq. (37) might be viewed also as a hint for a tiny minimum at finite ring length (≈ 30 sites for the parameter set under consideration) being deepest in the noninteracting case. Anyhow, the simple $1/N$- extrapolation of odd (even) rings is expected to yield a lower (upper) bound of d_∞.

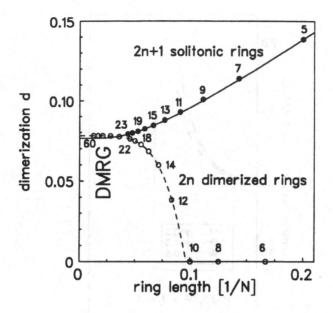

Figure 27. Size dependence of the dimerization amplitude d for even (o) and odd (•) periodic spin-Peierls rings. Even rings are treated by exact diagonalizations until $N=22$. The d for $N = 28$ to 60 are obtained by density matrix renormalization group computations. The full and dashed curves are the fits explained in the text. The parameter set used $\gamma_{sp} = 0.4, J = 1/3, K = 1$, corresponds to $U = 13, V = t_0 = 1, \eta = 0$, and $g = 0.6$ for the Peierls-one-band Hubbard model.

The initial enhancement of the dimerization amplitude as a function of the n.n. transfer integral or n.n. exchange is shown in Fig. 28. This mechanism is mainly responsible for the stabilization of the Spin-Peierls phase in CuGeO₃. The structural reason for significant frustration comes from the edge-sharing. Its practical absence in corner-sharing chain compounds might be one of the reasons for the lack of spin-Peierls states in these compounds.

In the case of the four-band model no sufficiently long rings can be considered to derive quantitatively reliable estimates of the lattice distortions in the infinite chain limit. However, the local symmetry of the lattice distortions points to the expected character of superstructure of long chains. For undoped CuO₃ chains and typical strong on-site Coulomb interactions we found a spin-Peierls ground state in the language of the Heisenberg model. In the language of the four-band model it corresponds to a Cu-site bond order wave (BOW) and a weak charge density wave (CDW) at the chain

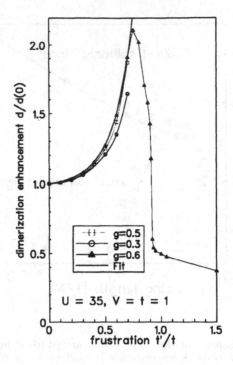

Figure 28. Dimerization amplitude d vs. frustration $J_2 = 4t'^2/U$ for fixed nearest neighbor spin-phonon interaction g_{sp} and nearest neighbor exchange $J_1 = 4t^2/(U - V)$.

oxygen sites $O(1)$. In the formally opposite limit[56] probably *not* realized in the cuprates, we find a CDW state involving the Cu and the side oxygens $O(2)$ and a BOW for the chain oxygens (see Fig. 29).

For high doping $\delta = 0.5$ in the strongly correlated regime we found a $4k_f$ mixed BOW-CDW solution. The actual parameter regime corresponding to a transition to the weaker coupled $2k_F$ solution like the observations reported for the CuO_3 chains in $YBa_2Cu_3O_7$ in [32] or for $PrBa_2Cu_3O_7$ in [30] remains unclear at present but these data clearly point to somewhat *reduced* correlation effects at these high doping level. Presumably the improved screening properties are responsible for significantly changed parameters.

period–4 superstructure

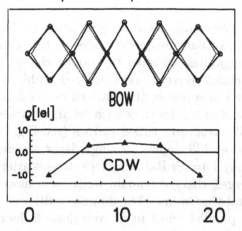

Figure 29. Schematical view of the lattice distortions in 1/4-filled CDW-BOW states and weak correlations as a simple model for the CuO_3 chains in $YBa_2Cu_3O_7$ [55].

8. 1D-Aspects of Superconductivity in 2D and 3D Cuprates

8.1. CHAIN-LIKE CHARGE RESERVOIRS

According to LDA-bandstructure calculation the COP in the CuO_3 chains are heavily doped with aproximately 1/2 hole. In the case of nearly perfect $YBa_2Cu_3O_7$ i.e. with long CuO_3 chains the latter participates in the superconductivity. This fact is very interesting since it is well known that the 2D CuO_2 planes lose their remarkable superconducting properties at $\delta > 0.3$. The plane induced chain superconductivity is usually considered as a consequence of a proximity-like effect. In this context one might also understand why insulating behaviour has been reported for *isolated* chains in $PrBa_2Cu_3O_7$ in [30] as well as in the static charged stripes at 1/8 doping in the CuO_2 planes (see below). Quite interestingly, the perfect double zigzag chains in $PrBa_2Cu_4O_8$ remain metallic but become not superconducting, too, down to very low temperatures.

8.2. CHAIN ASPECTS OF STRIPES IN DOPED CuO_2 PLANES

There is a rather interesting common phenomenon observed for doped, originally insulating, antiferromagnetic oxides such as cuprates, nickelates, vanadates, and manganites: to repel the doped charge. At low temperature this repelled charge is very often ordered within lower dimensional (compared with the dimension of the antiferromagnetic host) objects. For quasi-2D cuprates and nickelates they form "charged" rivers, more precisely coupled CDW-SDW structures called stripes (see Fig. 30). The simplest situation is realized at 1/8 doping in single layer superconductors of the $La_{2-x}A_xCuO_4$ family ($A=$ Sr,Ba). From neutron scattering data the following picture emerges: charged domain walls occur at every fourth line of vertical or horizontal Cu-Cu line. The domain walls separate magnetically ordered regions (3 parallel lines) in an antiphase fashion for neighboring regions. Obviously, there is a large similarity for the 0.5 hole doped corner-sharing CuO_3 chain considered above and this charged stripe. In particular, the insulating nature of this stripe can be understood as the combined CDW-BOW pinned to the lattice. There are two main possibilities for charge ordering with respect to the CuO_2 plane: vertical (or horizontal) as mentioned above and diagonal stripes connecting n.n.n. Cu-Cu sites (see e.g. Fig. 8.2 and Fig. 15 of [76]). The charge river of vertical stripes is built by corner-sharing CuO_3 chains while the diagonal charged river corresponds to edge-sharing CuO_2 chains. In the present context it is important that these chains are embedded into an active magnetic surrounding. Naturally, the interaction between magnetic and charged stripe regions leads to very complex new physics not present in the isolated chains. In particular, for 1/6 doping rate of the planes, i.e. just at or near the maximal T_c-value, there are 3/4-filled charge stripes and undoped magnetic, two-leg ladder like, stripes exhibiting a spin gap in the isolated state. According to Emery *et al.* [77] this might induce a spin gap for the doped holes in the charged rivers and this way superconductivity with d-wave symmetry of the order parameter. At 1/8 doping there is no spin gap and consequently also no superconductivity due to the three-leg like form of the magnetic stripes. This explains qualitatively the experimentally observed suppression of superconductivity at 1/8 doping ratio.

There is growing evidence for the presence of stripe related fluctuations in almost all cuprate superconductors. In particular, the recent observation of one-dimensional features in the ARPES-data for underdoped $La_{2-x}Sr_xCuO_4$ ($x=0.05$; 0.1) interpreted in terms of dynamical stripes [79] is noteworthy. On the other hand it is clear that static stripe structures which occur at the 1/8 doping ratio are absolutely detrimental to superconductivity. But at the same time the role of dynamical stripes in the

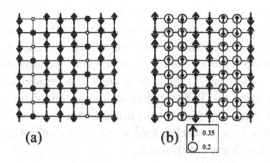

(a)　　　　　(b)

Figure 30. Schematical view of charge and spin order in stripe states in 1/8 hole doped CuO_2 planes (a) and near optimal hole doping $\approx 1/6$ in a two-legg ladder-like state (b); after Ref. 78.

high-T_c superconductivity mechanism remains unclear.

Irrespective of the final answer to this intriguing questions there is much physics in common between the magnetic flux line physics (see for instance the lecture of Brand [80] and the charged stripes especially with respect to their pinning and dynamics.

9. Summary and Outlook

Let us summarize what we have learned about general cuprate physics and specific problems of selected chain-like materials. The diversity of theoretical models allows in principle for each purpose an optimal description with respect to the desired physical information and with respect to the computational efforts. The down-mapping of Hamiltonians with a larger number of degrees of freedom onto reduced ones is often accompanied with the generation of an increased number of effective terms with unclear physical significance. The traditional slightly generalized pd-model with respect to nonequivalent oxygen sites and one or at most two orbitals per oxygen site is rich enough to describe most physics around the Fermi energy ($\pm$ 4 to 5 eV). However, for a microscopic explanation of certain phenomena such as magnetism where intermediate ranged couplings to other low-dimensional subsystems is necessary the inclusion of non-pd orbitals may be relevant.

The heavily doped state is not well described by the most popular approximations. In this regime double occupancy of Cu sites cannot be neglected. Typical corner-sharing linear chain compounds exhibit significantly enhanced transfer integrals and especially also substantially enhanced intersite Coulomb interactions. The former give rise to unusual strong in chain exchange integrals ensuring together with weak interchain coupling the magnetic ordering at very low temperature. The bandstructure derived dispersion reveals reasonable estimates for the antiferromagnetic interchain exchange interaction and this way about the effective magnetic dimension of various chain-like compounds. The analysis of the XAS, EELS, and other electron spectroscopic data reveals to which interactions different spectroscopies are most sensitive. Thus, a combined theoretical and experimental analysis starting from the LDA and exact diagonalization of model Hamiltonians on small clusters with a careful analysis of electron spectroscopy and magnetic data is a powerful tool in order to achieve a better understanding of cuprate chain compounds.

Acknowledgement

It is a pleasure to thank R. Hayn, J. Fink, J. Richter, and A.S. Moskvin, for valuable discussions and the Deutsche Forschungsgemeinschaft as well as the DAAD (individual grant H.R.) for financial support.

References

1. Rosner, H. (1999) Electronic Structure and exchange integrals of low-dimensional cuprates, *Ph-D Thesis* Technische Universität Dresden
2. Müller-Buschbaum, H.J. (1991) Zur Kristallchemie von Kupferoxometallaten, *Angew. Chem.* **103** 741–761; — (1991) The Crystal Chemistry of Copper Oxometallates, *Angew. Chem., Int. Ed. Engl.* **30**, 723–744
3. We shall not consider compounds such as CsCuO, $SrCu_2O_2$, and $BaCu_2O_2$ [2] exhibiting more or less pronounced zigzag Cu-O chains resulting from connected twofold coordinated O-Cu-O sticks where the Cu(I) state with almost filled 3d shell is involved.
4. Uehara, M., Nagata, T., Akimitsu, J., Takahashi, H., Mori, N., and Kinoshita, K. (1996) Superconductivity in the ladder material $Sr_{0.4}Ca_{13.6}Cu_{24}O_{41.84}$, *J. Phys. Soc. Jpn.* **65**, 2764–2767
   The recently reported [5] superconductivity in a closely related system at 80 K has not yet been confirmed by other experimental groups.
5. Szymczak, R., Szymczak, H., Baran, M., Mosiniewizz-Szablewska, E., Leonyuk, L., Babona, G.-J., Maltsev, V., Shvanskaya, L. (1999) Magnetic and superconducting properties of doped $(Sr,Ca)_{10}Cu_{17}O_{29}$-type single crystals, *Physica C* **311**, 187–196
6. Motoyama, N., Eisaki, H., Uchida, S. (1996) Magnetic Susceptibility of ideal Spin 1/2 Heisenberg Antiferromagnetic Chain System, Sr_2CuO_3 and Sr_2CuO_3, *Phys. Rev. Lett.* **76**, 3212–3215
7. Johnston, D.C. (1997) Normal-State Magnetic Properties of Single-Layer Cuprate High-Temperature Superconductors and Related Materials in K.H.J. Buschow (ed.) *Handbook of Magnetic Materials* **10** 1–237
8. Kim, C., Shen, Z.-X, Motoyama, N., Eisaki, H., Uchida, S., Tohyama, T., and Maekawa, S. (1997) Separation of spin and charge excitation in one-dimensional $SrCuO_2$, *Phys. Rev. B* **56**, 15589–15595
9. Fujisawa, H., Yokoya, T., Takahashi, T., Miyasaka, S., Kibune, M., and Takagi, H. (1999) Angle-resolved photoemission study of Sr_2CuO_3 *Phys. Rev. B* **59**, 7358–7361
10. Oseroff, S.B., Cheong, S.-W., Aktas, B., Hundley, M.F., Fisk, Z., and Rupp, Jr. W. (1995) Spin-Peierls State versus Néel State in Doped $CuGeO_3$, *Phys. Rev. Lett.* **74**, 1450–1453

11. Dagotto, E. (1999) Experiments on Ladders Reveal a Complex Interplay between a Spin-Gaped Normal State and Superconductivity, *Condensed Matter, abstract cond-mat/9908250*22pp. and 41 Figures

12. Meijer, G.I., Rossel, C., Henggeler, Keller, L., Fauth, F., Karpinski, J., Schwer, H., Kopnin, E.M., Wachter, P., Black, R.C., and Diederichs, J. (1998) Long-range antiferromagnetic order in quasi-one-dimensional $Ca_{0.83}CuO_2$ and $Sr_{0.73}CuO_2$, *Phys. Rev. B* **58**, 14452–14455

13. Except the delafossite derived compounds $RCuO_{2+\delta}$, R=La,Y and $0.5 < \delta < 1$, where the excess oxygen δ is introduced into a triangular (hexagonal) lattice of CuO_2 rods. For $\delta \leq 0.5$ and R=Y one arrives at an orthorhombic structure containing a plane with a "sawtooth" structure. For references see e.g. in:
    Ramirez, A.P., Cava, R.V., Krajewskii, J.J., and Peck, W.F. Jr., (1994) Groundstate of a triangular copper oxide system, *Phys. Rev. B* **49**, 16082–16085.

14. Fink, J., Golden, M.S., Knupfer, M., Böske, TH., Haffner, S., Neudert, R., Atzkern, S., Dürr, C., Hu, Z., Legner, S., Pichler, T., Rosner, H., Drechsler, S.-L., Málek, J., Hayn, R., Eschrig, H., Ruck, K., and G. Krabbes, G. (1999) The electronic structure of high-T_c superconductors: Introduction and recent studies of model compounds, *these Proceedings, 1-38*

15. Ong, E.W., Kwei, G.H., Robinson, R.A., Ramakrishna B.L., and von Dreele, R.B. (1990) Long-range antiferromagnetic ordering in $BiCuO_4$, *Phys. Rev. B* **42**, 4255–4262

16. Maltsev, V. (1999) *Growth of single crystals and crystal chemical features of Bi-containing alkaline-earth cuprates*, Ph-D Thesis, Lomonosov Moscow State University

17. Sasago, Y. Hase, M., Uchinokura, K., Tokunagai, M., and Miura, N., Discovery of a spin-singlet ground state with an energy gap in $CaCuGe_2O_6$, *Phys. Rev. B* **52**, 3533–3539

18. Zheludev, A., Shirane,G., Sasago, Y. Hase, M., Uchinokura, K. (1996) Dimerized ground state and magnetic excitations in $CaCuGe_2O_6$, *Phys. Rev. B* **52**, 3533–3539

19. Müller-Buschbaum, H.J. (1977) Oxometallate mit ebener Koordination, *Angew,. Chem.* **89**, 704-717

20. Leonyuk, L.I, Babonas, G.-J., Pushcharovskii, D.Ju, and Maltsev, V.V., (1998) Main Subdivisions of the Structural Systematization of Cuprates, *Crystallograph. Rep.* **43**, 256–270

21. Tsukuda, I., Sasago, Y., Uchinokarura, K., Zheludev, A., Maslov, S., Shirane, G., Kakurai, K., and Ressouche, E. (1999) $BaCu_2Si_2O_7$: A quasi-one-dimensional S=1/2 antiferromagnetic chain system, *Phys. Rev. B* **60**, 6601–6007

22. The crystal structures of compounds containing Ge or Si are characterized by the presence of a second structural element: the GeO_4 or SiO_4 tetrahedron, respectively. For this reason in the chemical literature the classification notation as germanates or silicates, respectively, is used. However, the band structure calculations for $GeCuO_3$ and $SiCuO_3$[23, 1] show clearly that the states near the Fermi energy which determine the electronic and magnetic properties are predominant of Cu and O character. The corresponding compounds are therefore strongly correlated charge transfer insulators in the sense of the classification of Sawatzky, Zaanen, and Allen. The Cu^{2+} ions are in a strongly elongated oxygen octahedral environment and hence again dominated by the COP states. For our purposes one could therefore write for instance $(GeO)CuO_2$ to emphasize the edge-shared COPs network instead of the standard $CuGeO_3$ notation. Strictly speaking, these compounds should be correctly denoted as germanyl (silicyl) cuprates.

23. Mattheiss, L.F., (1994) Band picture of the spin-Peierls transition in the spin-1/2 linear chain cuprate $GeCuO_3$, *Phys. Rev. B* **49**, 14050–14053

24. Rosner H., (1997) Electronic structure and magnetic properties of the linear chain cuprates Sr_2CuO_3 and Ca_2CuO_3, *Phys. Rev. B* **56**, 3402–3412

132

25. Rosner, H., (1998) Tight-binding parameters and exchange integrals of $Ba_2Cu_3O_4Cl_2$, *Phys. Rev. B* **57**, 13660–13666

26. Neudert, R., Knupfer, M., Golden, M.S., Fink, J., Stephan, W., Penc, K., Motoyama, N., Eisaki, H., and Uchida, S. (1998) Manifestation of Spin-Charge Separation in the Dynamic Dielectric Response of One-Dimensional Sr_2CuO_3, *Phys. Rev. Lett.* **81**, 657–660

27. Rosner, H., Hayn, R., and Drechsler, S.-L., (1999) The electronic structure of Li_2CuO_2, *Physica B* **259–261**, 1001–1002;
    Neudert. R., Rosner, H., Drechsler, S.-L., Kielwein, M., Sing, M., Hu, Z., Knupfer, M., Golden, M.S., Fink, J., Nücker, N., Merz, M., Schuppler, S., Motoyama, N., Eisaki, H., Uchida, S., Domke, M., and Kaindl, G. (1999) Unoccupied electronic structure of Li_2CuO_2, *Phys. Rev. B* **60**, 13 413 – 13 417;
    Rosner, H. Diviš, M., Koepernik, K., Drechsler, S.-L., and Eschrig, H. (2000) Comment on 'The electronic structure of $CaCuO_2$ and $SrCuO_2$', *J. Phys..: Condens. Matter* **12**, 5809–5812;
    Rosner, H., Johannes, M., and Drechsler, S.-L. (2001) Comment on: 'Energy band structures of the low-dimensional antiferromagnets Sr_2CuO_3 and $Sr_2CuO_2Cl_2$', *J. of Appl. Phys.*, 2 pp. submitted

28. Weht, R. and Pickett, W.E.; (1998) Extended moment formation and second neighbor coupling in Li_2CuO_2, *Phys. Rev. Lett.* **81**, 2502–2505

29. Müller, K.-H., (1998) Magnetic Properties of Low-Dimensional Cuprates, *These Proceedings* 51–80

30. Grévin, B., Bertier, Y., Collin, G., and Mendels, P., (1998) Evidence for Charge Instability in the CuO_3 chains of $PrBa_2Cu_3O_7$ from $^{63,65}Cu$ NMR, *Phys. Rev. Lett.* **80**, 2405–2408

31. Bonner, J.C. and Fisher, M., (1964) Linear Magnetic Chains with Anisotropic Coupling, *Phys. Rev.* **135**, A640-658

32. Edwards, H.L., Derro, A.L, Barr, A., Markert, J.T. and de Lozanne, A.L. (1994) Spatially Varying Energy in the CuO Chains of $YBa_2Cu_3O_{7-x}$ Detected by Scanning Tunneling Spectroscopy *Phys. Rev. Lett.* **75**, 1387–1390

33. Edwards, H.L., Barr, A., Markert, J.T. and de Lozanne, A.L. (1995) Modulations in the CuO Chain Layer of $YBa_2Cu_3O_{7-\delta}$: Charge Density Waves?, *Phys. Rev. Lett.* **73**, 1154–1157

34. de Jongh, L.J. and Miedema, A.R. (1974) Experiments on simple magnetic model systems *Advances in Physics* **23** , pp1–200

35. Winkelmann, M., Graf, H.A., and Andersen, N.H. (1994) Magnetic structure of $MgCu_2O_3$ and doping-induced spin reorientation in $Mg_{1-x/2}Cu_{2-x/2}O_3$, *Phys. Rev. B* **49**, 310–317

36. Feldkemper, S., Weber, W., Schulenburg, J. , and Richter, J. (1995) Ferromagnetic coupling in nonmetallic Cu^{2+} compounds, *Phys. Rev. B* **52**, 313–323

37. Shender, E.F., and Holdsworth P.C.W. (1996) Order by Disorder and Topology in Frustrated Magnetic Systems, in M. Millonas (ed.) *Fluctuation and Order: the new synthesis*, Springer-Verlag New York, 159 – 279
    Shender, E.F.,(1982) Sov. Phys. JETP **56** 178–200

38. Schüttler, H.-B. and Fedro, A.J. (1992) Copper-oxygen charge excitations and the effective-single-band theory of cuprate superconductors, *Phys. Rev. B* **45**, 7588–7591

39. Yushankhai, V.Y., Oudovenko, V.S., and Hayn, R. (1992) Proper reduction scheme to an extended t-J model and the hole dispersion in $Sr_2CuO_2Cl_2$, *Phys. Rev. B* **55**, 15 562–15 575

40. Eggert, S., Affleck, I., and Takahashi, M. (1994) Susceptibility of the Spin-1/2 Heisenberg antiferromagnetic Chain, *Phys. Rev. Lett.* **73**, 332–335

41. Ami, T., Crawford, M.K., Harlow, R.L., Wang, Z.R., Johnston, D.C., Huang, Q., and Erwin, R.W. (1995) Magnetic susceptibility and low-temperature structure of the linear chain cuprate Sr_2CuO_3, *Phys. Rev. B* **51**, 5994–6001

42. Eggert, S., (1996) Accurate determination of the exchange constant in Sr_2CuO_3 from recent theoretical result, *Phys. Rev. B* **53**, 5116–5118

43. Drechsler, S.-L., Málek, J., and Eschrig, H. (1997) Exact diagonalization study of the hole distribution in CuO_3 chains within the four-band dp model, *Phys. Rev. B* **55**, 606–620

44. Oleś, A.M. and Grzelka, W. (1991) Electronic structure and correlation of CuO_3 chains in $YBa_2Cu_3O_7$, *Phys. Rev. B* **44**, 9531–9540

45. Drechsler, S.-L., Málek, J., Zalis, S., and Rosćszewski, K. (1996) Exchange integral and the charge gap of the linear-chain cuprate Sr_2CuO_3, *Phys. Rev. B* **53**, 11 328–11 331

46. Estes, W.E., Gavel, D.P., Hatfield, W.E., and Godjson, D.J. (1978) *Inorg. Chem.* **17**, 1415–1421

47. Drechsler, S.-L., Málek, J., Zalis, S., and Rosćszewski, K. (1996) Superexchange, charge gap and lattice distortions of $Sr(Ca)_2CuO_3$- the interplay of e-e and e-ph interactions, *J. of Supercond.* **9**, 439–441

48. Suzuura, H., Yasuhura, H., Furusaki A.; Nagaosa N., and Y. Tokura, (1996) Singularities in Optical spectra of Quantum Spin Chains, *Phys. Rev. Lett.* **76**, 2579–2582

49. Lorenzana, J. and Eder, R., (1996) Dynamics of the 1D Heisenberg model and optical absorption of spinons in cuprate antiferromagnetic chains, *Phys. Rev. B* **55**, R3358–R3361

50. Takigawa, M., Starykh, O.A., Sandvik, A.V., and Singh, R.R.P. (1996) Nuclear relaxation in the spin-1/2 antiferromagnetic chain compound Sr_2CuO_3 - comparison with between theories and experiments, *Phys. Rev. B* **56**, 13681–13684

51. Drechsler, S.-L., Málek, J., Eschrig, H., Rosner, H., and Hayn, R., (1997) Electronic structure and BOW-CDW states of CuO_3 chains, *J. of Supercond.* **10**, 393–396

52. Málek, J., Drechsler, S.-L., Paasch, G., and Hallberg, K. (1997) Solitonic approach to the dimerization problem in correlated one-dimensional systems, *Phys. Rev. B* **56**, R8467–R8470

53. Drechsler, S.-L., Málek, J., and Eschrig, H. (1993) Bipolarons and Polarexcitons in CuO_3 chains, *Synthet. Met.* **57**, 4472–4477

54. Eschrig, H. and Drechsler, S.-L. (1991) A Microscopic Scenario for s-channel Superconductivity in layered Cuprates, *Physica C* **173**, 80–88

55. Paulsen, J., Eschrig, H., Drechsler, S.-L., and Málek, J. (1995) Electron-Lattice Interaction and Nonlinear Excitations in in Cuprate Structures, in K.A. Müller and D. Mihailovich (eds.), *Anharmonic Properties of High-T_c Cuprates* World Scientific Publ. Comp., Singapore, pp. 95–104

56. Drechsler, S.-L., Málek, J., Lavrentiev, M., and Köppel, H. (1994) On Optical Phonons in Alternating Chains with 3/4-filled Bands and Internal Degrees of Freedom – A simple Model for Ca_2CuO_3, *Phys. Rev. B* **49**, 233–243

57. Drechsler, S.-L., Málek, J., and Lavrentiev, M. (1993) Nonlinear excitations and phonons in alternating chains with mixed CDW-BOW ground state, *Journ. de Phys. IV, Colloque C2, supplément au Journ. de Physique I* **3**, 273–280

58. Horsch, P. (1981) Correlation effects on bond alternation in polyacetylene, *Phys. Rev. B* **24**, 7351 — 7360

59. Baeriswyl, D. and Maki, K. (1985) Electron correlations in polyacetylene, *Phys. Rev. B* **31**, 6633-6642; Horsch, P. (1981) *Phys. Rev. B* **24**, 7351 — 7360.

60. Kuprievich, V.A. (1994) Two-electron self-consistent-field study of $2k_F$ broken-symmetry states in correlated monatomic and diatomic Peierls-Hubbard chains, *Phys. Rev. B* **50**, 16872-16879

61. Dixit, S.N. and Mazumdar, S. (1984) *Phys. Rev. B* **29** 1824 – 1830

62. Hayden, G.W. and Soos, Z.G. (1988) Dimerization enhancement in one-dimensional Hubbard and Pariser-Parr-Pople models, *Phys. Rev. B* **38**, 6075–6083

63. Waas, V., Büttner, H., and Voit, J. (1990) Finite-size studies of phases and dimerization in one-dimensional extended Peierls-Hubbard models, *Phys. Rev. B* **41**, 9366 – 9376

134

64. Sandvik, A.W., (1999) Multi-chain Mean-field Theory of Quasi One-Dimensional Quantum Spin Systems, *Phys. Rev. Lett.*} **83**, *3069 -3072*

65. Schulz, H.J, (1996) Dynamics of Coupled Quantum Spin Chains, *Phys. Rev. Lett.* **77**, 539–543

66. Zaliznyak, I.A., Broholm, C., Kibune, M., Nohara, M., and Takagi, H. (1999) Anisotropic spin freezing in the S=1/2 zigzag chain compound $SrCuO_2$, *Phys. Rev. Lett.* **83**, 5370-5373

67. Inagaki, S., and Fukuyama, H., (1983) *J. Phys. Soc. Jpn.* **52**, 3620-3630

68. Nakano, T., and Fukuyama, H., (1980) *J. Phys. Soc. Jpn.* **49**, 1679-3630

69. Kojima, K.M., Fudamoto, Y., Larkin, M., Luke, G.M; Larkin, M., Luke, G.M., Merrin, J., Nachumi, B., Uemura, Y.J., Motoyama, N., Eisaki, H., Uchida, S., Yamada, K., Endoh, Y., Hosoya, S., Sternlieb, B.J., and Shirane, G. (1997) Reduction of Ordered Moment and Néel Temperature of Quasi-One-Dimensional Sr_2CuO_3 and Ca_2CuO_3 *Phys. Rev. Lett.* **77**, 1787–1790

70. Matsuda, M., Katsuma, K., Kojima, K.M., Larkin, M., Luke, G.M., Merrin, J., Nachumi, B., Uemura, Y.J., Eisaki, H., Motoyama, N., Uchida, S., and Shirane, G. (1997) Magnetic phase transition in the S=1/2 zigzag-chain compound $SrCuO_2$, *Phys. Rev. B* **55**, R11 953–R 11 956

71. Keren, A., Kojima, K.M., Le, L.P., Luke, G.M; Wu, W.D., Uemura, Y.J., Tajima, S., and Uchida, S. (1995) Muon-spin-rotation measurements in the 'infinite-chain' Ca_2CuO_3, *Journ. of Magnetism and Magnetic Materials* **140-144**, 1641–1642

72. van Oosten, A. and Mila, F. (1997) Ab initio determination of exchange integrals and Néel temperature in the chain cuprates, *Chem. Phys. Lett.* **359**, 359–365

73. Drechsler, S.-L., Málek, J., Rosner, H., Neudert, R., Golden, M.S., Knupfer, M., Fink, J., Eschrig, H., Waidacher, C., and Hetzel, M. (1999) Hole Distribution in Cuprate Chains, *Journ. of Low Temp. Phys.* **117**, 407–411

74. Okada, K., and Kotani, A. (1996) Intersite Coulomb Interactions in Quasi-One-Dimensional Copper Oxides, *J. Phys. Soc. Jpn.* **66** 341 – 344

75. Okada, K., Kotani, A., Maiti, K., and Sarma, D. (1996) Cu $2p$ Core-Level Photoemission Spectrum of Sr_2CuO_3, *J. Phys. Soc. Jpn.* **65** 1844 – 1848

76. Teplov, M.A., Bakharev, O.N., Dooglav, A.V., Egorov, A.V., Krjukov, E.V., Mukhamedshin, I.R., Sakhratov, Yu.A., Brom, H.B., and Witteveen, J. (1997) Tm MMR and Cu NQR studies of phase separation in TmBaCuO compounds, in E. Kaldis, E. Liarokapis, and K.A. Müller (eds.) *High-T_c superconductivity 1996: Ten years after the discovery*, NATO ASI Series, Series E: Applied Sciences - Vol. 343, Kluwer Academic Publishers, Dordrecht, pp. 531 – 562
    Dooglav, A.V., Alloul, H., Bakharev, O.N., Berthier, C., Egorov, A.V., Horvatic, M., Krjukov, E.V., Mendels, P., Sakhratov, Yu.A., and Teplov, M.A. (1998) Cu(2) nuclear resomnance evidence for a magnetic phase in aged 60-K superconductors $RBa_2Cu_3O_{6+x}$ (R=Tm, Y), *Phys. Rev. B* **57**, 11 792 – 11 797

77. Emery, V.J., and Kivelson, S.A. (1999) Electronic structure of doped insulators and high temperature superconductivity, *J. of Low Temp. Phys.* **117** 189 – 197

78. White, S.R., and Scalapino, D.J. (1997) Density Matrix Renormalization Group study of the striped phase in the 2D $t - J$ model, *Phys. Rev. Lett.* **80** 1272 – 1275.

79. Fujimori, A. Ino, A., Yoshida, T., Mizokawa T., Shen, Z.-X, aand Ueda, S. (2000) Fermi surface, pseudogaps, and dynamical stripes in *Preprint cond-mat/0011293*, 10 pp; Ino, A., Kim, C., Nakamura, M., Yoshida, T., Mizokawa, T., Shen, Z.-X., Fujimori, A., Kakeshita, T., Eisaki, H., and Uchida, S. (2000) Electronic structure of $La_{2-x}Sr_xCuO_4$ in the vicinity of the superconductor-insulator transition, *Phys. Rev. B* **62**, 4137 – 4141

80. Brandt, H., (1999) Pinning of vortices and linear and nonlinear ac susceptibilities in high-T_c superconductors, *These Proceedings* 455-486

Charge Density Waves with Complex Order Parameter in Bilayered Cuprates

S.Varlamov[1,2], M.Eremin[1] and I.Eremin[1]

[1] *Kazan State University, 420008 Kazan, Russia*
[2] *Cottbus Technical University, 03013 Cottbus, Germany*

Abstract. Lindhard response functions and momentum dependence of a CDW pseudogap have been calculated. In general case, the pseudogap possesses s+id symmetry with two different critical temperatures. This fact enabled us to explain the temperature behavior of spin susceptibility for the compound $YBa_2Cu_4O_8$ in the entire temperature interval $T > T_c$. The smooth development of the pseudogap formation temperature is explained from underdoped to overdoped states and the Fourier amplitudes $< s_q >$ (spin) and $< e_q >$ (charge) modulations have been calculated.

Keywords: pseudogap, charge density wave, Lindhard response function

1. Introduction

The problem of the electronic structure of layered cuprates is the focus of modern investigations. The various kind of the instabilities are possible in these compounds. Recent photoemission experiments have revealed especially clearly the presence of a pseudogap in the elementary excitation spectrum of normal phase of $Bi_2Sr_2CaCu_2O_8$ [1]. The nature of the pseudogap is not completely understood. We continue our previos examinations of the singlet correlated band model for layered cuprates [9] with respect to its ability to describe the observed temperature dependence of the Knight shift for Cu(2) [3], Fermi surface evolution [4] and the doping dependence of a pseudogap formation temperature [5]

2. Basic relations for one band approach

In our calculation we start from the Hamiltonian:

$$H = \sum t_{ij}\Psi_i^{pd,\sigma}\Psi_j^{\sigma,dp} + \sum J_{ij}[2(s_is_j) - \frac{n_in_j}{2}] + \sum g_{ij}\delta_i\delta_j + H_{CDW} \quad (1)$$

where $\Psi_i^{pd,\sigma}$, $\Psi_j^{\sigma,pd}$ are quasiparticle Hubbard-like operators for the copper-oxygen singlet, J_{ij} is the superexchange constant of the copper

135

spin coupling and g_{ij} is a screened Coulomb repulsion of the doped holes, $1 + \delta_i = \sum_\sigma \Psi_i^{\sigma,\sigma} + 2\Psi_i^{pd,pd}$. The quasiparticle interaction H_{CDW} mediated by the phonon field leads to a CDW transition [6].

In addition to the usuall mean field approach we have taken into account that anticommutators of Hubbard-like operators can be affected by the doping index level per one unit cell, spin magnetization and charge modulation (non-Fermi statistics effect (NFS)) [2]:

$$P_i^\sigma = \Psi_i^{pd,pd} + \Psi_i^{\sigma,\sigma} = \frac{1 + \delta_i}{2} + (-1)^{\frac{1}{2} - \sigma} s_i^z \qquad (2)$$

The appeearence of CDW and SDW we describe via the Fourier components:

$$e_{q_e} = \frac{1}{N} \sum \delta_i \exp(iqR_i), \qquad s_{q_s}^z = \frac{1}{N} \sum s_i^z \exp(iq_s R_i), \qquad (3)$$

where q_e and q_s are the instability wave vectors with respect to CDW and SDW formation, respectively. Generally the wave vectors of CDW (q_e) and SDW (q_s) can be different. Below we consider both as commensurate wave vectors $q_s = q_e = (\pi, \pi)$ and incommensurate wave vectors $q_s = q_e = (\pi \pm \varepsilon_x, \pi \pm \varepsilon_y)$. The equations of motion are written as:

$$i\hbar \frac{\partial}{\partial t} \Psi_k^{\sigma,pd} = \varepsilon_k \Psi_k^{\sigma,pd} + \eta_{k+q}^\sigma \Psi_{k+q}^{\sigma,pd}$$

$$i\hbar \frac{\partial}{\partial t} \Psi_{k+q}^{\sigma,pd} = \varepsilon_{k+q} \Psi_{k+q}^{\sigma,pd} + \eta_k^\sigma \Psi_k^{\sigma,pd} \qquad (4)$$

where

$$\eta_{k+q}^\sigma = [t_{k+q} - \frac{4}{(1 + \delta_0)^2} < s_i s_j > t_k] < \frac{e_q}{2} - (-1)^{\frac{1}{2} - \sigma} s_q^z > + G_{k+q}^\sigma + G_k^{ph} \qquad (5)$$

and ε_k is the energy dispersion in the normal state. The CDW order parameter is determined by the relations [7]:

$$G_{k+q}^\sigma = -\frac{2}{(1 + \delta_0)N} \sum_{k'} \{j_{k'-k} < \Psi_{k'+q}^{pd,\bar{\sigma}} \Psi_{k'}^{\bar{\sigma},pd} > + g_{k'-k} < \Psi_{k'+q}^{pd,\sigma} \Psi_{k'}^{\sigma,pd} > \}, \qquad (6)$$

$$G_k^{ph} = \sum_{\omega_q} [A(\omega_q) - B(\omega_q) \frac{(\hbar\omega_q)^2 \, \Theta \, (\hbar\omega_D - |\varepsilon_k|) \, \Theta \, (\hbar\omega_D - |\varepsilon_k - \varepsilon_{k+q}|)}{(\varepsilon_k - \varepsilon_{k+q})^2 - (\hbar\omega_q)^2}] \qquad (7)$$

where $\omega_q = 40 meV$ is the frequency of active phonon mode in the CDW transition.

The thermodynamic values of the Fourier component $< e_q >$ and $< s_q >$ are calculated self-consistently:

$$e_q = \frac{1}{2} \sum_k [< \Psi_{k+q}^{pd,\sigma} \Psi_k^{\sigma,pd} > + < \Psi_{k+q}^{pd,\bar{\sigma}} \Psi_k^{\bar{\sigma},pd} >]$$

$$s_q^z = \frac{1}{2} \sum_k [< \Psi_{k+q}^{pd,\bar{\sigma}} \Psi_k^{\bar{\sigma},pd} > - < \Psi_{k+q}^{pd,\sigma} \Psi_k^{\sigma,pd} >] \tag{8}$$

The correlation function is determined by

$$< \Psi_{k+q}^{pd,\sigma} \Psi_k^{\sigma,pd} > = P \frac{\eta_{k+q}^\sigma}{E_{1k}^\sigma - E_{2k}^\sigma} [f(E_{1k}^\sigma) - f(E_{2k}^\sigma)] \tag{9}$$

with

$$E_{1k,2k}^\sigma = \frac{\varepsilon_k + \varepsilon_{k+q}}{2} \pm \frac{1}{2} [(\varepsilon_k - \varepsilon_{k+q})^2 + 4 \mid \eta_{k+q}^\sigma \mid^2]^{1/2} \tag{10}$$

The spectral weight of the singlet band changes with doping level δ (hole concentration per one unit cell of bilayer) as $2\delta/(1 + \delta)$. Thus, the condition of half-filling (which approximately corresponds to the optimal doping level) yields $\delta_{opt} = 1/3$. Because the bilayer unit cell contains two copper sites we conclude that T_c has a maximum near hole concentrations $x \approx 1/6$ per one copper site.

The evolution of the Fermi surface and DOS for three doping levels are presented in Fig.1. White circles are taken from experimental data [8]. The one band approach is usually valid for a large enough doping level. Therefore if δ goes to zero we explored the two band model for layered cuprates [9]. In this case the energy dispersion is:

$$\varepsilon_k = \frac{E_k^{dd} + E_k^{pp}}{2} + \frac{1}{2} [(E_k^{dd} - E_k^{pp})^2 + 4 E_k^{dp} E_k^{pd}]^{1/2} \tag{11}$$

$$E_k^{dd} = \varepsilon_d + \sum [\frac{1 - \delta_0}{2} + \frac{2}{(1 - \delta_0)} < s_i s_j >] t_{ij} \exp(ikR_{ij}) \tag{12}$$

$$E_k^{pp} = \varepsilon_p + \sum [\frac{1 + \delta_0}{2} + \frac{2}{(1 + \delta_0)} < s_i s_j >] t_{ij} \exp(ikR_{ij}) \tag{13}$$

are dispersions of the lower Hubbard copper band and the singlet-correlated oxygen band, respectively and

$$E_k^{dp} = \sum [\frac{1 + \delta_0}{2} - \frac{2}{(1 - \delta_0)} < s_i s_j >] t_{ij}^{12} \exp(ikR_{ij}) \tag{14}$$

$$E_k^{pd} = \sum [\frac{1 - \delta_0}{2} - \frac{2}{(1 + \delta_0)} < s_i s_j >] t_{ij}^{12} \exp(ikR_{ij}) \tag{15}$$

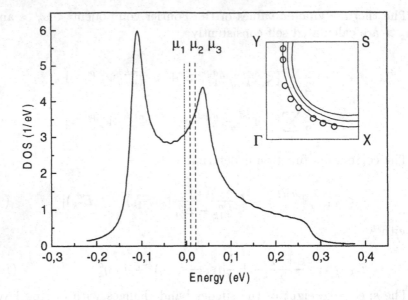

Figure 1. The calculated density of states. The vertical lines mean the three different positions of the chemical potential μ. In the inset the calculated Fermi surfaces has drown for three values of the chemical potencial.

describe their hybridization. If δ_0 goes to zero the spin dependent factors in E_k^{dd} and E_k^{pp} become zero due to antiferromagnetic correlations. At the same time, as one can see from the expressions for E_k^{pd} and E_k^{dp} the antiferromagnetic correlations enhance the interband coupling. So, at small doping level the energy dispersion will be determined by the expression

$$\varepsilon_k(\delta \to 0) = \frac{\varepsilon_d + \varepsilon_p}{2} + \frac{1}{2}\sqrt{(\varepsilon_p - \varepsilon_d)^2 + 16[t_1^{12}(\cos k_x + \cos k_y)]^2} \quad (16)$$

where t_1^{12} is a transfer integral between nearest neighbor copper sites. From (16) one can obtain that the bottom of the band corresponds to the point $(\frac{\pi}{2}, \frac{\pi}{2})$ of the Brillouin zone. The expression (16) is valid at $< s_i^z > = 0$, i.e. for the usual paramagnetic phase. With decreasing doping level the Fermi surface shrinks to the pockets near the point $(\frac{\pi}{2}, \frac{\pi}{2})$ of the Brillouin zone. This behavior qualitatively agrees with photoemission experimental data [1],[5] and provides a good basis for the analysis of Peierls like instabilities, which as it was pointed out by many authors [6],[10], is very sensitive to the topological properties of the Fermi surface.

3. Lindhard response function

Now we study the instability reason that is motored by the topological properties of the Fermi surface and the high density of states of the singlet correlated band. Generally, in the singlet band there are two peaks [9]. The one of them is a saddle singularity peak and the second which placed near the bottom of the band is so-called hybridization peak. In bilayer compounds such as $YBa_2Cu_4O_8$ the chemical potencial placed near the saddle peak.

The calculated shape of the Lindhard response function:

$$\chi(q) = \frac{1}{N} \sum_k \frac{f(E_k) - f(E_{k+q})}{E_k - E_{k+q}} \qquad (17)$$

along the Brillouin zone are given in Fig.2 for the different positions of the chemical potential μ_1, μ_2 and μ_3, respectively. At higher doping (μ_3) the response function has the main maximum near $Q_L = (\pi, \pi)$. When the doping level is decreased its peak goes down but new hills are appear. In particular, at the intermediate doping (μ_2) they are around $(0, \pm\pi/2), (\pm\pi/2, 0)$ and for μ_1 the hills are shifted towards $(\pi, 0)$ as it was pointed out earlier [6]. These hills hint the possible 1D instability wave vector at 1/8 holes doping [11], [12].

4. Results of calculations and discussion

The short range interactions (superexchange and screened Coulomb repulsion), as a rule, yield the d-wave order parameters of CDW and SDW, whereas NFL-effects, together with phonon mediated interaction, lead to a anisotropic s-wave pseudogap component. In general the calculated CDW and SDW order parameters have $s + id$ symmetry and the temperature dependence of s- and d-components are different. The critical temperature of the d-component (T_d^*) is always higher than T_s^* at all considered doping levels.

Recently, we have examined a charge density wave scenario near the optimal doping with $Q = (\pi, \pi)$ [7]. The momentum dependence of the CDW gap function was written as:

$$G_{k,T} = A(T) - B(T) \frac{(\hbar\omega_q)^2 \, \Theta \, (\hbar\omega_D - |\varepsilon_k|)}{(\varepsilon_k - \varepsilon_{k+q})^2 - (\hbar\omega_q)^2} + iD(T)[cosk_x - cosk_y],$$

$$(18)$$

where A(T) and B(T) parameters are calculated self-consistently following to the usual CDW theory [13]. The third term, D(T), has

140

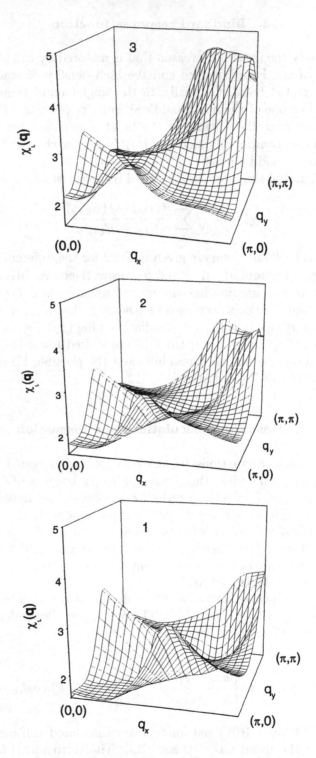

Figure 2. The calculated Lindhard response function for three different positions of the chemical potential μ_3, μ_2, μ_3, respectively.

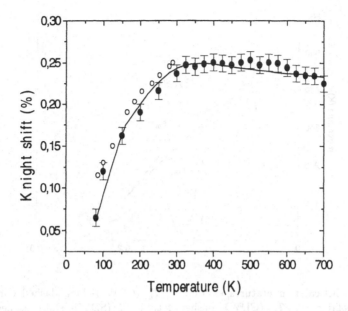

Figure 3. Temperature dependence of the Knight shift for Cu(2) in $YBa_2Cu_4O_8$. The experimental data were taken from [3]

been appeared due to a superexchange and a short-range Coulomb interactions. It is remarkable to note that both of them support each other in opening of the D(T)-component. Therefore, the critical temperature in a mean-field approximation is higher than T_c because in the case of superconducting transition temperature the Coulomb repulsion suppress superexchange pairing interaction. In particular, for $YBa_2Cu_4O_8$ using well-known experimental data for Cu(2) Knight shift data we have deduced T^* about 300 K and $J_0 + 2g \approx 210 meV$. The critical temperature for A and B components is 150-180 K. Fig.3 displays our computed value of the Knight shift on the Cu(2) copper nuclei in $YBa_2Cu_4O_8$ with $\delta = 0.15$.

Numerical solutions of the equations (6)-(9) are shown in Fig.4. Both types of solutions (CDW or SDW) display the correct doping dependence. In according to photoemission data [1] and NMR [14] the critical temperature of the pseudogap goes down when the doping decreases. From Eqs. (6), (9) it is clear that the critical temperature of the d- component CDW (T_d^*) is not sensitive to the external magnetic field. This result agrees well with recent experimental observation [15]. Calculated mean field critical temperatures of CDW (T_d^*) are higher than for SDW in the complete doping range. This result is also consistent with the widely accepted opinion that a transition towards the so called "stripe" phase in underdoped cuprates are charge rather than

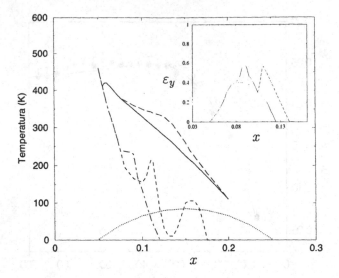

Figure 4. Critical temperatures vs. doping; T_d^* (CDW)- long dashed curve, T_d^* (SDW)- solid curve, T_s^* (CDW)- dashed-dotted, T_s^* (SDW)- short dashed. The parabola corresponds to the superconducting transition temperature T_c (schematically). Inset: The variation of incommensurability component ε_y vs. doping index; for CDW- solid and for SDW - dashed curve, respectively.

spin driven [16]. In Fig.4 we show the magnitude of the instability wave vector $Q = (\pi \pm \varepsilon_x, \pi \pm \varepsilon_y)$ vs. doping index. We have found the maximum of ε_y around $x_{CDW} \sim 0.09$ for CDW and $x_{SDW} \sim 0.11$ for SDW. It would be interesting to check our theoretical conclusion experimentally.

It is easy to see from Eq.(9) that for a d-wave order parameter with pure commensurate wave vector $q = Q = (\pi, \pi)$ both expectation values $< s_q >$ and $< e_q >$ vanish, but they are finite when the wave vector Q becomes incommensurate. Because in general the order parameters are complex, the expectation values $< s_q >$ and $< e_q >$ have real and imaginary components. Its values at the superconducting transition temperature are presented in Fig.4. The real component $< e_q >$ is about $0.1 - 0.15$ at temperatures $T \geq T_c$ and disappears at T_s^*. The imaginary parts of $< s_q >$ and $< e_q >$ in general are about $0.01 - 0.05$ and stay up to T_d^* and then drastically vanish.

5. Conclusion

To conclude we have calculated Lindhard response function and Fourier amplitudes of CDW and SDW order parameters in underdoped cuprates.

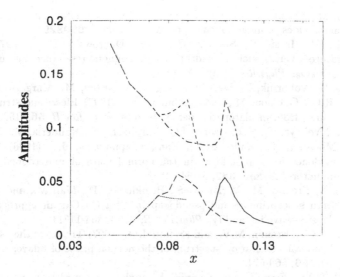

Figure 5. Calculated values $< e_q >$ and $< s_q >$ for temperatures around T_c;$Re < e_q >$(CDW)-dashed curve, $Re < e_q >$ (SDW) - dotted-dashed, $Im < sq >$ (SDW) - solid, $Im < e_q >$(CDW) -long dashed, $Im < e_q >$(SDW) -dotted.

Our calculation provides an explanation for the experimentally observed evolution of the Fermi surface and the doping dependence of the pseudogap formation temperature, the temperature dependence of the Knight shift. The charge instability preforms the spin instability and the incommensurability has the maximum at $x \sim 0.1$ holes per one copper site. We hope our present calculations will help in better understanding many of the strange features in the shape and linewidth of NMR in layered cuprates.

Acknowledgements

We would like to thank G.Seibold for helpful comments. This work is supported in part by INTAS (Grant 96-0393) and Russian Scientific Coincil on Superconductivity (Project N 98014). S.Varlamov would like to thank NATO for enabling to visit NATO Advanced Study Institute and discuss our work.

References

[1] Loeser, A.G., Shen, Z.-X., Dessau, D.S., Marshall, D.S., Park, C.H., Fournier, P., and Kapitulnik, A. (1996) Excitation gap in the normal state of underdoped $Bi_2Sr_2CaCu_2O_{8+\delta}$, *Science*, **273**, 325-329.

144

[2] Eremin, M., Varlamov, S., and Eremin, I. (1999) How large can be SDW and CDW amplitudes in underdoped cuprates, *cond.-mat.*, **9908297**, ?.

[3] Curro, N.J., Imai, T., Slichter, C.P., and Dabrowski, B. (1997) High-temperature ^{63}Cu(2) nuclear quadrupole and magnetic resonance measurements of YBa$_2$Cu$_4$O$_8$, *Phys.Rev.B*, **56**, 877-885.

[4] LaRosa, S., Vobornik, I., Zwick, F., Berger, H., Grioni, M., Margaritondo, G., Kelley, R.G., Onellion, M., and Chubukov, A. (1997) Electronic structure of CuO_2 planes: from insulator to superconductors, *Phys.Rev.B*, **56**, R525-528.

[5] Ding, H., Yokoya, T., Campuzano, J.C., Takahashi, T., Randeria, M., Norman, M.R., Mochiku, T., Kadowaki, K., and Giapintzakis, J. (1996) Spectroscopic evidence for a pseudogap in the normal state of underdoped high-Tc superconductors, *Nature*, **382**, 51-54.

[6] Eremin, I., Eremin, M., Varlamov, S., Brinkmann, D., Mali M., and Roos, J. (1997) Spin susceptibility and pseudogap in YBa$_2$Cu$_4$O$_8$: an approach via a charge-density-wave instability, *Phys.Rev.B*, **56**, 11305-11311.

[7] Varlamov, S.V., Eremin, M.V., and Eremin, I.M. (1997) Theory of the pseudogap in the elementary excitation spectrum of the normal phase of bilayer cuprates , *JETP Lett*, **66**, 569-574.

[8] Aebi, P. (1999) Fermi surface mapping by angle-scanned photoemission, *These Proceedings*.

[9] Eremin, I., Solov'yanov, S., and Varlamov, S. (1997) Theory of the electronic structure and spin susceptibility of La$_{2-x}$Sr$_x$CuO$_4$, *JETP*, **85**, 963-970.

[10] Markiewicz, R.S. (1997) Stripes, pseudogaps, and Van Hove nesting in the three-band t-J model, *Phys.Rev.B*, **56**, 9091-9105.

[11] Drechsler, S.-L., Rosner, H., and Eschrig, H. (1999) Electronic and magnetic properties of cuprate chains and related materials, *These Proceedings*.

[12] Bianconi, A. (1993) On the possibility of new high Tc superconductors by producing metal heterostructures as in the cuprate perovskites, *Proceedings of the workshop on Phase Separation In Cuprate Superconductors, World Scientific, Singapore*.

[13] Balseiro, C.A., and Falikov, L.M. (1979) Superconductivity and charge-density waves, *Phys.Rev.B*, **20**, 4457-4464.

[14] Williams, G.V.M., Tallon, J.L., Michalak, R., and Dupree, R. (1996) NMR evidence for common superconducting and pseudogap phase diagrams of YBa$_2$Cu$_3$O$_{7-d}$ and La$_{2-x}$Sr$_x$CaCu$_2$O$_6$, *Phys.Rev.B*, **54**, R6909-6912.

[15] Gorny, K., Vyaselev, O.M., Martindale, J.A., Nandor, V.A., Pennington, C.H., Hammel, P.C., Hults, W.L., Smith, J.L., Kuhns, P.L., Reyes, A.P., and Moulton, W.G. (1999) Magnetic field independence of the spin gap in $YBa_2Cu_3O_{7-\delta}$, *Phys.Rev.Lett*, **82**, 177-180.

[16] Zachar, O., Kivelson, S.A., and Emery, V.J. (1998) Landau theory of stripe phases in cuprates and nickelates, *Phys.Rev.B*, **57**, 1422-1426.

Coexistence of Superconductivity and Magnetism in Borocarbides

K.-H. MÜLLER[1], K. DÖRR[1], J. FREUDENBERGER[1],
G. FUCHS[1], A. HANDSTEIN[1], A. KREYSSIG[2],
M. LOEWENHAUPT[2], K. NENKOV[1], M. WOLF[1]
1. *IFW Dresden, POB 270016, 01171 Dresden, Germany*
2. *IAPD, TU Dresden, 01062 Dresden, Germany*

ABSTRACT. The interplay of rare-earth magnetism and superconductivity has been a topic of interest for many years. In cuprates as well as in the classical magnetic superconductors the superconducting state usually coexists with antiferromagnetic order on the rare-earth sublattice. Usually, their magnetic ordering temperature T_N is much below the superconducting transition temperature T_c. The discovery of superconducting borocarbides RT_2B_2C with R = Sc, Y, La, Th, U, Dy, Ho, Er, Tm or Lu and T = Ni, Pd or Pt (where not all of these combinations of R and T result in superconductivity) has reanimated the research on the coexistence of superconductivity and magnetic order. Most of these borocarbides crystallize in the tetragonal $LuNi_2B_2C$ type structure which is an interstitial modification of the $ThCr_2Si_2$ type. Contrary to the behavior of Cu in the cuprates Ni does not carry a magnetic moment in the borocarbides. Various types of antiferromagnetic structures have been found to coexist with superconductivity for R = Tm, Er, Ho and Dy in RNi_2B_2C. In the case of $HoNi_2B_2C$ three different types of antiferromagnetic structures have been observed (i) a commensurate one with Ho moments aligned ferromagnetically within layers perpendicular to the tetragonal c axis where consecutive layers are aligned in opposite directions, (ii) an incommensurate spiral along the c axis and (iii) an incommensurate a-axis modulated structure that was shown to be important for the reentrant behavior whereas the other two structures coexist with the superconducting state. The variation of T_N and T_c with the de Gennes factor can well be drawn on straight lines from Lu to Gd and from Lu to Tb, respectively. Consequently, $T_c > T_N$ holds for Tm, Er, Ho and $T_c < T_N$ for Dy. However, the study of various pseudoquaternary $(R,R')Ni_2B_2C$ compounds has shown that this so called de Gennes scaling is not universal for the borocarbides and it breaks down in some cases which is attributed to effects of crystalline electric fields, the difference in the R ionic radii or the effect of non-magnetic impurities in an antiferromagnetic superconductor. In an external magnetic field some of the RNi_2B_2C compounds show metamagnetic transitions combined with a large negative magnetoresistance.

S.-L. Drechsler and T. Mishonov (eds.), High-T_c Superconductors and Related Materials, 145–166.
© 2001 *Kluwer Academic Publishers. Printed in the Netherlands.*

1. Introduction

Superconducting rare-earth transition-metal borocarbides, RT_2B_2C (with R = Y, or a 4f element and T = Ni, Pd or Pt), have been discovered only few years ago [1-4]. They have attracted a great deal of attention because in this new class of materials the highest value of the superconducting transition temperature among all known intermetallic compounds ($T_c \approx 23K$ i.e. as high as in the A15 super-conductor Nb_3Ge) has been achieved. Typical resistance-versus-temperature transition curves for $LuNi_2B_2C$ are shown in Fig. 1. Since the rare-earth sites in solids often carry magnetic moments that are coupled by indirect exchange in-teraction the interplay between rare-earth magnetic long-range order and super-conductivity has been an active area of interest for many years. It was shown that superconductivity and ferromagnetism compete with each other and can hardly coexist [5] whereas superconductivity and antiferromagnetic order may coexist [6]. A striking feature distinguishing the superconducting borocarbides from other superconductors is that for certain combinations of elements R and T superconductivity and antiferromagnetic order have been found to coexist in RT_2B_2C where the values of the magnetic ordering temperature T_N are compa-rable with the T_c values (see Fig. 2). For comparison in rare-earth containing high-T_c cuprates, as e.g. $RBa_2Cu_3O_{7-\delta}$, the value of T_c is about 90K and, with the exception of R = Pr, it does practically not depend on the choice of R. On the other hand, the superconductivity remains even below T_N which may achieve few Kelvin (e.g. 2K for R = Gd). This means that there is no measur-able effect of the ordered magnetic moments on superconductivity. This sug-gests that exchange interaction between the conduction electrons and the rare-

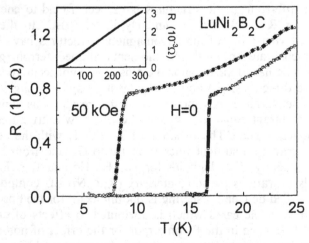

Figure 1. Resistance-vs.-temperature transition curves for a polycristalline $LuNi_2B_2C$ sample for two magnetic fields (0 and 50kOe). The zero field transition temperature T_c is 17K.

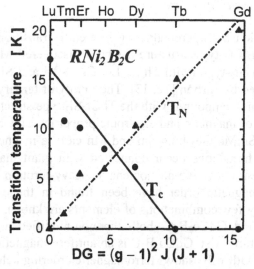

Figure 2. Critical temperatures for superconductivity, T_c, and for magnetic ordering, T_N, for RNi_2B_2C compounds with R = Lu, Tm, Er, Ho, Dy, Tb and Gd. DG is the de Gennes factor, g the Landé factor and J the total angular momentum of the R^{3+} Hund's rule ground state.

earth magnetic moments is minor and pair-breaking due to exchange scattering is weak. On the other hand, the relative high value of $T_N \approx 2K$ suggests that some small indirect exchange between the rare-earth magnetic moments operates across the CuO_2 layers [7]. The high-T_c superconductivity is based on strongly correlated copper-3d and oxygen-2p electrons in mixed valence cuprates and appears to be an alternative ground state instead of Néel-type antiferromagnetic order of Cu^{II} spin 1/2 magnetic moments in the corresponding (non-doped insulating) two-valent parent cuprates with ordering temperatures T_N as high as 520K [8]. With the exception of R = Pr, the two antiferromagnetic structures of the R and Cu^{II} sublattices are only weakly coupled to each other [7]. In the classical magnetic superconductors, the ternary borides RRh_4B_4 and the Chevrel phases RMo_6S_8, magnetic ordering is assumed to be mainly caused by magnetic dipolar interaction and T_N is typically $\leq 1K$ whereas T_c is about one order of magnitude larger. A key element of the rare-earth transition-metal borocarbides is that, contrary to the high-T_c cuprates and the classical magnetic superconductors the exchange coupling between the rare-earth magnetic moments is mediated by conduction electrons. Thus the interaction between the magnetic moments and the conduction electrons must be relatively strong. The magnetic energy is comparable with the superconducting condensation energy. Therefore the investigation of these compounds is expected to result in new insights into the interplay of superconductivity and magnetism [9-11].

Stoichiometry and crystal structure

With the investigation of superconducting rare-earth transition-metal borocarbides the new $LuNi_2B_2C$-type structure has been discovered which can be considered as the $ThCr_2Si_2$-type (with Th → Lu, Cr → Ni and Si → B in Fig. 3) interstitially modified by carbon [12, 13]. The family of ternary rare-earth transition-metal metalloid compounds with the $ThCr_2Si_2$-type structure is very large and a broad variety of magnetic and electronic properties have been observed in it. For example in $SmMn_2Ge_2$ both Sm and Mn carry a magnetic moment and two metamagnetic transitions occur connected with giant magneto-resistance effects [14]. Different collective phenomena as heavy-fermion behavior, superconductivity and magnetic order have been found in the exotic compound $CeCu_2Si_2$ [15]. Only few combinations of elements are known where, as in the case of $GdCo_2B_2$ and $GdCo_2B_2C$, both the non-modified and the modified $ThCr_2Si_2$-type structure exist. $GdCo_2B_2C$ is an antiferromagnetic compound (T_N = 6K) [16] whereas $GdCo_2B_2$ shows ferromagnetic ordering below 27K [17].

The $LuNi_2B_2C$-type structure has three open parameters, the two lattice constants a and c and the position z of the boron atom. For RNi_2B_2C, in a good approximation, the parameters c and z linearly decrease with increasing radius of R (where R is assumed to be in the trivalent oxidation state and c is taken as the unit for z) whereas a linearly increases with the radius of R, with the exception of Ce [13]. Thus while going through the series of R elements from Lu to

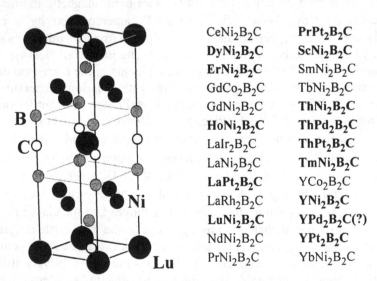

$CeNi_2B_2C$	$PrPt_2B_2C$
$DyNi_2B_2C$	$ScNi_2B_2C$
$ErNi_2B_2C$	$SmNi_2B_2C$
$GdCo_2B_2C$	$TbNi_2B_2C$
$GdNi_2B_2C$	$ThNi_2B_2C$
$HoNi_2B_2C$	$ThPd_2B_2C$
$LaIr_2B_2C$	$ThPt_2B_2C$
$LaNi_2B_2C$	$TmNi_2B_2C$
$LaPt_2B_2C$	YCo_2B_2C
$LaRh_2B_2C$	YNi_2B_2C
$LuNi_2B_2C$	$YPd_2B_2C(?)$
$NdNi_2B_2C$	YPt_2B_2C
$PrNi_2B_2C$	$YbNi_2B_2C$

Figure 3. Unit cell of the $LuNi_2B_2C$-type structure (space group I4/mmm) and some compounds crystallizing in this structure [12, 13]. The crystal structure of YPd_2B_2C is not yet finally identified. The bold compounds are superconductors.

La the structure shows a contraction along the tetragonal c-axis but an expansion perpendicular to it and the boron shifts away from the RC-layers more in the vicinity of the Ni layers. However, the radius variation of the rare-earth does not much affect the B-C distance and the B-Ni distance. Consequently, there is a remarkable reduction of the B-Ni-B tetrahedral angle from 108.8° for Lu to 102° for La which is expected to influence the variation of the electronic structure within the series of RNi_2B_2C compounds. The lattice parameters of $CeNi_2B_2C$ do not fit the derived linear relationship. Neither the trivalent nor the tetravalent radius for Ce falls on the corresponding straight line. The approximate valence $Ce^{+3.75}$ obtained by interpolation reveals possible mixed or intermediate valence behavior of cerium in this compound [13].

Magnetic ordering in RNi_2B_2C compounds may result in a structural distortion caused by magnetoelastic effects. Using high-resolution neutron scattering on a powder sample of $HoNi_2B_2C$ a tetragonal-to-orthorhombic distortion has been observed at low temperatures where the Ho magnetic moments order in a c-axis modulated antiferromagnetic structure (see Section 2, Fig. 8). As shown in Fig. 4 the distortion is a shortening of the tetragonal unit cell in [110] direction. At 1.5K this shortening is 0.19% [18]. A similar tetragonal-to-orthorhombic phase transition driven by magnetoelastic interaction has also been reported for $ErNi_2B_2C$ [19].

In spite of the discovery of the $LuNi_2B_2C$-type structure the crystal structures of superconducting R-T-B-C phases in the composition range near the stoichiometry 1:2:2:1 is far from being completely determined. This is particularly true fo R-Pd-B-C compounds where the highest value of T_c, 23K for Y-Pd-B-C, has

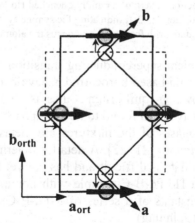

Figure 4. Orthorhombic distortion of tetragonal $HoNi_2B_2C$: a,b – original tetragonal axes; dashed line: tetragonal basal plane; $\rightarrow$, $\uparrow$: shift of the Ho atoms leading to the orthorhombic cell with the axes a_{orth}, b_{orth}. Thick arrows : Ho magnetic moments in the commensurate c-axis modulated structure [18].

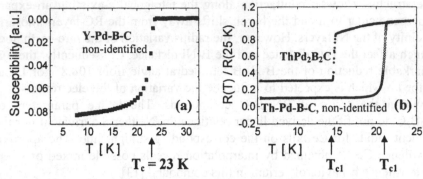

Figure 5. Superconducting transition curves of two arc melted crystallographically not yet identified compounds (a) Y-Pd-B-C and (b) Th-Pd-B-C (together with $ThPd_2B_2C$) [20].

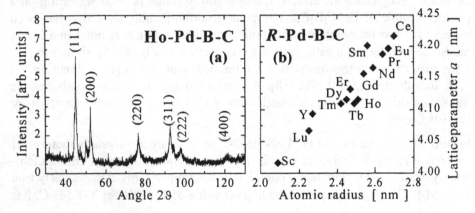

Figure 6. (a) X-ray diffraction pattern of a rapidly quenched Ho-Pd-B-C sample with the nominal composition 1:2:2:1, indicating fcc symmetry. (b) lattice constants a of R-Pd-B-C compounds with fcc symmetry versus the atomic radius of the R-atoms [22].

been reported [3]. Typical superconducting transition curves of such materials are presented in Fig. 5. Pd-based borocarbides have been prepared by arc melting [3, 20] and also by non-equilibrium routes as rapid quenching [21, 22]. In all cases the samples have been found to be multiphase where (at least some of) the superconducting phases of the mixtures are metastable and have a lattice structure with fcc symmetry [21, 22]. A detailed determination of that structure from x-ray diffraction data is difficult and has not yet been done. Fig. 6 shows the x-ray pattern of a Ho-Pd-B-C sample with nominal 1:2:2:1 stoichiometry and the fcc lattice constants of a series of R-Pd-B-C compounds prepared by rapid quenching (melt spinning).

As shown in Fig. 7 the $LuNi_2B_2C$-type structure has been successfully modified by incorporating an additional LuC layer resulting in LuNiBC [12, 13] or two additional (LaN instead of LuC) layers resulting in $La_3Ni_2B_2N_3$ [23, 24]. Both

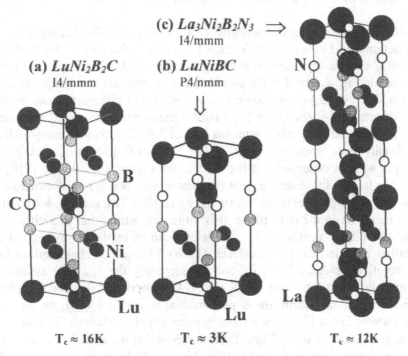

(c) *La₃Ni₂B₂N₃* ⟹
I4/mmm

(a) *LuNi₂B₂C*
I4/mmm

(b) *LuNiBC*
P4/nmm

⇓

N

B

C

Ni

Lu

Lu

La

$T_c \approx 16K$ $T_c \approx 3K$ $T_c \approx 12K$

Figure 7. Tetragonal rare-earth nickel borocarbides (nitrides) with single, double and triple RC(N)- layers and values of the superconducting transition temperature T_c [23-25].

compounds are superconducting [23-26]. Even compounds $(YC)_n(Ni_2B_2)$ with n = 3 and 4 have been prepared which however have not found to be superconducting [27]. The investigation of such multilayer compounds will help to understand the mechanisms for superconductivity and magnetism in the rare-earth transition-metal borocarbides [28]. However this study will be confined to magnetism and superconductivity in single-layer RNi_2B_2C compounds.

Electronic structure and dimensionality

Although the isomer shift of the Dy nucleus, determined by Mössbauer studies on $DyNi_2B_2C$, suggested that the Dy-C plane is insulating and the electrical conduction takes place in the Ni-B sheets [29] it is now generally accepted that the RNi_2B_2C compounds are three-dimensional in their behavior, and thus are in fact quite different than the layered cuprates [30]. This is supported by electronic structure calculations indicating that the 1:2:2:1 borocarbides are three-dimensional metals with all atoms contributing to the metallic character [31, 32, 28]. A clearly three-dimensional, yet anisotropic electronic structure has also been experimentally determined using x-ray absorption spectroscopy [33].

2. Magnetic ordering and superconductivity in RNi_2B_2C

The different types of antiferromagnetic order in the RNi_2B_2C compounds have been determined by neutron diffraction [30], and in the case of $SmNi_2B_2C$ and $GdNi_2B_2C$ using x-ray resonant exchange scattering [34, 35]. Four examples are presented in Fig. 8. For R = Dy and Ho the moments are ferromagnetically ordered within the (a,b)-plane, however the moments of different planes are oppositely directed (Fig. 8(a)). For $ErNi_2B_2C$ a transversely polarized spin density wave is observed (Fig. 8b)), whereas in $TbNi_2B_2C$ a longitudinally polarized spin density wave is found (Fig. 8(c)). For $TmNi_2B_2C$ also a transversely polarized spin wave is observed, but the modulation vector is parallel to [110] (Fig. 8(d)). Neglecting these details, one can distinguish two types of structure - (i) the moments are parallel to the tetragonal c axis (R = Tm, see Fig. 8(d)), and (ii) the moments lie in the (a,b)-plane. It is quite interesting, that this different behavior (with the exception of Er, see below) can be explained by second order crystalline electric field (CEF) effects. These CEF's can be characterized by the CEF coefficients A_{nm}. The CEF and consequently the A_{nm} are assumed approximately being the same in all RNi_2B_2C compounds. In lowest order the interaction of a rare-earth ion is proportional to $\alpha_J A_{20}$ with α_J as the second order Stevens factor [36] which roughly speaking characterizes the shape of the 4f charge density for the R^{3+} ion. Table 1 shows that for all R^{3+} ions with $\alpha_J < 0$ the moments are within the (a,b)-plane, for $\alpha_J > 0$ the moments are parallel to

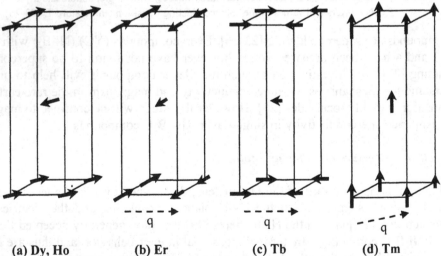

| (a) Dy, Ho | (b) Er | (c) Tb | (d) Tm |

Figure 8. Different types of magnetic structures in the ground state of RNi_2B_2C compounds. (a) For R = Dy or Ho a commensurate antiferromagnetic structure is found. (b), (c) and (d): for R = Er, Tb and Tm incommensurate structures with a propagation vector q in the (a,b)-plane are formed. (b) moments in the (a,b) plane and $\perp$ to q. (c) moments in the (a,b) plane and $\parallel$ q. (d) moments $\parallel$ c and $\perp$ to q.

the c axis. The exceptional case of Gd is obvious : within this approximation Gd^{3+} has a spherical charge density and should not be sensitive to CEF. The case of Er is more complicated and was discussed in detail by Cho et al. [37]: They argue that in the case of Er for the ground state higher order CEF coefficients can not be neglected. On the other hand, the susceptibility at high temperatures should be determined by A_{20} and α_J only. Indeed, their measurement of the susceptibility χ on $ErNi_2B_2C$ single crystals at high temperatures gave results compatible with Table 1. At temperatures below about 150K χ measured in fields parallel to the c-axis becomes larger than χ for fields perpendicular to the c-axis, which is interpreted as due to the influence of higher order CEF terms becoming important at low temperatures. This is also in accordance with the results of neutron scattering experiments.

The large variety of antiferromagnetic structures, and the fact that in most cases they are not simple commensurate structures is related to the competition of CEF's with the RKKY exchange interaction, the modulation of which is not commensurable with the lattice structure.

Table 1. Properties of free R^{3+} ions: n – number of 4f electrtons, S – total spin, L – total orbital angular momentum, J – total angular momentum, g – Landé factor, DG – de Gennes factor, α_J – second Stevens cocfficient, μ_p – paramagnetic moment, μ_s – saturation moment, $<\mu>$ - staggered magnetic moment in RNi_2B_2C [30]. The orientation of the moments with respect to the c-axis is given in the last column [30, 34, 35].

R^{3+}	n	Hund's rules quantum numbers			g	DG	α_J (10^{-2})	μ_p [μ_B]	μ_s [μ_B]	$<\mu>$ [μ_B]	$\parallel$ c $\perp$ c
		S	L	J							
Ce	1	1/2	3	5/2	6/7	0.18	-5.71	2.5	2.1		
Pr	2	1	5	4	4/5	0.80	-2.10	3.6	3.2	0.81	$\perp$
Nd	3	3/2	6	9/2	8/11	1.8	-0.64	3.6	3.3	2.10	$\perp$
Sm	5	5/2	5	5/2	2/7	4.56	4.13	0.9	0.7		$\parallel$
Gd	7	7/2	0	7/2	2	15.8	0	7.9	7		$\perp$, $\parallel$
Tb	8	3	3	6	3/2	10.5	-1.01	9.7	9	7.8	$\perp$
Dy	9	5/2	5	15/2	4/3	7.1	-0.64	10.7	10	8.5	$\perp$
Ho	10	2	6	8	5/4	4.5	-0.22	10.6	10	8.6	$\perp$
Er	11	3/2	6	15/2	6/5	2.6	0.25	9.6	9	7.2	$\perp$
Tm	12	1	5	6	7/6	1.2	1.01	7.6	7	3.4	$\parallel$
Yb	13	1/2	3	7/2	8/7	0.32	3.17	4.5	4		
Lu	14	0	0	0	-	0	0	0	0		

154

Antiferromagnetism and superconductivity in $Ho_xY_{1-x}Ni_2B_2C$

Because different types of antiferromagnetic structures occur in RNi_2B_2C compounds the interplay between antiferromagnetism and superconductivity is more complicated than for simple antiferromagnets. A phenomenon, some times observed, is the so called reentrant superconductivity i.e. for decreasing temperature first the material becomes superconducting and then a transition into the normal state is observed before, at lower temperature, the superconducting state is reached again. Those transition curves for $HoNi_2B_2C$ and $Ho_{0.85}Y_{0.15}Ni_2B_2C$ are quite similar (Figs. 9(a) and 9(b), [38]). For temperatures below 6 K neutron scattering experiments on $HoNi_2B_2C$ reveal the commensurate antiferromagnetic structure shown in Fig. 8(a). Additionally, incommensurate magnetization

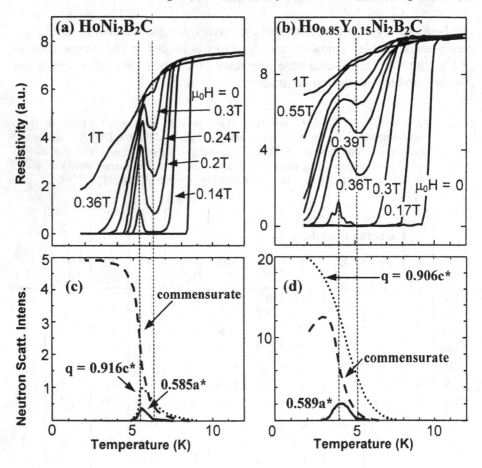

Figure 9. (a) and (b) Resistivity versus temperature curves for $HoNi_2B_2C$ and $Ho_{0.85}Y_{0.15}Ni_2B_2C$, respectively, showing reentrant behavior. (c) and (d) The comparison with the neutron diffraction pattern shows that the a* structure is related to the reentrant behavior.

structures characterized by the two wave vectors $q_1 = 0.916c^*$ and $q_2 = 0.585a^*$ were found between 6 and 4.5 K, which is in accordance with the results of other groups [39,40]. The most important result is that the temperature dependencies of the two incommensurate components shown in Figs 9(c) and 9(d) are quite similar for $HoNi_2B_2C$, but they are quite different for $Ho_{0.85}Y_{0.15}Ni_2B_2C$. Whereas for all investigated samples the a-axis modulated structure exists only in the narrow temperature range where the reentrant behavior is observed, the temperature dependence of the spiral magnetic structure along c-axis is quite different for the two materials of Fig. 9. For $HoNi_2B_2C$ its intensity is different from zero only in the same temperature range as the a-axis modulated structure. For $Ho_{0.85}Y_{0.15}Ni_2B_2C$ this c^*-structure exists down to the lowest measured temperatures. Thus only the incommensurate a-axis modulated magnetic structure seems to be closely related to the reentrant behavior i.e. to enhanced pair breaking. In Fig. 10 the three different types of antiferromagnetic order observed in $HoNi_2B_2C$ and $Ho_xY_{1-x}Ni_2B_2C$ are shown.

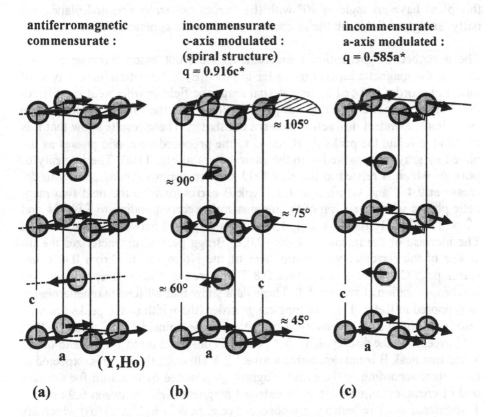

antiferromagnetic commensurate :

incommensurate c-axis modulated : (spiral structure) $q = 0.916c^*$

incommensurate a-axis modulated : $q = 0.585a^*$

Figure 10. The different magnetic structures of $Ho_xY_{1-x}Ni_2B_2C$ as determined by neutron scattering . (a) commensurate antiferromagnetic, (b) incommensurate c^*-structure and (c) proposal how the incommensurate a^*-structure looks like.

Metamagnetic transitions and magnetoresistance in $HoNi_2B_2C$

Two further phenomena closely related to each other and to the complicated antiferromagnetic structures of RNi_2B_2C are (i) metamagnetic transitions and (ii) the dependence of the resistance on an applied magnetic field (magnetoresistance). In the case of $HoNi_2B_2C$ angular dependent measurements of magnetization curves on a single crystal with an applied field in the (a,b)-plane showed up to three metamagnetic transitions i.e. a jump-like increase of the magnetization in a narrow range of field [41]. Analyzing the angular dependence of the critical fields and the corresponding magnetization values, Canfield et al. [41] concluded that in dependence on the field and its direction in the (a,b)-plane four different magnetic phases occur. As shown in Fig. 11(a), at low temperatures (2K) and low fields the structure of Fig. 8(a) is observed. At higher fields two of three succeeding (a,b) Ho planes are ferromagnetically coupled, and one plane is antiparallel to them. At still higher fields the moments of this plane have an angle of 90° with the ferromagnetically coupled planes. Finally, at highest fields, all the moments show a ferromagnetic order.

The described magnetization measurements yield not much information concerning the magnetic long-range order in $HoNi_2B_2C$. To determine the type of long-range order induced by an external magnetic field powder neutron diffraction experiments were carried out [42]. In Fig. 11(b) the low-angle part of the such field dependent diffraction patterns are shown. These results show that it is possible to relate the peaks A, B and C to the proposed magnetic phases as depicted by gray bars marked with the same letters in Fig. 11(a). The intensity of peak A, which is related to the zero field antiferromagnetic phase, starts to decrease at 0.4 T and vanishes at 1 T. Peak B corresponds to the next two magnetic phases with a stacking of the magnetization corresponding to ↑↑↓↑↑↓ and ↑↑→↑↑→ ($↑ \parallel$ [110]and → $\parallel$ [1,-1,0]) and is observed between 0.4 and 1.3 T. The increase of the intensity of the (0,0,2) Bragg peak C characterizes the increase of the ferromagnetic component of the Ho-moments. From 0.4 T upwards peak C increases, and above 0.8 T it increases with a steeper slope and reaches saturation at around 2 T. These data show that all four magnetic phases as proposed in Ref. [41] are of long-range order (the width of the peaks are only resolution limited). The observed transition fields coincide with those in Ref. [41] except for the discrepancy, that peak A is observed up to 1 T instead of 0.5 T and that peak B is not detectable above 1.2 T although the latter is expected to be present according to the phase diagram. A possible explanation for the second discrepancy might be that for external magnetic fields between 0.9 and 1.6 T additional weak reflections are observed (e.g. peak D in Fig. 11(b)) which are supposed to be related to an additional long-range ordered magnetic structure possibly being similar to the a*-type structure of Figs. 9 and 10.

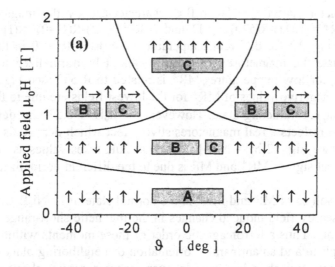

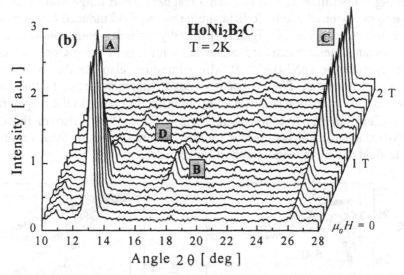

Figure. 11. (a) Magnetic phase diagram of HoNi$_2$B$_2$C at T = 2K proposed in [41]. The field is applied within the ab-plane with an angle ϑ with respect to the [110] direction. The arrows give the directions of the magnetic moments of the Ho-ions in the ferromagnetically ordered ab-planes: $\uparrow$ - [110], $\downarrow$ - [$\bar{1}\,\bar{1}$0], $\rightarrow$ - [$\bar{1}$10]. (b) The neutron diffraction pattern of HoNi$_2$B$_2$C depends on an applied field (T = 2K). This field induces four different magnetic structures. The peak A corresponds to $\uparrow\uparrow\uparrow$, B to $\uparrow\uparrow\downarrow$ and to $\uparrow\uparrow\rightarrow$, C to $\uparrow\uparrow\downarrow$, $\uparrow\uparrow\rightarrow$ and $\uparrow\uparrow\uparrow$. The length of the gray areas A, B, C in (a) describe the scattering intensity of the corresponding peaks at T = 2K [42].

The magnetoresistivity of $HoNi_2B_2C$ characterized by the magnetoresistance ratios $MR^* = [R(H)-R(5T)]/R(5T)$ and $MR = [R(H)-R(H=0)]/R(H=0)$ is shown in Fig. 12 for different temperatures. As $R(H=0) = 0$ in the superconducting state the magnetoresistance ratio cannot be normalized to that value. Therefore, for low temperatures MR^* is related to $R(5T)$ (see Fig. 12(a)). The very large negative values of MR^* for the lowest fields are due to the transition into the superconducting state. However, for $\mu_0H \approx 0.5T$ the dependence of MR^* on H reflects a real magnetoresistive effect which persist also in the normal state (see Fig. 12(b)) where MR even at 20K takes values as large as 20%. The different sign of MR^* and MR is due to the different normalization.

The large values of MR and MR^* in the normal state of $HoNi_2B_2C$ are assumed to be caused by field induced changes in the magnetic short-range order of the Ho moments. Thus a ferromagnetic order of these moments within the tetragonal basis plane and an antiparallel orientation of neighboring planes is assumed to be favored, on a short length scale, even for temperatures above the magnetic ordering temperature $T_N \approx 6$ K. Such a magnetic short range order may be sensitive to the external magnetic field similar as the field induced changes of long-range magnetic order of Fig. 11(a), usually called metamagnetic transitions [43]. Metamagnetic transitions connected with a large magnetoresistance have been reported for $TbNi_2B_2C$ [44]. This compound also has the $LuNi_2B_2C$-type crystal structure (see Fig. 3) but it is not a superconductor. In that case the metamagnetic transition is indicated by an inflection point in the magnetization-versus-field curve, for temperatures below $T_N \approx 15$ K. Also in the case of $TbNi_2B_2C$ a large MR is observed for temperatures above T_N which probably is due to field-sensitive magnetic short-range order.

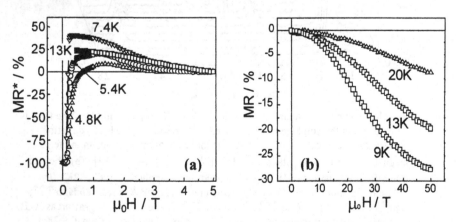

Figure 12. Magnetoresistivity characterized by $MR^* = [(R(H)-R(5T)])/(R(5T)$ and $MR = [(R(H)-R(0)]/R(0)$ of polycrystalline $HoNi_2B_2C$, measured at different temperatures.

3. Suppression of superconductivity in $(R,R')Ni_2B_2C$ compounds

According to Fig. 3, R = Sc, Y, La and all 4f elements form RNi_2B_2C compounds with the $LuNi_2B_2C$-type structure. In this series of quaternary compounds a large variety of superconducting and magnetically ordered states is realized. Therefore it's natural to investigate pseudoquaternary compounds $(R,R')Ni_2B_2C$ in order to realize intermediate states. This may help to better understand the intrinsic mechanisms for superconductivity and magnetism and their interplay in these compounds [45-48].

Fig. 13 shows the influence of dilution of R = Lu and Y by R' = Ho, Dy, or Gd on the superconducting transition temperature T_c. The dependence of T_c on the concentration x (or the effective de Gennes factor $\overline{DG} = xDG[R] + (1-x)DG[R']$ where DG[R] is the de Gennes factor of the R^{3+} ion) in $Gd_xY_{1-x}Ni_2B_2C$ can be well described within the classical theory of Abrikosov and Gor'kov [49] for magnetic impurities in a non-magnetic superconductor (solid line in Fig. 13(a)). For Dy or Ho impurities, the T_c-versus-$\overline{DG}$-curves in Fig. 13 become more flat i.e. the pair breaking effect of Dy and Ho is less pronounced than that of Gd. This is caused by the influence of crystalline electric fields acting on Dy^{3+} and Ho^{3+} thus reducing the magnetic degrees of freedom of these ions [50,45], as described by Fulde and Keller in a modified Abrikosov-Gor'kov-theory [51]. For medium and high concentrations of Dy in $(Y,Dy)Ni_2B_2C$ or $(Lu,Dy)Ni_2B_2C$ the T_c-versus-$\overline{DG}$-curves are strongly non-monotonic and even go to $T_c = 0$. It should be noted that the published experimental data for these pseudoquaternary compounds scatter much. This is particularly true for $(Y,Dy)Ni_2B_2C$ [52,53]. This uncertainty may be caused by unknown details of the metallurgical state of the samples as e.g. non-homogeneous stoichiometry or atomic interchange between the B and the C sublattices. In any case, the steep branches of the T_c-vs.-$\overline{DG}$ curves for high Dy concentrations (in Fig. 13(a) and (b)) can be interpreted as being based on electron scattering on non-magnetic (Y or Lu) impurities in the antiferromagnetic superconductor $DyNi_2B_2C$. A theory for this effect has been developed by Morozov [54].

As can be seen from Figs. 13 and 14, the decrease of T_c with increasing $\overline{DG}$ is stronger for $(Lu,R')Ni_2B_2C$ than for $(Y,R')Ni_2B_2C$ (R' = Ho, Dy, Gd). Obviously this observation is related to the considerably smaller ionic radius Lu^{3+} compared to Y^{3+}, Ho^{3+}, Dy^{3+} and Gd^{3+}. Thus in $(Lu,R')Ni_2B_2C$ considerably larger distortions in the rare-earth sublattice will occur than in $(Y,R')Ni_2B_2C$, which might result in enhanced pair breaking. The detailed mechanism for this effect is still unknown.

A very strong decrease of T_c is observed if $HoNi_2B_2C$ is diluted by La (see

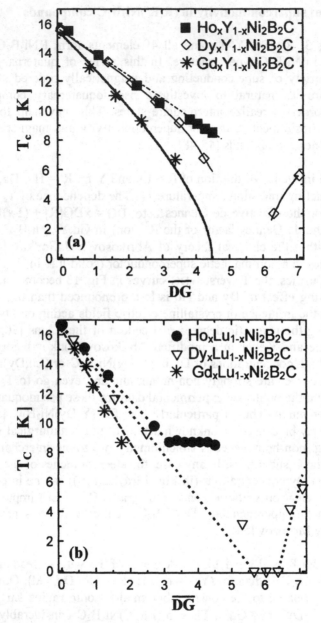

Figure 13. Dependence of the transition temperature T_c on the effective de Gennes factor $\overline{DG} = x\,DG[R] + (1-x)\,DG[R']$ for the non-magnetic superconductors YNi_2B_2C (a) and $LuNi_2B_2C$ (b), both diluted by magnetic rare earth elements Ho, Dy, Gd. The solid line in (a) corresponds to the theory of Abrikosov and Gor'kov [49].

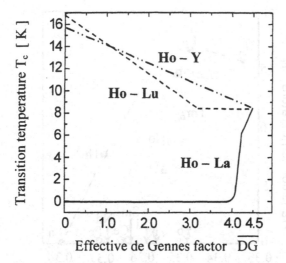

Figure 14. The different influence of the elements R = Y, Lu and La on the superconducting transition temperature T_c in the series $Ho_xR_{1-x}Ni_2B_2C$.

are in discussion. On the one hand, La has a much larger ionic radius than Ho. Hence large distortions will occur around the La impurities. Furthermore, below 6 K $HoNi_2B_2C$ is an antiferromagnetic superconductor. Therefore in a certain concentration range the La ions may act as non-magnetic impurities in an antiferromagnetic superconductor, similar as Lu and Y in $DyNi_2B_2C$ (see Fig. 13). A third possible reason for the decrease of T_c for $(Ho,La)Ni_2B_2C$ in Fig. 14 may be the tendency of RNi_2B_2C compounds to have small or zero values of T_c if the ionic radius or the lattice constant a is relatively large [55], as shown in Fig. 15. Superconductivity is absent in $LaNi_2B_2C$ because the electronic structure of that compound remarkably deviates from that in YNi_2B_2C and $LuNi_2B_2C$ [28]. It is obvious from Fig. 15 that the absence of superconductivity in RNi_2B_2C with light R elements is mainly due to the larger ionic radius of light R^{3+} ions (resulting in a large lattice constant a) but not to magnetic effects (as in the case of heavy R elements). An exception is $YbNi_2B_2C$ which, by de-Gennes-factor as well as lattice-constant arguments, should be a superconductor. Instead it is a non-magnetic non-superconducting heavy fermion compound because the Yb-4f electrons strongly hybridize with the conduction electrons [56].

We have also investigated the behavior of $(R,R')Ni_2B_2C$ compounds with both elements R and R' being non-magnetic as e.g. Y and Lu. Fig. 16 shows the x-dependence of T_c for $Y_xLu_{1-x}Ni_2B_2C$. For a "gray" system with a fictive R ion

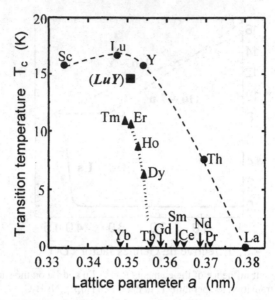

Figure 15. Transition temperature and lattice parameter *a* for RNi₂B₂C compounds with with LuNi₂B₂C type structure for non-magnetic [55] and magnetic R elements. The curves are guides to the eye.

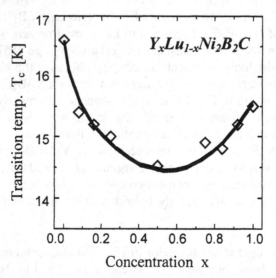

Figure 16. Transition temperature T_c in dependence on the concentration x for pseudoquaternary compounds $Y_xLu_{1-x}Ni_2B_2C$.

of averaged size resulting in an averaged lattice constant $\overline{a} = x a[Y] + (1-x) a[Lu]$ the value of T_c would be on the upper curve in Fig. 15 with a maximum value even higher than T_c of both parent compounds with Y or Lu. Thus the real quaternary system $Y_x Lu_{1-x} Ni_2 B_2 C$ behaves totally different from the fictive "gray" system. In Ref. 48 the non-monotonic curve in Fig. 16 with a minimum near $x = 0.5$ was attributed to disorder at the rare-earth lattice sites, where, however, the microscopic mechanism which mediates disorder to T_c and other physical quantities is not yet clarified. Typical scenarios for disorder effects are: the peak of the density of states may be broadened or the phonon spectrum may be modified by disorder.

4. Conclusions

The series of quaternary compounds $(R,R')Ni_2 B_2 C$ with R, R' = Sc, Y, La or 4f elements exhibit superconductivity and exchange-coupled magnetic order, which compete and, in some cases, coexist. An exception is $YbNi_2 B_2 C$ where a moderate heavy-fermion state has been detected. Experimental work showed that different types of magnetic order occur (e.g. in $HoNi_2 B_2 C$) and affect superconductivity to a different degree. In particular, future work has to clarify, details of the incommensurate a-axis modulated magnetic structure in $HoNi_2 B_2 C$ and other compounds and its role in reentrant superconductivity and other physical phenomena. In $(R,R')Ni_2 B_2 C$ various effects could be identified that reduce the superconducting transition temperature i.e. enhance pair breaking: (i) scattering on magnetic impurities in a non-magnetic superconductor according to the theory of Abrikosov/Gor'kov. This effect is weakened by crystalline electric field effects. (ii) Distortions caused by differences in the ionic sizes of R and R'. (iii) Effects of disorder on the R lattice site which are not yet understood in detail, (iv) scattering on nonmagnetic impurities in an antiferromagnetic superconductor and (v) change of the electronic density of states at the Fermi level, caused by the substitution of R by R'. Many problems of the crystal chemistry of the family of superconducting borocarbides have to be solved as e.g. the stoichiometry and crystal structure of Pd-based borocarbides or the degree of atomic interchange between the C and the B sublattices or details of the metallurgical state of the samples that control their superconducting and magnetic properties.

5. Acknowledgements

We would like to thank L. Schultz for his help and S.-L. Drechsler for stimulating discussions. This work was supported by NATO and by DFG/SFB 463.

164

6. References

1. Mazumdar, Ch., Nagarajan, R., Godart, C., Gupta, L.C., Latroche, M., Dhar, S.K., Levy-Clement, C., Padalia, B.D., and Vijayaraghavan, R. (1993) Superconductivity at 12K in Y-Ni-B System, *Solid State Commun.* **87**, 413-416.
2. Nagarajan, R., Mazumdar, Ch., Hossain, Z., Dhar, S.K., Gopalakrishnan K.V., Gupta, L.C., Godart, C., Padalia, B.D., and Vijayaraghavan, R. (1994) Bulk Superconductivity at an Elevated Temperature ($T_c \approx$ 12K) in a Nickel Containing Alloy System Y-Ni-B-C, *Phys. Rev. Lett.* **72**, 274-277.
3. Cava, R.J., Takagi, H., Batlogg, B., Zandbergen, H.W., Krajewski, J.J., Peck Jr, W.F., van Dover, R.B., Felder, R.J., Siegrist, T., Mizuhashi, K., Lee, J.O., Eisaki, H., Carter, S.A., and Uchida, S. (1994) Superconductivity at 23K in Yttrium Palladium Boride Carbide, *Nature* **367**, 146-148.
4. Cava, R.J., Takagi, H., Zandbergen, H.W., Krajewski, J. J., Peck Jr, W.F., Siegrist, T., Batlogg, B., van Dover, R.B., Felder, R.J., Mizuhashi, K., Lee, J.O,, Eisaki, H., and Uchida, S. (1994) Superconductivity in the Quaternary Intermetallic Compounds $LnNi_2B_2C$, *Nature* **367** 252-253.
5. Ginzburg, V.L. (1956) O Ferromagnetnykh Sverkhprovodnikakh. *Zh. Eksp. Teor. Phys.* **31**, 202-210 (engl. Transl. (1957) *Sov. Phys. JETP* **4**, 153).
6. Baltensperger, W., and Strässler, S. (1963) Superconductivity in Antiferromagnets, *Phys. kondens. Materie* **1**, 20-26.
7. Fischer, O. (1990) Magnetic Superconductors, in K.H.J. Buschow and Wohlfarth E.P. (eds.), *Ferromagnetic Materials*, vol. 5, Elsevier, Amsterdam p. 466.
8. Fulde, P. (1995) *Electron Correlations in Molecules and Solids*, Springer, Berlin.
9. Maple, M.B. (1995) Interplay between Superconductivity and Magnetism, *Physica B* **215**, 110-126.
10. Lynn, J.W., (1997) Rare Earth Magnetic Ordering in Exchange-Coupled Superconductors, *J. Alloys and Comp.* **250**, 552-558.
11. Gupta, L.C. (1998) Quaternary Borocarbide Superconductors, *Phil. Mag. B* **77**, 717-726.
12. Siegrist, T., Zandbergen, H.W., Cava, R.J., Krajewski, J.J., and Peck Jr, W.F. (1994) The Crystal Structure of Superconducting $LuNi_2B_2C$ and the Related Phase LuNiBC, *Nature* **367**, 254-256.
13. Siegrist, T., Cava, R.J., Krajewski, J.J., and Peck Jr, W.F. (1994) Crystal Chemistry of the Series LnT_2B_2C (Ln = Rare Earth, T = Transition Element), *J. Alloys and Comp.* **216**, 135-139.
14. Brabers, J.H.V.J., Bakker, K., Nakotte, H., de Boer, F.R., Lenczowski, S.K.J., and Buschow, K.H.J. (1993) Giant Magnetoresistance in Polycrystalline $SmMn_2Ge_2$, *J. Alloys & Comp.* **199**, L1-L3.
15. Steglich, F., Geibel, Chr., Modler, R. Lang, M., Hellmann, P., and Gegenwart, Ph. (1995) Classification of Strongly Correlated f-Electron Systems, *J. Low Temp. Phys.* **99**, 267-281.
16. Mulder, F.M., Brabers, J.H.V.J., Coehoorn, R., Thiel, R.C., Buschow, K.H.J., and de Boer, F.R. (1995) ^{155}Gd Mössbauer Effect and Magnetic Properties of Novel RT_2B_2C Compounds with T ≡ Ni, Co, *J. Alloys and Comp.* **217**, 118-122.
17. Rupp, B., Rogl, P., and Hulliger, F. (1987) Magnetism and Structural Chemistry of Ternary Borides $RECo_2B_2$ (RE ≡ Y, La, Pr, Nd, Sm, Gd, Tb, Dy, Ho, or Er) and Boron Substitution in $(Y,Ce)Co_2Si_{2-x}B_x$, *J. Less Comm. Met.* **135**, 113-125.
18. Kreyssig, A., Loewenhaupt, M. Freudenberger, J., Müller, K.-H., and Ritter, C. (1999) Evidence of Tetragonal to Orthorhombic Distortion of $HoNi_2B_2C$ in the Magnetically Ordered State, *J. Appl. Phys.* **85**, 6085-6060.

19. Detlefs, C., Islam, A.H.M.Z., Gu, T., Goldman, A.I., Stassis, C., Canfield, P.C., Hill, J.P., and Vogt, T. (1997) Magnetoelastic Tetragonal-to-Orthorhombic Distortion in $ErNi_2B_2C$, *Phys. Rev. B* **56**, 7843-7846.

20. Sarrao, J.L., de Andrade, M.C. Herrmann, J., Han, S.H., Fisk, Z., Maple, M.B., and Cava, R.J. (1994) Superconductivity to 21K in Intermetallic Thorium-Based Boride Carbides, *Physica C* **229**, 65-69.

21. Ström, V., Kim, K.S., Grishin, A.M., and Rao, K.V. (1996) Superconducting Metastable Phase in Rapid Quenched Y-Pd-B-C Borocarbides, *J. Appl. Phys.* **79**, 5860-5862.

22. Freudenberger, J. (2000) Paarbrechung in Seltenerd-Übergangsmetall-Borkarbiden, Thesis, TU Dresden.

23. Cava, R.J., Zandbergen, H.W., Batlogg, B., Eisaki, H., Takagi, H., Krajewski, J.J., Peck Jr, W.F., Gyorgy, E.M., and Uchida, S. (1994) Superconductivity in Lanthanum Nickel Boro-Nitride, *Nature* **372**, 245-247.

24. Zandbergen, H.W., Jansen, J., Cava, R.J., Krajewski, J.J., and Peck Jr, W.F. (1994) Structure of the 13-K Superconductor $La_3Ni_2B_2N_3$ and the Related Phase LaNiBN, *Nature* **372**, 759-761.

25. Gao, L., Qiu, X.D., Cao, Y., Meng, R.L., Sun, Y.Y., Xue, Y.Y., and Chu, C.W. (1994) Superconductivity in$(LuC)_2(Ni_2B_2)$ and $(LuC)(Ni_2B_2)$, *Phys. Rev. B* **50**, 9445-9448.

26. Michor, H., Krendelsberger, R., Hilscher, G., Bauer, E., Dusek, C., Hauser, R., Naber, L., Werner, D., Rogl, P., and Zandbergen, H.W. (1996) Superconducting Properties of $La_3Ni_2B_2N_{3-\delta}$, *Phys. Rev. B* **54**, 9408-9420.

27. Rukang, L., Chaoshui, X., Hong, Z., Bin, L., and Li, Y. (1995) The Preparation and Characterization of a New Layered Yttrium Nickel Borocarbide, *J. Alloys and Comp.* **223**, 53-55.

28. Drechsler, S.L., Shulga, S., and Rosner, H. (1998) Superconducting Transition Metal Boro-carbides – Challenging New Materials between Traditional Superconductors and HTSC's, *These Proceedings.*

29. Sanchez, J.P., Vulliet, P., Godart, C., Gupta, L.C., Hossain, Z., and Nagarajan, R. (1996) Magnetic and Crystal-Field Properties of the Magnetic Superconductor $DyNi_2B_2C$ from [161]Dy Mössbauer Spectroscopy, *Phys. Rev. B* **54**, 9421-9427.

30. Lynn, J.W., Skanthakumar, S., Huang, Q., Sinha, S.K., Hossain, Z., Gupta, L.C., Nagarajan, R., and Godart, C. (1997) Magnetic Order and Crystal Structure in the Superconducting RNi_2B_2C Materials, *Phys. Rev. B* **55**, 6584-6598.

31. Pickett, W.E., and Singh D.J. (1994) $LuNi_2B_2C$: A Novel Ni-Based Strong-Coupling Super-conductor, *Phys. Rev. Lett.* **72**, 3702-3705.

32. Mattheiss, L.F. (1994) Electronic Properties of Superconducting $LuNi_2B_2C$ and Related Boride Carbide Phases, *Phys. Rev. B* **49**, 13279-13282.

33. von Lips, H., Ilu, Z.,Grazioli, C., Drechsler, S.-L., Behr, G., Knupfer, M., Golden, M.S., Fink, J., Rosner, H., and Kaindl, G. (1999) Polarization-Dependent X-Ray-Absorption Spectroscopy of Single-Crystal YNi_2B_2C Superconductors, *Phys. Rev. B* **60**, 11444-11448.

34. Detlefs, C., Islam, A.H.M.Z., Goldman, A.I., Stassis, C., Canfield, P.C., Hill, J.P., and Gibbs, D. (1997) Determination of Magnetic-Moment Direction using X-Ray Resonant Ex-change Scattering, *Phys. Rev. B* **55**, R680-R683.

35. Detlefs, C., Goldman, A.I., Stassis, C., Canfield, P.C., Cho, B.K., Hill, J.P., and Gibbs, D. (1996) Magnetic Structure of $GdNi_2B_2C$ by Resonant and Nonresonant X-Ray Scattering, *Phys. Rev. B* **53**, 6355-6361.

36. Hutchings, M.T. (1964) Point-Charge calculations of Energy Levels of Magnetic Ions in Crystalline Electric Fields, in Seitz and Turnbull (eds.) *Solid State Physics* **16**, p. 227.

37. Cho, B.K., Canfield, P.C., Miller, L.L., Johnston, D.C., Beyerman, W.P., and Yatskar, A. (1995) Magnetism and Superconductivity in Single-Crystal $ErNi_2B_2C$, *Phys. Rev.* **B** 52, 3684-3695.

166

38. Müller, K.-H., Kreyssig, A., Handstein, A., Fuchs, G., Ritter, C., and Loewenhaupt, M. (1997) Magnetic Structure and Superconductivity in $(Ho_xY_{1-x})Ni_2B_2C$, J. Appl. Phys. **61**, 4240-4242.

39. Goldman, A.I., Stassis, C., Canfield, P.C., Zarestky, J., Dervenagas, P., Cho, B.K., Johnston, D.C., and Sternlieb, B. (1994) Magnetic Pair Breaking in $HoNi_2B_2C$, Phys. Rev. B **50**, 9668-9671.

40. Tomy, C.V., Chang, L.J., Paul, D.McK., Andersen, N.H., and Yethiraj, M. (1995) Neutron Diffraction from $HoNi_2B_2C$, Physica B **213&214**, 139-141.

41. Canfield, P.C., Bud'ko, S.L., Cho, B.K., Lacerda, A., Farrell, D., Johnston-Halperin, E., Kalatsky, V.A., and Pokrovsky, V.L. (1997) Angular Dependence of Metamagnetic Transitions in $HoNi_2B_2C$, Phys. Rev. B **55**, 970-976.

42. Kreyssig, A., Freudenberger, J., Sierks, C., Loewenhaupt, M., Müller, K.-H., Hoser, A., and Stuesser, N. (1999) Field and Temperature Dependence of Magnetic Order in $HoNi_2^{11}B_2C$, Physica B **259-261**, 590-591.

43. Gignoux, D. and Schmitt, D. (1995) Metamagnetism and Complex Magnetic Phase Diagrams of Rare Earth Intermetallics, J. Alloys and Comp. **225**, 423-431.

44. Müller, K.-H., Handstein,.A., Eckert, D., Fuchs, G., Nenkov, K., Freudenberger, J., Richter, M. and Wolf, M. (1998) Metamagnetismus and Large Magnetoresistance in $TbNi_2B_2C$, Physica B **246-247**, 226-229.

45. Freudenberger, J., Fuchs, G., Nenkov, K., Handstein, A., Wolf, M., Kreyssig, A., Müller, K.-H., Loewenhaupt, M., and Schultz, L. (1998) Breakdown of de Gennes Scaling in $Ho_xLu_{1-x}Ni_2B_2C$, J. Magn. Magn. Mater. **187**, 309-317.

46. Freudenberger, J., Fuchs, G., Müller, K.-H., Nenkov, K., Drechsler, S.-L., Kreyssig, A., Rosner, H., Koepernik, K., Lipp, D., and Schultz, L. (1999) Diluted and Concentrated Isoelectronic Substitutional Effects in Superconducting $R_xY_{1-x}Ni_2B_2C$ Compounds, J. Low Temp. Phys. **117**, 1623-1627.

47. Freudenberger, J., Kreyssig, A., Ritter, C., Nenkov, K., Drechsler, S.-L., Fuchs, G., Müller, K.-H., Loewenhaupt, M., and Schultz, L. (1999) Suppression of Superconductivity by Nonmagnetic Impurities, Structural properties and Magnetic Ordering in $Ho_xLu_{1-x}Ni_2B_2C$, Physica C **315**, 91-98.

48. Freudenberger, J., Drechsler, S.-L., Fuchs, G., Kreyssig, A., Nenkov, Shulga, S.V., K., Müller, K.-H., and Schultz, L. (1998) Superconductivity and Disorder in $Y_xLu_{1-x}Ni_2B_2C$, Physica C **306**, 1-6.

49. Abrikosov, A.A. and Gor'kov, L.P. (1961) Contribution to the Theory of Superconducting Alloys with Paramagnetic Impurities, Soviet Physics JETP **12**, 1243-1253.

50. Cho, B.K., Canfield, P.C., and Johnston, D.C. (1996) Breakdown of de Gennes Scaling in $(R_{1-x}R'_x)Ni_2B_2C$ Compounds, Phys. Rev. Lett. **77**, 163-166.

51. Fulde, P., and Keller, J. (1982) Theory of Magnetic Superconductors, in M.B. Maple and O. Fischer (eds.) Superconductivity in Ternary Compounds II, Springer, Berlin.

52. Hossain, Z., Nagarajan, R., Dhar, S.K., and Gupta, L.C. (1999) Depression of T_c by Non-Magnetic Impurities in Antiferromagnetic Superconductor with $T_c < T_N$, Physica B, **259-261**, 606-607.

53. Michor, H., El-Hagary, M., Hauser, R., Bauer, E., Hilscher, G. (1999) The Interplay of the Superconducting and Antiferromagnetic State in $DyNi_2B_2C$, Physica B **259-261**, 604-605.

54. Morozov, A.I. (1980) Impurities in Antiferromagnetic Superconductors, Soviet Physics Solid State **22**, 1974-1977.

55. Lai, C.C., Lin, M.S., You, Y.B., and Ku, H.C. (1995) Systematic Variation of Superconductivity for the Quaternary Borocarbide System RNi_2B_2C (R = Sc, Y, La, Th, U, or a lanthanide), Phys. Rev. B **51**, 420-423.

56. Dhar, S.K., Nagarajan, R., Hossain, Z., Tominez, E., Godart, C., Gupta, L.C., and Vijayaraghavan (1996) Anomalous Suppression of T_c and Moderate Heavy Fermion Behaviour in $YbNi_2B_2C$, Sol. State Commun. **98**, 985-989.

SUPERCONDUCTING TRANSITION METAL BOROCARBIDES

Challenging materials in between traditional superconductors and high-T_c cuprates

S.-L. DRECHSLER, H. ROSNER, S.V. SHULGA, H. ESCHRIG,
Institut für Festkörper- und Werkstofforschung Dresden,
D-01171 Dresden, Postfach 270116, Germany

1. Introduction

2. Crystal structures and related compounds

3. The electronic structure

4. The relationship between the electronic structure and the upper critical field $H_{c2}(T)$

5. Disorder and doping

6. LuNi$_2$B$_2$C and YNi$_2$B$_2$C: unconventional pairing?

7. A possible classification

Abstract. We present an overview of the present knowledge of the electronic structure of selected properties of quaternary intermetallic rare earth transition metal borocarbides and related boronitride compounds. The calculated highly anisotropic Fermi surfaces exhibit clear similarities such as nested regions but also significant distinctions. Electrons from the nested parts of the Fermi surface affect several properties in the superconducting state. We report theoretical calculations which emphasize the relevance of these electrons to the mechanism of superconductivity. Structural parameters for ScNi$_2$B$_2$C derived from total energy calculation allow to discriminate conflicting experimental structural reports. The importance of correlation effects due to the presence of the transition metal component in determining the electronic structure is discussed comparing the band structure calculation results with various electronic spectroscopies such as X-ray absorption spectroscopy (XAS). The excellent agreement with the XAS data suggests a minor importance of correlation effects compared with the cuprate superconductors. Thermodynamic properties of these systems

S.-L. Drechsler and T. Mishonov (eds.), High-T$_c$ Superconductors and Related Materials, 167–184.
© 2001 *Kluwer Academic Publishers. Printed in the Netherlands.*

are analyzed within the multi-band Eliashberg theory with special emphasis on the upper critical field $H_{c2}(T)$ and the specific heat. In particular, the unusual positive curvature of $H_{c2}(T)$ near T_c observed for high-quality single crystals, polycrystalline samples of YNi_2B_2C, $LuNi_2B_2C$ as well as to a somewhat reduced extent also for the mixed system $Y_{1-x}Lu_xNi_2B_2C$ is explained microscopically. The values of $H_{c2}(T)$ and of its positive curvature near T_c are intrinsic quantities generic for such samples. Both quantities decrease with growing disorder and thus provide a direct measure of the sample quality.

1. Introduction

Seven years after the discovery [1] of rare earth transition metal borocarbides (nitrides) (RTBC(N)) with T=Ni,Pd,Pt, the place of RTBC(N) compounds within the family of more or less exotic superconductors is still under debate. In contrast to first speculations of a strong similarity to quasi-2D cuprates (suggested by their reminiscent transition metal layered crystal structure), various LDA (local density approximation) band structure calculations performed in 1994/95 (see Refs. 2-14) clearly demonstrated their 3D electronic structure. Finally, the whole class has been classified as traditional intermetallic superconductors, more or less closely related to the A-15's. Only the interplay of antiferromagnetism and superconductivity for R=Ho, Er, Dy, Tm, Tb, Pr has been regarded as a challenging problem of general interest. However, during the last years the situation for non-magnetic borocarbides has been changed considerably as high-quality single crystals have become available revealing (i) more pronounced anisotropies for $LuNi_2B_2C$ [15] (some of them even being strongly temperature dependent) in contrast to the first experimental observations [16] for YNi_2B_2C and (ii) the clear multi-band character of the superconducting state in YNi_2B_2C and $LuNi_2B_2C$ [17]. The crucial importance of such details of the electronic structure, although known in principle from the first calculations, has not been fully appreciated so far. The value of the gap ratio $2\Delta/T_c$, initially fixed to the conventional BCS value 3.5, now varies from 0.45 [18] to 3.2 [19]. Finally, in the context of anisotropy and other unusual properties (see below), d-wave superconductivity has been proposed for YNi_2B_2C and $LuNi_2B_2C$ [20, 21]. Quite interestingly, other RTBC(N) superconductors such as $(LaN)_3(NiB)_2$ show more standard s-wave like behavior [22].

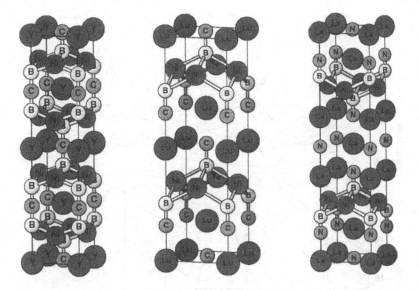

Figure 1. The crystal structure of the single-layer (YNi_2B_2C, left), double-layer (Lu-NiBC, middle), and triple-layer ($La_3Ni_2B_2N_3$, right) rare-earth transition metal borocarbides (nitrides), respectively.

2. Crystal structures and related compounds

2.1. CRYSTAL STRUCTURE AND CRITICAL TEMPERATURE

The tetragonal layered crystal structures of the I4/mmm or P4/nmm types resolved so far for all well characterized RTBC(N) compounds can be written schematically as $(RC(N))_n(TB)_2$ with $n = 1, 2, 3$ (see Fig. 1). There are empirically discovered systematic dependences of T_c on the T-T distance, the transition metal component, the non-isoelectronic dopants and the B-T-B bond angle. Finally, the number of *metallic* layers separating and doping the $(NiB)_2$ networks also has a profound effect on the actual T_c value. Thus, for the single RC-layer (T=Ni) compounds the highest $T_c \approx 14$ to 16.6 K values are obtained for R= Sc, Y, Lu, but superconductivity has not been observed so far for R= La. The double-layer Lu- and Y-compounds exhibit very small transition temperatures of 2.9 K and 0.7 K [22], respectively, which however can be increased considerably replacing Ni by Cu [23]. In the case of the two-layer boronitride $(LaN)_2(NiB)_2$, triple-layer, and quadro-layer $(YC)_n(NiB)_2$ ($n = 3, 4$) so far no superconductivity has been detected whereas the boronitride triple-layer compound exhibits a relatively high $T_c \approx 12K$.

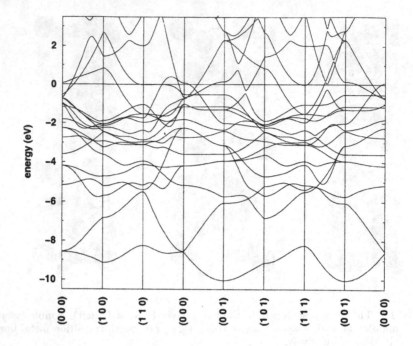

Figure 2. LDA-bandstructure of YNi$_2$B$_2$C. The Fermi level is at zero energy. The flat band near the Fermi level between the (110) and (000) points produces the narrow peak in the density of states (see Fig.3).

2.2. RELATED SUPERCONDUCTORS

The place of RTBC(N) compounds within the family of more or less exotic superconductors (see Fig. 2) is still under debate. The Problem under consideration has several aspects: (i) What is the pairing interaction once standard Cooper=pairs have been adopted? If this is the electron-phonon interaction, we may continue to ask which phonons, high or low-frequency? (ii) In this debate the character of the electronic structure plays a decisive role. We may ask: which orbitals are involved in the states near the Fermi energy E_F? How are they affected by the electron-electron correlation? Which details of the electronic structure near E_F affect the superconductivity?

3. The electronic structure

A typical bandstructure for a single-layer compound is shown in Fig. 2 for the case of YNi$_2$B$_2$C. Similarly, all bandstructure calculations [2-14] for RTBC(N) reveal sizeable dispersion in c-direction of the bands crossing the Fermi level. Electronically the coupling of the 2D-(TB)$_2$ networks is

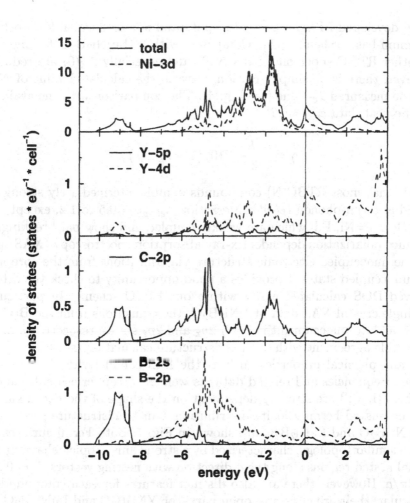

Figure 3. Total as well as partial DOS for YNi$_2$B$_2$C. The Fermi level is at zero energy.

mediated mainly by the carbon and boron $2p_z$ states. Further important issues are the peak of the density of states (DOS) $N(0)$ near the Fermi level $E_F=0$ (see Fig. 3) and the intermediate strength of correlation effects [24] which is in between the weakly correlated Ni-metal and the strongly correlated insulating NiO.

The electronic structure near $E_F=0$ of all RTBC(N) compounds is characterized by a special band complex containing three bands above about 1eV of the main group of the T-derived d-states. For Y(Lu)Ni$_2$B$_2$C there is a flat band near E_F (see Fig. 2) giving rise to a narrow asymmetric peak in the total DOS $N(E)$ (see Fig. 3) and to a large T-d-partial DOS $N_d(0) \approx 1/2N(0)$ which can be analyzed with high accuracy within our recently developed FPLO-scheme (full-potential non-orthogonal local-orbital minimum-basis scheme) [25]. Compared with Y(Lu)Ni$_2$B$_2$C for most of the other RTBC-superconductors $N(0)$ and especially $N_d(0)$ are reduced. However, there is no simple relation between the calculated value of $N(0)$ and the measured T_c-value (see Fig. 4). The comparison with the available specific heat data $c_{p,el} = \gamma T$

$$\gamma = \frac{\pi^2 k_B^2}{3} N(0) \left(1 + \lambda_{el-ph}\right), \tag{1}$$

reveals that most RTBC(N)-compounds exhibit intermediately strong averaged electron-phonon (el-ph) interaction $\lambda_{el-ph} \sim 0.5$ to 1.2, except LaT$_2$B$_2$C T=Ni, Pd; which are weakly coupled and show pairbreaking.

Since polarization dependent x-ray absorption spectroscopy (XAS) probes the unoccupied electronic structure via transitions from the core level into unoccupied states it provides a good opportunity to check the orbital resolved DOS calculated easily within our FPLO scheme. In particular, for single crystal XAS data of YNi$_2$B$_2$C, the transitions from the B(C) 1s core level into unoccupied the B(C) $2p_x$ and $2p_z$ states, respectively, are in reasonable agreement with our bandstructure calculation (see Fig. 5)[26].

Many physical properties such as the Hall conductivity, de Haas-van Alphen frequencies and related data, as well as the upper critical magnetical field H$_{c2}$(T) are strongly dependent on the shape of the Fermi surface and the related Fermi velocity distribution. Our bandstructure predictions for YNi$_2$B$_2$C and LuNi$_2$B$_2$C are shown in Fig. 7. Both Fermi surfaces exhibit a similar topology, characterized by a strong anisotropic behaviour and special nested regions along the a direction with nesting vectors $\vec{q}_n \sim 0.5$ to 0.6 $2\pi/a$. However, there are also distinct features for each compound. In particular, their are close and open parts for YNi$_2$B$_2$C and LuNi$_2$B$_2$C, respectively, near $k_z = 0.5\pi/c$. This might explain the qualitative differences between both compounds reported for the Hall data [27] which remained unexplained so far. Nested and anisotropic properties for the Fermi surface of LuNi$_2$B$_2$C close to our predictions [28] (see Fig. 6) have been observed by electron-positron annihilation radiation [29].

We emphasize that LDA bandstructure calculations are able to describe with high accuracy various structure related properties such as the lattice constants, the atomic positions, the bulk moduli as well as the frequencies of Raman activ phonon modes. Recently, for the case of ScNi$_2$B$_2$C, two significantly different structural parameter sets have been reported [30, 31] with c axis lattice constants differing by about 0.5 Å. Assuming the generally accepted I4/mmm space group for this compound, too, we were able to calculate the values for the three independent parameters (unit-cell volume V_0, c/a ratio and boron position) with respect to an minimum of the total en-

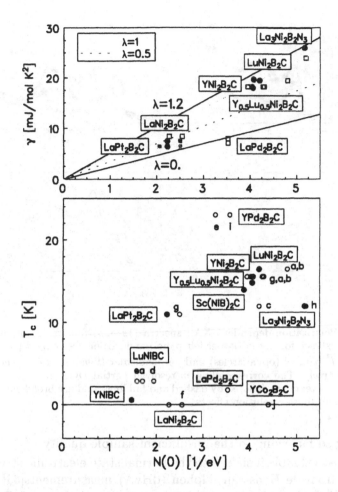

Figure 4. Experimental values of the Sommerfeld constant (top) and the superconducting transition temperature T_c (bottom) *vs.* calculated total density of states at the Fermi level $N(0)$ (states per formula unit) for various RTBC(N)-compounds and different LDA-calculational schemes: Ref. 3-(a), Ref. 2-(b), Ref. 7-(c), Ref. 12-(d), Ref. 10-(e), Ref. 5-(g), Ref. 9-(h), Ref. 6 (i), Ref. 4-(f), Ref. 13-(g), Ref. 9-(h), Ref. 14-(j). The straight lines in the upper picture denote various *el-ph* coupling constants $0 \le \lambda \le 1.2$ (dotted line: $\lambda=0.5$, dashed-dotted line: $\lambda=1.0$)

ergy. The result for V_0 lending strong support for the data of Freudenberger *et al.* [31] and disfavouring the data of Ku *et al.* [30] is shown in Fig. 8. To our knowledge, this is the second remarkable case in the history of the borocarbids research where experimentally determined structural data have been succesfully criticised by LDA bandstructure calculations [32]. In this context we mention that also the experimentally unknown boron position $(0,0,z)$ should be close to our prediction of $z = 0.362$. Experimental data

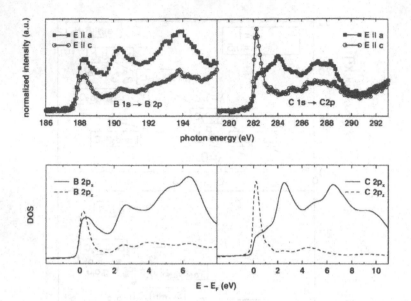

Figure 5. Polarization-dependent XAS spectra ($1s{\rightarrow}2p$ transitions) of single crystal YNi$_2$B$_2$C are shown for boron (upper left panel) and carbon (uper right panel) with the electric field $\vec{E}$ parallel (open circles) and perpendicular (filled sqares) to the tetragonal c-axis, respectively. The corresponding m-resolved partial DOS from our LDA-FPLO calculations (see text) are denoted by dashed and full lines, and are broadened by account for life time and finite resolution effects.

are lacking so far owing to the unsufficient sample quality.

The most valuable insight into the normal state electronic structure can be gained from de Haas-van Alphen (dHvA) measurements[33]. In high-quality YNi$_2$B$_2$C single crystals 6 cross sections are observed. The related Fermi velocities $v_{F,i}$, $i=1,...,6$, on extremal orbits can be grouped into two sets differing by a factor of 4. These observations and the sizable anisotropy of the H_{c2} for such crystals clearly indicate that they are nearly in the clean-limit regime.

To summarize, at the present status, there is a reasonable qualitative agreement between predictions of the LDA-results and experimental data. This points to a less relevance of correlation effects compared with the high-T_c cuprates.

4. The relationship between the electronic structure and the upper critical field $H_{c2}(T)$

A detailed analysis of the magnitude and the shape of $H_{c2}(T)$ is given in [17]. In particular, the failure of the standard isotropic band approach

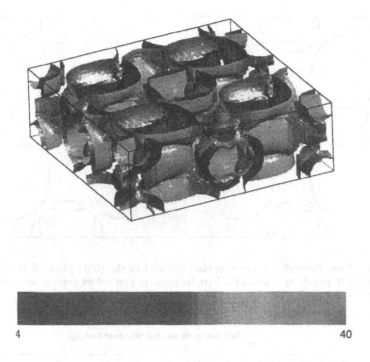

4   40

Figure 6. Fermi surface of YNi_2B_2C, the colors corresponds to the Fermi velocities given in atomic units.

points to a multiband description, where electrons with significantly smaller v_F compared with the Fermi surface average (see Fig. 6) $\sqrt{\langle v_F^2 \rangle_{FS}}$ and relatively strong *el-ph* coupling are mainly responsible for the superconductivity. This model explains also approximately the experimentally observed strong deviations of the shape of $H_{c2}(T)$ from the standard parabolic-like curve. In particular, the positive curvature of $H_{c2}(T)$ becomes maximal approaching T_c. All these peculiarities can be described with high accuracy by the simple expression

$$H_{c2}(T) \approx \frac{H_{c2}(0)z}{1 - (1 + \alpha)tz + L(tz)^2 + M(tz)^3},$$

$$t = T/T_c \qquad z = (1 - T/T_c)^{1+\alpha}.$$

$$(2)$$

The positive curvature near T_c (characterized by the value of the critical exponent α in Eq. 4.1) has been observed in resistivity, magnetization,

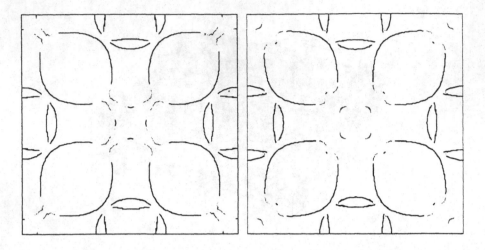

Figure 7. Cuts through the Fermi surface parallel to the (001) plane at the Γ-point for YNi_2B_2C (left panel) and $LuNi_2B_2C$ (right panel). The colors correspond to the Fermi velocities as in Fig. 6

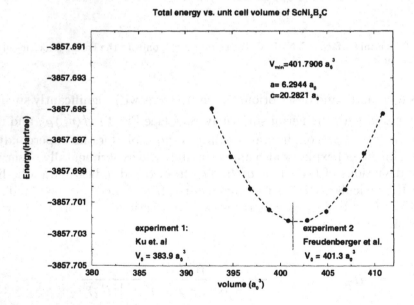

Figure 8. Calculated total energy vs. unit cell volume of $ScNi_2B_2C$ for a fixed c/a ratio and boron position corresponding to the minimum of the total energy found in a self-consistent procedure. The bars denote the two available experimental results.

and specific heat measurements [28]. This shows unambiguously that the positive curvature is an inherent thermodynamic property generic for all clean non-magnetic RTBC(N) superconductors.

The saturation (negative curvature) at low T described by the coefficient $L - (1 + \alpha)^2 > 0$ is sensitive to details of the Fermi surface and to the value of the smallest of the superconducting gaps.

Together with the second coefficient M it determines the inflection field near $\sim 0.3t$. Thus, a simple but instructive classification of various borocarbide compounds becomes possible. In particular, the nearly triangular shape of $H_{c2}(T)$ for $YC(NiB)_2$ described by large values of L, requires a significantly larger ratio of the fast and slow Fermi velocities v_f compared e.g. with $LuC(NiB)_2$ showing more pronounced parabolic behavior at low T.

In general, the detailed study of different anisotropies with respect to the Fermi surface, to the pairing interaction and possibly also to the order parameter symmetry is the most perspective way to elucidate the mechanism of superconductivity generic for the borocarbides and -nitrides. Obviously, this can be reached only by a strong collaboration of various experimental and theoretical techniques investigating the electronic structure of these fascinating compounds.

The inspection of the v_F distribution over the Fermi surface of $LuNi_2B_2C$ derived from our FPLO-LDA calculations depicted in Fig. 6. yields that the major part of the electrons (dominating the normal state transport properties) exhibit large v_F-values (red and green) differing up to factor of 6 from the slow electrons (blue). This is in accord with our phenomenological analysis [17] of $H_{c2}(T)$ requiring a factor of at least five in their relative magnitudes. The electrons with low v_F's are found near nested parts of the Fermi surface. Quite interestingly, vectors closely related to the nesting vector $q \approx 0.6\ 2\pi/a$, i.e. connecting the neighboring blue parts (compare Fig. 6) of Ref. 12), seem to occur also in low-frequency phonons exhibiting anomalously strong softening entering the superconducting state [34] as well as in the incommensurate a-axis modulated magnetic structure which partially suppresses the superconductivity in low magnetic fields [35].

5. Disorder and doping

The generic structure of RTBC(N) compounds exhibits, in spite of the different dimensionality discussed above, a remarkable similarity to the cuprate superconductors with respect to the specific role of the two subsystems comprising these compounds. They consist of the NiB-network which is the stage for the superconducting spectacle and the RC-subsystem playing the role of a charge reservoir. Its main function consists to adjust the Fermi level close to the sharp peak in the DOS (see Fig. 3). For the superconductivity itself it plays a minor active role because of its small admixture

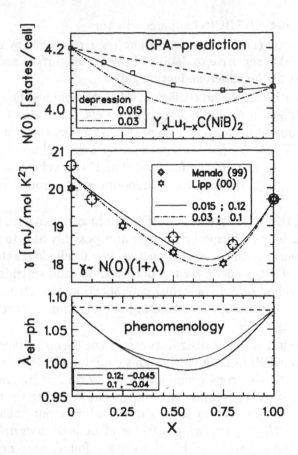

Figure 9. Composition dependence of the DOS at Fermi level $N(0)$ calculated by LDA-CPA (upper panel), the experimental value for the Sommerfeld constant γ (middle panel) and the phenomenologically extracted value of the electron-phonon coupling constant λ (lower panel) for the mixed quasi-quaternary compound $Y_xLu_{1-x}Ni_2B_2C$. The squares in the DOS are self-consistently calculated using the experimental structural data. The dashed-dotted line is an ad-hoc estimate taking into account possible local lattice distortions. The specific heat data are taken from Refs. [37, 38]. The value of the averaged electron phonon coupling constant λ has been derived combining the results for $N(0)$ and γ. The label numbers correspond to the disorder and asymmetry parameter, respectively, analyzed in terms of Eq. 3.

to the DOS peak near E_F. Therefore, it can be used to study the effect of disorder in a wide interval. This is of general interest, because these superconductors show anomalous disorder effects. So, the magnitude of $H_{c2}(0)$ and the curvature near T_c of $H_{c2}(T)$ are strongly reduced by disorder (see also Fig. 10).

The reduction of the multi-band effects discussed above can be studied by replacing partially some of the constituent atoms by entities with sim-

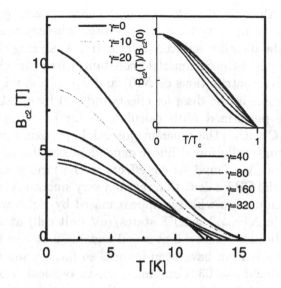

Figure 10. Upper critical field B_{c2} vs. temperature T within the two-band model for various degrees of disorder given by the impurity scattering rate γ (in cm^{-1}). Inset: the same in relative units.

ilar chemical and physical properties. Modest effects can be expected for chemically and magnetically equivalent substitutions R $\to$ R' in the RC(N)-layer(s). For this purpose the crystal structure, T_c, $H_{c2}(T)$, and the electronic specific heat $c_{p,el} = \gamma T$ of the mixed (polycrystalline) Y$_x$Lu$_{1-x}$Ni$_2$B$_2$C system have been studied recently [36, 37, 38]. Since the measured lattice constants a and c vary nearly linearly between the corresponding pure limits, a maximum of T_c' might be expected for $x=0.5$ according to the "universal" curve $T_c = T_c(d_{Ni-Ni})$ proposed by Lai *et al.* [39], where d_{Ni-Ni} denotes the Ni-Ni distance. Instead of the expected maximum with $T_c \approx$ 17 K a sample dependent dip in between $x=0.5$ to 0.7 with $T_c \approx$ 14.5 to 14.9 K has been found.

Several other physical quantities such as the the Sommerfeld constant γ (see Fig. 4), the curvature exponentα of H_{c2} exhibit similar dips as a function of the composition [40]. For any physical quantity A from a formal point of view it is convenient to analyze the deviations from the linear interpolation between the border compounds (the so called virtual crystal approximation) by [41] the Fourier components d_m

$$A(x) = (A(0)x + (1-x)A(1)) \left(1 - \sum_m d_m \sin(m\pi x)\right), \qquad (3)$$

where the strength of the disorder is mainly characterized by the odd components $d_{disorder} = d_1 + d_3 + \ldots$ characterizes whereas the even components describe the disorder induced asymmetry. Analyzing the empirically found dip behavior of the Sommerfeld constant γ (see Eq. 1), the problem about the relative contributions of $N(0)$ and λ_{el-ph} should be addressed. To get some insight in the disorder effects induced by substitutions at T- and R-sites, we performed $N(0)$ calculations for $Lu(Ni_{1-x}Co_xB)_2C$ and $Y_xLu_{1-x}(NiB)_2C$ within the coherent potential approximation (CPA) [42] using the experimentally found linear dependence of the lattice constants upon composition. We found strong effects in the former case whereas in the second case the CPA calculation reveal a very smooth $N(x)$ dependence with a shallow minimum which is approximated by Eq. 3 with $d_N \equiv d_1 = 0.015$ resulting in $N(x_{min}) = 4.058$ states/(eV unit cell) at $x_m = 0.73$ compared with the border values 4.2 at $x = 0$ and 4.072 at $x = 1$. Since in our approach local relaxation have been ignored so far, we admit also a twice as large total value $d_N = 0.03$. Combining our theoretical $N(x)$-dependence with the experimental γ-data we extract the averaged el-ph coupling constant λ. In both cases one arrives at much larger disorder suppressions of λ: $d_\lambda \sim 0.1$ (see Fig. 9, lower panel). Such a disorder suppression of λ is supported also, by the composition dependence of the specific heat jump at T_c [38] which shows a clear disorder induced suppression, too. In particular, even a shallow dip occurs near $x = 0.4$. From a microscopic point of view this should be attributed to a hardening of phonon modes. We speculate that the disorder affects the soft anomalous phonon near 7 meV probably closely related to the Fermi surface nesting discussed above.

The reduction of the positive curvature of H_{c2} points to a direct disorder effect at the level of the superconductivity. It can be described within the framework of the two-band Eliashberg theory admitting significantly increasing effective impurity scattering rates (see Fig. 10). Further experimental support for the presence of such direct disorder effects stems from the electronic specific heat data in the mixed state showing a nonlinear field dependence $c_p \propto T(H/H_{c2})^{1-\beta}$ which vanishes in the dirty limit [38]. In this context recent experimental results of Nohara et al. [43, 44] for $Y(Pt_xNi_{1-x})_2C$ single crystals are of considerable interest. According to this data both curvatures (α and β vanish already at $x \sim 0.1$ which clearly indicates the enhanced disorder sensitivity induced by even by isoelectronic substitutions at the T-site compared with our data for the R-site. There at $x = 0.2$ the dirty limit is reached in contrast to the R substitutional case (provided rather similar ionic radii of the substituted isoelectronic ions as in the Y-Lu case) where the system remains in the quasi-clean limit for any composition.

6. LuNi$_2$B$_2$C and YNi$_2$B$_2$C: unconventional pairing?

For these two compounds there are several properties which taken together might be interpreted also as hints for unconventional (d-wave or p-wave) superconductivity:

(i) a nonlinear $H^{1-\beta}$-dependence of the electronic specific heat c_{el} in the superconducting state instead of the standard linear dependence,

(ii) the very weak damping of the dHvA oscillations in the superconducting state related to the superconducting gap has been interpreted as strong evidence for a very small or vanishing gap at parts of the Fermi surface [45],

(iii) a non-exponential *non-universal* power-like T-dependence of the electronic specific heat in the superconducting phase $C_{el} \propto T^\beta$, $\beta \sim 3$ at low temperatures ($\beta \approx 2.75$ for YNi$_2$B$_2$C [46] and $\beta > 3$ for LuNi$_2$B$_2$C and LaPt$_2$B$_2$C). Strictly speaking, a pure one-band d-wave superconductor shows *quadratic* temperature dependence; a cubic dependence is expected for an order parameter with point-like nodes.

(iv) the anisotropy of $H_{c2}(T)$ within the basal plane of LuNi$_2$B$_2$C [15] has been ascribed to d-wave pairing [20],

(v) a quadratic flux line lattice at high fields has been observed not only for magnetic RTBC but also for the non-magnetic title compounds[19],

(vi) deviations from the Korringa behavior of the nuclear spin lattice relaxation rate $1/T_1T$=const have been ascribed to the presence of antiferromagnetic spin-fluctuations on the Ni site.

However, it should be noted that several of these unusual properties (i,ii,iii,iv,v) have been observed also for some more or less traditional superconductors such as V$_3$Si and NbSe$_2$ as well as for heavy fermion superconductors. The possibility of a square flux line lattice for the case of a two-band superconductor has been mentioned by Moskalenko in 1966 [47]. At present it is also unclear to what extent the observed anisotropies and unusual temperature dependences could be described alternatively by a full anisotropic (multi-band) extended s-wave theory. Phase-sensitive experiments [48] and/or the observation of the Andreev bound state near appropriate surfaces[49] must be awaited to confirm or disprove the d-wave scenario.

Most importantly, some of the unusual T-dependences ascribed sometimes to the presence of antiferromagnetic spin fluctuations and "non-Fermi liquid" effects might be caused by the rather strong energy dependence of the DOS near the Fermi level[13]. Finally, the observation of a weak Hebel-Slichter peak in the ^{13}C NMR data for the spin-lattice relaxation time T_1 ($1/T_1T$) [50] must be mentioned. This points to the presence of at least one s-wave pairing component in the multi-band order-parameter including also the C 2s-electrons. However, a strong electron-electron interaction is suggested by the twice as large so-called enhancement factor $\alpha_c = 0.6$ compared with those of conventional "s-band" metals Li and Ag.

In this context it should be mentioned that according to our FPLO calculations the C $2s$ contribution to $N(0)$ is very small ($\leq 1\%$).

7. A possible classification

Bearing the general similarities in the electronic and lattice structures of most RTB(N)-compounds in mind, we propose that essentially the same pairing mechanism should be responsible for all representatives. Since a substantial boron isotope effect has been measured for YNi_2B_2C and due to the absence of conclusive evidence for antiferromagnetic spin-fluctuations [13], the assumption of a dominating el-ph mechanism seems to be a quite natural one. In this case, the possible change of the order parameter symmetry (within a part of the multicomponent(band) order parameter) must be ascribed to the Coulomb repulsion. According to [51], increasing for instance the Coulomb pseudopotential μ^*, one approaches a critical ratio μ^*/λ: below and above of which s-wave or d-wave superconductivity occurs, respectively. Due to the minimal extent of the $3d$ Wannier functions, the largest value of the on-site Coulomb interaction U can be expected in the case of the single layer Ni-series. The presence of three intermediate LaN-layers in the triple-compound $(LaN_{1-\delta})_3(NiB)_2$ might contribute to an additional screening of the on-site Coulomb interaction less pronounced in the single-layer compounds. The disorder due to a significant amount of (charged !) N vacancies in the triple layer lanthanum boronitride $\delta \sim 0.05$ as well as in T-site substituted single- and two-layer borocarbides might kill any d-wave or highly anisotropic s-wave component present in the clean parent compounds. In this respect the observation of an disorder induced gap in the electronic specific heat of $YC(Pt_xNi_{1-x}B)_2$ in the superconducting state by Nohara et al. [44] is quite remarkable. All mentioned above "anomalous" properties are reduced or removed with increasing disorder.

Similarly, the presence of two Lu(Y)C-layers in Lu(Y)NiBC might be helpful to establish superconductivity at least at low temperatures absent in the single-layer compounds $LaNi_2B_2C$, YCo_2B_2C despite their larger $N(0)$-values (see Fig. 3).

In general the understanding of the absence of superconductivity in some of these compounds is of equal importance as the understanding of the representatives with the highest T_c-values. For the $4d$ and $5d$-members of the RTBC-family the DOS near the Fermi level $N(0)$ and especially the partial DOS with d-character $N_d(0)$ are reduced resulting possibly in a reduced Coulomb interaction, too.

The rich variety of superconducting compounds in the novel class of

RTBC(N) compounds under consideration and the growing knowledge of their electronic structure offers the fascinating possibility to work out a semi-microscopic quantitative description of these superconductors instead often theories of superconductivity with many adjustable parameters and quantities. This will be helpful to clarify obvious present differences and similarities with other exotic superconductors.

8. Acknowledgment

It is a pleasure to thank K.-H. Müller, G. Fuchs, J. Freudenberger, A. Kreyssig, C. Grazioli, O. Dolgov, and numerous other colleagues for fruitful discussions, collaboration and providing us with experimental data prior to publication. Without the synergetic interaction with them and their stimulating interest the present work would not be possible. We would like to thank also the Deutsche Forschungsgemeinschaft and the Sonderforschungsbereich 463 " Seltenerd-Übergangsmetallverbindungen: Struktur, Magnetismus und Transport" for financial support. Further support by the Graduiertenkolleg "Struktur und Korrelationseffekte in Festkörpern" der TU Dresden is acknowledged (H.R.).

References

1. Mazumdar, C., et al. (1993) *Solid State Commun.* **87**, 413; Nagarajan, R., et al., (1994) *Phys.Rev. Lett.* **72**, 274; Cava, R., et al., (1994) *Nature* **376** 146; (1994) *ibid.*, 252.
2. Pickett, W. and Singh, D. (1994) *Phys. Rev. Lett.* **72**, 3702.
3. Mattheiss, L.F. (1994) *Phys. Rev.* B **49**, 13 279.
4. Mattheiss, L.F., et al. (1994) *Solid State Commun.* **91** 587.
5. Singh, D. (1994) *Phys. Rev.* B **50**, 6486.
6. Pickett, W.E. and Singh, D. (1995) *J. of Supercond.* **8**, 425.
7. Singh, D. and Pickett, W.E. (1995) *Phys. Rev.* B **51** 8668.
8. Singh, D. (1995) *Solid State Commun.* **98**, 899.
9. Mattheiss, L.F. (1995) *Solid State Commun.* **94** 741.
10. Coehoorn, R. (1994) *Physica C* **228**, 331.
11. Rhee, R.Y. et al., (1995) *Phys. Rev.* B **51**, 15 585.
12. Kim, H., et al., (1995) *Phys. Rev.* B **52**, 4592.
13. Suh, B.J., et al. (1996) *Phys. Rev.* B **53**, R6022.
14. Ravindran, P., et al., (1998) *Phys. Rev.* B **58** 3381.
15. Metlushko, V. et al., (1997) *Phys. Rev. Lett.* **79**, 1738.
16. Xu, M. et al., (1994) *Physica C* **235**, 2533.
17. Shulga, S.V. et al. (1998), *Phys. Rev. Lett.* **80**, 1730.
18. Jacobs, T., et al., (1995) *Phys. Rev.* B **52**, R7022.
19. de Wilde, Y. et al. (1997) *Phys. Rev. Lett.* **78**, 4273; Eskildsen, R.,et al., (1997) *ibid.* **79** 487; Yethiraj, M., et al., (1997) *ibid.* **78**, 4849.
20. Wang, G. and K. Maki, K., (1998) *Phys. Rev.* B **58**, 6493.
21. Nohara, M. et al. (1997) *J. Phys. Soc. Jpn.* **66** 1888; Takagi, R., et al. (1997) *Physica B* **237-238**, 292.
22. Hilscher, G. and Michor, H. (1998) in "Studies of High Temperature Superconductors" (Ed. A.V. Narlikar), Nova Science Publishers, N.Y. v. **26/27**.

184

23. Gangopadhyay, A.K. and Schilling, J.S. (1996) *Phys. Rev. B* **54**, 10 107.
24. Böske, T. *et al.* (1996) Solid State Commun. **98**, 813.
25. Koepernik, K. and Eschrig, H. (1999) *Phys. Rev. B,* **59** 1743.
26. Lips, H., *et al.*, (1999) *Phys. Rev. B* **60**, 11144, Lips, H., *et al.*, *these Proceedings.*
27. Narozhnyi, V.N. *et al.* (1999) *Phys. Rev. B* **59**, 14762.
28. Drechsler, S.-L., *et al.*, (1999) *Physica C* **317-318**, 117.
29. Dugdale, S.B., *et al.*, Phys. Rev. Lett., **83**, (1999) 4824.
30. Ku, H.C., *et al.* (1994) *Phys. Rev. B* **50**, 351.
31. J. Freudenberger, "Paarbrechung in Seltenerd-Übergangsmetall-Borkarbiden" Dissertation, TU Dresden, (2000)
32. Singh, D. and Pickett, W.E. (1995) *Nature* **374**, 682.
33. Nguyen, L.H., *et al.*, (1996) *J. Low Temp. Phys.* **105**, 1653.
34. Stassis, C. *et al.*, (1999) *Physica C* **317-318**, 127, Bullock, M. *et al.* (1998) *Phys. Rev. B* **57**, 7916.
35. Müller, K.-H. *et al.*, (1997) *J. Appl. Phys.* **81** 4240.
36. Freudenberger, J. *et al.*, (1998) *Physica C* **306**, 1.
37. Manalo, S., (1999) *et al.*, *Preprint cond.-mat/9911305.*
38. Lipp, D., *et al.*, (2001) to be published.
39. Lai, C., *et al.*, (1995) *Phys. Rev. B* **51**, 420.
40. Drechsler, S.L., *et al.*, (2000) Physica C (in press).
41. Drechsler, S.-L., *et al.* (1999) *J. Low Temp. Phys.* **117**, 1617 .
42. Koepernik, K. *et al.*, (1997) *Phys. Rev. B* **55**, 5717.
43. Nohara, M., Isshiki, M., Sakai, F., and H. Takagi, H. (1999) *J. Phys. Soc. Jpn.* **68** 1078.
44. M. Nohara, H. Suzuki, N. Mangkorntong, and H. Takagi, Proceedings of M^{2}S-HTSC, Houston 20-25.2. 2000, Physica C, in press (2000).
45. T. Terashima *et al.*, Phys. Rev. B **56** (1997) 5120.
46. N.M. Hong *et al.*, Physica C **227** (1994) 85.
47. V.A. Moskalenko, ZhETF, **51**, (1966) 1163.
48. D.A. Wollman *et al.* Phys. Rev. Lett. **71** (1993) 2134.
49. Hu, C.-R. (1994) *Phys. Rev. Lett.* **72**, 1526.
50. Saito, T., *et al.*, (1998) *J. Magn. and Magnetic Materials,* **177-181**, 557.
51. Varelogiannis, G. (1998) *Phys. Rev. B* **57**, 13 743.

BCS ANALYSIS OF THE PHYSICAL PROPERTIES OF NONMAGNETIC BOROCARBIDES

What can we learn from the experimental data?

S. V. SHULGA

Institute of Spectroscopy, RAS, Troitsk, 142190, Russia

AND

S.-L. DRECHSLER

Institut für Festkörper- und Werkstofforschung Dresden e.V., Postfach 270116, D-01171 Dresden, Germany

Abstract. Magnetic properties of borocarbides are considered in the frameworks of the standard isotropic single band as well as within the effective two-band models. The unusual impurity dependence of the upper critical magnetic field $H_{c2}(T)$ is highlighted.

1. Introduction

Four years after the discovery [1] of superconductivity in rare earth transition metal borocarbides (RTMBC) those compounds have generated large interest due to their relatively high transition temperatures $T_c \sim 15$ to 23 K and due to the relationship between the mechanisms of superconductivity in RTMBC, in cuprates, and in ordinary transition metals and alloys. Another highlight is the coexistence of magnetism and superconductivity in some of those compounds which contain magnetic rare earth elements [2, 3]. A study of the nonmagnetic compounds such as RNi_2B_2C, with R=Lu,Y,Th,Sc [4], is a prerequisite for the understanding of their magnetic counterparts. Experimental data for $LuNi_2B_2C$ [5] demonstrate beside a maximal positive curvature (PC) of $H_{c2}(T)$ near T_c, observed also for YNi_2B_2C [2, 6, 7], a weak T-dependent anisotropy within the tetragonal basal plane and a T-independent out-of-plane anisotropy of the upper critical field H_{c2}. Both anisotropies have been described [5] in terms of nonlocal corrections to the Ginzburg-Landau (GL) equations. In that phenomenological picture the positive curvature of H_{c2} ($\vec{H}\|$ to the tetragonal c-axis)

185

S.-L. Drechsler and T. Mishonov (eds.), High-T$_c$ Superconductors and Related Materials, 185–192.
© *2001 Kluwer Academic Publishers. Printed in the Netherlands.*

is ascribed, almost purely, to the basal plane anisotropy. In that nonlocal GL-theory the upper critical field becomes $H_{c2} = h_1(t + h_2t^2) + ... +$, where $t = (1 - T/T_c)$. With $h_2 = 0.135$ a small positive curvature occurs, even if no basal-plane anisotropy exists. A fit to the anisotropic in-plane field data of $LuNi_2B_2C$ yields $h_2 = 1.736$. Thus the positive curvature is strongly enhanced by a factor of ≈ 12.87. However, it should be noted that the reported anisotropy of H_{c2} for YNi_2B_2C is significantly smaller than that for $LuNi_2B_2C$ [5, 7, 8] whereas its positive curvature is comparable or even larger.

A positive curvature was observed in underdoped cuprates [9], ternary compounds [10], layered materials [11] as well as in elemental niobium [12]. The standard GL theory which is valid within quite general assumptions, predicts a *negative* curvature below T_c. So, to describe the positive curvature of H_{c2} in borocarbides one has to go beyond the standard GL-theory. The bipolaronic scenario [13] as well as other non-Fermi liquid (FL) models can be disregarded due to observation of de Haas-van Alphen (dHvA) oscillations in these compounds [14]. In frame of the FL approach we have only two possibilities to generalize the standard GL-theory. The first scenario is a strong temperature or magnetic field dependence of material parameters just below T_c. Recently a dramatic phonon softening was observed [15] in $LuNi_2B_2C$ and YNi_2B_2C in low fields, but at $H = H_{c2}$ the phenomenon disappears. The remaining second possibility is the presence of several order parameters. The model system can be a superconductor with charge- or spin-density waves [16], superconducting-normal-metal multilayers [11], a strongly coupled two-band system [17] or even a conventional metal with large anisotropy of the Fermi velocity [12]. Hence, the standard Isotropic Single Band (ISB) model is not suitable for borocarbides and the physical properties of these compounds analysed in frame of this model conflict each other. The revealed inconsistency of the model contains important information which can be helpful to choose a more realistic non-ISB model.

2. Isotropic single band model

The standard isotropic single band (ISB) model [18] is the most well-developed part of the modern theory of superconductivity. It describes *quantitatively* the renormalization of the physical properties of metals due to the electron-phonon interaction. The input (material) quantities are the density of states at E_F, $N(0)$, the Fermi velocity v_F, the impurity scattering rate γ_{imp}, the paramagnetic impurity scattering rate γ_m, the Coulomb pseudopotential μ^*, and the electron-phonon spectral function $\alpha^2F(\omega)$. Note, that the T-dependencies of physical properties depend mainly on the first momenta of the spectral function - the coupling constant $\lambda = 2\int d\omega\alpha^2F(\omega)/\omega$

and the average boson energy $\omega_{ln} = \exp[2\int_0^\infty d\Omega \ \ln(\Omega)\alpha^2 F(\Omega)/\Omega\lambda_0]$ [18, 19].

If the shape of $\alpha^2 F(\Omega)$ can not be recovered from tunneling or optical data, the simplified list of material parameters ($N(0)$, v_F, γ_{imp}, γ_m, λ, ω_{ln}) can be determined from few experimental data: the normal state low-T electronic specific heat $\gamma_S \propto N(0)(1+\lambda)$, the plasma frequency $\omega_{pl} \propto N(0)v_F^2$ inferred from the optical conductivity, $H_{c2}(0) \propto (1+\lambda)^{2.4}/v_F^2$, T_c, as well as the normal state low-T dc resistivity $\rho(0) \approx \rho(T_c)$ which similarly to the Dingle temperature T_D, gives a direct measure of the sample purity.

Here we present and analyse theoretically the data of $H_{c2}(T)$ [17] in a broad interval 0.3K$\leq T \leq T_c$ for high purity LuNi$_2$B$_2$C and YNi$_2$B$_2$C single crystals. At first we adopt for the Coulomb pseudopotential μ^*=0.1 and $\hbar\omega_c$=600 meV for the energy cut-off in Eliashberg equations. The total el-ph coupling constant, $\lambda = 2\int d\omega\alpha^2 F(\omega)/\omega$, can be estimated from the boron isotope effect for T_c (its exponent $\alpha_B \approx 0.2$ [20]) and the measured phonon spectrum [21]. We first consider LuNi$_2$B$_2$C. To find a lower bound for λ, we accounted for only the high-energy carbon phonons centered at 50 meV and the special boron branch at 100 meV. Fitting the experimental α_B and T_c values, we obtained the partial coupling constants λ_{100}=0.31, λ_{50}=0.22, and $\lambda=\lambda_{100} + \lambda_{50}$=0.53, where the subscripts denote the corresponding phonon energies in meV. An upper bound of λ=0.77 has been found using the Lu phonons centered near 9 meV (λ_9=0.34) and the same B band (λ_{100}=0.43) as in the case before. In the following a wide averaged spectrum with λ=0.65 (λ_{100}=0.37, λ_{50}=0.12, λ_9=0.16) will be used which reproduces the experimental values of α_B and T_c. $N(0)$=11.8 mJ/mol k_B^2K^2 has been estimated from the experimental value [6] of $\gamma_S = 2\pi^2 k_B^2(1+\lambda)N(0)/3$= 19.5 mJ/mol K^2. The value of v_F=2.76$\cdot$10^7 cm/s follows from the experimental value [22] of the plasma frequency $\hbar\omega_{pl} = \sqrt{4\pi e^2 v_F^2 N(0)/3}$ =4.0 eV. The analogous values for YNi$_2$B$_2$C are λ=0.637, $N(0)$=11.1 mJ/mol k_B^2K^2, v_F= 3$\cdot$10^7 cm/s and $H_{c2}(0)$=2T, where the data of Refs. [6] and [23] have been used. Since there is no experimental evidence [24, 25] for the presence of magnetic impurities in high quality samples, we neglect the magnetic scattering rate $\gamma_{imp,m}$. For the quantification of the non-magnetic counterpart $\gamma_{imp} \approx 2\pi T_D$, the Dingle temperatures, T_D, measured by the de Haas-van Alphen (dHvA) effect are very suitable [14]. The experimental values T_D=2.8 K and 4K reveal γ_{imp}=18K and 25K for our YNi$_2$B$_2$C and LuNi$_2$B$_2$C single crystals, respectively, indicating that the clean limit is reached since $\gamma_{imp} \leq 2\Delta_0 \approx 51K$ holds for both samples, where $2\Delta_0$ denotes the smaller of the two gaps. Hence, the scattering by impurities can be neglected setting γ_{imp}=0.

We solved the equations for $H_{c2}(T)$ [26] with these parameter sets for two types of spectral densities $\alpha^2 F(\omega)$: (i) a wide spectrum and (ii) a single

Einstein mode peaked at $\hbar\omega_E=42.4$ meV chosen to yield the experimental $T_c=16.5$ K for LuNi$_2$B$_2$C using the same value of $\lambda=0.65$ as in the first case. The results are shown in Fig. 1. Comparing the LuNi$_2$B$_2$C data with the ISB curves one clearly realizes strong deviations. In particular, there is a discrepancy of about 3 between experimental and ISB model predicted values of $H_{c2}(0)$. For YNi$_2$B$_2$C the discrepancy reaches even a factor of 5.

An approximate formula for the low temperature value of the orbital upper critical magnetic field $H_{c2}(0)$ was derived by Marsiglio and Carbotte [19, 18] which includes the first correction for strong-coupling characterized by the single strong-coupling parameter (T_c/ω_{ln}). For our goal it is more convenient to obtain a more simple, factorable, expression. The clean limit orbital field $H_{c2}(T = 0, 1 + \lambda)$ was calculated using several spectral functions. We found that in the clean limit the results obtained can be approximated by the following simple formula

$$H_{c2}(T = 0)[\text{T}] \approx 0.02 T_c^2[\text{K}](1 + \lambda)^{2.4}/v_F^2[10^7 \text{cm/s}]. \quad (1)$$

As a result, in the clean limit the following combination of the basic quantities depend on the coupling constant λ, only. It gives the simple ISB criterium

$$\frac{\omega_{pl}^2\ [\text{eV}^2]\ H_{c2}(0)[\text{T}]}{\gamma_S[\text{mJ/mol} \cdot \text{K}^2]T_c^2[\text{K}]} \approx \frac{(1 + \lambda)^{1.4}}{3.6}. \quad (2)$$

For YNi$_2$B$_2$C with $\hbar\omega_{pl}=4.2$ eV, $H_{c2}=10.6$ T, $\gamma_S=17.3$ mJ/mol K^2, Eq. 2 gives a unrealistic value of the coupling constant $\lambda =4.6$, which is a direct manifestation of the inapplicability of the ISB model to borocarbides. At the same time, if one is able to estimate the value of λ, the departure of the quantity

$$V = \frac{3.6\omega_{pl}^2\ [\text{eV}^2]\ H_{c2}(0)[\text{T}]}{\gamma_S[\text{mJ/mol} \cdot \text{K}^2]T_c[\text{K}](1 + \lambda)^{1.4}} \quad (3)$$

from unity is a measure of "unISBility". Assuming $\lambda = 1.2$ for YNi$_2$B$_2$C one gets the large value $V=4.2$.

The second useful relation is the following: the T-dependence of $H_{c2}(\text{T})$ of the ISB system with ordinary magnetic impurities exhibits approximately the same shape as the $H_{c2}(\text{T})$ of the ISB clean limit system which have the same (suppressed) value of T_c due to reduction of the coupling constant. Fig. 1b presents the T dependencies of the clean limit upper critical magnetic field $H_{c2}(T)$ calculated for the Einstein coupling function $\Omega=100\ cm^{-1}$ and $\lambda=1.3, 1.0, 0.7$. They have to be compared with the curves calculated for the same Einstein spectral function $\Omega=100\ cm^{-1}$, $\lambda=1.3$ containing magnetic impurities with $\gamma_m=5, 11\ cm^{-1}$. The discussed above useful relation allows to simplify the formula (1) for the case of the strong pear-breaking, since it

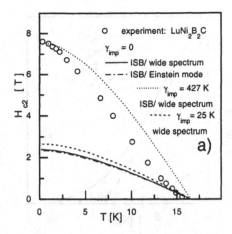
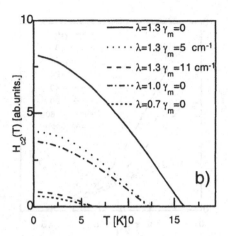

Figure 1. Panel (a) shows the experimental data for $H_{c2}(Y)$ of LuNi$_2$B$_2$C ($\vec{H}\|$ the c-axis) compared with ISB model curves. Panel (b) presents the temperature dependencies of the clean limit upper critical magnetic field $H_{c2}(T)$ calculated for the Einstein coupling function $\Omega=100$ cm^{-1} and $\lambda=1.3$, 1.0, 0.7 (solid lines) in comparison with the $H_{c2}(T)$ of the system containing magnetic impurities with $\gamma_m=5$, 11 cm^{-1} (dashed and dotted lines, respectively).

is equivalent to the case of the weak coupling clean limit ISB system with $\lambda=0.4$-0.5. Substituting $\lambda=0.45$ into (1) one gets

$$H_{c2,m}(T=0)[\mathrm{T}] \approx 0.05 T_c^2[\mathrm{K}]/v_F^2[10^7 \mathrm{cm/s}]. \tag{4}$$

Similarly, in the case of strong pear-breaking Eq. (3) takes the form

$$V_m = \frac{6\omega_{pl}^2\ [\mathrm{eV}^2]\ H_{c2}(0)[\mathrm{T}](1+\lambda)}{\gamma_S[\mathrm{mJ/mol \cdot K}^2]T_c[\mathrm{K}]} \tag{5}$$

For ErNi$_{1.93}$Co$_{0.07}$B$_2$C [3] $V_m=0.$(A possible scenario for the suppression of the Fermi velocity anisotropy will be discussed below in terms of an effective two-band model.

3. Effective two-band model

The effective N-band model for an anisotropic system is obtained by dividing the Fermi surface into N parts and approximating the k-dependent quantities in each part by their mean values [27]. The number of material parameters of the N-band model equals $3N^2+6N$. With respect to the coupling, the Fermi surface of borocarbides can be roughly divided into 3 parts:

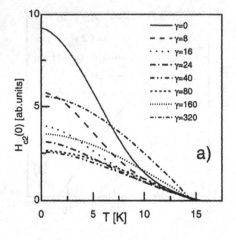
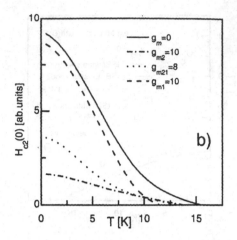

Figure 2. Panel (a) shows the temperature dependencies of the upper critical magnetic field $H_{c2}(T)$ of our two-band model adopting $\lambda_2=1.17$, $\lambda_1=1$, $\lambda_{21}=0.8$, $\lambda_{12}=0.1$, and $v_{F1}/v_{F2}=4.5$ The lines from top to bottom correspond to the impurity scattering rates $\gamma = \gamma_1 = \gamma_{21} = 0$, 10, 20, 40, 80, 160, 320 cm^{-1}. Panel (b) illustrates the outcome of the partial magnetic impurity pair-breaking in the same model system. The case when the magnetic scattering acts in the first band, only (dashed line), in the second band, only (dash-dotted line), or as an inter-band scatter (dotted line). The clean limit data are denoted by the solid lines.

a strongly coupled (nested) part, a moderately coupled (main) one, and a weakly coupled (F_β - dHvA cross-section) disjoined Fermi sheets. Fortunately, the isolated F_β group does not contribute essentially to $H_{c2}(T)$. Hence, we may restrict ourselves to $N=2$.

In this section we consider a completely anisotropic two band model with $\lambda_2=1.17$, $\lambda_1=1.$, $\lambda_{21}=0.8$, and $\lambda_{12}=0.1$ and $v_{F1}/v_{F2} >1$. The adopted set of parameters was chosen to illustrate the possible role of a small subgroup of quasiparticles ($N_2(0)/N_1(0)=8$) from nested parts of the Fermi surface with a large coupling constant λ_2, and and small v_F. A special feature of this set is a significant interband coupling (described by the constant λ_{21}) acting between the small-v_F bands and the large-v_F band. This case differs from the situation considered in Ref. [27]. There, the stronger coupling is in the large-v_F band and the curvature of $H_{c2}(T)$ near T_c is *negative*. A positive curvature would only appear at intermediate T if the interband coupling and the impurity scattering are both weak. In contrast, in this region $H_{c2}(T)$ shows almost no curvature in our model [17]. In other words, the result of Refs. [27] can be understood as an average over two weakly coupled superconductors, the first with a high $H_{c2}(0)$ but with a low T_c and the second one with a small $H_{c2}(0)$ but high T_c. In our case the isolated small-v_F subsystem would have high values of λ, $H_{c2}(0)$, and T_c. The values of $H_{c2}(0)$ and T_c of the coupled system are reduced by the second large-v_F subsystem with weak interaction. In this case, the positive curvature of the resulting $H_{c2}(T)$ near T_c becomes a direct manifestation of that interband coupling. In our two-band model the upward curvature of $H_{c2}(T)$, $H_{c2}(0)$, as well as T_c are suppressed by growing impurity content and the positive curvature vanishes upon reaching the dirty limit (see Fig. 2a). The in-band impurity scattering acts traditionally and increases $H_{c2}(T)$, but at low γ_{imp} the interband merge by impurities dominates showing opposite behaviour.

In contrast, in-band as well as interband magnetic scattering suppress $H_{c2}(0)$ and T_c (see Fig. 2b). One can see, that the most pronounced decrease of $H_{c2}(0)$ takes place for pairbreaking in the strongly coupled (second) band, while the transition temperature exhibits a minor reduction in this case. In contrast, the paramagnetic impurity scattering acting on the first (weakly coupled) band affects strongly T_c and hardly $H_{c2}(0)$. The interband scattering affects in a moderate way both quantities. Thus we may suggest that the dramatic decrease of $H_{c2}(0)$ in ErNi$_{1.93}$Co$_{0.07}$B$_2$C [3] is predominantly due to magnetic scattering of quasiparticles from the nested parts of the Fermi surface.

4. Acknowledgements

We express our deep gratitude to O. V. Dolgov, E. G. Maksimov, M. Kulić, K.-H. Müller, G. Fuchs, J. Freudenberger, L. Schultz, H. Eschrig, M.S. Golden, H. von Lips, J. Fink, V.N. Narozhnyi, H. Rosner, A. Gladun, D. Lipp, A. Kreyssig, M. Loewenhaupt, K. Koepernik, K. Winzer, K. Krug, and M. Heinecke for numerous discussions, fruitful collaboration, and generous support of this work. This work was supported by the SFB 463.

References

1. Nagarajan, R., Mazumdar, Ch., Hossain, Z., Dhar, S.K., Gopalakrishnan, K.V., Gupta, L.C., Godart, C. , Padalia, B.D., and Vijayaraghavan R. (1994) Bulk superconductivity at an elevated temperature ($T_c \approx 12$ K) in a nickel containing alloy system Y-Ni-B-C, *Phys. Rev. Lett.* **72**, 274-277.
2. Eversmann, K., Handstein, A., Fuchs, G., Cao, L., and Müller, K.-H. (1996) Superconductivity and magnetism in Ho$_x$Y$_{1-x}$Ni$_2$B$_2$C, *Physica* **C266**, 27-32.
3. Schmidt, H., and Braun, H.R. (1997) Superconductivity, magnetism, and their coexistence in R(Ni$_{1-x}$Co$_x$)$_2$B$_2$C (R = Lu, Tm, Er, Ho, Dy), *Phys. Rev.* **55**, 8497-8505.
4. Ku, H.C., Lai, C.C., You, Y.B., Shieh, J.H., and Guan, W.Y. (1995) Superconductivity at 15 K in the metastable ScNi$_2$B$_2$C compound, *Phys. Rev.* **B50**, 351-353.
5. Metlushko, V., Welp, U., Koshelev, A., Aranson, I., Crabtree, G.W., and Canfield, P.C. (1997) Anisotropic upper critical field of LuNi$_2$B$_2$C, *Phys. Rev. Lett.* **79**, 1738-1741.
6. Michor, H., Holubar, T., Dusek, C., and Hilscher G. (1995) Specific-heat analysis of rare-earth transition-metal borocarbides: an estimation of the electron-phonon

coupling strength, *Phys. Rev.* **B52**, 16165-16175.

7. Rathnayaka, K.D.D., Bhatnagar, A.K., Parasiris, B.A., Naugle, D.G., Canfield, P.C., and Cho B.K. (1997). Transport and superconducting properties of RNi_2B_2C (R = Y, Lu) single crystals, *Phys. Rev.* **B55**, 8506-8519.

8. De Wilde, Y., Iavarone, M., Welp, U., Metlushko, V., Koshelev, A.E., Aranson, I., and Crabtree, G.W. (1997) Scanning tunneling microscopy observation of a square Abrikosov lattice in $LuNi_2B_2C$, *Phys. Rev. Lett.* **78**, 4273-4276.

9. Mackenzie, A.P., Julian, S.R., Lonzarich, G.G., Carrington, A., Hughes, S.D., Liu, R.S., and Sinclair, D.S. (1993) Resistive upper critical field of $Tl_2Ba_2CuO_6$ at low temperatures and high magnetic fields, *Phys. Rev. Lett.* **71** , 1238-1241.

10. Maple, M., Hamaker, H., and Woolf, L. (1982) in *Superconductivity in Ternary Compounds II*, Edited by Maple, M., and Fisher, O., Springer-Verlag, N.Y., Berlin.

11. Biagi, K.R., Kogan, V.G., and Clem J.R. (1985) Perpendicular upper critical field of superconducting-normal-metal multilayers, *Phys. Rev.* **B32**, 7165-7172.

12. Weber, H.W., Seidl, E., Laa, C., Schachinger, E., Prohammer, M., Junod, A., and Eckert, D. (1991) Anisotropy effects in superconducting niobium, *Phys. Rev.* **B44**, 7585-7600.

13. Alexandrov, A.S. (1993) Bose-Einstein condensation of charged bosons in a magnetic field, *Phys. Rev.* **B48**, 10571-10574.

14. Goll, G., Heinecke , M., Jansen, A.G.M., Joss, W., Nguyen, L., Steep, E., Winzer, K., and Wyder, P. (1996) de Haas-van Alphen study in the superconducting state of YNi_2B_2C, *Phys. Rev.* **B53**, R8871-R8874.

15. Dervenagas, P., Bullock, M., Zarestky, J., Canfield, P., Cho, B.K., Harmon, B., Goldman, A.I. and Stassis, C. (1995) Soft phonons in superconducting $LuNi_2B_2C$, *Phys. Rev.* **B52**, R9839-R9842.

16. Gabovich, A.M., and Shpigel, A.S. (1988) Upper critical magnetic field of super-conductors with a dielectric gap on the Fermi-surface sections, *Phys. Rev.* **B38**, 297-306.

17. Shulga, S.V., Drechsler, S.-L., Fuchs, G., Müller, K.-H., Winzer K., Heinecke, M., and Krug K. (1998) Upper critical field peculiarities of superconducting YNi_2B_2C and $LuNi_2B_2C$ *Phys. Rev. Lett.* **80**, 1730-1733.

18. Carbotte, J.P. (1990) Properties of boson-exchange superconductors, *Rev. Mod. Phys.* **62**, 1027-1158.

19. Marsiglio, F., and Carbotte, J.P. (1990), Dependence of the second upper critical field on coupling strength, *Phys. Rev.* **B41**, 8765-8771.

20. Lawrie, D.D., and Franck, J.P. (1995) Boron isotope effect in Ni and Pb based borocarbide, *Physica* **C245**, 159-163.

21. Gompf, F., Reichardt, W., Schober, H., Renker, B., and Buchgeister M. (1997) Lattice vibrations and electron-phonon coupling in superconducting quaternary boro-carbides: an inelastic neutron scattering investigation, *Phys. Rev.* **B55**, 9058-9066.

22. Bommeli, F., Degiorgi, L., Wachter, P., Cho, B.K., Canfield, P.C., Chau, R., and Maple, M.B. (1997) Optical Conductivity of the Superconductors LNi_2B_2C (L = Lu and Y), *Phys. Rev. Lett.* **78**, 547-550.

23. Widder, K., Berner, D., Zibold, A., Geserich, H.P., Knupfer, M., Kielwein, M., Buchgeister, M., and Fink J. (1995) Dielectric function of YNi_2B_2C between 10 meV and 50 eV, *Europhys. Lett.* **30**, 55-60.

24. Schmidt, H., Müller, M., and Braun H.F. (1994) Superconductivity in the pseudo-quatternary system $Y(Ni_{1-x}Co_x)_2B_2C$, *Physica* **C235-240**, 779-780.

25. Suh, B.J., Torgeson, D.R., Cho, B.K., Canfield, P.C., Johnston, D.C., Rhee, J.Y., and Harmon B.N. (1996) Absence of antiferromagnetic correlations in YNi_2B_2C, *Phys. Rev.* **B53**, R6022-R6025.

26. Prohammer, M., and Schachinger, E. (1987) Upper critical field of superconducting anisotropic polycrystals, *Phys. Rev.* **B36**, 8353-8359.

27. Langmann, E. (1992) Fermi-surface harmonics in the theory of the upper critical field, *Phys. Rev.* **B46**, 9104-9115.

POLARIZATION-DEPENDENT X-RAY ABSORPTION SPECTROSCOPY ON SINGLE CRYSTAL YNI$_2$B$_2$C SUPERCONDUCTORS

H. VON LIPS, Z. HU, C. GRAZIOLI, G. BEHR, M. KNUPFER,

M.S. GOLDEN, AND J. FINK
Institut für Festkörper- und Werkstofforschung Dresden
D-01171 Dresden, Postfach 270016, Germany

H. ROSNER
Institut für Theoretische Physik, TU Dresden, Germany

U. JAENICKE-ROESSLER
Institut für Kristallographie und Festkörperphysik, TU Dresden, Germany

AND

G. KAINDL
Institut für Experimentalphysik, FU Berlin, Germany

1. INTRODUCTION

The discovery of superconductivity in the intermetallic borocarbides systems RM_2B_2C (R=Y, rare earth; M=Ni, Pd, Pt) [1, 2] has given a new impulse to superconductivity research in intermetallics. These quaternary systems exhibit a variety of interesting physical properties. Non-magnetic borocarbides are either non-superconducting or are 'good' superconductors with transition temperatures T_c of up to ca. 23 K. Even though these T_c's are quite modest in comparison with the cuprates and fullerides, they are high for intermetallic compounds and comparable to the highest T_c-values in the A15 superconductors. Another noteworthy property of the borocarbides is the coexistence of magnetism and superconductivity when R is a magnetic rare earth element (see for instance Müller [3]). Also a peculiar behavior of the upper critical field H_{C2} is found which is explained in more datail in the contribution by Drechsler [4].

193

S.-L. Drechsler and T. Mishonov (eds.), High-T$_c$ Superconductors and Related Materials, 193–198.
© *2001 Kluwer Academic Publishers. Printed in the Netherlands.*

194

Band structure calculations [5] predict the Fermi energy to be situated in a peak maximum of the density of states (DOS). This similarity and the comparable T_c-values to the A15 compounds give rise to the reasonable assumption that for these new systems the connection of T_c with the DOS can be explained within a conventional model of electron-phonon coupling. Moreover according to the calculations, the constitual atoms form a strongly bond network, resulting in an essentially three-dimensional electronic structure with significant contributions of all elements of the system, whereby the $M\,d$ states are dominating the DOS at E_F. With photoemission spectroscopy on polycrytalline YNi_2B_2C samples the dominance of Ni $3d$ states in the valence band could be shown [6].

In this work we present polarization-dependent x-ray absorption spectroscopy (XAS) on high quality single crystals of YNi_2B_2C and compare our results with orbital-resolved LDA bandstructure calculations. The absorption spectra are well described by the calculated DOS, supporting the scenario of a peaked-DOS-governed superconductor.

2. EXPERIMENT

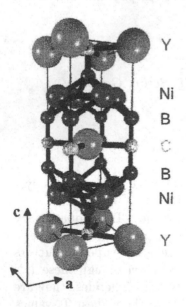

Figure 1. Crystal structure of YNi_2B_2C.

For all RNi_2B_2C, a body-centered tetragonal structure was found [7] (see Fig. 1).

Single crystals of YNi_2B_2C have been obtained by the floating zone method [8], using an inductive heating coil. The starting material is polycrystalline YNi_2B_2C of high purity. Large crystals with a size of 10 mm length and a diameter of about 4 mm can be grown by this method. The crystals were prepared with a surface of [100]-orientation, where the surface plane contains the crystallographic a and c-axes. The lattice parameter of our YNi_2B_2C samples were determined by x-ray diffraction to be $a = 3.53\,\text{Å}$ and $c = 10.57\,\text{Å}$.

The XAS experiments were performed at the synchrotron light source BESSY I in Berlin using the SX700-II monochromator [9] operated by the Freie Universität Berlin. As the samples were [100]-oriented and mounted on a rotable sample holder, the absorption spectra were measured at normal incidence with the polarization vector **E** of the synchrotron radiation set parallel to one of the crystallographic axes **a** or **c**.

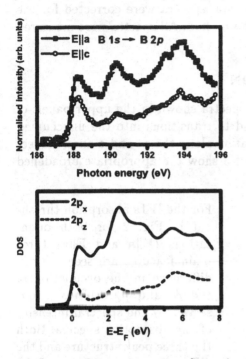

Figure 2. Boron 1s x-ray absorption spectra of single crystal YNi$_2$B$_2$C for **E**||**a** and **E**||**c** (upper panel); broadened orbital-projected unoccupied DOS for the B $2p_x/2p_z$ orbitals (lower panel).

With XAS-measurements the unoccupied electronic structure is probed. By absorbing x-ray photons, core electrons are excited into unoccupied states. The system thus set into an excited state can thereafter relax via photon or electron emission. In dependence of the initial photon energy the absorption cross section can be indirectly monitored either by counting the fluorescence photons (FY) using a solid state Ge-detector or the emitted electrons (TEY) by a channeltron. Whereas the latter method is surface sensitive, the first method is considered bulk sensitive and is therefore more appropriate for the borocarbides which are difficult to cleave.

XAS is site selective and is subject to dipol selection rules. Additionally, the use of polarized x-rays and single crystalline samples results in an orbitally resolved picture of the unoccupied states. Thus with the polarization vector **E** aligned along the **a**-axis B 1s and C 1s core level electrons are excited into unoccupied states with B $2p_x/2p_y$ and C $2p_x/2p_y$ character, respectively. With the polarization vector aligned along the **c**-axis, the electrons are excited into the respective $2p_z$ orbitals.

For comparision with the experimental LDA calculations of the partial DOS for YNi$_2$B$_2$C were performed using the linear combination of atomic-like orbitals method (LCAO). These scalar relativistic calculations employed a minimal basis set consisting of Y($5s,5p,4d$), Ni($4s,4p,3d$), B($2s,2p$), and C($2s,2p$) orbitals. All lower lying states were treated as core states and a contraction potential has been used at each site to optimize the local basis [10]. In order to take the quasi-particle and core-hole lifetimes, as well as the instrumental resolution broadening into account, the calculated unoccupied DOS are convoluted with a Lorentzian of FWHM according to $\Gamma + 0.2^*(E\text{-}E_F)$ and with a Gaussian with a FWHM equivalent to the experimental resolution. The experimental spectra were corrected for the energy dependence of the incoming photon flux. All spectra were measured at room temperature.

3. RESULTS AND DISCUSSION

The B 1s and C 1s x-ray absorption spectra shown in the upper panels of Figs. 2 and 3, respectively, correspond to transitions into the unoccupied B/C $2p_x$- and $2p_z$-orbitals for polarization along the a- and c-axes, respectively. The lower panels of Figs. 2 and 3 show the appropriate broadened orbitally-resolved unoccupied DOS.

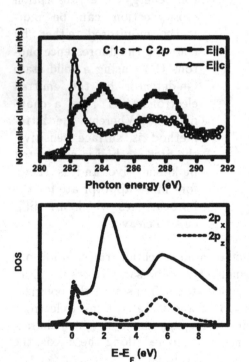

Figure 3. Carbon 1s absorption edge for $E\|a$ and $E\|c$ and the corresponding unoccupied states $2p_x$ and $2p_z$ from LDA calculations.

For the B 1s absorption threshold in Fig. 2 for both orientations ($E\|a$ and $E\|c$) three main features are seen. Also a difference in the occupation of the p_x and p_z orbitals is observed by the spectral intensity of the absorption spectra. Both the three peak structure and the difference in occupaction of the orbitals are well described by the calculated orbital-resolved DOS.

The C 1s excitation spectra in Fig. 3 exhibit a strong peak at the threshold for $E\|c$, whereas for $E\|a$ mainly two broad features are seen. These features are also quite well reproduced in the LDA calculations. However the ratio of the peak at the thesholds between $E\|a$ and $E\|c$ and the calculated p_x and p_z unoccupied DOS differ considerably.

A reasonable explanation for the difference in the spectral intensity of the XAS and the calculated DOS data for boron and carbon can be found in the short bond length in the c-direction between the boron and carbon atoms. The p_z-orbital overlap of the B $2p$ and C $2p$ states due to the short B-C bond length results in significant Coulomb repulsion between the B $2p$- and C $2p$-orbitals, which is not adequately treated in the calculations. Thus the electron occupancy of the B $2p_z$- and C $2p_z$-orbitals is lower than predicted in the LDA calculations.

The short bonds in the B–C–B link which joins the Ni_2B_2 layer-like units in the crystal structure results not only in a strong hybridization of B $2p_z$ and C $2p_z$ orbitals. An analysis of the peak positions relative to the absorption thresholds of the B $1s$ and C $1s$, as well as Ni $2p$ (shown in Ref. [11]) excitation data, reveals a strong hybridization among all the constituent elements, which is consistent with the predicted three-dimensional electronic structure for these systems.

For the absorption spectra for boron and carbon shown here, as well as for the nickel edge, a peak is seen at the absorption threshold. If a peak at E_F could also be detected for the occupied states, the scenario of a peaked DOS with its maximum close to the Fermi energy as predicted by LDA calculations would be proved. However, so far no peak could be found in the occupied states by photoemission spectroscopy (PES) [6], which is a surface sensitive method. To obtain a clean surface by either cleaving or scraping, strong chemical bonds are broken, as the intermetallic borocarbides are strongly covalent, three-dimensional systems. Therefore a possible reason for the absence of the peak might be due to possible surface reconstruction saturating dangling bonds by which the DOS can be reduced in the near surface region in comparison to the bulk [11]. So far we can only state that a peak of the DOS at E_F leading to the ralatively high T_c for intermetallic compounds is supported by our results. However a step-like DOS with a tail extending into the unoccupied states can not be ruled out.

The good agreement presented here between the LDA calculations and the experimental data give rise to the conclusion that correlation effects between the $3d$ electrons play a less important role in the intermetallic borocarbides as in the cuprates IITSC. This statement is supported by the value determined for the Ni d-d on-site Coulomb energy of $U_{dd} \approx 4.4$ eV[12] for YNi_2B_2C, which is smaller than the conduction bandwidth of about 6 eV.

4. SUMMARY

The examination of the unoccupied electronic states of YNi_2B_2C single crystals using polarization-dependent XAS reveal a three-dimensional, yet anisotropic electronic structure. In particular, between B and C orbitals along the c-axis a strong hybridization is observed. Furthermore, orbitally-

198

resolved LDA calculations describe the x-ray absorption spectra of B $1s$ and C $1s$ quite well, indicating that correlation effects play a minor role in this system in comparision to the cuprates. Furthermore the features in the unoccupied B $2p$ and C $2p$ states at the absorption thresholds are consitent with a DOS peak centered at E_F, as also predicted by LDA calculatons.

We are grateful to the DFG for financial support both as part of SFB 463 and the Graduiertenkolleg *Struktur und Korrelationseffekte in Festkörpern* of the TU Dresden.

References

1. Nagarajan, R., Mazumdar, C., Hossain, Z., Dhar, S.K., Gopalakrishnan, K.V., Gupta, L.C., Godart, C., Padalia, B.D., and Vijayaraghavan, R. (1994) Bulk Superconductivity at an elevated Temperature ($T_c \approx 12$ K) in a Nickel containing Alloy Sytem Y–Ni–B–C *Phys. Rev. Lett.* **72**, 274–277.
2. Cava R.J., Takagi, H., Zandbergen, H.W., Krajewski, J.J., Peck Jr., W.F., Siegrist, T., Batlogg, B., van Dover, R.B., Felder, R.J., Mizuhashi, K., Lee, J.O., Eisaki, H., and Uchida, S. (1994) Superconductivity in the quaternary intermetallic compounds $LnNi_2B_2C$, *Nature (London)* **367**, 252–253.
3. Müller, K.-H. (1999) Coexistence of Superconductivity and Magnetism in Rare Earth Transition Metal Borocarbides *these proceedings*.
4. Drechsler, S.-L. (1999) Siperconducting Transition Metall Borocarbides – Challenging New Materials in Between Traditional Superconductors and HTSC's *these proceedings*.
5. Mattheiss, L.F., Electronic properties of superconducting $LuNi_2B_2C$ and related carbide phases (1994) *Phys. Rev. B* **49**, 13 279–13 282; Pickett, W.E. and Singh, D.J. (1994) $LuNi_2B_2C$: A Novel Ni-Based Strong-Coupling Superconductor *Phys. Rev. Lett.* **72**, 3702–3705.
6. Golden, M.S., Knupfer, M., Kielwein, M., Buchgeister, M., Fink, J., Teehan, D., Pickett, W.E., and Singh, D.J. (1994) A Resonant-Photoemission Study of YNi_2B_2C, *Europhys. Lett.* **28**, 369–374. Fujimori, A., Kobayashi, K., Mizokawa, T., Mamiya, K., Sekiyama, A., Eisaki, H., Takagi, H., Uchida, S., Cava, R.J., Krajewski, J.J., and Peck Jr., W.F. (1994) Photoemission and inverse-photoemission study of superconducting YNi_2B_2C: Effects of electron-electron and electron-phonon interaction, *Phys. Rev. B* **50**, 9660–9663.
7. Siegrist, T., Zandbergen, H.W., Krajewski, J.J., and Peck Jr., W.F. (1994) The crystal structure of superconducting $LuNi_2B_2C$ and the related phase LuNiBC, *Nature (London)* **367**, 254–256; Siegrist, T., Cava, R.J., Krajewski, J.J., and Peck Jr., W.F. (1994) Crystal chemistry of the series LnT_2B_2C (Ln=rare earth, T=transition element), *J. Alloys and Compounds* **216**, 135–139.
8. Takeya, H., Hirano, T., and Kadowaki, K. (1996) Single crystal growth of quaternary superconductor YNi_2B_2C by a floating zone method, *Physica C* **256**, 220–226.
9. Domke, M., Mandel, T., Puschmann, A., Xue, C., Shirley, D.A., Kaindl, G., Petersen, H., and Kuske, P. (1992) Performance of the high-resolution SX700/II monochromator, *Rev. Sci. Instrum.* **63**, 80–89.
10. Eschrig, H. 1989 *Optimized LCAO Method*, Springer-Verlag, Berlin.
11. von Lips, H., Hu, Z., Grazioli, C., Behr, G., Knupfer, M., Golden, M.S., Fink, J., Rosner, H., and Kaindl, G. (1999) Polarization-dependent x-ray absorption spectroscopy on single crystal YNi_2B_2C superconductors *Phys. Rev. B* **60**.
12. Böske, T., Kielwein, M., Knupfer, M., Barman, S.R., Behr, G., Buchgeister, M., Golden, M.S., Fink, J., Singh, D.J., and Pickett, W.E. (1996) Core electron spectroscopic studies of YNi_2B_2C, *Solid State Commun.* **99**, 23–27.

P-WAVE SUPERCONDUCTIVITY IN Sr_2RuO_4

K. MAKI, E. PUCHKARYOV AND G. F. WANG
Department of Physics and Astronomy,
University of Southern California Los Angeles, CA 90089-0484,
USA

Abstract. There is accumulating evidence that the superconductivity of recently discovered Sr_2RuO_4 is of p-wave and nonunitary. After a brief introduction on experiments so far done on Sr_2RuO_4, we will describe the effect of impurities and some properties of the vortex state in p-wave superconductors. A possible relevance of the present model for the super-conductivity of organic superconductors in Bechgaard salts $(TMTSF)_2X$ with $X = ClO_4, PF_6$, etc. is also briefly described.

1. Introduction

The recent identification of d-wave symmetry of the superconductivity in the hole doped high T_c cuprates is probably the most important single event in the history of not only the high T_c cuprates but also of superconductivity [1, 2]. At least we know there is a large class of unconventional supercon-ductors arising from the strong electron correlation (i.e. nonphonic origin) [3].

The case in point is the recently discovered superconductivity in Sr_2RuO_4 [4]. This material without Cu has the same crystal structure as La_2CuO_4. But unlike La_2CuO_4, Sr_2RuO_4 is metallic and behaves like a standard Fermi liquid below 20 K, though the system is quasi-two dimensional (i.e. the cylindrical Fermi surface). Also it becomes superconducting around T = 1.2 – 1.5K.

Now there is accumulating evidence that the superconductivity is of isotropic p-wave and perhaps nonunitary [5, 6]. The nonunitary means that the superconducting state breaks not only the usual gauge symmetry but also the time reversal symmetry. In other words, the quasi-particle energy of the up spin state is different from the down spin state. Two models

199

S.-L. Drechsler and T. Mishonov (eds.), High-Tc Superconductors and Related Materials, 199–213.
© *2001 Kluwer Academic Publishers. Printed in the Netherlands.*

have been proposed for Sr_2RuO_4: the single band model of Sigrist and Zhitomirsky [7] and the three band model of Agterberg, et.al. [8]. The single band model requires not only that the superconductivity is of p-wave but also nonunitarity, since the T-linear coefficient of the specific heat at low temperatures γ_0 is always somewhat larger than $\frac{1}{2}\gamma_N$. The origin of the nonunitarity state is still puzzling, though we have some evidence of possible time-reversal symmetry breaking [9]. On the other hand the three-band model takes care of three separate bands observed by de Haas van Alphen effect [10] and in this sense more realistic. Also there is a theoretical reason to believe that the superconductivity involves only the γ band, while both the α and β bands are inert from the point of view of superconductivity. Therefore we do not need nonunitarity state in order to describe Sr_2RuO_4. The key to this second possibility is the ratio γ_0/γ_N. If $\gamma_0/\gamma_N \leq 0.5$, there will be no chance for the single band model. Indeed a very recent experiment by Nishizaki et al. [11] appears to suggest $\gamma_0/\gamma_N \approx 0.25$ in the purest crystal of Sr_2RuO_4. If it is the case, we have to adopt the three-band model. In this paper, however, we limit ourselves to the single band model for simplicity. In this case we have to assume the time reversal symmetry is maximally broken; we assume $\Delta_\uparrow(\vec{k}) = \Delta(\vec{k})$ while $\Delta_\downarrow(\vec{k}) = 0$. On the other hand we take $\Delta_\gamma(\vec{k}) = \Delta(\vec{k})$ and $\Delta_\alpha(\vec{k}) = \Delta_\beta(\vec{k}) = 0$ for the three-band model. Also, $\Delta(\vec{k})$ has 2D representation:

$$\Delta(\vec{k}) = \Delta e^{\pm i\phi} = \Delta(\hat{k}_1 \pm i\hat{k}_2), \tag{1}$$

where $\hat{k}$ is the unit vector within the a-b plane.

Since $|\Delta(k)| = \Delta$, this p-wave state behaves in many respects like the s-wave state described by the BCS (Bardeen, Cooper, and Schrieffer) theory [12] as far as the one spin component is concerned. On the other hand if you measure the spin dependent behavior like Knight shift and T^{-1} in NMR, the response will be completely different.

Before entering into the main subject we point out that the present model may apply to the organic superconductors in Bechgaard salts $(TMTSF)_2X$ with $X = ClO_4, PF_6$, etc. [13]. In particular, recent measurements of the upper critical field H_{c2} with $\vec{B} \parallel \vec{a}$ and $\vec{B} \parallel \vec{b'}$ indicate that H_{c2} far exceeds the Pauli limiting field $H_p = \Delta_0/\sqrt{2}\mu_b \sim 2$ Tesla [14]. Since these systems are in the clean limit it appears that the only escape from the Pauli limiting is possible if the superconductivity is of spin triplet pairing. On the other hand, it appears that the superconductivity recovers the full energy gap in the absence of magnetic field as seen by tunneling spectroscopy [15] and more recently by the thermal conductivity measurement [16]. Therefore, the only order parameter which satisfies the above 2 constraints appears to be the p-wave superconductor given in Eq(1).

In the following we summarize our recent work on the impurity scattering [17] and on the vortex state [18]. Surprisingly, the p-wave superconductor behaves in many respects very similarly to the d-wave superconductor in the hole doped high T_c cuprates.

2. Impurity Scattering

For simplicity we limit ourselves to the impurity scattering in the unitarity limit [18]. The resonance is on the Fermi surface. Then the renormalized frequency in the quasi-particle Green's function in given by

$$\tilde{\omega} = \omega + \Gamma \frac{\sqrt{\tilde{\omega}^2 + \Delta^2}}{\tilde{\omega}} \tag{2}$$

where $\Gamma = n_i \pi N_0$ and n_i is the impurity concentration, and N_0 is the quasi-particle density of states in the normal state on the Fermi surface. We note that the same equation (1) has been considered earlier [19]. But Buckholtz and Zwicknagl limit their consideration to the density of states and the boundary effect. Perhaps no realistic system existed to compare with their results at that time. Let us first consider the gap equation, which is given by

$$1 = \frac{2\pi T}{\Delta} \sum_n{}' \frac{1}{\sqrt{1 + u^2}}, \tag{3}$$

where the prime in the sum means that sum over ω_n (the Matsubara frequency) is cut off at $\omega_n = E_c$. Then putting $\Delta \to 0$, Eq(3) gives the well-known Abrikosov-Gor'kov result

$$-\ln\left(\frac{T_c}{T_{c0}}\right) = \psi\left(\frac{1}{2} + \frac{\Gamma}{2\pi T_c}\right) - \psi\left(\frac{1}{2}\right), \tag{4}$$

where $\psi(z)$ is the di-gamma function and T_c (T_{c0}) is the superconductivity transition temperature in the presence (absence) of impurities. T_c vanishes at $\Gamma = \Gamma_c = \frac{1}{2}\Delta_{00}$ where Δ_{00} is the superconducting order parameter at $T = 0K$ and in the absence of impurities. At $T = 0K$ Eq(3) simplifies also and we obtain

$$-\ln\left(\frac{\Delta(\Gamma, 0)}{\Delta_{00}}\right) = \ln\left(C_0 + \sqrt{1 + C_0^2}\right) - \zeta\left(C_0^{-1} - \arctan\left(C_0^{-1}\right)\right), \tag{5}$$

where $\zeta = \Gamma/\Delta$, and C_0 is defined as $C_0 = \lim_{\omega \to 0} \frac{\tilde{\omega}}{\Delta}$. Also we can obtain C_0 as

$$C_0 = \left(\frac{\zeta^2}{2} + \zeta\sqrt{1 + \frac{\zeta^2}{4}} \right)^{\frac{1}{2}} \tag{6}$$

The residual density of states (i.e. the quasi-particle density of states on the Fermi surface) is then given as

$$\frac{N(0)}{N_0} = \frac{C_0}{\sqrt{1 + C_0^2}} = \sqrt{\frac{\zeta}{\frac{\zeta}{2} + \sqrt{1 + \frac{\zeta^2}{4}}}}, \tag{7}$$

In particular, $N(0)/N_0$ increases as $\sqrt{\zeta}$ for small ζ and approaches unity as $(1 + \zeta^{-2})^{-\frac{1}{2}}$ when $\zeta = \Gamma/\Delta$ diverges. We show in Fig. 1 T_c/T_{c0}, Δ/Δ_{00}, and $N(0)/N_0$ vs. Γ/Γ_c. The similarity of these behaviors with the ones in d-wave superconductors is striking.

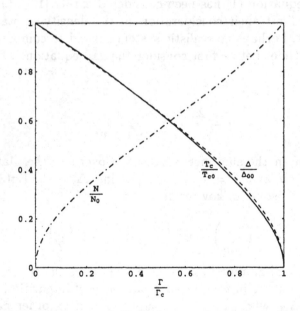

Figure 1. $\Delta(0,\Gamma)/\Delta_{00} (- - -)$, $T_c/T_{c0} (\text{———})$ and the residual density of states $N(0)/N_0 (- \cdot - \cdot -)$ are shown as functions of Γ/Γ_c, where $\Gamma_c = 0.5\Delta_{00}$.

On the other hand, the difference between p-wave superconductors and d-wave superconductors becomes obvious if one looks at the quasi-particle density of states, which is given by

$$\frac{N(E)}{N_0} = Re \frac{u}{\sqrt{u^2 + 1}}\bigg|_{u=-i\frac{E}{\Delta}}, \tag{8}$$

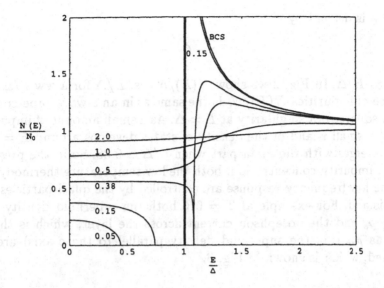

Figure 2. The density of states $N(E)/N_0$ is shown as a function of the reduced energy E/Δ for several values of the reduced scattering rate Γ/Δ.

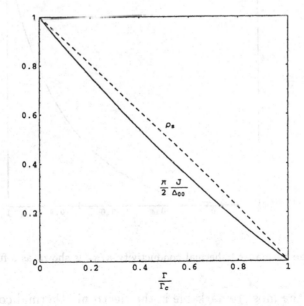

Figure 3. The superfluid density $\rho_s(\Gamma, 0)$ and the normalized Josephson current $\pi J(\Gamma, 0)/2\Delta_{00}$ are shown as functions of Γ/Γ_c.

where u is defined by

$$u = x - \zeta \frac{\sqrt{1-u^2}}{u} \qquad (9)$$

and $x = E/\Delta$. In Fig. 2 we show $N(E)/N_0$ vs. E/Δ for a few Γ/Δ. In the absence of impurities $N(E)/N_0$ is the same as in an s-wave superconductor with a square-root singularity at $E = \Delta$. As a small amount of impurities is added, a small island of the density of states develops around $E = 0$. This island merges with the main part when $\Gamma/\Delta \simeq 0.38$. So in the presence of a small impurity concentration, both the low temperature thermodynamics and the low frequency response are controlled by the quasi-particles in this small island. For example at $T = 0K$ both the superfluid density (in the plane) ρ_s and the Josephson current across the layer, which is the same thing as ρ_{sc} (i.e. the superfluid density parallel to the c axis) are easily obtained, which is shown in Fig. 3.

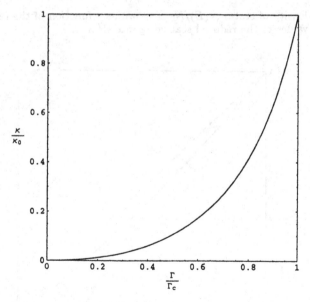

Figure 4. The normalized thermal conductivity κ/κ_n is shown as a function of Γ/Γ_c.

Perhaps the most remarkable is the electronic thermal conductivity for $T \ll T_{c0}$. At low temperatures the thermal conductivity decreases linearly with temperature T as in d-wave superconductors [20]. Further, the Wiedemann-Franz law is reestablished with the real part of the electric conductivity in the superconducting state. Now the thermal conductivity takes a very simple form

$$\frac{\kappa}{\kappa_n} = \left(\frac{\zeta}{\frac{\zeta}{2} + \sqrt{1 + \frac{\zeta^2}{4}}}\right)^2 = \left(\frac{N(E)}{N_0}\right)^4, \tag{10}$$

where $\kappa_n = \frac{\pi^2 nT}{3m\Gamma}$ is the thermal conductivity in the normal state. In the p-wave superconductors the thermal conductivity at low temperatures increases very rapidly with impurities as shown in Fig. 4. Also, we show

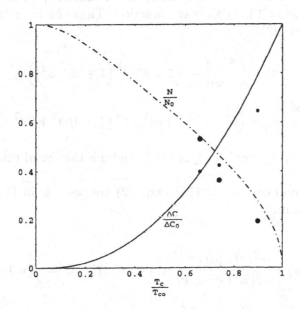

Figure 5. The normalized jump in the specific heat at $T = T_c$, $\Delta C/\Delta C_0$ and the density of states $N(0)/N_0$ are shown as functions of T_c/T_{c0}. Large dots are the · ·imental points for $N(0)/N_0$; smaller dots are the experimental points for $\Delta C/\Delta C_0$.

the jump in the specific heat in the presence of impurities in Fig. 5. Making use of the present theory we analyzed the experimental data in [21]. It is very important to choose the correct T_{c0} (the superconducting transition temperature of the clean system). The analysis is made by assuming $T_{c0} = 1.275K$ and we obtain a rather reasonable fit as seen in Fig. 5. On the other hand, a recent report suggests $T_{c0} = 1.5K$ [22]. Clearly, a further work on the thermodynamics is necessary.

3. Vortex state with $B \parallel c$

From our experience with d-wave superconductors, we know that the vortex state will tell much about the underlying symmetry of the superconducting order parameter.

Assuming that the order parameter in the vortex state is given by

$$\Delta(\vec{r}, \vec{k}) = \left(e^{i\phi} + Ce^{-i\phi}(a^+)^2 \right) |0\rangle, \tag{11}$$

where $|0\rangle$ is the Abrikosov solution for s-wave superconductors and a^+ is an analogy of the raising operator in the harmonic oscillator [18], we first determine the upper critical field for $\vec{B} \parallel \vec{c}$. Therefore, the first term is written as sum of the $N = 0$ Landau wave function, while the second term is that of $N = 2$ Landau wave function. Then the linearized Gor'kov equation is given [23]

$$-\ln t = \int_0^\infty \frac{du}{\sinh u} \left(1 - e^{-\rho u^2} \left(1 + 2C\rho u^2 \right) \right), \tag{12}$$

$$-C\ln t = \int_0^\infty \frac{du}{\sinh u} \left(C - e^{-\rho u^2} \left[\rho u^2 + C \left(1 - 4\rho u^2 + 2\rho^2 u^4 \right) \right] \right), \tag{13}$$

where $t = T/T_c$, $\rho = v^2 eH_{c2}(T)/2(2\pi T)^2$ and v is the Fermi velocity within the a-b plane.

Analytical solutions of Eqs (11) and (12) are possible in limiting cases. At $t = 0$, we obtain

$$\rho_0 = \lim_{t \to 0}(t^2 \rho) = \frac{v^2 eH_{c2}(0)}{2(2\pi T_c)^2} \frac{e^{\sqrt{3}-1}}{4\gamma} = 0.291878, \quad C = \frac{1}{2}(\sqrt{3} - 1) \tag{14}$$

with $\ln \gamma$ the Euler constant and for $t \to 1$,

$$\rho = \left(\frac{7\zeta(3)}{2}(3 - \sqrt{6}) \right)^{-1} (-\ln t), \quad C = \sqrt{\frac{3}{2}} - 1 \tag{15}$$

Indeed, the relatively large ρ compared with other solutions indicate Eq (11) is the correct choice. On the other hand $C \simeq 0.2247$ at $T = T_c$ is rather surprising. This means that there is an important admixture of the $N = 2$ Landau wave function even at $T = T_c$. All other cases except 3D axial (p-wave) and hybrid (d-wave) cases we have looked at so far [23, 24], C tends to zero as T approaches T_c. However, until now both 3D axial and hybrid state appear not to correspond to the real physical electronic systems. So this appears to be the unique feature of the p-wave superconductors. $C(t)$ in the intermediate temperature is given by

$$C(t) = \sqrt{\frac{1}{2} + \left(1 - \frac{1}{2}x \right)^2} - 1 + \frac{1}{2}x \tag{16}$$

with

$$x = \rho \int_0^\infty du \frac{u^4 e^{-\rho u^2}}{\sinh u} \bigg/ \int_0^\infty du \frac{u^2 e^{-\rho u^2}}{\sinh u} \qquad (17)$$

We show in Figs. 6 and 7, $h(t) = H_{c2}(t) / \left(-\frac{\partial H_{c2}}{\partial t}\big|_{t=1}\right)$ and $C(t)$ respectively.

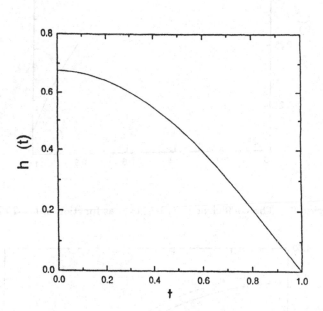

Figure 6. The normalized upper critical field $h(t)$ is shown as function of $t = T/T_c$.

Making use of the solution we find, we can construct the Abrikosov parameter β_A for arbitrary vortex lattice as [23, 25]

$$\beta_A = \sqrt{\frac{R}{2\pi}} \sum_{n,m} (-1)^{nm} e^{-\frac{\pi}{2} R(n^2 + m^2)} \left(1 + 2C^2\right)^{-2}$$

$$\times \int_{-\infty}^\infty dx e^{-x^2} \left(1 + C^2 (f(x + x_1) + f(x - x_1))(f(x + x_2) + f(x - x_2))\right.$$

$$\left. + C^4 f(x + x_1) f(x - x_1) f(x + x_2) f(x - x_2)\right), (18)$$

where $f(x) = x^2 - 1$, $x_1 = \sqrt{\frac{\pi R}{2}}(n - m)$, $x_2 = \sqrt{\frac{\pi R}{2}}(n + m)$ and $R = 1$ corresponds to the square lattice, while $R = \sqrt{3}$ to the hexagonal lattice. Here β_A is defined as

$$\beta_A = \left\langle |\Delta(r)|^4 \right\rangle \bigg/ \left\langle |\Delta(r)|^2 \right\rangle^2 . \qquad (19)$$

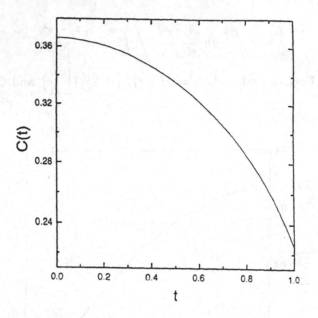

Figure 7. The coefficient $C(t)$ is shown as function of $t = T/T_c$.

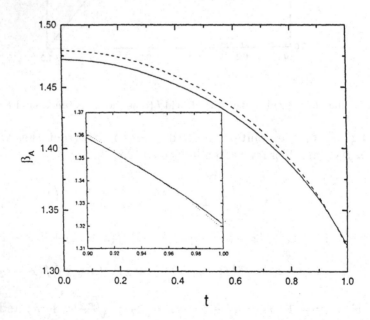

Figure 8. β_A for the square (solid line) and the triangular (dashed line) lattices are shown as functions of $t = T/T_c$.

We show in Fig. 8 β_A for the square lattice and for the hexagonal lattice. In the vicinity of $T = T_c$, β_A for the square lattice and the one for the hexagonal lattice are very close, but β_A for the hexagonal lattice is slightly smaller. This means at $T = T_c$ the hexagonal lattice is more stable. But two curves cross around $t = 0.95$ and for $t \leq 0.95$ the square vortex lattice becomes more stable. Indeed, the present theory is very consistent with a recent SANS experiment [26] where they saw the square vortex lattice oriented parallel to the a axis in the most part of the B-T phase diagram. In fact, the phase diagram is very similar to the one proposed for a d-wave superconductor recently [27]. On the other hand, the present theory does not contain the orientation energy for the square lattice. Therefore, we need an extra fourfold energy to fix the orientation of the square lattice.

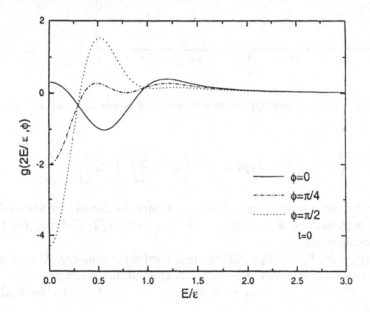

Figure 9. $G(2E/\epsilon, \phi)$ versus E/ϵ for $\phi = 0$ (solid), $\phi = \pi/4$ (dotted), and $\phi = \pi/2$ (dashed) at $t = 0$

For small $\Delta = \langle |\Delta(r)|^2 \rangle^{\frac{1}{2}}$, the quasi-particle density of states is given by

$$\frac{N(E, \phi)}{N_0} = 1 + 2 \left(\frac{\Delta}{\epsilon} \right)^2 G \left(\frac{2E}{\epsilon}, \phi \right);$$

$$G(x, \phi) = \frac{1}{\sqrt{\pi}(1 + 2C^2)} \int_{-\infty}^{\infty} du e^{-u^2}$$

210

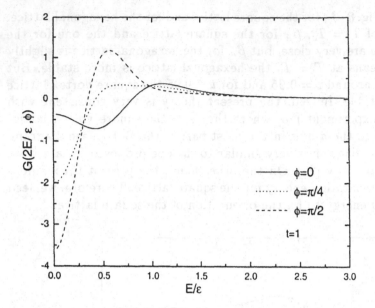

Figure 10. $G(2E/\epsilon, \phi)$ versus E/ϵ for $\phi = 0$ (solid), $\phi = \pi/4$ (dotted), and $\phi = \pi/2$ (dashed) at $t = 1$

$$\times \left(1 + 4C\left(u^2 - \frac{1}{2}\right)\cos(2\phi) + 4C^2\left(u^2 - \frac{1}{2}\right)^2\right) Re\left[\frac{1}{(x-u)^2}\right] \quad (20)$$

and $x = 2E/\epsilon$ and $\epsilon = v\sqrt{2eB}$. The quasi-particle density of states for a few $(\Delta/\epsilon)^2$ is shown vs. x in Figs. 9, 10 for $\phi = 0$, $\pi/4$, and $\pi/2$ for $t = 0$ and $t = 1$ respectively.

Surprisingly, $N(E, \phi)/N_0$ exhibits clear twofold symmetry. Also, the angular averaged $N(E, \phi)/N_0$ is the same as the one for $\phi = \pi/4$.

Similarly, the extra quasi-particle damping constant due to the Andreev scattering is given by

$$\Gamma(E, \phi) = 2\sqrt{\pi}\frac{\Delta^2}{\epsilon}\left(1 + 2C^2\right)^{-1} e^{-x^2}$$

$$\times \left(1 + 4C\left(x^2 - \frac{1}{2}\right)\cos(2\phi) + 4C^2\left(x^2 - \frac{1}{2}\right)^2\right). \quad (21)$$

Indeed, this extra damping constant is accessible through the de Haas van Alphen effect in the vortex state as an extra Dingle temperature [28]. Of course in this case we measure only $\langle\Gamma(E, \phi)\rangle = \Gamma(E, \pi/4)$. $\Gamma(E, \phi)$ for $\phi = 0$, $\pi/4$, $\pi/2$ is shown in Fig. 11. In this analysis we took $C = \frac{1}{2}(\sqrt{3} - 1) \simeq 0.366$ appropriate for $t = 0$.

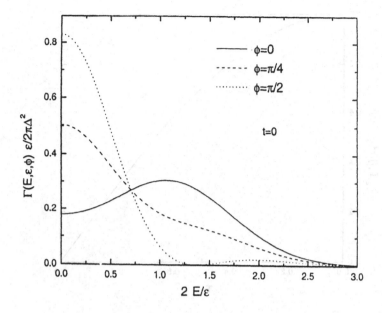

Figure 11. $\Gamma(E,\phi)$ is shown for $\phi = 0$ (solid), $\phi = \pi/4$ (dotted), and $\phi = \pi/2$ (dashed)

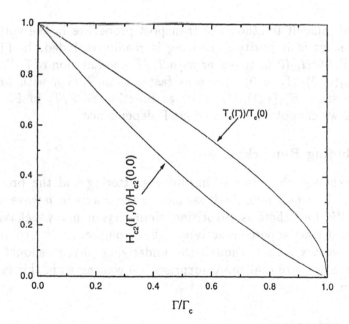

Figure 12. $H_{c2}(\Gamma,0)/H_{c2}(0,0)$ and T_c/T_{c0} as functions of Γ/Γ_c.

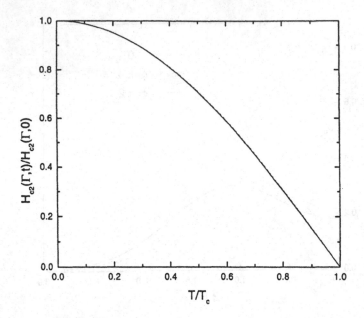

Figure 13. $H_{c2}(\Gamma, t)/H_{c2}(\Gamma, 0)$ as function of $t = T/T_c$

It is not difficult to study the transport properties in the vortex state. Also, the effect of impurity scattering is readily included. In Fig. 12 we show $H_{c2}(\Gamma, 0)/H_{c2}(0, 0)$ together with T_c/T_{c0} as function of Γ/Γ_c. As seen readily $H_{c2}(\Gamma, 0)/H_{c2}(0, 0)$ decreases faster than T_c/T_{c0} with impurities. Finally we show $H_{c2}(\Gamma, t)/H_{c2}(\Gamma, 0)$ as function of T/T_c in Fig. 13. For $\Gamma/\Gamma_c \leq 0.2$ we cannot see any obvious Γ-dependence.

4. Concluding Remarks

We have explored the effect of impurity scattering and the properties of vortex state in a magnetic field parallel to the c axis in p-wave superconductivity. We find there is surprising similarity in many behaviors with those in the d-wave superconductivity. The strong sensitivity to impurities, the square vortex lattice, though the underlying physics should be quite different. We are sure that more surprises are waiting us in this fascinating new system.

Acknowledgements

One of us (KM) thanks the hospitality of Peter Fulde at Max-Planck Institut für Physik Komplexer Systeme where the present work started. Also he has benefited from discussions and correspondence with Daniel Agterberg,

Tanya Riseman and Yoshiteru Maeno.

Present work is supported by National Science Foundation under grant number DMR95-31720.

References

1. Tsuei, C.C. and Kirtley, J.R. (1997) *Physica C*, **282-287**, 4.
2. Van Harlingen, D.J. (1997) *Physica C*, **282-287**, 128.
3. Maki, K. and Won, H. (1996) *J. Physique I* (Paris), **6**, 2317; (1998), in *Symmetry and Pairing in Superconductors* edited by M. Ausloos (Kluwer, Dordrecht).
4. Maeno, Y. Hashimoto, H., Yoshida, K., Nishizaki, S., Fujita, T., Bednorz, J.G. and Lichtenberg, F. (1994) *Nature* (London), **374**, 532; Maeno, Y. (1997) *Physica C*, **282-287**, 206.
5. Kitaoka, Y. Ishida, K. Asayama, K. Ikeda, S., Nishizaki, S. Maeno, Y., Yoshida, K. and Fujita, T. (1997) *Physica C*, **282-287**, 210.
6. Nishizaki, S., Maeno, Y., Farmer, S. Ikeda, S. and Fujita, T. (1997) *Physica C*, **282-287**, 1413.
7. Sigrist, M. and Zhitomirsky, M.E. (1996) *J. Phys. Jpn*, **67**, 3452.
8. Agterberg, D.F., Rice, T.M. and Sigrist, M. (1997) *Phys. Rev. Lett.*, **76**, 3374.
9. See for example Mota, A.C., Dumont, E., Amann, A., and Maeno, Y. (1999) in *SCES*, **98**, in press.
10. Mackenzie, A.P., Julian, S.R., Diver, A.J., McMullan, G.J., Ray, M.P., Lonzarich, G.G., Maeno, Y., Nishizaki, S. and Fujita, T. (1996) *Phys. Rev. Lett.*, **76**, 3786.
11. Nishizaki, S., Mao, Z. Q. and Maeno, Y. (preprint)
12. Bardeen, J., Cooper, L.N., and Schrieffer, J.R. (1957) *Phys. Rev.*, **108**, 1175.
13. Ishiguro, T. and Yamaji, K. (1990) *Organic Superconductors* (Springer, Berlin).
14. Naughton, M.J., Lee, I.J., Chaikin, P.M. and Danner, G.M. (1997) *Synth. Metals*, **85**, 1481; Lee, I.J., Naughton, M.J., Danner, G.M. and Chaikin, P.M. (1997) *Phys. Rev. Lett.*, **78**, 3555.
15. Bando, H., Kajimura, K., Anzai, H., Ishiguro, T. and Saito, G. (1985) *Mol. Cryst. Liqu. Crystal*, **119**, 41.
16. Belin, S. and Behnia, K. (1997) *Phys. Rev. Lett.*, **70**, 2125.
17. Maki, K. and Puchkaryov, E. (1998) preprint.
18. Wang, G.-f. and Maki, K. (1998) preprint.
19. Buchholtz, L.J. and Zwicknagl, G. (1981) *Phys. Rev. B*, **23**, 5788.
20. Sun, Y. and Maki, K. (1995) *Phys. Rev. B*, **53**, 6059; *Europhys. Lett.*, **32**, 355.
21. Nishizaki, S., Maeno, Y., Farmer, S., Ikeda, S. and Fujita, T. (1998) *J. Phys. Soc. Jpn.*, **67**, 560.
22. Mackenzie, A.P., Haselwimmer, R.K.W., Tyler, A.W., Lonzarich, G.G., Mori, Y., Nishizaki, S. and Maeno, Y. (1998) *Phys. Rev. Lett.*, **80**, 161.
23. Won, H. and Maki, K. (1995) *Europhys. Lett.*, **30**, 421; (1996) *Phys. Rev. B*, **53**, 5927.
24. Sun, Y. and Maki, K. (1993) *Phys. Rev. B*, **47**, 9108.
25. Kleiner, W.H., Roth, L.M. and Autler, S.H. (1964) *Phys. Rev. A*, **133**, 1226.
26. Riseman, T.M., Kealey, P.C., Forgan, E.M., Mackenzie, A.P., Galvin, L.M., Tyler, A.W., Lee, S.L., Ager, C., Paul, D.McK., Aegerter, C.M., Cubitt, R., Mao, Z.Q., Akima, S. and Maeno, Y. (1998) *Nature* (in press).
27. Shiraishi, J., Kohmoto, M. and Maki, K. (1998) *Korean J. of Physics* (in press); in *Symmetry and Pairing in Superconductors* edited by M. Ausloos (Kluwer, Dordrecht).
28. Maki, K. (1991) *Phys. Rev. B*, **44**, 2861; Stephen, M.J. (1992) *Phys. Rev. B*, **45**, 5481.

PLASMA MODES IN HTC SUPERCONDUCTORS WITHIN THE ANISOTROPIC LONDON MODEL

OLIVIER BUISSON

Centre de Recherches sur les Très Basses Températures, laboratoire associé à l'Université Joseph Fourier, C.N.R.S., BP 166, 38042 Grenoble-cedex 9, France.

AND

MAURO M. DORIA

Instituto de Física, Universidade Federal do Rio de Janeiro, C.P. 68528, Rio de Janeiro, 21945-970 RJ, Brazil.

1. Introduction

Up to the discovery of HT_c superconductors, it was generally believed [1] that plasma modes could not exist below the superconducting gap. In contrast to He superfluid, where many collective modes were observed, for superconductors only one mode, the Carlson-Goldman mode, was observed [2] very near the critical temperature. In superconductors, the Coulomb interaction shifts all the density oscillations modes to the plasma frequency [3], thus at an energy much higher than the superconducting gap. An exception to this rule are the layered superconductors, whose penetration depth is huge along the c-axis (perpendicular to the CuO_2 planes) and the static dielectric constant along the c-axis is estimated very large. These two properties make the plasma frequency along the c-axis strongly reduce. For highly anisotropic superconductors, such as BiSrCaCuO or LaSrCuO, these c-axis collective excitations were observed well below the gap and correspond to plasma oscillations of the superconducting electrons [4, 5, 6, 7, 8, 9, 10, 11].

Theoretically, plasma waves have been studied by several authors. Fertig *et al* [17] derived plasma dispersion working in the BCS approximation. Mishonov [15] discussed plasma oscillations in the highly anisotropic BiSrCaCuO superconductors. Van der Marel *et al* [18], Tachiki *et al* [29] and

S.-L. Drechsler and T. Mishonov (eds.), High-Tc Superconductors and Related Materials, 215–228.

Doria *et al* [26, 27] discussed plasma properties in the superconducting state using the London or Drude approximation. Bulaevski *et al* [20] obtained charge oscillations in layered superconductors using the Laurence-Doniach model. Pokrovsky *et al* [31] presented a microscopic theory of plasma resonance and provides an interesting insight into the role played by the elastic scattering of electrons by impurities. Artemenko *et al* [16, 30] discussed the temperature dependence of plasma oscillations using the kinetic equations of Green functions generalized to the case of layered superconductors with weak interlayer coupling. They predict that as the temperature approaches the critical one, these oscillations transform continuously into the Carlson-Goldman mode.

This paper is organized as follows. In section 2, the London-Maxwell model is applied to describe wave propagation in an anisotropic superconductor. Propagating plasma modes in highly anisotropic films with the c-axis perpendicular to the interfaces are presented in the next section. Two situations are considered, below and above the c-axis plasma frequency. In section 4, we study plasma mode propagation and the optical properties for oblique incidence on the film. The interaction between plasma modes and vortices are discussed in section 5, and finally, we conclude in section 6.

2. The London-Maxwell model

We are interested in the system's response to a fixed frequency, and so, a time dependence $\exp(i\omega t)$ is assumed for all fields. Hereafter we take that $\partial/\partial t \to i\omega$ in all equations governing the system. Current transport inside the anisotropic superconductor is described through the first London equation,

$$i\omega\mu_0\lambda_{\parallel}^2 \mathbf{J}_{\parallel} = \mathbf{E}_{\parallel}, \qquad i\omega\mu_0\lambda_{\perp}^2 J_{\perp} = E_{\perp}, \tag{1}$$

where $\mathbf{E}_{\parallel}$ and $\mathbf{E}_{\perp}$ are the field components parallel and perpendicular to the CuO_2 planes, respectively. $\lambda_{\parallel}$ and $\lambda_{\perp}$ are the London penetration depth along these two directions. The electromagnetic coupling, given by Maxwell's equations,

$$\nabla \cdot \mathbf{D} = 0 \tag{2}$$
$$\nabla \cdot \mathbf{H} = 0 \tag{3}$$
$$\nabla \times \mathbf{E} = -i\omega\mu_0\mathbf{H} \tag{4}$$
$$\nabla \times \mathbf{H} = i\omega\mathbf{D} \quad \text{where} \quad \mathbf{D} = \epsilon_0\epsilon_s \cdot \mathbf{E} - i\mathbf{J}/\omega. \tag{5}$$

shows that the superconductor dielectric constant is tensorial, $\mathbf{D} = \epsilon_0\epsilon(\omega) \cdot \mathbf{E}$.

$$\epsilon(\omega) = \begin{pmatrix} \varepsilon_{\perp} & 0 & 0 \\ 0 & \varepsilon_{\parallel} & 0 \\ 0 & 0 & \varepsilon_{\parallel} \end{pmatrix} \quad , \varepsilon_{\perp}(\omega) = \varepsilon_c - \frac{1}{(k\lambda_{\perp})^2} \quad , \varepsilon_{\parallel}(\omega) = \varepsilon_{ab} - \frac{1}{(k\lambda_{\parallel})^2}, \tag{6}$$

where $k \equiv \omega\sqrt{\mu_0\epsilon_0}$ is the wavenumber and ϵ_c and $\epsilon_{a,b}$ are components of the tensor ε_s, namely the dielectric constants in the insulating phase along the c-axis and the a,b-plane, respectively.

Phenomenological theories, such as the present one, only describe the superconductor at energies much lower than the pair breaking threshold. ω is much smaller than the superconducting gap, and consequently, it is also much smaller than the plasma frequency along the CuO_2 planes: $k\lambda_{\parallel} \ll 1$. This renders a negative, and large in modulus, dielectric tensor component parallel to the surfaces:

$$|\varepsilon_{\parallel}| \gg 1, \quad \varepsilon_{\parallel} \approx -\frac{1}{(k\lambda_{\parallel})^2} \tag{7}$$

In the present model the c-axis charge oscillation is reached when the c-axis dielectric tensor component vanishes,

$$\varepsilon_{\perp}(\omega_p) = 0, \quad \omega_p = \frac{c}{\sqrt{\varepsilon_c}\lambda_{\perp}} \tag{8}$$

Hence $\varepsilon_{\perp}$ changes sign as ω crosses ω_p.

We remark a rather curious property due to anisotropy: the presence of a volumetric charge density inside the superconductor, $\rho = -i\nabla \cdot \mathbf{J}/\omega = \epsilon_0(1 - \varepsilon_{\perp}/\varepsilon_{\parallel})\partial E_{\perp}/\partial x$. But this charge density is not responsible for the low frequency plasma modes discussed here.

The temperature dependence and the losses due to quasiparticle scattering can be introduced into the dielectric function by considering a bi-fluid model or Drude model [18, 29]. This correction to Eq.(1) leads to,

$$\varepsilon_{\perp} = \varepsilon_c(1 - \frac{1}{(k\lambda_{\perp})^2} - \frac{\omega_{pn}^2}{\omega(\omega + i\gamma)}) \tag{9}$$

where $\omega_{pn}^2 = e^2n/(\epsilon_c m^*)$, $n(T)$ is the density of thermally excited electrons, m^* their effective mass along the c-axis and γ their scattering rate. Since γ is usually very large ($\gamma \approx 10^{14}sec^{-1}$) and $n(T)$ is small in the superconducting state at low temperature, the London dielectric function is a good approximation. More elaborated studies have been recently performed to better describe the dynamics and the dielectric function along the c-axis in layered superconductors [33, 32, 30, 34, 35].

3. Plasma modes in anisotropic superconducting films

We study an anisotropic superconducting film, grown with the c-axis perpendicular to the surfaces, and choose a coordinate system where the two film-dielectric interfaces are at $x = d/2$ and $x = -d/2$. Propagation is along

218

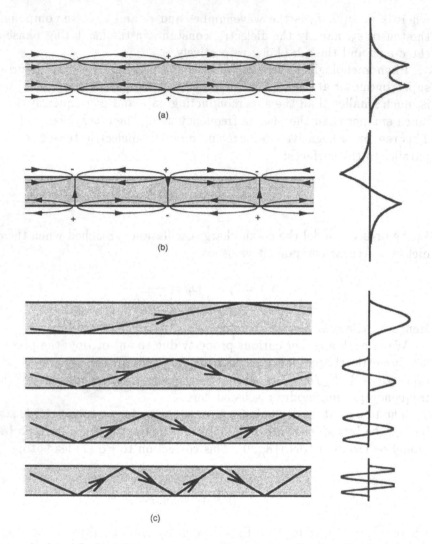

Figure 1. A pictorial view of wave propagation in a superconducting film surrounded by two equivalent non-conducting media. For $\omega < \omega_p$ the instantaneous electric field and the superficial charges are shown for symmetric (a) and antisymmetric (b) modes. For $\omega > \omega_p$ the optical ray associated to the plane wave travelling inside the film is shown here (c) for M=1,2,3 and 4. The symmetry of each state is also shown here, for both cases (a), (b) and (c), through the diagram E_z versus x.

the z-axis and all fields are plane waves of the kind, $\exp[-i(q_x x + q_z z - \omega t)]$ inside the superconducting film, and $\exp[-i(\tilde{q}_x x + \tilde{q}_z z - \omega t)]$ in the external dieletric medium.

From the above London-Maxwell's theory, we obtain Fresnel's equation of wave normals:

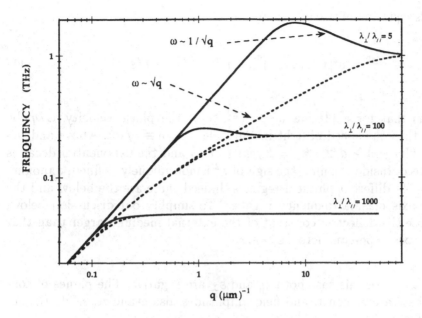

Figure 2. The symmetric and antisymmetric $\omega < \omega_p$ dispersion relations are shown for three anisotropies. A very thin superconducting film, $d = 10nm$-thick, surrounded by $SrTiO_3$, is considered ($\varepsilon_c \approx 30$, $\lambda_\parallel = 0.15\mu m$, and $\bar{\varepsilon} \approx 20000$).

$$\frac{q_z{}^2}{\varepsilon_\perp} + \frac{q_x{}^2}{\varepsilon_\parallel} = k^2 \text{(inside the superconducting film)},$$

$$\tilde{q}_z^2 + \tilde{q}_x^2 = \bar{\varepsilon}k^2 \text{(in the external dieletric medium)} \qquad (10)$$

We are interested here in the so-called p or TM (Transverse Magnetic) polarization, the only one that really probes properties of the dielectric component $\varepsilon_\perp$. For this polarization the oscillation of the superficial charge density at the interfaces yields interesting low frequency properties [27]. The field components are (E_x, H_y, E_z) and the transverse current ($J_x, 0, J_z$), not present in the external insulating dielectric media, leads to a superficial charge density at the interface.

The plasma mode corresponds to a propagation along the interfaces and to an exponential decay in the dielectric media perpendiculary to the film. This condition is $\tilde{q}_x = -(\pm)i\tilde{\tau}$ above and below the film, according to Eq.(3). We also define $q_x = -i\tau$, and then added to the condition of continuity of the phase at the interfaces, we also define $\tilde{q}_z = q_z \equiv q$. Therefore

Fresnel's equation of wave normals,

$$\tilde{\tau}^2 = q^2[1 - (\omega/qv)^2], \quad \tau^2 = \frac{\varepsilon_{\parallel}}{\varepsilon_{\perp}}q^2 - k^2\varepsilon_{\parallel} \tag{11}$$

demands that for a plasma mode ($\tilde{\tau}^2 > 0$) the phase velocity, ω/q, be smaller than the speed of light in the dielectric, $v \equiv c/\sqrt{\tilde{\varepsilon}}$. Above and below the film ($|x| \geq d/2$), $E_z = \tilde{E}_o \exp(-\tilde{\tau}\,x)$ and the exponential decay is guaranteed. Inside the film, the sign of τ^2 is not uniquely defined, announcing possibly different physical regimes. Indeed these are the below and the above c-axis plasma frequency regimes. To simplify our discussion below, we choose the dielectric constant of the external medium larger than that of the c-axis superconductor, $\tilde{\varepsilon} > \varepsilon_c$.

$\omega < \omega_p$:   In this case both $\varepsilon_{\parallel}$ and $\varepsilon_{\perp}$ are negative. The planes of constant phase are $z=$const. and field amplitudes also evanesce inside the superconductor:

$$E_z = E_o \exp(-\tau\,x) + F_o \exp(\tau\,x), \quad \tau^2 > 0 \tag{12}$$

$\omega > \omega_p$:   In this case $\varepsilon_{\parallel} < 0$ and $\varepsilon_{\perp} > 0$. Even within the Eq. (7) approximation, $\tau^2 \approx -|\varepsilon_{\parallel}/\varepsilon_{\perp}|q^2 + 1/\lambda_{\parallel}^2$ has no definite sign. The mode may be evanescent or propagative inside the superconductor. But for $\tilde{\varepsilon} > \varepsilon_c$, the evanescent modes disappear and τ^2 is negative. This is the case of confined propagation, with oblique incidence at the interfaces and planes of constant phase $q\,z \pm \tau'\,x = $ const.:

$$E_z = E_o \exp(-i\tau'\,x) + F_o \exp(i\tau'\,x), \quad \tau'^2 = -\frac{\varepsilon_{\parallel}}{\varepsilon_{\perp}}q^2 + k^2\varepsilon_{\parallel} > 0 \tag{13}$$

In summary, for the present purposes, the sign of τ^2 is uniquely determined below and above ω_p by the sign of the ratio $\varepsilon_{\parallel}/\varepsilon_{\perp}$.

The dispersion relations follow from the continuity of the ratio H_y/E_z at a single interface, say $x = d/2$, once assumed that the longitudinal field E_z is either an even or an odd function with respect to the $x = 0$ plane: a simplifying property that results from the choice of identical dielectric media above and below the film.

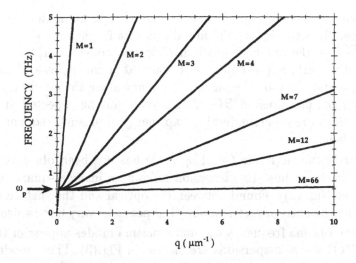

Figure 3. The dispersion relations above ω_p is shown for a 100 nm-thick film and anisotropy $\lambda_\perp/\lambda_\parallel = 100$. The M=1 mode propagates with speed v and the M=66 mode is the upper limit for the validity of the present theory ($\tilde{\varepsilon} = \varepsilon_c = 30$, $\lambda_\parallel = 0.15\mu m$).

Table I: The four possible dispersion relations and E_z.

E_z	Dispersion Relation
$E_{oz}\ \cosh{(\tau\ x)}$	$\frac{\tau}{\tilde{\tau}}\frac{\tilde{\varepsilon}}{\varepsilon_\parallel} = -\tanh{(\tau\ \frac{d}{2})}$
$E_{oz}\ \sinh{(\tau\ x)}$	$\frac{\tau}{\tilde{\tau}}\frac{\tilde{\varepsilon}}{\varepsilon_\parallel} = -\frac{1}{\tanh{(\tau\ \frac{d}{2})}}$
$E_{oz}\ \cos{(\tau'\ x)}$	$\frac{\tau'}{\tilde{\tau}}\frac{\tilde{\varepsilon}}{\varepsilon_\parallel} = -\tan{(\tau'\ \frac{d}{2})}$
$E_{oz}\ \sin{(\tau'\ x)}$	$\frac{\tau'}{\tilde{\tau}}\frac{\tilde{\varepsilon}}{\varepsilon_\parallel} = \frac{1}{\tan{(\tau'\ \frac{d}{2})}}$

The symmetric mode, shown in Fig. (1a), corresponds to the already observed plasma modes in granular aluminium [24] and YBCO thin films [25]. The dispersion follows a square-root law which is independent on the perpendicular penetration depth (see Fig. (2)). For larger wavevector, an asymptotic regime appears, which is strongly dependent on the anisotropy. The dispersion relation becomes nearly independent of the wave number, reaching the asymptotic value ω_{ps}. This is the surface plasmon that for an interface metal-dielectric is just the metal's plasma frequency divided by the factor $\sqrt{1 + \tilde{\varepsilon}}$, $\tilde{\varepsilon}$ the external medium dielectric constant. Here for large anisotropy the surface plasmon takes place at a much lower frequency, namely ω_{ps} is just given by the plasma frequency along the c-axis: $\omega_{ps} = \omega_p$. Thus the surface plasmons in $Bi - 2212$ crystal may be expected at energy much lower than the superconducting gap energy or plasma frequency along the CuO_2 planes!

The antisymmetric mode (see Fig. (1b)) has not been observed so far in superconducting films. Its dispersion follows an inverse square root dependence (see Fig. (2)). Found between the optical and the surface plasma modes regime the antisymmetric mode disappears for very large anisotropies.

Above the plasma frequency ω_p, many plasma modes appear in the thin films (Fig.(1c)) whose dispersions are plotted in Fig.(3). These modes were also predicted by Artemenko et al [30] using a microscopic theory. Notice that the London model only applies for modes whose wavelengths along the c-axis are larger than the separation between two consecutive CuO_2 planes.

4. Optical properties for oblique incidence

Far infrared measurements of highly anisotropic superconductors have shown striking properties and have initiated the discussion about plasma modes along the c-axis [4, 5, 6, 7, 8, 9, 10, 11]. An edge was observed in reflectance spectra below the critical temperature of the superconducting transition. This edge was interpreted by a charge oscillations for the first time, at our knowledge, by D.A. Bonn et al [4]. It was later observed and discussed in much more detailed [6, 7, 8, 9, 10, 11]. Van der Marel et al [18] interpreted the reflectance measurements using the perpendicular dielectric function of Eq. (9). Tachiki et al[29] derived the reflection and transmission of an incident electromagnetic wave polarized along the c-axis. They predict interference of Josephson plasma waves in HTc cuprates films [33]. Artemenko et al [30] and Hollauer et al [28] derived the reflection coefficient of a superconducting thin film for light incident at an angle θ using microscopic and London theory, respectively. Below the plasma frequency, light is perfectly reflected for any θ angle but not above ω_p, where it may propagate inside the superconductor.

In the following, these unusual properties of the anisotropic superconducting film are derived using the London-Maxwell approach. Consider an oblique incident light on an superconducting anisotropic film polarized in the incident plane (see Fig.(4)). In the dielectric media, $\tilde{q}_x = -k\sqrt{\tilde{\varepsilon}}\cos\theta$, and $\tilde{q}_z = k\sqrt{\tilde{\varepsilon}}\sin\theta$. Snell's law states that the wavenumber parallel to the surface is the same for the incident, reflected and transmitted waves:

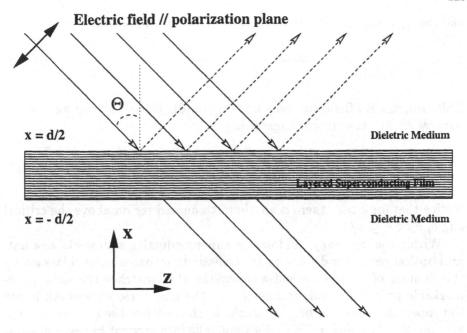

Electric field // polarization plane

Θ

x = d/2   Dieletric Medium

Layered Superconducting Film

x = - d/2   Dieletric Medium

X

Z

Figure 4. A pictorial view of p polarized wave incidence. The multiple reflections occuring at the interfaces suggest that an above ω_p situation is being represented here.

$q_z = \tilde{q}_z$. In the approximation of Eq.(7), Fresnel's equation (Eq.(10)) gives for the transverse wavenumber inside the superconductor,

$$q_x{}^2 = \frac{1}{\lambda_{\parallel}{}^2}\left(\frac{\tilde{\varepsilon}}{\varepsilon_{\perp}(\omega)} \sin^2 \theta - 1\right) \tag{14}$$

This equation is a valuable tool to understand many of the unusual properties of light interaction with the anisotropic superconductor.

Below ω_p: For $\omega \ll \omega_p$, the transverse dielectric constant is approximately given by $\varepsilon_{\perp} \approx -1/(k\lambda_{\perp})^2$ and $q_x{}^2 \approx -[(k\lambda_{\perp})^2 \sin^2 \theta + 1]/\lambda_{\parallel}{}^2$ is always negative. The light is always exponentially attenuated and decays over a finite length inside the superconductor which is given by $1/|q_x|$. At $\theta = 0$, this length is independent on the frequency and is given by $\lambda_{\parallel}$. For $\theta \neq 0$, as the frequency increases from 0 to ω_p, this lenght slightly decreases.

Above ω_p: For normal incidence ($\theta = 0$), the light is again exponentially attenuated over the distance $\lambda_{\parallel}$. For $\theta \neq 0$, there exists a frequency window sufficiently close to the plasma frequency for what the light propagates inside the superconducting film. This frequency window starts at ω_p and ends at $q_x = 0$ (Eq.(14)), that corresponds to the frequency,

$$\omega_0 = \frac{\omega_p}{\sqrt{1 - \frac{\tilde{\varepsilon}}{\varepsilon_s} \sin^2 \theta}} \tag{15}$$

This propagative frequency window, $(\omega_p, \omega_0(\theta))$ exists only for an angular window $(0, \theta_c)$, the critical angle being,

$$\theta_c = \begin{cases} \arcsin \sqrt{\varepsilon_c/\tilde{\varepsilon}}, & \tilde{\varepsilon} > \varepsilon_c \\ \pi/2, & \tilde{\varepsilon} < \varepsilon_c \end{cases} \tag{16}$$

Notice that for $\tilde{\varepsilon} > \varepsilon_c$ there is another attenuated region above the critical angle, $\theta_c < \theta \leq \pi/2$.

Within the frequency window, the superconducting slab works as a natural optical resonator displaying transmissivity resonances, each labeled by the number of transverse half-wavelengths that matches the slab. A remarkable property of this regime is that the transverse wavelength inside the superconductor can be very small, to the order of the inter-plane separation [28]. The present work does not take into account losses, and so, is restricted to extremely low temperature superconductors.

5. Plasma modes in presences of vortices

Up to this point the superconductor was just a dispersionless conductor. It was described by the London equation which is equivalent to the lossless free electron model at $q = 0$. Therefore many plasma properties are expected to be the same as other systems, such as the two-dimensional electron gas or thin metallic films for high enough frequencies $\omega\tau \gg 1$ where τ is the electron relaxation time. However such analogy is no more correct in a constant magnetic field. The presence of vortices strongly differentiate the plasma properties of a superconductor from a normal metal.

Experimental and theoretical studies have shown that a constant magnetic field can influence considerably the plasma modes. For a magnetic field parallel to the c-axis, T. Mishonov [15] has predicted a magnetoplasmon mode similar to that observed in the 2D electron gas. O. Tsui *et al* [12, 13] and Y. Matsuda *et al* [14] have observed collective Josephson plasma resonances in the vortex state in Bi-2212 when plasma oscillations are parallel to the DC applied magnetic field. The presence of vortices strongly reduces the plasma energy. M. Tachiki *et al* [29] explained such a behavior due to the reduction of superconducting electrons density in the mixed state. Bulaevskii *et al* [20] took into account the structure of vortices in such high anisotropic superconductors (pancake vortices) in order to interprate the experimental result.

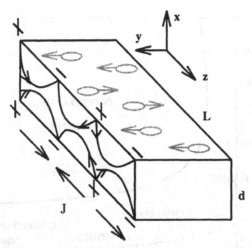

Figure 5. A pictorial view of a plasma wave in presence of vortices. Scales are out of proportion in order to enhance some of features, such as the superficial charge densities and the motion of the vortex lines. The instantaneous electric field is shown here only inside the film.

For magnetic field perpendicular to the c-axis, vortices are positioned between the layers. The plasma wave with the c-axis polarisation carries current parallel to the c-axis. Tachiki *et al* [29] derived the plasma frequency shift due to the vortex dynamic using the approximation of Gittleman and Rosenblum [37].

Hereafter we review how a perpendicular magnetic field modifies the symmetric plasma mode in a thin film for frequencies much below ω_p. The current associated to the charge oscillation exerts a Lorentz force on the vortices, which causes an oscillatory motion under the influence of vortex pinning (α) and vortex viscous drag (η) forces [37]:

$$\eta\frac{\partial \mathbf{u}}{\partial t} + \alpha\mathbf{u} = \mathbf{\Phi}_0(\mathbf{J} \times \hat{\mathbf{n}}) \tag{17}$$

The vortex displacement from its equilibrium position (see Fig.(5)) is described by the field $\mathbf{u}(\mathbf{x})$. Vortex motion creates an electrical field that adds to the inertial field. Indeed the London equation is completed by a new term associated to the vortex motion [36]:

$$\mathbf{E} = \mu_0\lambda^2 \cdot \frac{\partial \mathbf{J}}{\partial t} - \mu_0 H_0\left(\frac{\partial \mathbf{u}}{\partial t} \times \hat{n}\right) \tag{18}$$

The above equations lead to a parallel dielectric function given by,

$$\varepsilon_{\parallel} = \varepsilon_{ab} - \frac{1}{(k\bar{\lambda}_{\parallel})^2}, \quad \text{where} \quad \bar{\lambda}_{\parallel}^2 = \lambda_{\parallel}^2 + \left(\frac{B\,\Phi_0}{\mu_0\,\alpha}\right)\frac{1}{1 + i(\omega/\omega_d)} \tag{19}$$

226

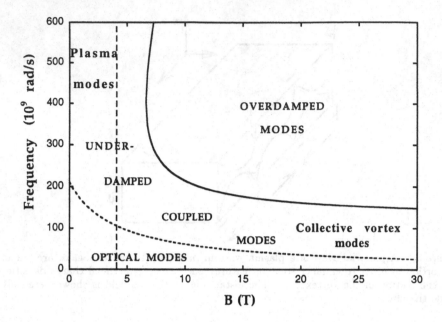

Figure 6. The diagram B vs. ω for a very thin YBCO superconducting film, d=10nm-thick, surrounded by the dielectric material $SrTiO_3$ shows three regions: optical, underdamped coupled and overdamped modes. The vertical dashed line separates the plasma oscillations (below) to the vortex oscillations (above). ($\lambda_\parallel$=0.15 μm, $\alpha_0 = 3.0 \times 10^5$ N/m^2, $\eta_0 = 1.2 \times 10^{-6}$ Ns/m^2, hence $\omega_0 = 250 \times 10^9$ rad/s)

where $\omega_d = \alpha/\eta$ is the depinning frequency. Hence vortices and superelectrons contribute additively to the parallel penetration depth. The perpendicular dielectric function $\varepsilon_\perp$ is unchanged by the perpendicular magnetic field if the vortex lines are assumed parallel to the c-axis.

At low magnetic field ($B \ll \frac{\mu_0 \alpha}{\Phi_0}(k\lambda_{parallel})^2 (1 + i(\omega/\omega_d))$), we retreive the usual dielectric constant (Eq. (6)) and the symmetric mode is unchanged. For high magnetic field, the dielectric constant is strongly modified and the magnetic term is dominant. Collective modes still exist but they are described by oscillation of vortices instead of oscillation of superconducting electrons. This oscillation is underdamped in the low frequency regime ($\omega \ll \omega_d$) and corresponds to a collective vortex mode. At higher frequency ($\omega \geq \omega_d$), these modes becomes overdamped due to the vortex viscosity. In Fig. (6), a diagram B versus ω is plotted indicating the different regime of the collective modes in thin films.

6. Conclusion

In highly anisotropic superconductors, plasma oscillations of the superconducting electrons polarized along the direction perpendicular to the CuO_2 planes were observed inside the superconducting state. The effects of such modes on wave propagation properties were studied in anisotropic superconducting films, immersed on a dielectric media, such that its uniaxial c-axis is perpendicular to the surfaces. Plasma modes are waves that propagate along the surfaces of the superconducting film and evanesce in the dielectric media perpendicularly to the surfaces. While only *two* branches of propagating modes are predicted for $\omega < \omega_p$, there are many for $\omega > \omega_p$, similar to propagation in optic fibers. Oblique incidence on the anisotropic superconducting film is also studied and the below and the above ω_p transmissivity patterns have several of its properties determined. Below, the regime is attenuated for any incident angle, and there is a transmissivity maximum, quite pronounced in case of a very high external dielectric constant. Above, a propagative regime exists where the superconductor is a natural optical resonator producing light with an extremely small transverse wavelength inside the superconductor.

Finally the effect of a c-axis parallel magnetic field, onto the plasma mode is considered. Vortices, known to cause dissipation and to change the penetration depth, affect the symmetric mode. A region in the B versus ω diagram is determined where vortex contribution is dominant and dissipation is small. At large magnetic field, a collective vortex mode is predicted.

Acknowledgements

We thank G. Hollauer, F. D'Almeida and F. Parage for helpful discussions. This work was done under a CNRS(France)-CNPq(Brasil) collaboration program.

References

1. P.C. Martin in *Superconductivity* ed. R.D (1970). Parks.
2. R.V. Carlson and A.M. Goldman, Phys. Rev. Lett. **34**, 11 (1975).
3. P.W. Anderson, Phys. Rev. **112**, 1900 (1958).
4. D. A. Bonn *et al*, Phys. Rev. B **35**, R8843 (1987).
5. B. Koch, M. Dueler, H.P. Geserich, Th. Wolf, G. Roth, and G. Zachman, in *Electronic Porperties of High-Tc Superconductors*, ed. by H. Kuzmany, M. Mehring and J. Fink, Springer 1990.
6. K. Tamasaku, Y. Nakamura, and U. Uchida, Phys. Rev. Lett. **69**, 1455 (1992).
7. A.M. Gerrits, A. Wittlin, V.H.M. Duijn, A.A. Menovsky, J.J.M. Franse, and P.J.M. van Bentum, Physica C **235-240**, 1117 (1994).
8. J.H. Kim, H.S. Somal, M.T. Czyzyk, D. van der Marel, A. Wittlin, A.M. Gerrits, V.H.M. Duijn, N.T. Hien, and A.A. Menovsky, Physica C **247**, 297 (1995).
9. C.C. Homes, T. Timusk, R. Liang, D.A. Bonn, and W.N. Hardy, Phys. Rev. Lett. **71**, 1645 (1993).

228

10. S. Tajima, G.D. Gu, S. Miyamoto, A. Odagawa, and N. Koshizuka, Phys. Rev. B **48**, 16164 (1993).
11. S. Uchida, K. Tamasaku, and S. TajimaPhys. Rev. B **53**, 14558 (1996).
12. Ophelia K.C. Tsui, N.P. Ong, Y. Matsuda, Y.F. Yan, and J.B. Peterson, Phys. Rev. Lett., **73**, 724 (1994).
13. Ophelia K.C. Tsui, N.P. Ong, and J.B. Peterson, Phys. Rev. Lett., **76**, 819 (1995).
14. Y. Matsuda, M.B. Gaifullin, K. Kumagai, K. Kadowaki, and T. Mochiku, Phys. Rev. Lett., **75**, 4512 (1995).
15. T. Mishonov, Phys. Rev. B **44**, 12033 (1991).
16. S.N. Artemenko and A.G. Kobelkov, JETP Lett. **58**, 445 (1993).
17. H.A. Fertig and S. Das Sarma, Phys. Rev. Lett. **65**, 1482 (1990), and Phys. Rev. B **44**, 4480 (1991).
18. D. Van der Marel, H. U. Habermeier, D. Heitmann, W. König and A. Wittlin, *Physica C* **176**, 1 (1991).
19. Y.B. Kim and X.G. Wen, Phys. Rev. B **48**, 6319 (1993).
20. L.N. Bulaevskii, M. Zamora, D. Baeriswyl, H. Beck, and J.R. Clem, Phys. Rev. B **50**, 12831 (1994).
21. J.E. Mooij and G. Schön, Phys. Rev. Lett. **55**, 114 (1985).
22. B. Mirhashem and R. Ferrell, *Physica C* **161**, 354 (1989).
23. T. Mishonov and A. Groshev, Phys. Rev. Lett. **64**, 2199 (1990).
24. O. Buisson, P. Xavier, and J. Richard, Phys. Rev. Lett. **73**, 3153 (1994); *Phys. Rev. Lett.* **74E**, 1493 (1995).
25. F.J. Dunmore, D.Z. Liu, H.D. Drew, and S. Das Sarma, Phys. Rev. B **52**, R731 (1995).
26. M.M. Doria, F. Parage, and O. Buisson, Europhys. Lett. **35**, 445 (1996).
27. M.M. Doria, G. Hollauer, F. Parage and O. Buisson, Phys. Rev. B **56**, 2722 (1997).
28. Private communication.
29. M. Tachiki, T. Koyama, and S. Takahashi, Phys. Rev. B **50**, 7065 (1994).
30. S.N. Artemenko and A.G. Kobelkov, Physica C **253**, 373 (1995).
31. S.V. Pokrovsky and V.L. Pokrovsky, J. of Superconductivity **8**, 183 (1995).
32. T. Koyama and M. Tachiki, Phys. Rev. B **54**, 16183(1996).
33. S. Takahashi and M. Tachiki, Physica C **282-287**, 2425 (1997).
34. S. N. Artemenko and A. G. Kobelkov, Phys. Rev. Lett. **78**, 3551 (1997)
35. To be published in SPIE proceeding, San Diego (1998).
36. M. M. Doria, F. M. R. d'Almeida and O. Buisson , Phys. Rev. B **57**, 5489 (1998).
37. J.I. Gittleman and B. Rosenblum, Phys. Rev. **39**, 2617 (1968).

PLASMA MODES IN THIN SUPERCONDUCTING FILMS

OLIVIER BUISSON
*Centre de Recherches sur les Très Basses Températures,
laboratoire associé à l'Université Joseph Fourier,
C.N.R.S., BP 166, 38042 Grenoble cedex 9, France.*

1. Introduction

It was largely considered that density oscillations have nothing to do with superconductivity [1]. This follows the argument [2] that the Coulomb interaction is very effective and shifts all these modes to the plasma frequency which is well above the frequency of the superconducting gap. Thus even in the superconducting state, plasma oscillations were identical to ones observed in the normal metal.

Recently two different plasma oscillations of superconducting electrons were predicted and observed bothly in low dimensional superconductors, mainly, in layered high-Tc superconducting crystal [3, 4, 5, 6, 7] or in thin films [8, 9, 10, 11, 12]. In the former, the observed plasma modes [13, 14, 15, 16, 18, 19] are analogous to those measured in a Josephson tunel junction [20] and are strongly dependent on the interlayer coupling. These bulk plasma modes are extensively discussed in this book. In thin films plasma oscillations may also appear regardless of the anisotropy. The electrical field created by the charge modulation is mainly concentrated outside the film. The Coulomb interaction is therefore strongly softened. A dispersion relation, ω_p versus the wavevector q, is predicted for these later modes and depends on the superconductor dimensionality:

$$\omega_p^2 \approx \frac{1}{\epsilon_m \, \lambda^2(T)} \, q^\alpha \tag{1}$$

where $\alpha=2$ and $\alpha=1$ for 1D- and 2D-superconductor respectively. ϵ_m is the dielectric constant of the dielectric medium surrounding the superconducting thin film and $\lambda^2(T)$ is the effective penetration depth.

Although many theoretical works have discussed plasma oscillations in superconducting films, very few experiments were performed to directly

S.-L. Drechsler and T. Mishonov (eds.), High-Tc Superconductors and Related Materials, 229–242.
© *2001 Kluwer Academic Publishers. Printed in the Netherlands.*

observe them. In this paper, we describe two different experiments which have recently shown these collective modes.

In a first section, we will detail experiments realized on superconducting films deposited onto SrTiO$_3$ substrates. After a description of the experimental set-up (section 2), measured plasma dispersions for superconducting film [21], wire [24] and 2D wire networks [25] are presented and compared to theoretical ones. In section 6, we will briefly present the observation made on YBCO films using far-infrared measurements by Dunmore et al.[22].

2. "Superconducting film/SrTiO$_3$": an experimental model system

As discussed in the introduction, plasma modes in thin films depend on the surrounding dielectric media. Indeed from Eq. 1 the squared plasma frequency ω_p^2 is inversely proportional to the dielectric constant of the surrounding medium. By choosing strontium titanate (SrTiO$_3$) as substrate of the superconducting film, whose dielectric constant reaches values as high as $\epsilon_m \approx 20000$ at $T \approx 4$ K [23, 28], the plasma frequency is lowered down to the GHz range. The system "superconducting film/SrTiO$_3$" turns to be an experimental model to study plasma modes in superconducting thin films. Using this experimental model system, we have studied plasma modes of superconducting electrons in different structures such as films, thin wires and wire networks.

We notice that SrTiO$_3$ substrate is not useful to study plasma modes in normal metal thin films. Indeed plasma resonances appear if the relaxation time τ of the electron is long enough to verify $\omega_p \tau \gg 1$. This condition is verified in a superconductor in the GHz range since the relaxation of superconducting electrons diverges below the critical temperature. In normal metal thin films, the relaxation time is very short (about 10^{-14} sec) which limits the observation of plasma resonances only for frequencies higher than τ^{-1}.

2.1. SAMPLES AND EXPERIMENTAL SET-UP

For all the samples, the superconducting films (granular aluminum or niobium) are deposited onto a (110) SrTiO$_3$ crystal whose features are the following: 5 mm along the [001] direction , 0.3 mm and 2.4 mm along the [1$\bar{1}$0] and [110] directions. In the study of thin films and arrays, the deposition is realized onto the edge of the crystal previously polished (see Fig. 1a). This particular geometry has been chosen to have a very thick dielectric medium below the superconducting film: a 5 mm thick is sufficient to consider the SrTiO$_3$ dielectric medium as semi-infinite space. The thin wire was evaporated on the face of the (110) SrTiO$_3$ substrate (Fig. 1b). Hereafter, the y-axis and z-axis corespond to the thin film plane and the

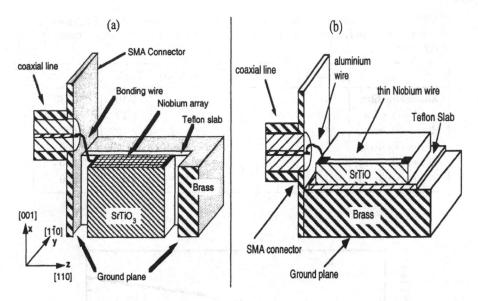

Figure 1. Schematic view of the two experimental configurations: (a) for thin films and wire networks; (b) for thin wires.

x-axis to its perpendicular direction. The plasma modes propagate along the z-axis.

These two experimental configurations were schematically described in Fig. 1a and Fig. 1b . The main features are the following: a 50 Ω coaxial line guides the microwave between the vector analyzer and the sample which is at low temperature; the electrical contact between the superconducting thin film and the inner conductor of the coaxial line was made by using a 20 μm-diameter aluminum bonding wire; the "superconducting film/SrTiO$_3$" block is isolated from the electrical ground plane by a Teflon slab.

Table 1 resumes the main physical properties of the samples. The critical temperature of the superconducting transition and the film resistivity were measured by transport measurements. The penetration depth are derived from the BCS dirty model.

2.2. MICROWAVE MEASUREMENTS

The reflection coefficient amplitude of the samples is measured by an HP8720B vector analyzer in the 130 MHz to 4 GHz range (Fig. 2). In the superconducting state below the critical temperature, several plasma resonances are observed. Their amplitude and their frequency are strongly temperature dependent. By approaching the superconducting transition from

TABLE 1. Characteristics of the different niobium arrays: d the thickness, w the wire width, $\rho(T_c)$ the resistivity at T_c, $\lambda(0)$ the penetration depth using the dirty limit.

	d (nm)	w (μm)	$\rho(T_c)$ ($\mu\Omega$.cm)	T_c (K)	$\lambda(0)$ (nm)
aluminum film	10		310	2.10	1 530
niobium wire	10	5	55	5.19	330
15 μm Nb array	13	3.2	49.3	3.56	820
31 μm Nb array	10	2.7	34.2	5.66	300
46 μm Nb array	10	2.8	66	4.27	520
76 μm Nb array	10	1.6	32.3	4.80	360

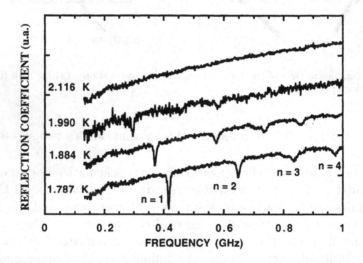

Figure 2. Frequency dependence of the reflexion coefficient for different temperatures. The curves are shifted on the vertical axis for clarity. The $T = 2.116$ K-curve is above T_c. Standing wave resonances of the plasma modes are indexed by n.

below, the resonant frequency is lowered, losses increase, and the number of observed resonances decreases. In the normal state, no plasma resonance is observed.

We obtain the dispersion relation of these collective oscillations in the following way. Since the resonances correspond to standing waves of plasma modes along the z-axis [21], the wave vector q, associated to each of them, verifies the $q L = n\pi$ condition where L is the film length and n the resonance index. Therefore for each resonances frequency, it corresponds the indexed wavevector $q(n)$.

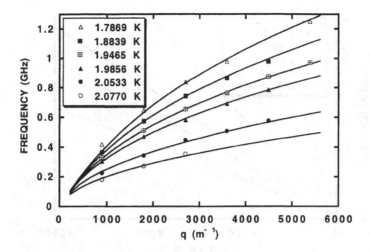

Figure 3. Dispersion relation of a granular aluminium film for different temperatures. Experimental data (points) are fitted by theoretical dispersions obtained from Eq. 2 (solid lines)

3. Thin film dispersion relation

The experimental plasma dispersion relations, ω versus q, of a granular aluminium film are plotted in Fig. 3 for different temperatures. The wave vector ranges between about $1000 \, m^{-1}$ and $6000 \, m^{-1}$ corresponding to large wavelenghts (6 mm to 1 mm). The dispersion relation of plasma modes is strongly temperature dependent. The mode is softened as the temperature approaches the transition. Moreover the dispersion relation is non-linear with the wavevector.

According to previous theories [8, 10, 11, 26], the plasma dispersion for a superconducting film deposited onto a semi-infinite $SrTiO_3$ dielectric and taking into account retardation effects is given by:

$$\omega_p^2 = \frac{d}{\mu_0 \, \epsilon_0 \, \sqrt{\epsilon_x \epsilon_z} \, \lambda^2(T)} \sqrt{q^2 - \mu_0 \, \epsilon_0 \, \epsilon_x \omega^2} \qquad (2)$$

where d is the film thickness. In this dispersion relation, the $SrTiO_3$ anisotropy is taken into account by the different values of dielectric constant along the two axis x and z. In our measurements ϵ_z is extracted from the dielectric resonances which appear at higher frequency [21] and ϵ_x is derived using the anisotropy ratio: $\epsilon_z/\epsilon_x = 2.1$ [23].

For each temperature, the experimental dispersion curves are fitted using this equation (Eq. 2) with the effective penetration depth $\lambda(T)$ as the only adjustable parameter. Very good agreement is observed between the

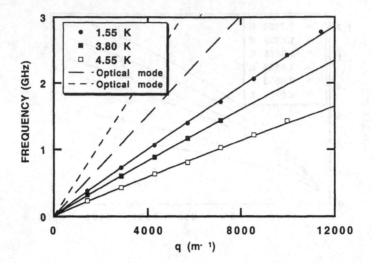

Figure 4. Dispersion relation of a thin niobium wire for different temperature. Experimental data points are fitted by theoretical law obtained from Eq. 3 (solid lines). Dashed lines are the two optical modes polarized along the x-axis and z-axis

measured plasma dispersion relations and the theoretical prediction for the wave vector dependence at all the measured temperature (Fig. 3). Moreover the temperature dependence of the penetration depth follows the Gorter-Casimir law [21]. The experimental penetration depth ($\lambda(0) = 1.47\mu$m) extracted from the adjustable parameter is perfectly consistent to the value obtained by the dirty limit model (see Table 1).

In conclusion, bothly the temperature dependence and the dispersion laws of plasma modes are very well described by theoretical models in the case of thin superconducting films.

4. Thin wire dispersion relation

Dispersion relation of a thin niobium wire is plotted in Fig. 4 for three different temperatures. Again the dispersion is strongly temperature dependent and the mode is softened as the temperature increases near T_c. Instead of the square root dependence observed in thin films, the dispersion relation in a thin wire presents a linear dependence over the entire wave vector range and for all the temperatures.

The experimental dispersion relation is very well fitted by the 1D law [8, 9, 12, 24]:

$$\omega_p^2 = \frac{r_0^2}{\epsilon_0\,\mu_0\,\epsilon_m\,\lambda^2(T)}\,\tilde{q}^2\ln(1/(\tilde{q}r_0)) \tag{3}$$

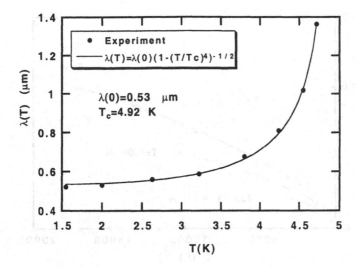

Figure 5. Experimental penetration depth deduced by the fit as function of temperature (data points). Solid line is obtained by the Gorter-Casimir law with the two adjustable parameters $\lambda_{Nb}(0)$ and T_c given in the graph.

where $\tilde{q} = \sqrt{q^2 - \mu_0\,\epsilon_0\,\epsilon_m\omega^2}$ and r_0 the wire radius, ϵ_m the dielectric constant. This dispersion relation is derived in the case of an isotropic dielectric medium and a cylindrical superconducting wire. Thus the only free parameter is again the penetration depth $\lambda(T)$.

In order to obtain this fit, we have made the two following simplifications: the dielectric medium is assumed isotropic with $\epsilon_m = (\epsilon_{001}\epsilon_{110}\epsilon_{110})^{1/3}$; the rectangular wire is assumed circular with the same cross section $(\pi r_0^2 = w\ d)$.

The temperature dependence of the extracted penetration depth is plotted in Fig. 5. It is very well fitted by the Gorter Casimir law. From this fit, the penetration depth and the critical temperature are deduced: $\lambda(0) = 0.53$ μm and $T_c = 4.92$ K. The comparison with the critical temperature defined by transport measurements is not precise because the resistive transition width is large between about 4.7 K to 5.4 K. Moreover the penetration depth derived from the BCS dirty limit, which is about 0.3 μm, gives only an approximate value.

We can now try to answer the following question: why does the plasma dispersion of the wire follow the 1D law? The wire is not 1D in the sense of the superconductivity because the estimated coherence lenght using the dirty limit $(\xi(0) \approx 8nm)$ is very small compared to the wire width. It is no more 1D for charge distribution since the Thomas-Fermi lenght (about 0.4 nm for niobium) is always very small compared to the thickness and the width of the wire. The charge modulation is therefore only localized near the

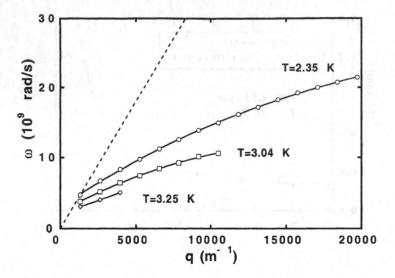

Figure 6. Dispersion relation of the 15 μm period array at three different temperatures compared to plane waves polarized along the x-direction. Drawn lines connect experimental data points.

interface between the superconductor and the dielectrics. The 1D behaviour is explained by the two following properties: first of all the supercurrent distribution is homogeneous in the wire section ($w\,d \ll \lambda_{Nb}^2(T)$); secondly the Coulomb interaction which generates the observed plasma modes is due to its very long range part ($q^{-1} \approx 1\ mm$). For such distances the wire width can be considered as 1D and the interaction described by its one dimensional Coulomb potential: $U_{Coulomb}^{1D} \approx ln(1/(qr_0))$.

As a conclusion, the dispersion relation of an isolated wire was measured for the first time, at our knowledge. The predicted linear dispersion has been verified as well as the temperature dependence. However the predicted logarithmic correction was not yet observed.

5. 2D wire networks

Four niobium networks with different lattice size p are considered ($p = 15\ \mu$m, 31 μm, 46 μm and 76 μm). The networks are all obtained from very thin niobium films of about 10 nm-thickness. The patterning of the arrays used a new lithography process [25, 27].

For the four different arrays, the dispersion is obtained as discussed in section 2.2. As an example, the experimental dispersion relation of the 15 μm lattice size is plotted in Fig. 6 for three different temperatures. The modes are observed in the frequency range 100 MHz to 3 GHz while

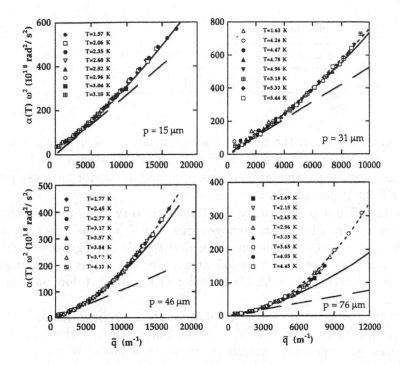

Figure 7. Dispersion relation for the four arrays, ω^2 versus $\tilde{q}$, at different temperatures scaled by the dimensionless factor $\alpha(T)$. Experimental data (points) are compared to the dispersion relation of an homogeneous film (large dashed lines) and an array of independent parallel wires (continuous lines). Dotted lines are obtained by the phenomenological law $\omega^2 \sim \tilde{q}^b$ with b given in Table 2.

the wave-vector ranges over about one decade, between 1500 m^{-1} and 20000 m^{-1}.

The temperature dependence of the dispersion curves shows scaling properties for the four different arrays in the new axis plot, ω^2 vs. $\tilde{q}$ (Fig. 7), where $\tilde{q} = \sqrt{q^2 - \mu_0 \, \epsilon_0 \, \epsilon_m \omega^2}$. Indeed all different temperature curves collapse into a single one provided that each curve is appropriately multiplied by a dimensionless factor $\alpha(T)$. Thus the experimental diagrams of Fig. 7 show that $\tilde{q}$ and T are the independent variables. The dispersion law does not follow the usual dispersion of an homogeneous thin film (see Eq. 2). Indeed the law is not linear in the ω^2 vs. $\tilde{q}$ graph as it should be for a thin film. This deviation from linearity increases with the array period. The deviation appears for surprisingly small wavevectors ($\tilde{q}p < 0.1$) which correspond to very long wavelength of nearly sixty times larger than the array period!

The absence of models able to describe these curves throughout their full measured wave vector region, made us introduce a new fitting into

TABLE 2. Characteristics of the different niobium arrays.

arrays	15 μm	31 μm	46 μm	76 μm
b	1.05	1.23	1.48	1.95

the diagram ω^2 vs. $\tilde{q}$: the phenomenological law $\omega^2 \sim \tilde{q}^b$, with b an adjustable parameter, whose optimal found values are listed in Table 2. This phenomenological parameter increases from one to two for an array period varying from 15 μm to 76 μm.

For the 15 μm period array, $b = 1.05$, thus very close to the behavior predicted by Eq. (2) for an homogeneous film. Thus the wires are close enough to sum up to a plane, meaning that the 2D homogeneous behavior is dominant. For the 76 μm period array, the best fitting is obtained for $b = 1.95$, indicating that the 1D behavior of Eq. 4 [9] is dominant in this case.

To our knowledge, no theory of plasma modes propagation for fully 2D conducting networks has been developed. Therefore, in order to interpret the measured dispersion relations, a very crude model is introduced by considering the square array just as an array of parallel cylindrical wires of radius r_0, thus neglecting the effects of the transverse branches, which are perpendicular to the propagation. Within the approximation of very small radius, $r_0 \ll p$ and $r_0 \ll \lambda_{Nb}(T)$, an analytical expression for the dispersion relation [24, 25] is obtained using the London model and Maxwell equations, in terms of the modified Bessel functions K_0 and K_1. It describes a TM mode propagating along the wires such that all wires oscillate in phase:

$$\omega_p^2 = \frac{1}{\epsilon_0 \, \mu_0 \, \epsilon_m \, \lambda^2(T)} \, \frac{\tilde{q}r_0}{K_1[\tilde{q}r_0]} \left\{ K_0[\tilde{q}r_0] + 2 \sum_{n=1}^{+\infty} K_0[\tilde{q}np] \right\} \tag{4}$$

The cylindrical wires are assumed to be half immersed in the semi-infinite dielectric medium of high constant ϵ_m ($\epsilon_m \gg 1$). $\tilde{q} = \sqrt{q^2 - \epsilon_0 \, \epsilon_m \mu_0 \, \omega^2}$ is the parameter characterizing evanescence inside the dielectric taking into account retardation effects: $\tilde{q}$ reduces to the wave vector q when the mode becomes very slow, $\omega/q \ll c/\sqrt{\epsilon_m}$. The first term of Eq. (4) is the contribution of a single wire to the dispersion relation and shows 1D behaviour [8, 9] for a wire radius much smaller than the penetration depth and the wavelength. The remaining sum corresponds to the Coulomb interaction between wires. In the limit $\tilde{q}p \gg 1$, this interaction is exponentially evanescent and the dispersion law of a single wire is dominant. In the other limit,

$\tilde{q}p \ll 1$, many wires are coupled and we retrieve the dispersion of an homogeneous thin film [8, 21], corrected by $\cdot w/p$ due to the dilution of the electronic concentration:

$$\omega_p^2 = \frac{w}{p} \frac{d}{\epsilon_0 \mu_0 \epsilon_m \lambda^2(T)} \tilde{q} \tag{5}$$

where w and d are respectively the branch width and the film thickness.

In order to discuss the experimental plasma modes dispersions, the dispersion curves of the homogeneous (Eq. (5)) and parallel wires model (Eq. (4)) are also plotted in Fig. 7. Comparison between experimental data and our theoretical model is done through the adjustment of one single parameter, $\lambda(0)$, whose optimal fit values are given in Table 2. In our experiments, instead of cylinders we have wires of rectangular cross section. Therefore, we have defined the radius as $\pi r_0^2 = w t$, thus conserving the film cross section. Since the dielectric medium under consideration, $SrTiO_3$, is anisotropic with constants ϵ_{001} and ϵ_{110}, we assume in the above expression that $\epsilon_m = \sqrt{\epsilon_{001}\epsilon_{110}}$ ($\epsilon_m \approx 15000$ and remains constant below about 6 K [23, 28]).

For the 15 μm period array, the homogeneous model fits the data over a large $\tilde{q}$ regime, up to $\tilde{q} = 6000\, m^{-1}$ corresponding to $\tilde{q}p \approx 0.1$. For the larger period arrays, the homogeneous model is valid up to the identical limit, $\tilde{q}p \approx 0.1$. Below this limit, long range interaction between the wires is dominant and the network can be essentially regarded as an homogeneous 2D thin film (Eq. (2)).

Above $\tilde{q}p \approx 0.1$, an increasing curvature in the experimental dispersion relation shows that the homogeneous model fails to fit the data for all the four different arrays. Plasma modes become sensitive to the artificial array structure and can no longer be explained by the homogeneous model. In comparison, the 2D dielectric lattice [29] follows the microwave dispersion relation of a homogeneous medium up to $\tilde{q}p \approx 1$. The same crossover value is obtained for an electron in a 2D periodic potential.

The model of parallel wires (Eq. (4)) succeeds to describe the unexpected deviation of plasma modes at very long wavelength. It predicts the curvature and fits very well the 15 μm, 31 μm and 46 μm arrays over all the investigated $\tilde{q}$ range. The observed curvature found in these two arrays is then clearly understood as the beginning of the crossover between the 2D regime and the wire regime (1D). Indeed above $\tilde{q}p = 0.1$, the first term of Eq. (4) related to the 1D interaction becomes of the same order than the infinite sum term which describes the interwire interaction. This crossover appears at an unexpectedly small $\tilde{q}p$ value because of the two following reasons: the 1D Coulomb interaction term, divergent at the origin, is dominant for a small radius ($K_0[\tilde{q}r_0] \approx -log[\tilde{q}r_0] \approx 8$ for $\tilde{q}r_0 \approx 10^{-4}$); the interwire interaction is exponentially evanescent, turning the sum term contribution into an extremely small contribution.

For the larger period array, the model of parallel wires (Eq. (4)) applies until another limit is met at $\tilde{q}p \approx 0.3$. Beyond $\tilde{q}p \geq 0.3$, the observed dispersion relation has a steeper curvature than predicted by Eq. (4) (see the 76 μm-period dispersion). While the kinetic energy was mainly localized inside the longitudinal branches in the low $\tilde{q}p$ regime, the superconducting current becomes significant in the transverse branches for larger $\tilde{q}p$. Propagation along the transverse branches must then be taken into account. Moreover the defects lines which appear at the sample edges, were neglected in our model. As the array period increases, the importance of these defects relatively increases and should be evaluated.

6. Far-infrared experiments

For usual experimental parameters, the plasma energy in thin superconducting films is expected in the THz range. Using the standard technique to observe plasma modes in 2DEG expriments, Dunmore et al. [22] have realized transmission measurements of a YBCO thin film in the superconducting state.

In order to couple the IR radiation to in-plane plasma modes, a grating was patterned above the superconducting film. The grating periodicity a defines the wavelength of the excited plasma waves by $q = n2\pi/a$. In Fig. 8, the measured transmission ratios for three selected grating periods are shown with their calculated representations from the London bi-fluid model. The transmission for both samples was found to have London behavior at low frequency both with and without grating: transmission vanishes as frequency goes to zero. The first peak is the plasmon antiresonance which is due to a zero-crossing in the imaginary part of the total effective conductivity, the dip is due to the plasma resonance and the second peak is due to diffraction from the grating.

From their measurements, the authors have deduced the dispersion relation. The law is not clearly square-root dependence but is well explained by a model taking into account the coupling of the plasma modes to the diffraction modes of the grating. Therefore this experiment shows also clearly the existence of plasma modes associated to superconducting electrons.

7. Conclusion

Plasma modes of superconducting electrons have been observed in thin superconducting films as different as granular aluminum, niobium or YBCO films. The system "superconducting film/SrTiO$_3$" appears as an experimental model to study plasma modes in superconducting thin films.

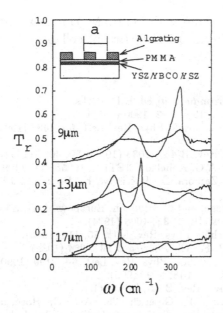

Figure 8. The transmission ratios of three selected grating periods are plotted along with theoretical predictions (thin lines). The curves are shifted on the T_r axis for clarity. As the grating period increases from 9 to 17 μm, the frequency position of the plasmon decreases. The inset figure is a schematic of the sample with grating period a, from Dunmore *et al* [22].

The plasma modes are strongly temperature dependent. Far below the critical temperature of the superconducting transition, they are weakly damped. As the temperature approaches T_c, the modes are softened due to the decrease of the superconducting electrons density. Above T_c plasma resonances were no more detected.

Different experimental dispersion relations were obtained for different shape of the superconductors. A square-root dependence was measured for homogeneous thin films and corresponds to the well-known dispersion of 2D electron gas. In thin wires the dispersion follows a linear dependence.

In superconducting wire networks, plasma modes displays two regimes, ranging from a 2D behavior, initially like a thin homogeneous film, and then as a set of interacting parallel wires. Their frequency is strongly lowered by the low effective carrier density associated to the network. A parallel wire model describes very well the low wave vector limit of the experimental data.

Acknowledgements

Much of the work described here was carried out in Grenoble and I wish to express my grateful thanks to Franck Parage, Thomas Reuss, Bertram Meyer and Benedetta Camarota for their direct contributions. I have profited greatly from discussions with Mauro M. Doria. A part of this work was done under a CNRS(France)-CNPq(Brazil) collaboration program.

References

1. P.C. Martin in *Superconductivity* ed. R.D. Parks.
2. P.W. Anderson, Phys. Rev. **112**, 1900 (1958).
3. H.A. Fertig and S. Das Sarma, Phys. Rev. Lett. **65**, 1482 (1990), and Phys. Rev. B **44**, 4480 (1991).
4. T. Mishonov, Phys. Rev. B **44**, 12033 (1991); **50**, 4004 (1994).
5. S.N. Artemenko and A.G. Kobelkov, JETP Lett. **58**, 445 (1993).
6. L.N. Bulaevskii, M. Zamora, D. Baeriswyl, H. Beck, and J.R. Clem, Phys. Rev. B **50**, 12831 (1994).
7. M.M. Doria, F. Parage, and O. Buisson, Europhys. Lett. **35**, 445 (1996).
8. I.O. Kulik, Sov. Phys.-JETP **38**, 1008 (1974).
9. J.E. Mooij and G. Schön, Phys. Rev. Lett. **55**, 114 (1985).
10. B. Mirhashem and R. Ferrell, *Physica C* **161**, 354 (1989).
11. T. Mishonov and A. Groshev, Phys. Rev. Lett. **64**, 2199 (1990).
12. M.V. Simkin, Physica C **168**, 279 (1990).
13. D. A. Bonn *et al*, Phys. Rev. B **35**, R8843 (1987).
14. B. Koch, M. Dueler, H.P. Geserich, Th. Wolf, G. Roth, and G. Zachman, in *Electronic Porperties of High-Tc Superconductors*, ed. by H. Kuzmany, M. Mehring and J. Fink, Springer 1990.
15. K. Tamasaku, Y. Nakamura, and U. Uchida, Phys. Rev. Lett. **69**, 1455 (1992).
16. C.C. Homes, T. Timusk, R. Liang, D.A. Bonn, and W.N. Hardy, Phys. Rev. Lett. **71**, 1645 (1993).
17. S.N. Artemenko and A.G. Kobelkov, Physica C **253**, 373 (1995).
18. J.H. Kim, H.S. Somal, M.T. Czyzyk, D. van der Marel, A. Wittlin, A.M. Gerrits, V.H.M. Duijn, N.T. Hien, and A.A. Menovsky, Physica C **247**, 297 (1995).
19. Y. Matsuda, M.B. Gaifullin, K. Kumagai, K. Kadowaki, and T. Mochiku, Phys. Rev. Lett., **75**, 4512 (1995).
20. A.J. Dahm, A. Denenstein, T. F. Finnegan, D. N. Langenberg, and D. J. Scalapino, Phys. Rev. Lett, **20**, 859 (1968).
21. O. Buisson, P. Xavier, and J. Richard, Phys. Rev. Lett. **73**, 3153 (1994); *Phys. Rev. Lett.* **74E**, 1493 (1995).
22. F.J. Dunmore, D.Z. Liu, H.D. Drew, and S. Das Sarma, Phys. Rev. B **52**, R731 (1995).
23. T. Sakudo and H. Unoki, Phys. Rev. Lett. **26**, 851 (1971).
24. F. Parage. PhD thesis (1997).
25. F. Parage, M.M. Doria and O. Buisson, Phys. Rev. B - Rapid Comm. (October 1998).
26. M.M. Doria, G. Hollauer and O. Buisson, Phys. Rev. B **56**, 2722 (1997).
27. Nover: Raith Gmbh, Hauert 18, Technologiepark, D44227 Dortmund, RFA.
28. K.A. Müller, W. Berlinger, M. Capizzi, H. Gränicher, Solid State Commun. **8**, 549 (1970).
29. S. L. McCall, P M. Platzman, R. Dalichaouch, D. Smith, S. Schultz , *Phys. Rev. Lett.* **67**, 2017 (1991).

COUPLING OF INTRINSIC JOSEPHSON OSCILLATIONS IN LAYERED SUPERCONDUCTORS

C. PREIS, C. HELM, J. KELLER, A. SERGEEV
Institute of Theoretical Physics, University of Regensburg
D-93040 Regensburg, Germany

AND

R. KLEINER
Physical Institute III, University of Erlangen-Nürnberg
D-91058 Erlangen, Germany

1. Introduction

The superconducting properties of the highly anisotropic cuprate–superconductors $Tl_2Ba_2Ca_2Cu_3O_{10+\delta}$ (TBCCO) and $Bi_2Sr_2CaCu_2O_{8+\delta}$ (BSCCO) are well described by a stack of Josephson junctions coupling the superconducting CuO_2 layers in c-direction. In particular the multiple branch structure observed in the current-voltage characteristics by several groups [1, 2, 3, 4, 5, 6] can be explained by this model. Due to the low value of the critical current in c-direction the system has a small Josephson plasma frequency ω_p and a large value of the McCumber parameter β_c. Therefore the current-voltage characteristics are highly hysteretic. The low value of the Josephson plasma frequency manifests itself in the transparency of the stack with respect to THz radiation polarized in c-direction [7]. The longitudinal and transversal plasma oscillations have also been observed directly [8, 9].

Recently Koyama and Tachiki [10] proposed a coupling of Josephson oscillations in different barriers due to charge fluctuations. In systems with weakly coupled superconducting layers the charges on different layers need not to be constant, which is in contrast to ordinary superconductors where charge neutrality can be assumed. Koyama and Tachiki [10] show that the gauge-invariant scalar potential

$$\mu_l = \Phi_l - (\hbar/2e)\dot{\chi}_l \tag{1}$$

S.-L. Drechsler and T. Mishonov (eds.), High-T_c Superconductors and Related Materials, 243–248.

does not vanish in general. Here Φ_l is the electric scalar potential and χ_l is the phase of the superconducting order parameter $\Delta_l = |\Delta_l| \exp(i\chi_l)$ on layer l. On the other hand the Josephson current density $j_c \sin \gamma_{l,l+1}$ between layers l and $l+1$ depends on the gauge-invariant phase difference

$$\gamma_{l,l+1}(t) = \chi_l(t) - \chi_{l+1}(t) - \frac{2e}{\hbar} \int_l^{l+1} dz \, A_z(z,t) \tag{2}$$

where A is the vector potential. Its time derivative (2nd Josephson relation)

$$\frac{\hbar}{2e} \dot{\gamma}_{l,l+1}(t) = \int_l^{l+1} dz \, E_z + \mu_{l+1} - \mu_l \tag{3}$$

then not only depends on the voltage between the layers, but also on the potential difference $\mu_{l+1} - \mu_l$. This finally leads to a coupling between Josephson oscillations in different barriers which is also present in stacks of short contacts. A non-vanishing generalized scalar potential which is related to quasi-particle charge imbalance can also be obtained without breaking charge neutrality. This effect has been investigated by Artemenko and Kobelkov [11] and by Ryndyk [12, 13].

In the following we give a short outline of the microscopic derivation of the coupling effect and discuss the influence of the coupling on the current voltage characteristics. More details and results can be found elsewhere [14].

2. General outline of the theory

We consider a model where the current between superconducting layers across the insulating barrier is described by a tunneling Hamiltonian

$$H_T(t) = \sum_{l,k,k',\sigma} T_{k,k'} c_{l+1,k',\sigma}^\dagger c_{l,k,\sigma} e^{-\frac{ie}{\hbar} \int_l^{l+1} dz \, A_z(t)} + h.c. \tag{4}$$

which depends on the vector potential $A_z(t)$ in the barrier. Moreover the current is driven by the difference of scalar potentials $\Phi_l(t)$ on the different layers. Thus the total time-dependent Hamiltonian is

$$H = \sum_l [H_l - e\Phi_l(t) N_l] + H_T(t) \tag{5}$$

where H_l is the Hamiltonian of the electrons in layer l including superconducting interactions and pairing. At last we have to take into account phase fluctuations of the superconducting order parameter induced by charge fluctuations. The final results of the calculation of physical quantities depend

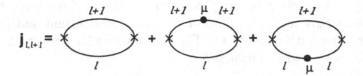

Figure 1. Graphs for the current density. Symbols: left $\times$ = current operator, right $\times$ = H_T, $\bullet$ = density vertex. Each cross corresponding to a hopping $T_{kk'}$ between layers l and $l+1$ is combined with a phase factor $\exp\left(\pm i\gamma_{l,l+1}(t)\right)$.

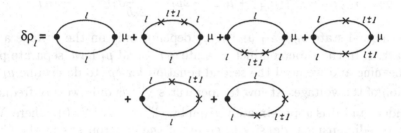

Figure 2. Graphs for the density response in layer l. Symbols: $\times$ = H_T, $\bullet$ = density vertex.

only on the gauge-invariant combinations $\gamma_{l,l+1}$ and μ_l of the electromagnetic potentials with the phase of the order parameter. Formally the same results are obtained if we replace in the Hamiltonian the vector potential $\int_l^{l+1} dz\, A_z(t) = A_{l,l+1}(t)$ by $(-\hbar/2e)\gamma_{l,l+1}(t)$ and the scalar potential $\Phi_l(t)$ by $\mu_l(t)$ and assume the order parameter to be real in the BCS-treatment of H_l. The phase $\gamma(t)$ can be written as

$$\gamma(t) = \gamma_0 + \omega t + \delta\gamma(t) \tag{6}$$

where the constant phase γ_0 is determined by the dc-current and ω is the Josephson frequency which is related to the dc-voltage. The oscillating part $\delta\gamma(t)$ is small for large McCumber parameter β_c.

By calculating the current response $j_{l,l+1}$ with respect to both the tunneling Hamiltonian H_T and the generalized scalar potential μ_l we restrict ourselves to second order tunneling processes and linear response with respect to μ_l. The Feynman graphs for the current density $j_{l,l+1}$ and the charge density response ρ_l within this approximation are shown in fig. 1 and fig. 2 respectively. Note that the equations corresponding to these graphs are non-linear in $\gamma(t)$.

For a systematic evaluation one has to insert the ansatz for $\gamma(t)$ into these expressions and has to separate different harmonics in the Josephson frequency ω. A considerable simplification is obtained if one keeps the non-linear effects only in the sin-term of the Josephson current and linearizes the other terms with respect to γ. Then one obtains for the current density across the barrier with thickness b

$$j_{l,l+1} = j_c \sin \gamma_{l,l+1} + \sigma_0 \frac{\hbar}{2eb} \dot{\gamma}_{l,l+1} + \sigma_1 (\mu_l - \mu_{l+1})/b \qquad (7)$$

and for the charge density response

$$\delta\rho_l = \chi^{(0)}_{\rho\rho} \mu_l + \chi^{(2)}_{\rho\rho} (\mu_{l+1} + \mu_{l-1} - 2\mu_l) + \frac{\hbar}{2eb} \sigma_1 (\gamma_{l-1,l} - \gamma_{l,l+1}) . \qquad (8)$$

In the normal state $\sigma_0 = \sigma_1$ and $j_{l,l+1}$ depends only on the voltage across the barrier. In the superconducting state $\gamma_{l,l+1}$ and μ_l have separate physical meaning and we need the second equation for $\delta\rho_l$ to determine μ_l as a function of the voltage. At low temperatures $\chi^{(0)}_{\rho\rho}$ is only weakly frequency dependent and it is approximately given by $\chi^{(0)}_{\rho\rho} \simeq -2e^2 N_2(0)$, where $N_2(0)$ is the two-dimensional density of states of the electron gas in the CuO_2-layers. The conductivities $\sigma_{0,1}$ as well as $\chi^{(2)}_{\rho\rho}$ describe the charge exchange with the neighboring layers and are proportional to $t_\perp^2$. Adding the displacement current we obtain

$$j = j_{l,l+1} + \epsilon\epsilon_0 \dot{E}_{l,l+1} \qquad (9)$$

where j is the external current density. With help of the 2nd Josephson equation and Maxwell's equations we finally arrive at one differential equation for the phase:

$$\frac{j}{j_c} = \left(1 - \alpha\Delta^{(2)}\right) \sin\gamma_{l,l+1} + \frac{1}{\omega_c}\left(1 - \eta\Delta^{(2)}\right) \dot{\gamma}_{l,l+1} + \frac{1}{\omega_p^2}\left(1 - \zeta\Delta^{(2)}\right) \ddot{\gamma}_{l,l+1} \qquad (10)$$

where $\omega_p^2 = j_c 2eb/(\epsilon\epsilon_0\hbar)$ and $1/\omega_c = \sigma_0/(\epsilon\epsilon_0\omega_p^2)$. α, η, ζ describe the coupling of the phase-difference in different layers via the derivative operator $\Delta^{(2)}$, which is defined as $\Delta^{(2)} f_l = f_{l+1} + f_{l-1} - 2f_l$. In particular

$$\alpha = -\epsilon\epsilon_0/(b\chi^0_{\rho\rho}) + O(t_\perp^2) , \qquad (11)$$

$$\eta = -\epsilon\epsilon_0/(b\chi^0_{\rho\rho})(1 - 2\sigma_1/\sigma_0) + O(t_\perp^2) . \qquad (12)$$

ζ is proportional to $t_\perp^2$. For $\omega \ll \Delta$ the quantity α is only weakly frequency dependent and it is $\alpha < 1$. If we neglect η and ζ we arrive back at the theory of Koyama and Tachiki [10]. In fact at $\omega \ll T \ll \Delta$ we find $\sigma_1 \simeq \sigma_0/2$ for d-wave superconductors. Thus neglecting η and ζ seems to be a good approximation for small values of ω.

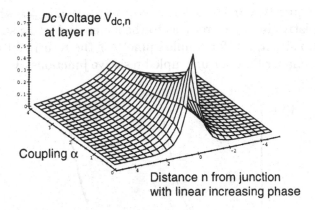

Figure 3. Distribution of the dc-voltage in the neighborhood of one resistive junction as function of the distance n and the coupling constant α.

3. Influence on the current-voltage characteristics

In the following we restrict ourselves to the approximation (10) with $\alpha = const, \eta = \zeta = 0$ supplemented by

$$\frac{\hbar}{2e}\dot{\gamma}_{l,l+1} = V_{l,l+1} \quad \alpha(V_{l+1,l+2} + V_{l-1,l} - 2V_{l,l+1}) \tag{13}$$

where we have defined the voltage $V_{l,l+1} = \int_l^{l+1} E_z dz = bE_{l,l+1}$. A junction is called to be in the resistive state if $< \dot{\gamma}_{l,l+1}(t) > \neq 0$. In the case of one junction in the resistive state there is a finite voltage-drop also in the neighboring junctions due to the coupling α. This is shown in Fig. 3. As in the absence of the coupling the total voltage is given by

$$V = \sum_l V_{l,l+1} = \frac{\hbar}{2e}\sum_l \dot{\gamma}_{l,l+1}. \tag{14}$$

For the total dc-voltage only the values of $< \dot{\gamma}_{l,l+1} >$ from the resistive barriers contribute. In the case of two or more resistive barriers in the stack (corresponding to the 2nd or higher order branch in the I-V characteristics) it makes a difference whether these resistive barriers are next neighbors or are separated by one or more non-resistive junctions. We find by both numerical simulations and analytical calculations that in the case of two resistive barriers next to each other the Josephson frequency of both junctions is the same, but (at the same current) is slightly higher than in the case of well-separated junctions. For two uncoupled junctions, of course, the total voltage is just the double of one resistive junction (the first branch). This is shown schematically in Fig. 4. The effect is more pronounced near

248

the plasma frequency. Particularly the return voltage ω_{return}, where at least one of the resistive junctions returns to the superconducting state is different for the two situations: for coupled junctions the return-voltage and the dc-current is smaller than for uncoupled resistive junctions. Our numerical

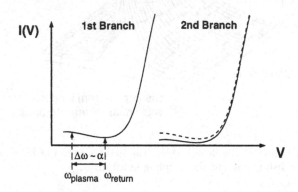

Figure 4. Schematic plot of the current-voltage characteristics for two coupled junctions (solid line) in comparison with two uncoupled junctions (dashed line)

simulations also show that the phases in two neighboring resistive junctions rotate coherently even if the critical currents of the junctions are slightly different. This phase-locking is very important for microwave application.

This work was supported by Bayerische Forschungsstiftung (C.P.), the German Science Foundation (C.H.), and the Humboldt-Foundation (A.S.).

References

1. R. Kleiner, F. Steinmeyer, G. Kunkel, and P. Müller. *Phys. Rev. Lett.*, 68(14):2394–2397, 1992.
2. K. Schlenga, G. Hechtfischer, R. Kleiner, W. Walkenhorst, and P. Müller. *Phys. Rev. Lett.*, 76:4943–4946, 1996.
3. A. Yurgens, D. Winkler, N.V. Zavaritsky, and T. Claeson. *Phys. Rev. B*, 53:R8887, 1996.
4. F.X. Régy, J. Schneck., J.F. Palmier, and H. Savary. *J. Appl. Phys.*, 76:4426, 1994.
5. K. Tanabe, Y. Jidaka, S. Karimoto, and M. Suzuki. *Phys. Rev. B*, 53:9348, 1996.
6. M. Itoh, S. Karimoto, K. Namekawa, and M. Suzuki. *Phys. Rev. B*, 55:R12001, 1997.
7. S. Tajima, G.D. Gu, S. Miyamoto, A. Odagawa, and N. Koshizuka. *Phys. Rev. B*, 48:16164, 1993.
8. S. Uchida and K. Tamasuka. *Physica C*, 297:1–7, 1997.
9. K. Kadowaki, I. Kakeya, K. Kindo, S. Takahashi, T. Koyama, and M. Tachiki. *Physica C*, 293:130, 1997.
10. T. Koyama and M. Tachiki. *Phys. Rev. B*, 54:16183–16191, 1996.
11. S.N. Artemenko and A.G.Kobelkov. *Phys. Rev. Lett.*, 78:3551–3354, 1997.
12. D.A. Ryndyk. *JETP Lett.*, 65:791, 1997.
13. D.A. Ryndyk. *Phys. Rev. Lett.*, 80:3376–3379, 1998.
14. C. Preis, C. Helm, J. Keller, K. Schlenga, A. Sergeev, and R. Kleiner. *to be published.*

NONSTATIONARY JOSEPHSON AND QUASIPARTICLE TUNNEL CURRENTS THROUGH JUNCTIONS BETWEEN SUPERCONDUCTORS WITH SPIN-DENSITY WAVES

A. M. GABOVICH AND A. I. VOITENKO

Crystal Physics Department, Institute of Physics, NASU
prospekt Nauki 46, 252650 Kiev, Ukraine

Abstract. The nonstationary Josephson, interference, and quasiparticle currents through junctions involving partially-dielectrized superconductors with spin-density waves were calculated. The existence of both superconducting and dielectric energy gaps leads to a large set of peculiarities for current-voltage characteristics. The new regime of broken symmetry is predicted for all currents. The results agree with tunnel and point-contact data for junctions including URu_2Si_2.

Key words: superconductivity, spin-density waves, Josephson effect, tunnel junction, dielectric gap

1. Introduction

The spin-singlet Cooper pairing and the magnetic ordering are considered with good reason as antagonists due to the diamagnetic currents induced by the molecular field and the paramagnetic effects consisting in the electron spin-flip [1]. Thus, a coexistence of ferromagnetism and superconductivity within the same volume element of a substance seems very improbable. At the same time, the antiferromagnetic (AFM) ordering of different rare-earth ions R in RMo_6S_8, RMo_6Se_8, $R(Rh_{1-x}Ir_x)_4B_4$, and RNi_2B_2C compounds is more merciful to the superconducting (SC) pairing of the host itinerant electrons [1, 2]. For such substances AFM and SC correlations tending to destroy each other often lead to reentrant phenomena, and the Néel temperatures T_N may be either higher or lower than SC critical ones T_c.

Another type of the AFM order is possible, namely, commensurate or

249

S.-L. Drechsler and T. Mishonov (eds.), High-T_c Superconductors and Related Materials, 249–257.
© 2001 *Kluwer Academic Publishers. Printed in the Netherlands.*

incommensurate spin-density waves (SDW's) [3]. This state originates from the parent high-temperature, T, phase due to the nesting of the Fermi surface (FS) sections. For strictly one-dimensional metal, the FS consists only of two parallel planes and the low-T state is insulating. Unfortunately, the standard solution of the mean-field equation for the dielectric order parameter is of a limited value because higher-order vertex corrections, i. e., all parquet diagrams, should be taken into account. For Peierls insulator unstable with respect to the charge-density wave (CDW) formation it was shown in Ref. [4]. Of course, the same is true for the SDW's and superconductivity coexistence problem.

Nevertheless, for real systems the problem seems rather academic. In organic superconductors the interchain interaction is so strong that they become quasi-two-dimensional [5]. In the model of the SDW state as a three-dimensional excitonic insulator, the quasi-one-dimensionality is not anticipated from the very beginning [6]. Here the excitonic instability is a consequence of the congruency between electron and hole pockets rather than between different branches of one band as in the Peierls-Hubbard insulator case [3]. The existence of the magnetic excitonic insulator state was first directly experimentally confirmed for the intermediate valent semiconductor $TmSe_{1-x}Te_x$ [7].

Having in view two kinds of systems mentioned above, one may neglect all fluctuation phenomena with a great accuracy and consider the mean-field picture of SDW's. It was done previously both for normal and SC metals [8, 9] in the framework of the partial dielectrization (partial gapping) scheme introduced in Ref. [10] for the CDW case. The nesting features can be preserved in quasi-two- and quasi-three-dimensional models, as was shown for dichalcogenides [11] and phosphate bronzes [12].

In this paper we consider tunnel junctions where one or both electrodes are SDW superconductors. Three current components are calculated, namely, nonstationary Josephson, pair-quasiparticle interference ("cosine"), and quasiparticle ones. Stationary Josephson currents between SDW superconductors and quasiparticle current through junctions with SDW partially-gapped non-SC metals have been calculated earlier [13, 14]. The results for CDW metals in the absence of superconductivity are identical. The tunneling between CDW superconductors was treated elsewhere [15].

2. Theory

The model mean-field Hamiltonian of the SDW superconductor has the form [8, 9, 10, 13, 14, 15]

$$\mathcal{H} = \mathcal{H}_0 + \mathcal{H}_{BCS} + \mathcal{H}_{SDW}, \tag{1}$$

where $\mathcal{H}_0$ is the free-electron Hamiltonian, $\mathcal{H}_{BCS}$ is the original Bardeen-Cooper-Schrieffer (BCS) Hamiltonian, and

$$\mathcal{H}_{SDW} = -2\Sigma \sum_{i=1}^{2} \sum_{\mathbf{p}\alpha} \alpha a_{i\mathbf{p}\alpha}^{\dagger} a_{i,\mathbf{p}+\mathbf{Q},\alpha} + \text{H.c.} \qquad (2)$$

is the SDW Hamiltonian describing the dielectric pairing. The operator $a_{i\mathbf{p}\alpha}^{\dagger}$ ($a_{i\mathbf{p}\alpha}$) is the creation (annihilation) operator of a quasiparticle with a quasimomentum $\mathbf{p}$ and spin projection $\alpha = \pm\frac{1}{2}$ from the ith FS section. Specifically, $i = 1$ and 2 for the nested sections where the electron spectrum is degenerate $\xi_1(\mathbf{p}) = -\xi_2(\mathbf{p} + \mathbf{Q})$, $\mathbf{Q}$ being the SDW vector, while $i = 3$ for the rest of the FS. The dielectric order parameter Σ emerges only on the nested FS sections. On the other hand, the *single* superconducting order parameter Δ appears on the whole FS. The ratio $\nu = N_{nd}(0)/N_d(0)$ describes the gapping degree of the primordial FS. Here $N_{d(nd)}(0)$ is the density of states for the dielectrized (nondielectrized) part of the FS.

Our approach based on the seminal work [10] explicitly comprises conjectures that the Fermi level lies inside the dielectric gap and SDW's impede superconductivity. The alternative point of view adopts the concept of complete dielectrization, i. e., Σ existing on the whole FS and the intrinsic doping of the system [16]. Then the Fermi level is outside the gap region and superconductivity may be even stimulated by the high density of electron states at dielectric gap edges [17]. There are also other scenarios of the positive influence of AFM correlations on superconductivity: (i) the spin-fluctuation induced d-wave pairing [18], (ii) unified SO(5)–SO(n) pictures of coexisting SDW's and d-wave Cooper pairing [19], (iii) spin-triplet p-wave pairing [20]. First two scenarios are usually applied to cuprates [21], and the last one to UPt$_3$ [22] and Sr$_2$RuO$_4$ [23].

Below we restrict ourselves to s-wave-like superconductors where the Cooper and electron-hole pairings compete for the FS. The quantity $\Sigma(T)$ is taken independent of Δ [15]. The order parameter Σ is pinned, real and can be of either sign [14]. $\Sigma(T)$ is considered to be of the BCS form.

The normal $G_{ij}^{\alpha\beta}(\mathbf{p};\omega_n)$ and anomalous $F_{ij}^{\alpha\beta}(\mathbf{p};\omega_n)$ Matsubara Green's functions (GF's) were found from the Dyson-Gor'kov equations [9, 13]. To calculate the tunnel currents we obtained in the usual manner [24] the temporal GF's $F(\omega)$ and $G(\omega)$ from their Matsubara counterparts. The equivalence of the dielectrized FS sections 1 and 2 reduces the number of independent GF's to six for each electrode. They depend not only on Δ but also on the combinations $D_{\pm} = \Delta \pm \Sigma$.

The total tunnel current I through the junction is calculated with the use of the tunnel Hamiltonian, $\mathcal{T}$, approach in the lowest order of the perturbation theory [24]. In the spirit of the conventional Ambegaokar-

Baratoff theory [25], we consider all matrix elements of T equal and independent of Δ and Σ. Besides, the current I is assumed independent of the relative spatial orientation of the junction plane and the SDW vector $\mathbf{Q}$, so that the universal tunnel resistance R may be introduced [14, 15]. Then, in the adiabatic approximation $V^{-1}\frac{dV}{d\tau} \ll T_c$ for the ac bias voltage $V(\tau) \equiv V_{\text{right}}(\tau) - V_{\text{left}}(\tau)$ across the Josephson junction, we obtain the expression for I in the general case of two different SDW superconductors as electrodes:

$$I[V(\tau)] = I^1(V)\sin 2\phi + I^2(V)\cos 2\phi + J(V), \qquad (3)$$

where $\phi = \int^\tau eV(\tau)d\tau$, τ denotes time, and e is the elementary charge. The explicit cumbersome expressions for amplitudes $I^{1,2}$ and J are given elsewhere. They are sums of functionals including different GF's inherent to each electrode, and therefore are multicomponent (nine terms for each amplitude in the general case of a junction between two different SDW superconductors):

$$I^{1,2}(V) = \sum_{i=1}^{9} I_i^{1,2}(V), J(V) = \sum_{i=1}^{9} J_i(V). \qquad (4)$$

3. Results

First, let us consider the case of nonsymmetrical (ns) junctions when an SDW superconductor is, say, the left-hand-side (l.h.s.) electrode and a BCS superconductor with a gap Δ_{BCS} is the right-hand-side (r.h.s.) one. Then, $\Sigma_{\text{right}} = 0, \nu_{\text{right}} = \infty$, some terms in Eqs. (4) vanish, and we can group the rest into three components for each amplitude. For two of them the symmetry relations appropriate to the BCS case hold:

$$I_{ns1,2}^1(-V) = I_{ns1,2}^1(V), I_{ns1,2}^2(-V) = -I_{ns1,2}^2(V), J_{ns1,2}(-V) = -J_{ns1,2}(V). \qquad (5)$$

At the same time, the third term has nonconventional properties, opposite to those given by Eq. (5):

$$I_{ns3}^1(-V) = -I_{ns3}^1(V), I_{ns3}^2(-V) = I_{ns3}^2(V), J_{ns3}(-V) = J_{ns3}(V). \qquad (6)$$

Thus, the current-voltage characteristics (CVC's) for all three currents are nonsymmetrical with respect to the V polarity, contrary to the well-known polarity independence for BCS superconductors [25]. Since Σ may be of either sign with equal probability due to the thermodynamic equivalence of both states [9, 13], two nonsymmetrical CVC's are equally probable, being a mirror reflection of each other with reference to the current axis. Of

TABLE 1. Types and positions of CVC peculiarities inherent to components of currents through ns-junctions.

$|eV|$	Type	Components
$K_\pm = |D_\pm| + \Delta_{\text{BCS}}$	a	1,3
$K = \Delta + \Delta_{\text{BCS}}$	a	2
$L_\pm = ||D_\pm| - \Delta_{\text{BCS}}|^c$	b	1,3
$L = |\Delta - \Delta_{\text{BCS}}|^c$	b	2

[a]Logarithmic singularities for Josephson current components and jumps for interference and quasiparticle current components.
[b]Jumps for Josephson current components and logarithmic singularities for interference and quasiparticle current components.
[c]For $T \neq 0$.

course, in real experiments a certain CVC, corresponding to a definite phase of Σ, may become preferential because of additional coincidental factors. The analysis shows that Riedel-like logarithmic singularities and/or jumps should occur at several voltages determined by specific combinations of $D_\pm, \Delta$, and Δ_{BCS} (see Table 1).

Another usual experimental setup involves symmetrical junctions when both electrodes are *thermodynamically* identical SDW superconductors. As follows, these junctions ought to be further classified as genuinely symmetrical (s) or symmetrical with broken symmetry (bs).

In the s-case, $\Sigma_{\text{left}} = \Sigma_{\text{right}} = \Sigma, \nu_{\text{left}} = \nu_{\text{right}}$, and the nonvanishing terms in Eqs. (4) can be combined into four components for each amplitude. The conventional symmetry relations [24, 25] hold

$$I_{si}^1(-V) = I_{si}^1(V), I_{si}^2(-V) = -I_{si}^2(V), J_{si}(-V) = -J_{si}(V), i = 1 \ldots 4, \quad (7)$$

so the CVC's for all currents are symmetrical with reference to V polarity. Also they do not depend on the sign of Σ. The feature points are collected in Table 2.

The *symmetry breaking* takes place when, say, the l.h.s. electrode is characterized by $\Sigma_{\text{left}} = \Sigma > 0$ and the r.h.s. one has negative $\Sigma_{\text{right}} = -\Sigma < 0$, or vice versa (the condition $\nu_{\text{left}} = \nu_{\text{right}}$ holds true). In both cases the junction is nonsymmetrical in reality, although $|\Sigma_{\text{left}}| = |\Sigma_{\text{right}}|$ and all macroscopical properties of each separate electrode are *identical* [9, 13]. However, if the junction concerned is a part of an electric circuit, it will manifest itself through the nonsymmetricity of the resulting CVC's as a phase-sensitive indicator of the symmetry difference between the electrodes [26]. Such a phenomenon comprises a new macroscopic manifestation of the symmetry breaking in many-body systems.

TABLE 2. Types and positions of CVC peculiarities inherent to components of currents through s- and bs-junctions.[a]

$|eV|$	Type	Position $\Delta < \Sigma$	$\Delta > \Sigma$	Components[b]
2Δ	c	2Δ	2Δ	4
$2|D_\pm|$	c	$2(\Sigma \pm \Delta)$	$2(\Delta \pm \Sigma)$	1,2,5
$H_+ = \|\,|D_+| + |D_-|\,\|$	c	2Σ	2Δ	1,2
$H_- = \|\,|D_+| - |D_-|\,\|^e$	d	2Δ	2Σ	1,2,5
$M_+ = |D_+| + \Delta$	c	$2\Delta + \Sigma$	$2\Delta + \Sigma$	3,6
$M_- = |D_-| + \Delta$	c	Σ	$2\Delta - \Sigma$	3,6
$N_+ = \|\,|D_+| - \Delta\,|^e$	d	Σ	Σ	3,6
$N_- = \|\,|D_-| - \Delta\,|^e$	d	$|2\Delta - \Sigma|$	Σ	3,6

[a]In the s-case the choice $\Sigma > 0$ is made. In the bs- case $\Sigma > 0$ in the l. h. s. electrode and $\Sigma < 0$ in the r. h. s. one. See details in the text.
[b]Components 5 and 6 are inherent to currents through bs- junctions only.
[c]Logarithmic singularities for Josephson current components and jumps for interference and quasiparticle current components.
[d]Jumps for Josephson current components and logarithmic singularities for interference and quasiparticle current components.
[e]For $T \neq 0$.

In the bs-case, each current amplitude has two more terms as compared to the s-case. The symmetry relations for the first four ones are the same as for the s-junction

$$I_{bsi}^1(-V) = I_{bsi}^1(V), I_{bsi}^2(-V) = -I_{bsi}^2(V), J_{bsi}(-V) = -J_{bsi}(V), i = 1 \ldots 4,$$
(8)

but the rest two terms are similar to the third components of currents through the ns-junction [see Eq. (6)]

$$I_{bsi}^1(-V) = -I_{bsi}^1(V), I_{bsi}^2(-V) = I_{bsi}^2(V), J_{bsi}(-V) = J_{bsi}(V), i = 5, 6.$$
(9)

These components are equal and enhance each other. As a consequence, the total current amplitudes $I_{bs}^{1,2}(V)$ and $J_{bs}(V)$ are neither symmetrical nor antisymmetrical in V. They are also sensitive to the sign of Σ, so that the different V-polarity branches are interchanged with changing Σ sign. In contrast, for the s-case, the components 5 and 6 are of different signs and *exactly compensate* each other. The CVC peculiarities of the bs-junctions are shown in Table 2.

We see that bs-junctions are somewhat in between s- and ns-ones. Hence, depending on generally unidentified factors including the electric

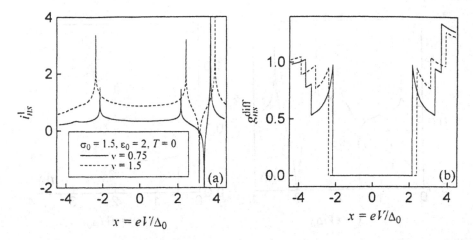

Figure 1. Dimensionless amplitude of the nonstationary Josephson current $i^1_{ns} \equiv I^1_{ns} eR/\Delta_0$ through nonsymmetrical junctions (a) and quasiparticle conductance $g^{\mathrm{diff}}_{ns} \equiv RdJ_{ns}/dV$ (b) versus $x = eV/\Delta_0$ for different ν, where $\Delta_0 = \Delta(T = 0, \Sigma = 0), \sigma_0 = \Sigma(T = 0)/\Delta_0, \epsilon_0 = \Delta_{\mathrm{BCS}}(T = 0)/\Delta_0$. See other notations in the text.

and thermal prehistories of a given formally symmetrical sample, the latter may reveal either s- or ns-properties. Thus, our classification deals with states rather than junctions.

The dependences of the dimensionless nonstationary Josephson current amplitudes $i^1_{ns,s,bs}(x) = I^1_{ns,s,bs}(V)eR/\Delta_0$ and differential quasiparticle conductivities $g_{ns,s,bs}(x) = RdJ_{ns,s,bs}(V)/dV$, where $x = eV/\Delta_0$ and Δ_0 is the SC gap in the absence of dielectrization are shown in Figs. 1–3. All feature points for $T = 0$ assembled in Tables 1 and 2 are seen. The structure of the CVC's is much more involved than for BCS [25]and even CDW superconductors [15].It is also important that the *dielectric* gap influences strongly the coherent Josephson *supercurrent*, although the adopted conditions of pinning ruled out the coherent link between two SDW's across the junction of the type considered for the CDW case [27].

4. Discussion

The compounds where superconductivity and SDW coexist are not as numerous as CDW superconductors [15]. Further still, it is hard to distinguish experimentally between objects with a dielectric order parameter Σ of the collective origin appearing due to the nesting FS and the AFM superconductors of other nature. Nevertheless, we can indicate at least the organic metals $(TMTSF)_2X$ ($X = AsF_6$, PF_6, ClO_4) [5], alloys $Cr_{1-x}Re_x$ [28], and the heavy-fermion compound URu_2Si_2 [29, 30]. The latter has

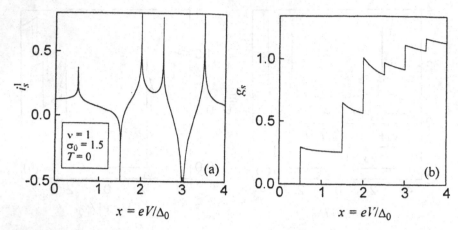

Figure 2. Dependences on x of the dimensionless Josephson current amplitude i_s^1 (a) and the differential quasiparticle conductance g_s (b) for symmetrical tunnel junction.

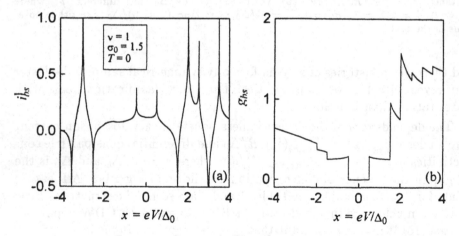

Figure 3. Same as in Fig. 2 but for bs-junction.

$T_N \approx 17\,\text{K}, T_c \approx 1.4\,\text{K}$, and is usually treated as a partially dielectrized metal, although the values of Σ and ν inferred from various specific heat, point-contact, and tunnel measurements differ substantially.

It is natural to apply our results to URu_2Si_2 as the most studied SDW superconductor. Point-contact measurements [31, 32, 33, 34] for ns-junctions reveal asymmetric CVC's with complex structure possessing numerous feature points for differential conductivity $G^{\text{diff}}(V)$. This contradicts the traditional BCS picture but agrees qualitatively with our results. However, it is impossible to compare directly point-contact CVC's and the calculated tunnel currents, although the relevant positions of the peculiarities should coincide.

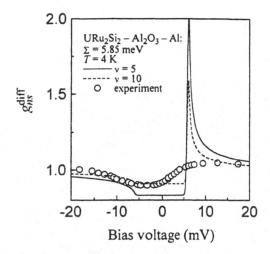

Figure 4. Dependences $g_{ns}^{\text{diff}}(V)$ for the ns-junction $URu_2Si_2-Al_2O_3-Al$ in the normal state taken from the experiment [32] and calculated for various ν's.

As for the tunnel measurements, they were made for $URu_2Si_2-Al_2O_3-Al$ junctions at $T \approx 4\,\text{K}$ [32], so we can check our calculations only for $G_{ns}^{\text{diff}}(V)$ in the normal state. Such a comparison is shown in Fig. 4. The fitting procedure leads to a satisfactory result but with the parameter ν exceeding considerably those inferred from specific heat data, namely, $\nu = 0.4$ [29] and $\nu = 1.5$ [30]. At the same time, because of the great discrepancy in the latter values, these data can not serve as the basis to prove or disprove the validity of our approach.

At last, it is important to emphasize that the predicted symmetry breaking seems to be confirmed by the point-contact measurements for homocontacts $URu_2Si_2-URu_2Si_2$ [33], where both symmetrical and nonsymmetrical CVC's were observed in different conditions for various samples.

Acknowledgements We are grateful to S.-L.Drechsler and K.Maki for useful discussions. This work was supported, in part, by the Ukrainian State Foundation for Fundamental Researches (Grant No. 2.4/100).

References

1. L. N. Bulaevskii, A. I. Buzdin, M.-L. Kulic, and S. V. Panjukov, Adv. Phys. **34**, 175 (1985).
2. P. C. Canfield, P. L. Gammel, and D. J. Bishop, Phys. Today **51**, 40 (October 1998); S. V. Shulga *et al.*, Phys. Rev. Lett. **80**, 1730 (1998).
3. G. Grüner, Rev. Mod. Phys. **66**, 1 (1994).
4. J. Sólyom, Adv. Phys. **28**, 201 (1979).
5. J. Friedel and D. Jerome, Contemp. Phys. **23**, 583 (1982).
6. A. N. Kozlov and L. A. Maksimov, Zh. Eksp. Teor. Fiz. 48, 1184 (1965) [Sov. Phys.

SUPERCONDUCTING FLUCTUATIONS AND THEIR ROLE IN THE NORMAL STATE ANOMALIES OF HIGH TEMPERATURE SUPERCONDUCTORS

A.A. VARLAMOV
Unità "Tor Vergata" di INFM,
Dipartimento STFE, Università "Tor Vergata",
via Tor Vergata 110, 00133 Roma, Italy
and
Department of Theoretical Physics
Moscow Institute for Steel and Alloys
Leninskii Prospect, 4 Moscow 117936 Russia

Abstract The role of the DOS fluctuations in the normal properties anomalies of HTS is analyzed. It is shown how, taking into account this effect, many puzzling and long debated properties of HTS materials (e.g. the steep increase of the electrical resistivity along the c axis just above T_c, the anomalous magnetoresistance, the effects of the magnetic field on the resistive transition along the c-axis, the c-axis far infrared absorption spectrum, NMR spectra around the critical temperature etc.) can be understood, leading to a simple consistent description in terms of fluctuation theory.

The lectures are based on the review article [1].

1. Introduction

There are no doubts that the puzzling anomalies of the normal state properties of high temperature superconductors (HTS) are tightly connected with the physical origin of superconductivity in these materials. Among them are:

- a peak in the c-axis resistivity above T_c followed by a decrease to zero as temperature is decreased [2, 3];

- the giant growth of this peak in the presence of an external magnetic field applied along c-axis and its shift towards low temperatures [4];

- the giant magnetoresistance observed in a wide temperature range above the transition [5, 6, 7];

S.-L. Drechsler and T. Mishonov (eds.), High-T_c Superconductors and Related Materials, 259–288.

- the deviation from the Korringa law in the temperature dependence of the NMR relaxation rate above T_c [8];

-the opening of a large pseudo-gap in c-axis optical conductivity at temperatures well above T_c [9, 10] ;

- the anisotropic gap observed in the electron spectrum by angular resolved photo-emission experiments [11].

- the gap-like tunneling anomalies observed already above T_c [12, 13, 14, 15, 16, 17].

- the anomalies in the thermoelectric power above T_c [18, 19].

These effects have been attributed by many authors to the opening of a "pseudo-gap". Naturally this has led to numerous speculations about the physical origin of such a gap.

During the last decade a lot of efforts have been undertaken to explain the unusual normal state properties of HTS materials using unconventional theories of superconductivity based on ideas of spin-charge separation, preformed Cooper pairs, polaron mechanism of superconductivity, etc.(see for instance [21]). They have been widely discussed and we will not overview them here. In the case of HTS with a well developed Fermi surface (i.e. optimally doped or overdoped part of the phase diagram) one can approach this problem from another side, namely to develop the perturbation theory for interacting electrons in the normal phase of a strongly anisotropic superconductor. We will not specify the origin of this interaction: for our purposes it is enough to assume that this interaction is attractive and leads to the appearance of superconductivity with Cooper pairs of charge $2e$ at temperatures below T_c. Of course, the smallness of the effects magnitude (necessary for the applicability of the perturbative approach) is a serious limitation of the proposed theory.

The Cooper channel of interelectron interaction is equivalent to taking into account superconducting fluctuations which are unusually strong in HTS [22]. The reasons for this strength are the effective low dimensionally of the electron spectrum, the low density of charge carriers and the high values of critical temperature of HTS. We will show that these peculiarities lead not only to the increase of the magnitude of the fluctuation effects, but frequently change the hierarchy of the different fluctuation contributions, leading to the appearance of competition among them and even to the change of the habitual (in conventional superconductivity) sign of the overall correction. As a result, the fluctuation effects can manifest themselves in very unusual form, so that their origin cannot be identified at the first glance.

The first well-known result is that in the metallic phase one automatically has the non-equilibrium analogue of pre-formed Cooper pairs above T_c. Indeed, taking into account thermal fluctuations (or interelectron interaction in the Cooper channel) leads to the appearance of some non-zero density of fluctuation Cooper pairs (in contrast to pre-formed pairs with finite lifetime) in the superconducting layers without the establishment of the long range order in the system. It is important that in the 2D case, typical for HTS materials, the density of Cooper pairs decreases with temperature extremely slowly: $\sim \ln \frac{T_c}{T-T_c}$. One should therefore not be surprised that precursor effects can often be detected in the normal phase well above T_c (especially in underdoped samples).

The formation of fluctuation Cooper pairs of normal electrons above T_c has an important though usually ignored consequence: the decrease of the density of one-electron states (DOS) at the Fermi level [23, 24]. This circumstance turns out to be crucial for the understanding of the aforementioned effects and it will constitute the quintessence of this review. In this way can be at least qualitatively (and in many cases quantitatively as well) explained the following phenomena:

The behavior of the c-axis resistance [4] which has been explained in terms of the suppression of the one-electron DOS at the Fermi level and the competition of this effect with the positive Aslamazov-Larkin (AL) paraconductivity [25, 26, 27, 28, 29].

The giant growth of the c-axis resistance peak in the presence of an external magnetic field applied along the c-axis is explained using the same approach [30] which was shown to fit well the experiments [29, 31, 32].

The anomalous negative magnetoresistance observed above T_c in BSCCO samples [33, 6, 35] was again explained by the same DOS fluctuation contribution [36]. Moreover, its competition with the positive Aslamazov-Larkin magnetoresitance gave good grounds for the prediction of a sign change in the magnetoresistance as temperature decreases towards T_c [36]. The latter effect was very recently confirmed experimentally on YBCO samples [7].

The decrease of the thermoelectric power at the edge of transition turns out to be the result of the DOS fluctuation contribution which dominates over the AL term [37] previously assumed to play the crucial role [38, 39, 40, 41].

The temperature dependence of the NMR rate $\frac{1}{T_1 T}$ can be explained as the result of the competition between the positive Maki-Thompson (MT) correction to the Korringa law and the negative DOS contribution at the edge of the transition [42, 43].

The observed pseudo-gap-like structure in the far infra-red optical conductivity along c-axis can also be attributed to the suppression of the one-electron DOS at the Fermi level. This leads to the appearance of a sizable negative contribution in optical conductivity, which shows up in a wide range of frequencies (up to $\omega_{DOS} \sim \tau^{-1}$), exceeding the positive AL and MT contributions in magnitude and range of manifestation [44].

We believe that the fact that even the simple approach proposed here was able to explain most of the anomalies of the normal state properties of HTS mentioned above is not accidental. It shows the importance of the interelectron interaction in the problem discussed and demonstrates that even the way of "up-grading" the traditional BCS theory to include the HTS peculiarities is creative in the explanation of the HTS properties. Further, the approach considered provides clear results which can be compared with those obtained from the alternative viewpoints and an attempt to match them in the region of the intermediate strengths of interaction may be undertaken.

2. Excursus to superconducting fluctuation theory

During the first half of the century, after the discovery of superconductivity by Kammerlingh-Onnes, the problem of fluctuations smearing the superconducting transition was not even considered. In bulk samples of traditional superconductors the critical temperature T_c sharply divides the superconducting and the normal phases. It is worth mentioning that such behavior of the physical characteristics of superconductors is in perfect agreement both with the Ginzburg-Landau phenomenological theory (1950) and the BCS microscopic theory of superconductivity (1957). Nevertheless, at the same time, it was well known that thermodynamic fluctuations can cause strong smearing of other second -order phase transitions, such as the λ-point in liquid helium.

As already mentioned, the characteristics of high temperature and organic superconductors, low dimensional and amorphous superconducting systems studied today, differ strongly from those of the traditional superconductors discussed in textbooks. The transitions turn out to be much more smeared here. The appearance of thermodynamically nonequilibrium Cooper pairs (superconducting fluctuations) above critical temperature leads to precursor effects of the superconducting phase occurring while the system is still in the normal phase, often far enough from T_c. The conductivity, the heat capacity, the diamagnetic susceptibility, the sound attenuation etc. may increase considerably in the vicinity of the transition temperature.

The first numerical estimation of the fluctuation contribution to the heat capacity of superconductors in the vicinity of T_c was done by Ginzburg in 1960 [45]. In that paper he showed that superconducting fluctuations increase the heat capacity even above T_c. In this way the fluctuations smear the jump in the heat capacity which, in accordance with the phenomenological Ginzburg-Landau theory of second order phase transitions (see for instance [46]), takes place at the transition point itself. The range of temperatures where the fluctuation correction to the heat capacity of a bulk

clean conventional superconductor is relevant was estimated by Ginzburg
as

$$\frac{\delta T}{T_c} \sim \left(\frac{T_c}{E_F}\right)^4 \sim \left(\frac{a}{\xi}\right)^4 \sim 10^{-12} \div 10^{-14} \tag{1}$$

where a is the interatomic distance, E_F is the Fermi energy and ξ is the
superconductor coherence length at zero temperature. It is easy to see that
this is many orders of magnitude smaller than the temperature range accessible in real experiments. This is why fluctuation phenomena in superconductors were considered experimentally inaccessible for a long time.

In the 1950s and 60s the formulation of the microscopic theory of superconductivity, the theory of type-II superconductors and the search for
high-T_c superconductivity attracted the attention of researchers to dirty
systems, and the properties of superconducting films and filaments began
to be studied. In 1968, in the well known paper of L. G. Aslamazov and A. I.
Larkin [47] , the consistent microscopic theory of fluctuations in the normal
phase of a superconductor in the vicinity of the critical temperature was formulated. This microscopic approach confirmed Ginzburg's evaluation [45]
for the width of the fluctuation region in a bulk clean superconductor, but
much more interesting results were found in [47] for low dimensional or dirty
superconducting systems. The exponent ν of the ratio (a/ξ_0), which enters
in (1), drastically decreases as the effective dimensionality of the electron
motion diminishes: $\nu = 4$ for 3D, but $\nu = 1$ for 2D electron spectrum (in
the clean case) which will be the most interesting for HTS materials.

Another source of the effective increase of the strength of fluctuation
effects is the decrease of the coherence length, which occurs in dirty superconductors because of the diffusive character of the electronic motion. This
means that fluctuation phenomena are mainly observable in amorphous
materials with removed dimensionality, such as films and whiskers, where
both factors mentioned above come into play. HTS is of special interest in
this sense, because their electronic spectrum is extremely anisotropic and
their coherence length is very small. As a result the temperature range in
which the fluctuations are important in HTS may reach tens of degrees.

3. Fluctuation Cooper pairs above T_c

Let us start from the calculation of the density of fluctuation Cooper pairs in
the normal phase of superconductor. We restrict ourselves to the region
of temperatures near the critical temperature, so we can operate in the
framework of the Landau theory of phase transitions [46].

When we consider the system above the transition temperature, the
order parameter $\Psi(\vec{r})$ has a fluctuating origin (its mean value is equal to

zero) and it depends on the space variables even in the absence of magnetic field. This is why we have to take into account the gradient term in the Ginzburg-Landau functional for the fluctuation part of the thermodynamical potential Ω_{fl}:

$$\Omega_{fl} = \Omega_s - \Omega_n = \alpha \int dV \left\{ \varepsilon |\Psi(\vec{r})|^2 + \frac{b}{2\alpha} |\Psi(\vec{r})|^4 + \eta_D |\nabla \Psi(\vec{r})|^2 \right\} \quad (2)$$

where $\varepsilon = \ln(T/T_c) \approx \frac{T-T_c}{T_c} \ll 1$ for the temperature region discussed and $\alpha = \frac{1}{4m\eta_D}$. The positive constant η_D of the phenomenological Ginzburg-Landau may be expressed in terms of microscopic characteristics of metal:

$$\eta_D = -\frac{v_F^2 \tau^2}{D} \left[\psi \left(\frac{1}{2} + \frac{1}{4\pi\tau T} \right) - \psi \left(\frac{1}{2} \right) - \frac{1}{4\pi\tau T} \psi' \left(\frac{1}{2} \right) \right] \rightarrow \quad (3)$$

$$\frac{\pi v_F^2 \tau}{8DT} \begin{cases} 1 & \text{for } \tau T \ll 1, \\ 7\zeta(3)/(2\pi^3 \tau T) & \text{for } \tau T \gg 1, \end{cases}$$

where v_F is the Fermi velocity τ is quasiparticle scattering time, D is the space dimensionality, $\psi(x)$ and $\psi'(x)$ are the digamma function and its derivative respectively, and $\zeta(x)$ is the Riemann zeta function [1]. Dealing mostly with $2D$ case we will often use this definition omitting the subscript "2": $\eta_2 \equiv \eta$.

Dealing with the region of temperatures $\varepsilon > 0$, in the first approximation we can neglect the fourth order term in (2). Then, carrying out the Fourier transformation of the order parameter

$$\Psi_{\vec{k}} = \frac{1}{\sqrt{V}} \int \Psi(\vec{r}) \exp^{-i\vec{k}\vec{r}} dV \quad (4)$$

one can easily write the fluctuation part of the thermodynamic potential as a sum over Fourier components of the order parameter:

$$\Omega_{fl} = \alpha \sum_{\vec{k}} \left(\varepsilon + \eta_D k^2 \right) \left| \Psi_{\vec{k}} \right|^2. \quad (5)$$

Here

$$\vec{k} = \frac{2\pi}{L_x} n_x \vec{i} + \frac{2\pi}{L_y} n_y \vec{j} + \frac{2\pi}{L_z} n_z \vec{l},$$

where $L_{x,y,z}$ are the sample dimensions in appropriate directions; $\vec{i}, \vec{j}, \vec{l}$ are unit vectors along the axes; $n_{x,y,z}$ are integer numbers; V is the volume of the sample.

[1] We will mostly use the system $\hbar = c = k_B = 1$ everywhere, excluding the situations where the direct comparison with experiments is necessary.

In the vicinity of the transition the order parameter Ψ undergoes equilibrium fluctuations. The probability of the fluctuation realization of some concrete configuration $\Psi(\vec{r})$ is proportional to [46]:

$$\mathcal{P} \propto \exp\left[-\frac{\alpha}{T}\sum_{\vec{k}}\left(\varepsilon + \eta_D k^2\right)\left|\Psi_{\vec{k}}\right|^2\right], \tag{6}$$

Hence the average equilibrium fluctuation of the square of the order parameter Fourier component $\left|\Psi_{\vec{k}}^{(fl)}\right|^2$ may be calculated as

$$\left\langle\left|\Psi_{\vec{k}}^{(fl)}\right|^2\right\rangle = \frac{\int \left|\Psi_{\vec{k}}\right|^2 \exp\left[-\frac{\alpha}{T}\left(\varepsilon + \eta_D k^2\right)\left|\Psi_{\vec{k}}\right|^2\right] d\left|\Psi_{\vec{k}}\right|^2}{\int \exp\left[-\frac{\alpha}{T}\left(\varepsilon + \eta_D k^2\right)\left|\Psi_{\vec{k}}\right|^2\right] d\left|\Psi_{\vec{k}}\right|^2}$$

$$\tag{7}$$

$$= \frac{T}{\alpha\left(\varepsilon + \eta_D k^2\right)},$$

The concentration of Cooper pairs $\mathcal{N}_{c.p.}$ is determined by the average value of the square of the order parameter modulus [46]. For the two-dimensional case, which is of most interest to us, one finds:

$$\mathcal{N}_{c.p.}^{(2)} = \left\langle\left|\Psi_{(fl)}\right|^2\right\rangle = \int \frac{d^2\vec{k}}{(2\pi)^2}\left|\Psi_{\vec{k}}^{(fl)}\right|^2 \exp i(\vec{k}\cdot\vec{r})\Big|_{\vec{r}\to 0} =$$

$$= \frac{T}{\alpha}\int \frac{1}{\varepsilon + \eta_2\vec{k}^2}\frac{d^2\vec{k}}{(2\pi)^2} = 2\mathcal{N}_e^{(2)}\frac{T_c}{E_F}\ln\frac{1}{\varepsilon} \tag{8}$$

where $\mathcal{N}_e^{(2)} = \frac{m}{2\pi}E_F$ is the one-electron concentration in 2D case, and η_2 is defined by the expression (3)

We see that in the 2D case the density of fluctuation Cooper pairs decreases very slowly as temperature increases : logarithmically only. Of course these are nonequilibrium pairs, their lifetime being determined by the Ginzburg-Landau time $\tau_{GL} = \frac{\pi}{8(T-T_c)}$ and there is no long range order in the system. Nevertheless, one can see that even in the normal phase of a superconductor at each moment there is a non-zero density of such pairs which may participate in charge transfer, anomalous diamagnetism, heat capacity increase near transition. In this sense we can speak about the existence of the average modulus of the order parameter (which is defined as the square root of the average square of modulus (8)).

The participation of normal electrons in nonequilibrium Cooper pairing above T_c is an inelastic process leading to some decay of the phase coherence between initial and final quasiparticle states. This means that fluctuations themselves act as a source of some phase-breaking time $\tau_\phi(\varepsilon)$ side by side with paramagnetic impurities and thermal phonons. The consequence of this fact is the shift of the transition temperature toward lower temperatures with respect to its mean field value T_{c0}.

This shift is easy to estimate by taking into account the next order correction ($\sim |\Psi|^4$) in the Ginzburg-Landau functional. We make the Hartree approximation by replacing the $|\Psi(\vec{r})|^4$ term in (2) by $\langle|\Psi_{(fl)}|^2\rangle|\Psi(\vec{r})|^2$. This leads to the renormalization of the reduced critical temperature value

$$\varepsilon^* = \varepsilon + \frac{b}{2\alpha}\langle|\Psi_{(fl)}|^2\rangle \tag{9}$$

and to appropriate reduction of the critical temperature T_c^* with respect to its BCS value T_{c0}. Using the microscopic values of α and b and also the results (7)-(8) for $\langle|\Psi_{(fl)}|^2\rangle$, one can easily find within logarithmic accuracy:

$$\frac{\delta T_c}{T_c} = \frac{T_c^* - T_{c0}}{T_{c0}} \sim -Gi_{(2)}\ln\frac{1}{Gi_{(2)}} \tag{10}$$

where $Gi_{(2)}^{(d)} \sim \frac{1}{p_F^2 l d}$ for dirty, $Gi_{(2)}^{(cl)} \sim \frac{1}{p_F d}\frac{T_c}{E_F}$ for clean film of thickness d (l is the electron mean free path). In the case of HTS single crystal $Gi_{(2)}^{(cl)} \sim \frac{T_c}{E_F}$.

The account of fluctuations results in the deviation of $\sqrt{\langle|\Psi_{(fl)}|^2(T)\rangle}$ below BCS curve $\Psi_0(T, T_{c0})$ (solid line) due to the growth of the order parameter modulus fluctuations with the increase of temperature.

The second effect of fluctuations consists in the decrease of the critical temperature with respect to T_{c0}, so we have to terminate our consideration at T_c^*. One can see that the value of $\langle|\Psi_{(fl)}|^2(T_c^*, T_{c0})\rangle$ is of the order of $\mathcal{N}_1\mathcal{G}\rangle_{(\varepsilon)}\ln\frac{\infty}{\mathcal{G}\rangle_{(\varepsilon)}}$ and it matches perfectly with the logarithmic tail calculated for the temperatures above T_c. The full curve $\sqrt{\langle|\Psi_{(fl)}|^2(T)\rangle}$ is presented in Fig. 1 by the solid line.

One comment has to be done at this stage. In practice the temperature T_{c0} is a formal value only, T_c^* is measured by experiments. So instead of $\Psi_0(T, T_{c0})$ the curve $\Psi_0(T, T_c^*)$ more naturally has to be plotted (dashed line at Fig. 1). It starts at T_c^* and finishes at zero temperature a little bit below the BCS value $\Psi_0(0, T_{c0})$, the shift of $\delta\Psi(0)$ due to quantum fluctuations turns out to be relatively the same as δT_c. One can see that the renormalized by fluctuations curve $\sqrt{\langle|\Psi_{(fl)}|^2(T)\rangle}$ passes above with respect to this, more natural parametrization of the BCS $\Psi_0(T, T_c^*)$ temperature dependence.

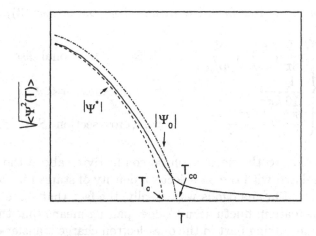

Figure 1. The renormalized by fluctuations averaged order parameter as the function of temperature.The dashed-dot line presents the unperturbed BCS curve. The dashed line is the BCS curve with renormalized by fluctuations T_c^*. The solid line presents the renormalized by fluctuations function $\langle |\Psi_{(fl)}|^2(T) \rangle$.

4. Qualitative consideration of different fluctuation contributions

The first and evident effect of the appearance of fluctuation Cooper pairs above transition is the opening of a new channel for the charge transfer. Cooper pairs can be treated as carriers with charge $2e$ and lifetime $\tau_{GL} = \frac{\pi}{8(T-T_c)}$. Instead of the electron concentration $\mathcal{N}_e$, in the Drude formula for paraconductivity has to be used calculated above density of Cooper pairs $\mathcal{N}_{c.p.}$.

$$\delta\sigma_{AL}^{(2)} \sim \frac{\mathcal{N}_{c.p.}(2e)^2\tau_{GL}}{2m} = \frac{\pi e^2}{4m}\frac{1}{(T-T_c)}\mathcal{N}_{c.p.} \tag{11}$$

In 2D case $\mathcal{N}_{c.p.}^{(2)} = \frac{p_F^2}{2\pi d}\frac{T_c}{E_F}\ln\frac{1}{\varepsilon}$

$$\delta\sigma_{AL}^{(2)} = \frac{e^2}{16d}\frac{T_c}{T-T_c} \tag{12}$$

The 2D result (12) remains surprisingly the same for clean and dirty cases. In the case of other dimensions of the effective electron spectrum this universality is loosen, the paraconductivity becomes to be dependent on the electron mean free path (by means of η_D parameter (3)):

$$\delta\sigma_{AL}^{(D)} = \begin{cases} \dfrac{1}{8\pi}\left(\dfrac{e}{\hbar}\right)^2\left(\dfrac{1}{4\eta_3\epsilon}\right)^{1/2} & \text{three}-\text{dimensional case,} \\[2ex] \dfrac{1}{16}\dfrac{e^2}{\hbar d\epsilon} & \text{film, thickness}: d \ll \xi, \\[2ex] \dfrac{\pi\eta_1^{1/2}}{16}\dfrac{e^2}{\epsilon^{3/2}S} & \text{wire, cross section}: S \ll \xi^2. \end{cases} \qquad (13)$$

The correction to the normal phase conductivity above the transition temperature related with the one-electron density of states renormalization can be reproduced in analogous way. Really, the fact that some amount of electrons participate in fluctuation Cooper pairing means that the effective number of carries taking part in the one-electron charge transfer diminishes leading to the decrease of conductivity:

$$\delta\sigma_{DOS} = -\frac{\Delta\mathcal{N}_e e^2 \tau_{imp}}{m} = -\frac{2\mathcal{N}_{c.p.} e^2 \tau_{imp}}{m} \qquad (14)$$

or in the 2D case

$$\delta\sigma_{DOS}^{(2)} = -\frac{e^2}{2\pi d}(T_c\tau)\ln\frac{1}{\varepsilon} \qquad (15)$$

The exact diagrammatic consideration of DOS fluctuation effect on conductivity fully confirms the estimate obtained in its sign and temperature dependence. The impurity scattering time dependence of (15) results to be correct in the dirty case ($\xi \ll l$) but in the clean case the exact calculations show a stronger dependence on τ (τ^2 instead of τ, see the next section).

Finally, let us discuss the anomalous Maki-Thompson contribution. This anomalous term has the same singularity in ε as the AL one (within logarithmic accuracy), but has a purely quantum nature and does not appear in the usual Time Dependent Ginzburg-Landau approach at all. Its physical nature has remained mysterious since 1968, when it was calculated in the diagrammatic approach by Maki. This contribution is related certainly with the coherent electron scattering, manifests itself in transport properties only and turns out to be strongly phase sensitive. All these facts give a hint to treat the MT contribution in the same way as it was done by Altshuler and Khmelnitski (see [46]) with weak localization and interaction corrections to conductivity.

Namely, let us consider possible types of electron-electron Cooper pairing above T_c in real space. The simplest one is the appearance of Cooper correlation between two electrons with momenta $\vec{p}$ and $-\vec{p}$ moving along the straight lines in opposite directions. (see Fig. 2 (a)). Such type of pairing

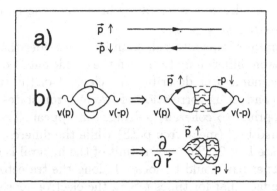

Figure 2. The self-intersecting trajectory treatment of the MT contribution in the real space

does not have the characteristics mentioned above and has to be attributed to classic AL process.

Nevertheless, another, much more sophisticated pairing process is imaginable: one electron with spin up and momentum $\vec{p}$ can move along some self-intersecting trajectory; simultaneously another electron with spin down and the opposite momentum $-\vec{p}$ can move in the opposite direction along the same trajectory (see Fig. 2(b)). The interaction of such pair of electrons during their motion along the trajectory leads to the appearance of some special contribution similar to localization and Coulomb interaction corrections to conductivity [48], but it has to be evidently singular in the vicinity of T_c. One can easily see that such type of pairing is possible only in the case of diffusion motion (necessary for the realization of the self-intersecting trajectory). Finally any phase-breaking scattering leads to the lost of the coherence and destruction of the Cooper correlation. So all properties of Maki-Thompson contribution coincide with the properties of the process proposed. The proper contribution to conductivity of such process has to be proportional to the ratio of the number of such interfering Cooper pairs $\delta\mathcal{N}_{s.i.}$ to the full concentration of fluctuation Cooper pairs (in the most interesting for our purposes 2D case):

$$\frac{\delta\mathcal{N}_{s.i.}}{\mathcal{N}_{c.p.}} \sim \int_{\tau_{GL}}^{\tau_\phi} \frac{l v_F dt}{(\mathbf{D}t)} = \ln\frac{\varepsilon}{\gamma} \tag{16}$$

where τ_ϕ is the one-electron state phase-breaking time and $\gamma = \frac{\pi}{8T_c\tau_\phi}$ is appropriate phase-breaking rate.

The denominator of this integral, as in [46], describes the volume available for the electron diffusion motion with the coefficient $\mathbf{D}$ during the time t is $(\mathbf{D}t)^{d/2}$. The numerator describes the element of the tube volume in which the superconducting correlation of two electron states with opposite momenta, appropriate to coherent pairing, can appear. The width of the tube is determined by l (mean free path) while the differential of the tangential is $v_F dt$ (see Fig. 2). The lower limit of the integral is chosen so that at least one Cooper pair could be located along the trajectory. The upper limit reflects the fact that for times $t > \tau_\phi$ the electron loose its phase and coherent Cooper pairing of two electron states above T_c is impossible.

The proper contribution of the Cooper pairs generated by coherent electrons moving along self-intersecting trajectories is

$$\Delta\sigma_{s.i.}^{(2)} = \mathcal{N}_{c.p.}^{(2)} \ln\left(\frac{\varepsilon}{\gamma}\right) \frac{(2e)^2 \tau_{GL}}{2m} \sim \sigma_{AL}^{(2)} \ln\left(\frac{\varepsilon}{\gamma}\right) \tag{17}$$

One can see that this result coincides with the result of microscopic calculations of the anomalous Maki-Thompson contribution [49]:

$$\sigma_{MT}^{(2)} = \frac{e^2}{8d} \frac{1}{\varepsilon - \gamma} \ln\frac{\varepsilon}{\gamma} \tag{18}$$

In similar way the correct temperature dependence of all contributions in all dimensions can be obtained.

5. The effect of superconducting fluctuations on the one-electron density of states.

As was already mentioned, the appearance of non-equilibrium Cooper pairing above T_c leads to the redistribution of the one-electron states around the Fermi level. A semi-phenomenological study of the fluctuation effects on the density of states of a dirty superconducting material was first carried out while analyzing the tunneling experiments of granular Al in the fluctuation regime just above T_c. The second metallic electrode was in the superconducting regime and its well developed gap gave a bias voltage around which a structure, associated with the superconducting fluctuations of Al, appeared. The measured density of states has a dip at the Fermi level [2], reaches its normal value at some energy $E_0(T)$, show a maximum at an energy value equal to several times E_0, finally decreasing towards its normal value at higher energies (Fig. 3). The characteristic energy E_0 was found

---

[2]Here we refer the energy E to the Fermi level, where we assume $E = 0$.

to be of the order of the inverse of the Ginzburg-Landau relaxation time $\tau_{GL} \sim \frac{1}{T-T_c} = T_c \varepsilon^{-1}$ introduced above.

The presence of the depression at $E = 0$ and of the peak at $E \sim (1/\tau_{GL})$ in the density of states above T_c are the precursor effects of the appearance of the superconducting gap in the quasiparticle spectrum at temperatures below T_c. The microscopic calculation of the fluctuation contribution to the one-electron density of states is a nontrivial problem and can not be carried out in the framework of the phenomenological Ginzburg-Landau theory. It can be solved within the diagrammatic technique by calculating the fluctuation correction to the one-electron temperature Green function with its subsequent analytical continuation to the real energies [23, 24]. We omit here the details of the cumbersome calculations and present only the results obtained from the first order perturbation theory for fluctuations. They are valid near the transition temperature, in the so-called Ginzburg-Landau region, where the deviations from the classical behavior are small. The theoretical results reproduce the main features of the experimental behavior cited above. The strength of the depression at the Fermi level is proportional to different powers of τ_{GL}, depending on the effective dimensionality of the electronic spectrum and the character of the electron motion (diffusive or ballistic). In a dirty superconductor for the most important cases of dimensions D=3,2 one can find the following values of the relative corrections to the density of states at the Fermi level [23]:

$$
\delta N_{fl}^{(d)}(0) \sim - \begin{cases} \dfrac{T_c^{1/2}}{\mathbf{D}^{3/2}}(T_c\tau_{GL})^{3/2}, & D = 3 \\[2ex] \dfrac{1}{\mathbf{D}}(T_c\tau_{GL})^2, & D = 2 \end{cases} \tag{19}
$$

where $\mathbf{D} = \frac{v_F l}{D}$ is the diffusion coefficient. At large energies $E \gg \tau_{GL}^{-1}$ the density of states recovers its normal value, according to the same laws (19) but with the substitution $\tau_{GL} \to E^{-1}$.

It is interesting that in the case of the density of states fluctuation correction the critical exponents change when moving from a dirty to a clean superconductor [24]:

$$
\delta N_{fl}^{(cl)}(0) \sim - \begin{cases} \dfrac{1}{T_c\xi_0^3}(T_c\tau_{GL})^{1/2}, & D = 3 \\[2ex] \dfrac{1}{T_c\xi_0^2}(T_c\tau_{GL}), & D = 2 \end{cases} \tag{20}
$$

(the subscripts *(cl)* and *(d)* stand here for clean and for dirty cases respectively). Nevertheless, as it will be seen below, due to some specific properties of the corrections obtained, this difference between clean and dirty systems does not manifest itself in the physically observable quantities (tunneling current, NMR relaxation rate etc.) which are associated with the density of states by means of some convolutions.

Another important respect in which the character of the density of states renormalization in the clean and dirty cases differs strongly is the energy scale at which this renormalization occurs. In the dirty case ($\xi_0 \gg l$) this energy turns out to be [23]:

$$E_0^{(d)} \sim T - T_c \sim \tau_{GL}^{-1}, \tag{21}$$

while in the clean one ($\xi_0 \ll l$) [24]:

$$E_0^{(cl)} \sim \sqrt{T_c(T - T_c)}, \tag{22}$$

To understand this important difference one has to study the character of the electron motion in both cases discussed [24]. Let us recall that the size of the fluctuating Cooper pair is determined by the coherence length

$$\xi(T) = \xi_0 \left(\frac{T_c}{T - T_c} \right)^{1/2} \tag{23}$$

of the Ginzburg-Landau theory. The zero-temperature coherence length ξ_0 differs considerably for the clean and dirty cases:

$$\xi_{0,cl}^2 = \frac{7\zeta(3)}{12\pi^2 T_c^2} \frac{E_F}{2m} \tag{24}$$

$$\xi_{0,d}^2 = \frac{\pi D}{8T_c} \tag{25}$$

To pass from the dirty to the clean case one has to make the substitution

$$D \sim \frac{p_F l}{m} \sim \frac{E_F \tau}{m} \to \frac{E_F}{m T_c}. \tag{26}$$

The relevant energy scale in the dirty case is the inverse of the time necessary for the electron to diffuse over a distance equal to the coherence length $\xi(T)$. This energy scale coincides with the inverse relaxation time τ_{GL}:

$$t_\xi^{-1} = D\xi^{-2} \sim \tau_{GL}^{-1} \sim T - T_c. \tag{27}$$

In the clean case, the ballistic motion of the electrons gives rise to a different characteristic energy scale

$$t_\xi^{-1} \sim v_F \xi^{-1} \sim (T_c \tau_{GL}^{-1})^{1/2} \sim \sqrt{T_c(T - T_c)}. \tag{28}$$

The fluctuation corrections to the density of states may be presented as a function of the energy and the temperature in a general form, for any dimensionality of the isotropic electron spectrum and any impurity concentration [24], but the relevant expressions are very cumbersome and we restrict ourselves to report the 2D clean case only

$$\delta N_{fl(2)}^{cl}(E) = -N_{(2)}\frac{8aT_c}{\pi E_F}\frac{T_c^2}{(4E^2 + aT_c\tau_{GL}^{-1})} \times$$

$$\times \left\{1 - \frac{2E}{(4E^2 + aT_c\tau_{GL}^{-1})^{1/2}}\ln\left[\frac{2E + (4E^2 + aT_c\tau_{GL}^{-1})^{1/2}}{(aT_c\tau_{GL}^{-1})^{1/2}}\right]\right\}. \tag{29}$$

where $N_{(2)} = \frac{m}{2\pi}$ is the 2D density of electron states in the normal metal and a is some number of the order of unity. One can check that the integration of this expression over all positive energies gives zero. This is a consequence of conservation of the number of particles: the number of quasiparticles is determined by the number of cells in the crystal and cannot be changed by the interaction. So the only effect which can be produced by the interelectron interaction is the redistribution of energy levels near the Fermi energy. This statement can be written as the "sum rule" for the fluctuation correction to the density of states:

$$\int_0^\infty \delta N_{fl}(E)dE = 0 \tag{30}$$

This sum rule plays an important role in the understanding of the manifestation of the fluctuation density of states renormalization in the observable phenomena. As we will see in the next Section the singularity in tunneling current (at zero voltage), due to the density of states renormalization, turns out to be much weaker than that in the density of states itself ($\ln \varepsilon$ instead of ε^{-1} or ε^{-2}, see (19)-(20)). The same features occur in the opening of the pseudo-gap in the c-axis optical conductivity, in the NMR relaxation rate etc. These features are due to the fact that we must always form the convolution of the density of states with some slowly varying function: for example, a difference of Fermi functions in the case of the tunnel current. The sum rule then leads to an almost perfect cancellation of the main singularity at low energies. The main non-zero contribution then comes from the high energy region where the DOS correction has its 'tail'.

Another important consequence of the conservation law (30) is the considerable increase of the characteristic energy scale of the fluctuation pseudo-gap opening with respect to E_0: this is $eV_0 = \pi T$ for tunneling and $\omega \sim \tau^{-1}$ for c-axis optical conductivity.

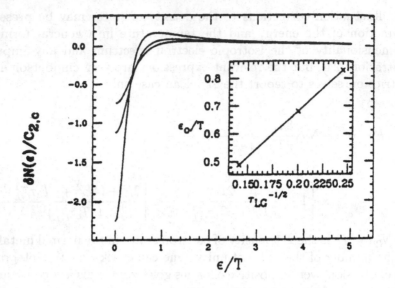

Figure 3. The theoretical prediction of Eq. (29) for the normalised correction $\delta N(E)$ to the single-particle density of states vs energy E (measured in units of T_c) for a clean two-dimensional superconductor above T_c. τ_{GL}^{-1} assumes the values $0.02T_c$, $0.04T_c$ and $0.06T_c$. In the inset the dependence of the energy at which $\delta N(E)$ is a maximum, E_0, on τ_{GL}^{-1} is shown.

6. The effect of fluctuations on the tunnel current

One of the most currently discussed problems of the physics of high temperature superconductivity is the observation of pseudo-gap type phenomena in the normal state of these materials [3]. Here we deal with the "pseudo-gap" type behaviour of the tunneling characteristics of HTS materials in the metallic phase (slightly under-, optimally or over-doped compounds, where the Fermi surface is supposed to be well developed). By this we mean the observation at temperatures above T_c critical one of non-linear $I - V$ characteristics usual for the superconducting phase where a real gap in the quasiparticle spectrum occurs. It is quite evident that the renormalization of the density of states near the Fermi level, even of only one of the electrodes, will lead to the appearance of anomalies in the voltage-current

[3]It is worth to mention that some confusion with the concept of "pseudo-gap" takes place in literature. It is used often as synonym for the spin gap (even being far from the anti-ferromagnetic phase), the same definition is used in the description of a variety of HTS normal state anomalies observed by means of transport and photo-emission measurements, NMR relaxation, optical conductivity and tunneling at temperatures above T_c in all range of oxygen concentrations without any proof that it has the same origin in the different experiments.

characteristics of a tunnel junction. So-called zero-bias anomalies, which are the increase of the differential resistance of a junction with amorphous electrodes at zero voltage and low temperatures, have been observed for a long time. They have been explained in terms of a density of states depression in an energy range of the order of $E_{am} \sim \tau^{-1}$ around the Fermi level due to the electron-electron interaction in the diffusion channel.

The quasiparticle current flowing through a tunnel junction may be presented as a convolution of the densities of states with the difference of the electron Fermi distributions in each electrode (L and R):

$$I_{qp} = \frac{1}{eR_n N_L(0)N_R(0)} \times \qquad (31)$$
$$\int_{-\infty}^{\infty} \left(\tanh \frac{E+eV}{2T} - \tanh \frac{E}{2T} \right) N_L(E)N_R(E+eV)dE,$$

where R_n is the Ohmic resistance per unit area and $N_L(0)$, $N_R(0)$ are the densities of states at the Fermi levels in each of electrodes in the absence of interaction. One can see that for low temperatures and voltages the expression in parenthesis is a sharp function of energy near the Fermi level. The characteristic width of it is $E_{\mathrm{ker}} \sim \max \{T, V\} \ll E_F$. Nevertheless, depending on the properties of densities of states functions, the convolution (31) may exhibit different properties.

If the energy scale of the density of states correction is much larger than E_{ker} , the expression in parenthesis in (31) acts as a delta-function and the zero-bias anomaly in the tunnel conductivity strictly reproduces the anomaly of the density of states around the Fermi level:

$$\frac{\delta G(V)}{G_n(0)} = \frac{\delta N(eV)}{N(0)}, \qquad (32)$$

where $G(V)$ is the differential tunnel conductance and $G_n(0)$ is the background value of the Ohmic conductance supposed to be bias independent, $\delta G(V) = G(V) - G_n(0)$.

This situation occurs in a junction with amorphous electrodes [52]. In the amorphous metal, the electron-electron interaction with small momentum transfer (diffusion channel) is retarded and this fact leads to a considerable suppression of the density of states in the vicinity of the Fermi level, within an energy range $E_{am} \sim \tau^{-1} \gg T \sim E_{\mathrm{ker}}$. At zero temperature for the 2D case one has:

$$\delta N_2(E) = \frac{\lambda}{4\pi^2 \mathbf{D}} \ln(E\tau), \qquad (33)$$

where the constant λ is related to the Fourier transform of the interaction potential. In the 3D case the correction to the density of states turns out to be proportional to $|E|^{1/2}$.

In the framework of this approach Altshuler and Aronov [52, 53] analyzed the experimental data obtained in studies of the tunneling resistance of $Al - I(O_2) - Au$ junctions and showed it to be proportional to $|V|^{1/2}$ at $eV \ll T$. The identification of the "wings" in the $I - V$ characteristics of such junctions with (33) was a key success of the theory of the electron-electron interaction in disordered metals [48].

It is worth stressing that the proportionality between the tunnel current and the electron DOS of the electrodes is widely accepted as an axiom, but generally speaking this is not always so. As one can see from the previous Section, the opposite situation occurs in the case of the DOS renormalization due to the electron-electron interaction in the Cooper channel: in this case the DOS correction varies strongly already in the scale of $E_0 \ll T \sim E_{ker}$ and the convolution in (31) with the density of states (29) has to be carried out without the simplifying approximations assumed to obtain (32).

We now derive the explicit expression for the fluctuation contribution to the differential conductance of a tunnel junction with one thin film electrode close to its T_c. To do this we differentiate (31) with respect to voltage, and insert the density of states correction given in (29). This gives (see [50]):

$$
\frac{\delta G_{fl}(V, \varepsilon)}{G_n(0)} = \frac{1}{2T} \int_{-\infty}^{\infty} \frac{dE}{\cosh^2 \dfrac{E + eV}{2T}} \delta N_{fl}^{(2)}(E, \varepsilon) =
$$

$$
= Gi_{(2)} (4\pi T \tau) \ln \left(\frac{1}{\varepsilon} \right) \operatorname{Re} \psi'' \left(\frac{1}{2} - \frac{ieV}{2\pi T} \right),
$$

(34)

where $\psi(x)$ is the digamma function, and τ is the electron's elastic scattering time.

It is important to emphasize several nontrivial features of the result obtained. First, the sharp decrease ($\varepsilon^{-2(1)}$) of the density of the electron states generated by the inter-electron interaction in the immediate vicinity of the Fermi level surprisingly results in a much more moderate growth of the tunnel resistance at zero voltage ($\ln \frac{1}{\varepsilon}$). Second, in spite of the manifestation of the density of states renormalization at the characteristic scales $E_0^{(d)} \sim T - T_c$ or $E_0^{(cl)} \sim \sqrt{T_c(T - T_c)}$, the energy scale of the anomaly development in the $I - V$ characteristic is much larger: $eV = \pi T \gg E_0$ (see Fig. 4).

This departure from the habitual idea of the proportionality between the tunnel conductance and the so-called tunneling density of states (32) is

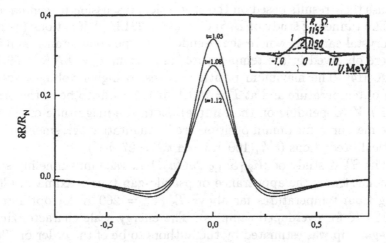

Figure 4. The theoretical prediction of Eq. (34) for the fluctuation-induced zero-bias anomaly in tunnel-junction resistance as a function of voltage for reduced temperatures $t = 1.05$ (top curve), $t = 1.08$ (middle curve) and $t = 1.12$ (bottom curve). The insert shows the experimentally observed differential resistance as a function of voltage in an Al-I-Sn junction just above the transition temperature [51].

a straightforward result of the convolution calculated in (31) with the difference of Fermi-functions as a kernel. As already explained in the previous section the physical reason is that the presence of inter-electron interaction cannot create new electron states: it can only redistribute the existing states.

In the inset of Fig. 4 the measurements of the differential resistance of the tunnel junction $Al - I - Sn$ at temperatures slightly above the critical temperature of Sn electrode are presented. This experiment was accomplished by Belogolovski, Khachaturov and Chernyak in 1986 [51] with the purpose of checking the theory proposed by Varlamov and Dorin [50] which led to the result (34). The non-linear differential resistance was precisely measured at low voltages which permitted the observation of the fine structure of the zero-bias anomaly. The reader can compare the shape of the measured fluctuation part of the differential resistance (the inset in Fig. 4) with the theoretical prediction. It is worth mentioning that the experimentally measured positions of the minima are $eV \approx \pm 3T_c$, while the theoretical prediction following from (34) is $eV = \pm \pi T_k$. For the HTS samples this means a scale of $20 - 40\ meV$, considerably larger than in the case of conventional superconductors.

The observations of the pseudo-gap type anomalies of $G(V, T)$ at temperatures above T_c obtained by a variety of experimental techniques on $BiSrCaCuO - 2212$ samples were reported very recently [14, 15, 16] . Let us discuss their results based on the theoretical discussion presented above.

1. The tunneling study of $BiSrCaCuO - 2212/Pb$ junctions (as grown single crystal samples) where the pseudo-gap type conductance nonlinearities were observed in the temperature range from $T_c = 87\ K - 89\ K$ up to $110K$ [14]. The maximum position moves to higher voltages with the growth of temperature and at $T = 100\ K$ it turns out to be of the order of $30 - 35\ mV$ (depending on the sample). Both the magnitude of the effect and the measured maximum position are in qualitative agreement with the theoretical predictions $(eV_m(100\ K) = \pm\pi T = 27\ mV)$.

2. The STM study of $Bi_{2.1}Sr_{1.9}CaCu_2O_{8+\delta}$ vacuum tunneling spectra [15], demonstrates the appearance of pseudo-gap type maxima in $G(V,T)$ starting from temperatures far above T_c ($T^* = 260\ K$ for optimum and $T^* = 180\ K$ for overdoped samples). The energy scale characterizing the pseudo-gap dip was estimated by the authors to be of the order of $100\ mV$ which again is consistent with the predicted maximum position $eV_m(100\ K)$ $= \pm\pi T = \pm27mV$.

3. Impressive results on tunneling pseudo-gap observations are presented in [16]. They were carried out by interlayer tunneling spectroscopy using very thin stacks of intrinsic Josephson junctions fabricated on the surface of $Bi_2SrCaCu_2O_8$ single crystal. The opening of the pseudo-gap was found in dI/dV characteristics at temperatures below $180K$. The data presented permitted us to fit them by means of formula (34) with $Gi = 0.008$ and tunneling critical temperature $T_c = 87\ K$ [54]. The fit reproduces not only the magnitude of the effect and the maxima positions, but also the shape of the experimental curves.

7. c-axis resistivity peak

Let us try to understand in the same qualitative manner the effect of fluctuations on the transverse resistance of layered superconductor. As we just have demonstrated above, the in-plane component of paraconductivity is determined by the Aslamazov-Larkin formula (12). To modify this result for c-axis paraconductivity one has to take into account the hopping character of the electron motion in this direction. One can easy see that, if the probability of one-electron interlayer hopping is $\mathcal{P}_1$, that one of the coherent hopping for two electrons during the virtual Cooper pair lifetime τ_{GL} is calculated as the conditional probability of these two events:

$$\mathcal{P}_2 = \mathcal{P}_1 \cdot (\mathcal{P}_1 \cdot \tau_{GL}) \tag{35}$$

and the transverse paraconductivity may be estimated as

$$\sigma_\perp^{AL} \sim \mathcal{P}_2 \cdot \sigma_\parallel^{AL} \sim \mathcal{P}_1^2 \frac{1}{\varepsilon^2}. \tag{36}$$

It is easy to see that the temperature singularity of $\sigma_{\perp}^{AL}$ turns out to be even stronger than that one in $\sigma_{\parallel}^{AL}$ because of the hopping character of the electron motion (the critical exponent "2" in the conductivity is characteristic of the zero-dimensional band motion), but in the case of a strongly anisotropic layered superconductor $\sigma_{\perp}^{AL}$, is considerably suppressed by the square of the small probability of inter-plane electron hopping which enters in the prefactor.

Namely, this suppression determines the necessity of taking into account the DOS contribution to the transverse conductivity, which is less singular in temperature but, in contrast to paraconductivity, manifests itself at the first, not the second, order in the interlayer transparency. One can easily estimate it in the same way as it was done above multiplying the in plane result (15) by the one-electron hopping probability:

$$\Delta\sigma_{\perp}^{DOS} \sim -\mathcal{P}_1 \ln \frac{1}{\varepsilon}. \tag{37}$$

It is important that, in contrast to the paraconductivity, the DOS fluctuation correction to the one-electron transverse conductivity is obviously negative and, being proportional to the first order of $\mathcal{P}_1$ only, can completely change the traditional picture about fluctuations just rounding the resistivity temperature dependence around transition. Excluding temporarily from consideration the anomalous Maki-Thompson contribution (which is strongly suppressed in HTS by strong pair-breaking effects), one can say that the shape of the temperature dependence of the transverse resistance is determined by the competition of two contributions of the opposite signs: the paraconductivity, that is strongly temperature dependent but is suppressed by the square of the barrier transparency ($\sim J^4$) and by the first order in transparency ($\sim J^2$) DOS contribution, which depends on the reduced temperature only logarithmically:

$$\sigma_{fl}^{\perp} \sim \alpha\mathcal{P}_1^2 \frac{1}{\varepsilon^2} - \beta\mathcal{P}_1 \ln \frac{1}{\varepsilon} \tag{38}$$

where α and β are coefficients which will be calculated in the framework of the exact microscopic theory presented in the next section. Namely these circumstances determine the possible competition between the contributions of different orders in transparency and lead to the formation of the maximum with the right temperature dependence.

8. Magnetoresistance above T_c

While the ab-plane magnetoresistance (MR) above T_c is well known to have a positive sign, due to the suppression of the AL paraconductivity the

280

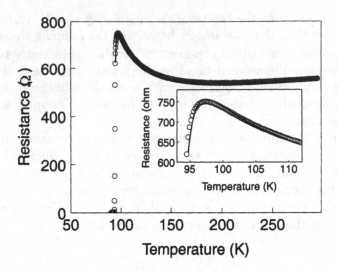

Figure 5. Fit of the temperature dependence of the transverse resistance of an under-doped BSCCO c-axis oriented film with the results of the fluctuation theory [28]. The inset shows the details of the fit in the temperature range between T_c and $110K$.

c-axis MR of BSCCO single crystals has been found to be negative and growing anomalously with the decrease of temperature in the range of $95 - 140K$ [5]. This effect was attributed [36] to the manifestation of the DOS fluctuations in MR. Really, the suppression of the DOS term, in contrast to the AL one, leads to the decrease of the resistivity, so in the region of temperatures where the DOS contribution dominates a negative c-axis MR is conceivable. The fitting of the experimental results mentioned above with formula of [30] gave the excellent results both in its magnetic field ($\sim H^2$) and temperature dependencies. Moreover, the analysis of the temperature dependence of the total fluctuation correction to MR at fixed magnetic field led to the non-monotonous, sign alternating function [36]. The physical origin of this change of sign is the same as for the appearance of the peak discussed above: relatively far from T_c the positive AL MR is suppressed by its proportionality to the square of the interlayer transparency and the negative DOS contribution dominates, while close enough to T_c the very singular temperature dependence of the negative AL contribution ($\sim \epsilon^{-4}$) makes it prevail on the less singular DOS one ($\sim \epsilon^{-2}$), in spite of the linear dependence on the transparency of the latter.

Shortly after the problem of the transverse magnetoresistivity was examined [7] basing on the idea of its fluctuation origin. In all cases a good agreement between the fluctuation theory and the measured data has been found, both for BSCCO and YBCO samples. In particular, in [7] the transverse magnetoresistance was measured down to temperatures close to T_c. Its temperature dependence was found to follow very closely the behavior predicted in [36] and the existence of a sign reversal temperature was experimentally confirmed.

It is important to stress that the suppression of DOS contribution by magnetic field takes place very slowly. Such robustness with respect to the magnetic field is of the same physical origin as the slow logarithmic dependence of the DOS-type corrections on temperature. This DOS contribution differs strongly from the Aslamazov-Larkin and Maki-Thompson ones, making the former noticeable in the wide range of temperatures (up to $\sim 2-3T_c$) and magnetic fields ($\sim H_{c2}(0)$). The scale of the suppression of DOS contribution can be treated as the value of the pseudogap observed in the experiments mentioned above [1]. It has the order of $\Delta_{pseudo} \sim 2-3T_c$ for magnetoconductivity and NMR, $\Delta_{pseudo} \sim \pi T_c$ for tunneling and $\Delta_{pseudo} \sim \tau^{-1}$ in optical conductivity.

9. Optical conductivity

The most puzzling experiments widely discussed recently are related with the observation of the pseudogap-like behavior in different properties of HTS materials above the critical temperature. One of them is the observation of the pseudogap opening in the c-axis far infrared conductivity which we discuss here basing on the same idea of the importance of Cooper channel interelectron interaction. Let us start from the analysis of each fluctuation contribution separately and then we will discuss their interplay in Re $[\sigma_\perp(\omega)]$.

The AL contribution describes the fluctuation condensate response to the electromagnetic field applied. The component of the current associated with it can be treated as the precursor phenomenon of the screening currents in the superconducting phase. Above T_c the virtual Cooper pairs binding energy is of the order of $T - T_c$, so it is not surprising that at higher frequencies the AL contribution decreases fast with the further increase of ω. Actually $\omega_{AL} \sim T - T_c$ is the only relevant scale for σ^{AL}: its frequency dependence doesn't contain T, τ_ϕ and τ. The independence from the latter

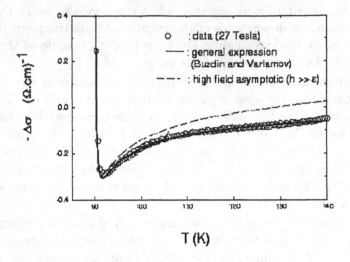

Figure 6. Magnetoconductivity versus temperature at 27 T for an underdoped Bi-2212 single crystal. The solid line represents the theoretical calculation. The symbols are the experimental magnetoconductivity $\Delta\sigma_{zz}(BcI)$

manifests mathematically the fact that elastic impurities do not represent obstacles for the motion of Cooper pairs.

Another effect related with the formation of the fluctuation Cooper pairs, but on the self-intersecting trajectories (like the weak localization correction), is described by the MT *anomalous* contribution. Being the contribution related with the Cooper pairs electric charge transfer it doesn't depend on the elastic scattering time but it turns out to be extremely sensitive to the phase-breaking mechanisms. So two characteristic scales turn out to be relevant in the frequency dependence of MT contribution: $T - T_c$ and τ_ϕ^{-1}. In the case of HTS, where τ_ϕ^{-1} has to be estimated as at least $0.1T_c$, for temperatures up to $5 \div 10\,K$ above T_c the MT contribution is overdamped, but in 2D regime it is still dependent on temperature.

The density of states fluctuation renormalization gives quite different contribution to $\mathrm{Re}\,[\sigma(\omega)]$ with respect to the others. At low frequencies ($\omega \ll \tau^{-1}$) the lack of electron states on the Fermi level leads to an opposite effect in comparison with AL and MT contributions: $\mathrm{Re}\left[\sigma^{\mathrm{DOS}}(\omega)\right]$ turns out to be negative and this means the increase of the surface impedance, or, in other words, the decrease of reflectance. Nevertheless, the electromagnetic field applied affects the electron distribution and at high frequencies $\omega \sim \tau^{-1}$ the DOS contribution changes its sign. It is interesting that DOS contribution, as one-electron effect, depends on the impurity scattering sim-

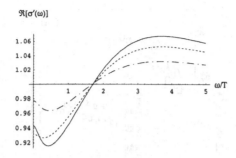

Figure 7. The plot shows the dependence of the real part of conductivity, normalized on the Drude normal conductivity, on ω/T, $\Re[\sigma'(\omega)] = \mathrm{Re}[\sigma(\omega)]/\sigma^n$. The dashed line refers to the ab-plane component of the conductivity tensor whose Drude normal conductivity is $\sigma_{\parallel}^n = N(0)e^2\tau v_F^2$. The solid line refers to the c-axis component whose Drude normal conductivity is $\sigma_{\perp}^n = \sigma_{\parallel}^n J^2 s^2/v_F^2$. In this plot we have put $T\tau = 0.3, E_F/T = 50, r = 0.01, \varepsilon = 0.04, T\tau_\varphi = 4$.

ilarly to the normal Drude conductivity. The decrease of $\mathrm{Re}\left[\sigma^{\mathrm{DOS}}(\omega)\right]$ starts already at frequencies $\omega \sim \min\{T, \tau^{-1}\}$ which for HTS are much higher than $T - T_c$ and τ_φ^{-1}.

The scenario of $\mathrm{Re}[\sigma_{\perp}^{\mathrm{tot}}]$ ω-dependence may be presented as following. The positive AL and MT effects, in their ω-dependence, are well pronounced at low frequencies on the background of the DOS contribution which remains, in this region, a negative constant. Then at $\omega \sim T - T_c$ the former decays and the $\mathrm{Re}\sigma_{\perp}$ remains negative up to $\omega \sim \min\{T, \tau^{-1}\}$. The DOS correction changes its sign at $\omega \sim \tau^{-1}$ and the following high frequency behavior is governed by the Drude law. So one can see that the characteristic pseudogap-like behavior in the frequency dependence of the optical conductivity takes place in the range $\omega \in [T - T_c, \tau^{-1}]$. The depth of the window increases logarithmically with ϵ when T tends to T_c.

10. NQR-NMR relaxation rate

One more example of the possible manifestation of the DOS is related to the NMR-NQR relaxation rate. The decrease of the electron density of states on the Fermi level evidently leads to the decrease of the relaxation rate with respect to Korringa law [42]. The opposite sign effect could be expected from paraconductivity-like contribution, but the necessity of the spin flip in the process of the nucleus-electron interaction excludes the AL

284

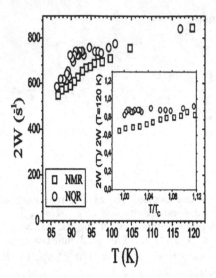

Figure 8. ^{63}Cu relaxation rates in zero field $2W(0)$ (from NQR relaxation) and $2W(H)$ in a field of 5.9 T (from NMR relaxation of the $-1/2 \rightarrow 1/2$ line) in the oriented powders of YBCO, with $T_c(0) = 90.5$ K and $T_c(H) = 87.5$ K. In the inset the relaxation rates, normalized with respect to $W(H) = W(0)$ for $T \gg T_c$, are reported as a function of T/T_c.

term from consideration (after the unique act of such interaction one of the fluctuation Cooper pair electrons should change its spin orientation, which destroyed the pair). Nevertheless the positive competitor for the DOS contribution appears as one sophisticated version of the anomalous MT process with the spin flip. Electron can move along the self-intersecting trajectory with the nucleus in the vicinity of the intersection point. After the passing along the trajectory once, electron can interact with the nucleus, change the spin orientation and the momentum on the opposite ones and then to pass the same trajectory again in the opposite direction, interacting with itself in "the past" (what is possible due to the retarded character of superconducting attraction). This, purely quantum, process exist in the s-wave scenario of pairing. It leads to the increase of the relaxation but, typically for MT processes, turns out to be very sensitive for the phase-breaking and is considerably suppressed in HTS. The concurrence of two mentioned terms was observed on experiment on YBCO samples. Moreover, the positive MT contribution was suppressed by the additional magnetic field applied [43]. The last fact could be treated as the indirect evidence of the presence of the s-wave component in the order parameter.

11. Thermoelectric power

Another example of the domination of the DOS contribution on the other fluctuation terms, even in the case of the isotropic electron spectrum, is the electron part of the thermoelectric power at the edge of superconducting transition which can be demonstrated from the following qualitative consideration (we will refer to the most interesting 2D case) [37].

The thermoelectric coefficient may be estimated through the electrical conductivity σ as $\beta \sim (\epsilon^*/eT)f_{as}\sigma$, where ϵ^* is the characteristic energy involved in thermoelectric transport and f_{as} is the electron-hole asymmetry factor, which is defined as the ratio of the difference between numbers of electrons and holes to the total number of particles. Conductivity can be estimated as $\sigma \sim e^2\mathcal{N}\tau^*/m$, where $\mathcal{N}$, τ^* and m are the density, lifetime and mass of charge (and heat) carriers, respectively. In the case of AL contribution the heat carriers are non-equilibrium Cooper pairs with energy $\epsilon^* \sim T - T_c$ and density $\mathcal{N} \sim p_F^d \frac{T}{E_F} \ln \frac{T_c}{T-T_c}$ and characteristic lifetime is given by the Ginzburg-Landau one. Thus in 2D case $\Delta\beta^{AL} \sim (T - T_c)/(eT_c)f_{as}\Delta\sigma^{AL} \sim ef_{as} \ln \frac{T_c}{T-T_c}$. One can easily get that the fluctuation correction due to AL process is less singular (logarithmic in 2D case) with respect to the corresponding correction to conductivity and does not depend on impurity scattering.

The analogous consideration of the single-particle DOS contribution ($\epsilon^* \sim T$, $\tau^* \sim \tau$) evidently results in the estimate of the same singularity as AL one: $\beta^{DOS} \sim ef_{as}T_c\tau \ln \frac{T_c}{T-T_c}$. The detailed comparison of the coefficients shows that the DOS contribution exceeds AL by the large parameter both in dirty and clean cases.

The fluctuation contribution provides a positive correction $\Delta\beta$, resulting in the decrease of the absolute value of Seebeck coefficient S at the edge of superconducting transition ($\Delta\beta/\beta_0 < 0$). To the result of the same sign the account of fluctuation correction to conductivity leads. So one can conclude that the thermodynamical fluctuations always reduce the overall S as temperature approaches T_c. So the very sharp maximum in the Seebeck coefficient experimentally observed in few papers seems to be unrelated to fluctuation effects within our simple model.

12. Conclusions

Several comments should be made in the conclusion. We have demonstrated that the strong and narrow in energy scale renormalization of the one-electron density of states in the vicinity of the Fermi level due to the Cooper channel interelectron interaction manifests itself in experiments as the wide enough pseudo-gap-like structures. The scale of these anomalies, as we have seen above, is different for various phenomena ($eV = \pi T$ for tunnel conductance, $\omega \sim \tau^{-1}$ for optical conductivity, $T \sim T_c$ in NMR and conductivity measurements).

The results presented above are based on the Fermi liquid approach which is formally expressed through the presence of a small parameter of the theory $Gi_2 \ln \frac{1}{Gi_2} \approx \frac{T_c}{E_F} \ln \frac{E_F}{T_c} \ll 1$ in obtained results. Moving along the phase diagram of HTS from the metal region (overdoped or optimally doped samples) to poor metals (underdoped compounds) one can see that the small parameter of the perturbation theory grows ($E_F \to 0$) causing the effects discussed to be more pronounced. This correlates with the effective increase of the interelectron interaction constant g.

Nevertheless, at some stage, the proposed approach becomes no more valid because of its restricted perturbation character. It would be natural to expect matching for the perturbative results with those of the strongly correlated theories in the unknown land *"hic sunt leones"* where $g \sim 1$. Such crossover qualitatively was discussed by M.Randeria [55] , appears in the preformed pairs models.

In conclusion, we have presented here what we hope to be a comprehensive overview of the theoretical and experimental facts which allow us to attribute an important role in the behaviour of HTS (the metal part of its phase diagram) to the fluctuation theory, and especially, to the frequently neglected DOS contribution. We did show that a large number of nontrivial experimental observed anomalies of HTS normal state properties can be qualitatively, and often quantitatively, described under this model. Of course, as a rule, this is not the only explanation which can be given for these behaviours. However, in our opinion, alternative approaches often lack the internal coherence and self consistency of the picture based on the fluctuation theory. In particular, it is important that all the experimental facts reviewed here can be explained by the fluctuation theory alone, without need for any other *ad hoc* assumption or phenomenological parameters, and that the values of all physical parameters extracted from the fits are always in a good agreement with independent measurements. Even more important is the fact that within the same approach one can explain so many different properties of HTS, in the wide temperature range from tens of degrees below to tens of degree above the critical temperature. It is very unlikely that these circumstances are purely fortunate.

Therefore we believe that, although almost for all experiments discussed the alternative explanations can be proposed, no other model, besides the fluctuation theory, has the same appeal embracing in a single microscopic view the wide spectrum of unusual properties which have been discussed in this review.

13. Acknowledgments

This work was partially supported by NATO Collaborative Research Grant # CRG 941187 and INTAS Grant # 96-0452.

References

1. A.A.Varlamov, G.Balestrino, E.Milani, D.Livanov (1999), *Advances in Physics*.
2. T.Penney, S.von Holnar, D.Kaiser, F.Holtzeberg, and A.W.Kleinsasser (1988), *Phys. Rev. B* **38**, 2918.
3. S.Martin, A.T.Fiory, R.M.Fleming, L.F.Schneemeyer and J.V.Waszczak (1988), *Phys. Rev. Lett.* **60**, 2194.
4. G.Briceno, M.F.Crommie, A.Zettle (1991), *Phys.Rev. Lett.*, **66**, 2164.
5. Y. F. Yan, P. Matl, J. M. Harris, and N. P. Ong (1995), *Phys. Rev.* **B 52**, R751.
6. K.Hashimoto, K.Nakao, H.Kado and N.Koshizuka (1996), *Phys. Rev.* **B 53**, 892.
7. J.Axnas, W.Holm, Yu.Eltsev, and O.Rapp (1996), *Phys Rev.Lett.* **77**, 2280.
8. C. H.Pennington and C. P.Slichter (1990), in *Physical Properties of High Temperature Supeconductors*, ed. by D. M. Ginzberg World Scintific, Singapore.
9. D.N.Basov, T.Timusk, B.Dabrowski and J.D.Jorgensen (1994), *Phys. Rev.* **B 50**, 3511.
10. D.N.Basov, T.Timusk, B.Dabrowski, H.A.Mook (1995), *Phys. Rev* **B 52**, 13141.
11. A.G.Loeser, Z.-X. Shen, D.S.Dessau (1996), *Physica* **C 263** , 208.
12. M.Gurvitch, J.M.Valles jr, A.M.Cucolo, R.S.Dynes, J.P.Garno, L.F.Schneemeyer, and J.V.Waszczak (1989), *Phys. Rev. Lett.* **63**, 1008.
13. A.M.Cucolo, C Noce, and A.Romano (1992),*Phys. Rev.* **B 46**, 5864.
14. H.J.Tao, Farun Lu, E.L.Wolf (1997), *Proceedings of* $M^2 SHTSC-Y$, Beijing, China.
15. A.Matsuda, S.Sugita, T.Watanabe (1998), submitted to *Phys. Rev. Lett.*.
16. M.Suzuki, S.Karimoto, K.Namekawa (1998), *J. of Phys. Soc. of Japan* **67**, v.3.
17. Ch.Renner, B.Revaz, J-Y. Genoud, K.Kadowaki and O.Fischer (1998), *Phys. Rev. Lett.*.
18. M.A.Howson *et al* (1990), *Phys. Rev.* **B 41**, 300.
19. N.V.Zavaritsky, A.A.Samoilov (1992), and A.A.Yurgens, *JETP Letters* **55**, 127.
20. J.M. Harris, Y.F.Yan, N.P.Ong (1992), *Phys.Rev.* **B 46**, 14293.
21. B.G.Levi (1996), *Physics Today*, **49**, p.17.
22. J.L.Tholence et al. (1994), *Physica* **C 235-240**, 1545.
23. E.Abrahams, M.Redi, and C.Woo (1970), *Phys. Rev.* **B 1**, 218.
24. C.Di Castro, C.Castellani, R.Raimondi, A.Varlamov (1990), *Phys. Rev.* **B 49**, 10211.
25. L.B.Ioffe, A.I.Larkin, A.A.Varlamov, L.Yu (1993), *Phys.Rev.* **B 47**, 8936.
26. K.E.Gray, D.H.Kim (1993), *Phys. Rev. Lett.* **70**, 1693.
27. G.Balestrino, M.Marinelli, E.Milani, A.Varlamov, L.Yu (1993), *Phys. Rev.* **B 47**, 6037.
28. G.Balestrino, E.Milani, A.Varlamov (1993), *Physica* **C 210** , 386.
29. G.Balestrino, E.Milani, C.Aruta, and A.Varlamov (1996), *Phys. Rev B* **54**, 3628.
30. V.V.Dorin, R.A.Klemm, A.A.Varlamov, A.I.Buzdin, D.V.Livanov (1993), *Phys. Rev.* **B 48**, 12951.
31. A.S.Nigmatulin, A.Varlamov, D.Livanov, G.Balestrino and E.Milani (1996), *Phys. Rev,* **B 53**.
32. D.Livanov, E.Milani, G.Balestrino and C.Aruta (1997), *Phys. Rev,* **B 55** R8701.
33. Y.F.Yan, P.Matl, J.M.Harris and N.P.Ong, *Phys. Rev.* **B 52**, R751 (1995) ; N.P.Ong, Y.F.Yan, and J.M.Harris (1994), *procedings of CCAST Symposium on High Tc Superconductivity and the C60 Family*, Beijing.
34. T. Kimura, S. Miyasaka, H. Takagi, K. Tamasaku, H. Eisaki, S. Uchida, M. Hiroi, M. Sera, K. Kobayashi (1996), *Phys.Rev.* **B53**, 8733.

35. W.Lang (1997), *Proceedings of M^2SHTS* Beijing.
36. G.Balestrino, E.Milani, and A.A.Varlamov (1995), *Soviet JETP Letters* **61**, 833.
37. A.Varlamov, D.Livanov, F.Federici (1997), *JETP Letters* **65**, 182.
38. K.Maki (1974), *J. Low Temp. Phys.* **14**, 419.
39. A.A.Varlamov and D.V.Livanov (1990), *Sov. Phys. JETP* **71**, 325.
40. M.Yu.Reizer and A.V.Sergeev (1994), *Phys. Rev.* **B 50**, 9344.
41. J.Mosqueira, J. A.Veira, and Felix Vidal (1995), *Physica* **C 229**, 301 (1994); J.Mosqueira, A.Veira, J.Maza, O.Cabeza, and Felix Vidal, *Physica* **C 253**, 1.
42. M.Randeria and A.A.Varlamov (1994), *Phys. Rev.* **B 50**, 10401.
43. P.Carretta, A.Rigamonti, A.A.Varlamov, D.V.Livanov (1996), *Phys.Rev.* **B 54**, R9682.
44. F.Federici, A.A.Varlamov (1996), *JETP Letters* **64**, 497.
45. V.L.Ginzburg (1960), *Soviet Solid State*, **2**, 61.
46. A.A.Abrikosov (1988), *Fundamentals of the Theory of Metals*, North-Holland, Elsevier, Groningen.
47. L.G.Aslamazov, A.I.Larkin (1968), *Soviet Solid State*, **10**, 875.
48. B.L.Altshuler and A.G.Aronov (1985), *Electron-Electron Interaction in Disordered Conductors* in Efros, A.L. and Pollak, M. (eds.), Elseiver Scientific Publishing.
49. R.S.Thompson (1970), *Phys. Rev.* **B 1**, 327.
50. A.A.Varlamov and V.V.Dorin (1983), *Soviet JETP* **57**, 1089.
51. M.Belogolovski, O. Chernyak and A.Khachaturov (1986), *Soviet Low Temperature Physics* **12**, 630.
52. B.L.Altshuler and A.G.Aronov (1979), *Solid State Communications* **30**, 115.
53. B.L.Altshuler and A.G.Aronov (1979), *Soviet JETP* **50**, 968.
54. A.M.Cucolo, M.Cuoco, A.Varlamov (1998), to be published in *Phys. Rev. B.*
55. M.Randeria (1994), *Physica* **B 199-200**, 373.

ON THE INTERPLAY BETWEEN T_C-INHOMOGENEITIES AT LONG LENGTH SCALES AND THERMAL FLUCTUATIONS AROUND THE AVERAGE SUPERCONDUCTING TRANSITION IN CUPRATES

FÉLIX VIDAL, JOSÉ ANTONIO VEIRA, JESÚS MAZA,
JESÚS MOSQUEIRA AND CARLOS CARBALLEIRA
Laboratorio de Bajas Temperaturas y Superconductividad,
Fac. de Física, U. Santiago de Compostela, E-15706 Spain

ABSTRACT. We review at an introductory and pedagogical level some aspects of the interplay between inhomogeneities at long length scales (at length scales much bigger than any characteristic length for superconductivity, in particular than the superconducting coherence length amplitude, $\xi(T)$, even at temperatures relatively close to T_C) and the intrinsic fluctuations of Cooper pairs above T_C in high temperature cuprate superconductors (HTSC). These inhomogeneities at long length scales do not directly affect the thermal fluctuations, but they may deeply affect, together (and entangled!) with the thermal fluctuations, the measured behaviour of any observable around the transition. The emphasis is centered on the role played by the presence of T_C-inhomogeneities, as those associated with oxygen content inhomogeneities, at these long length scales and uniformly or non-uniformly distributed in the samples, on the in-plane transport properties in inhomogeneous HTSC crystals. For completeness, we will also summarize some results on this interplay when various types of inhomogeneities (i.e., structural and stoichiometric, uniformly and non-uniformly distributed) may be simultaneously present.

Contents

1. Introduction

2. T_C-inhomogeneities uniformly distributed
    2.1. T_C-inhomogeneity effects on the in-plane magnetoconductivity above the average transition temperature.
    2.2. Splitting of the resistive transition: Intrinsic double superconducting transitions versus extrinsic effects.

S.-L. Drechsler and T. Mishonov (eds.), High-Tc Superconductors and Related Materials, 289–322.
© 2001 Kluwer Academic Publishers. Printed in the Netherlands.

3. T_C-inhomogeneities non-uniformly distributed
    3.1. Anomalous peaks of the in-plane magnetoresistivity around T_C.
    3.2. Negative in-plane longitudinal voltages around T_C.
    3.3. Anomalous peaks of the thermopower around T_C: Comparison with the electrical resistivity peaks.

4. Two examples of other inhomogeneity effects: How the crossing point of the magnetization and the full critical behaviour of the paraconductivity may be affected by the presence of inhomogeneities.
    4.1. The crossing point of the magnetization in highly anisotropic HTSC in presence of structural and stoichiometric inhomogeneities.
    4.2. Full critical behaviour of the paraconductivity versus T_C-inhomogeneities uniformly and non-uniformly distributed.

5. Conclusions.

6. Acknowledgements.

7. References.

1. Introduction

The interplay between the intrinsic thermal fluctuation effects around the superconducting transition and the extrinsic inhomogeneity effects associated with stoichiometric and structural inhomogeneities at different length scales, was already an important problem in low temperature superconductors (LTSC). For instance, in summarizing in 1978 the effects of the thermal fluctuations of Cooper pairs on the electrical resistivity, $\rho(T)$, above the superconducting transition in metallic films, effects that had been actively studied in the last ten years, Kosterlitz and Thouless concluded that the onset of the observed rounding of $\rho(T)$ "*may alternatively be a result of film inhomogeneities*". [1] In high temperature superconducting cuprates (HTSC), the dilemma between sample inhomogeneities and thermodynamic fluctuations above T_C was earlier stated by Bednorz and Müller in their seminal work, [2] although they formulated the alternative in an opposite way to that done by Kosterlitz and Thouless for LTSC's: After having indicated that the observed rounding of $\rho(T)$ around T_C in their LaSCO compounds may be due to inhomogeneities, Bednorz and Müller concluded that "*the onset (of the $\rho(T)$ drop) can also be due to fluctuations in the superconducting wave functions*". In fact, mainly due to the smallness of the superconducting coherence length amplitudes (at 0 K), $\xi(0)$, which are anisotropic but in all directions of the order of the interatomic distances, both effects, those associated with the intrinsic thermal fluctuations and those with the extrinsic inhomogeneities, may be very important in the HTSC. This is mainly due, in the case of the thermal fluctuations, to the fact that a small $\xi(T)$ leads to a small coherent volume, which will contain very few strongly correlated Cooper pairs. These fluctuation effects are also enhanced by the layered nature of the HTSC, which may lead these materials to behave as quasi bi-dimensional superconductors (still reducing, then, their superconducting coherent volume) and their high T_c, which increase then the available (gratis!) thermal agitation energy, of the order of $k_B T_c$ (where k_B is the Boltzmann constant) around their superconducting transition. [3]

In the case of the inhomogeneities, the smallness of $\xi(T)$ makes the different superconducting properties of these materials very sensitive to the presence of inhomogeneities, even when they have very small characteristic length, of the order of $\xi(T)$. In addition, their layered nature and the complexity of their chemistry enhance the probability of the presence of extrinsic inhomogeneity effects in real HTSC compounds. When they are present at long length scales (i.e., at length scales much bigger than any characteristic length in the system, as the magnetic field penetration length or, mainly, the superconducting coherence length, $\xi(T)$, even for temperatures relatively close to T_c), these inhomogeneities will not *directly* affect the thermal fluctuations themselves, but still they may deeply affect, together with the thermal fluctuations, the measured behaviour of any observable around the superconducting transition. [3]

In this paper, we will summarize some of our results on the interplay between the inhomogeneities of T_c at long length scales and the thermal fluctuations around the average superconducting transition. In particular, we will indicate through some examples how to disentangle the intrinsic effects (associated with the thermal fluctuations) from the extrinsic ones (associated with inhomogeneities). But, in any case these examples are aimed also to illustrate the fact that in analyzing an anomalous critical behaviour of any observable it is important to carefully check the possible presence of extrinsic inhomogeneity effects.

2. T_c-inhomogeneities uniformly distributed

Probably one of the most common types of inhomogeneities in HTSC are the critical temperature (T_c) inhomogeneities at long length scales. These inhomogeneities may be produced by, for example, oxygen content inhomogeneities at these long length scales. But, in addition to stoichiometric inhomogeneities, there are other possible causes for T_c-inhomogeneities in LTSC and HTSC compounds, such as local strains, [4] or low dimensionality effects. [5] In fact, these last effects may appear when, for instance, the sample dimension in one direction is smaller than the coherence length amplitude in that direction (so, in this case, the inhomogeneities will be in the so called small-length-scale regime, and the fluctuations will be also directly affected). [5]

An expected but non-trivial effect of these T_c-inhomogeneities at long length scales and uniformly distributed in the samples is that they round the critical behaviour of different observables around the superconducting transition, in competition with the intrinsic rounding effects associated with thermal fluctuations. [6] In the first part of this

Section, we summarize some of our results on this type of T_c-inhomogeneity effects on the in-plane resistivity and on the magnetoresistivity in some inhomogeneous HTSC crystals. Also, we will indicate how to disentangle the intrinsic and the extrinsic effects, around the average T_c, on both observables. In the second part of this subsection, we will summarize some of our results on the apparent double resistivity transition, effects that may be easily explained in terms of inhomogeneities. [7]

2.1. T_c-INHOMOGENEITY EFFECTS ON THE MAGNETOCONDUCTIVITY ABOVE THE AVERAGE TRANSITION TEMPERATURE

An illustrative example of the interplay between the intrinsic fluctuation effects and the extrinsic effects associated with inhomogeneities of characteristic lengths much bigger than the superconducting coherence length, $\xi(T)$, in all directions, is provided by the in-plane magnetoconductivity, $\sigma_{ab}(T,H)$, in presence of T_c-inhomogeneities at these long length scales and *uniformly distributed* in the sample. In the case of the electrical conductivity (in absence of applied magnetic field) such an interplay was first studied by Maza and Vidal (MV)[6] by using an effective medium approach. These results were recently extended by Pomar et al. [7] to study the influence of the T_c inhomogeneities on the in-plane magnetoconductivity in inhomogeneous $Bi_2Sr_2Ca_1Cu_2O_8$ (Bi-2212) crystals. We will summarize here these last results.

The effective (measured) in-plane magnetoconductivity $\sigma_{ab}^e(T,H)$, may be related to the intrinsic one (the conductivity measured in an ideal, homogeneous, crystal), $\sigma_{ab}(T,H)$, through the MV expression, based on the Bruggeman effective medium approach, [6,8]

$$\int_0^\infty \frac{\sigma_{ab}(T,H) - \sigma_{ab}^e(T,H)}{\sigma_{ab}(T,H) + 2\sigma_{ab}^e(T,H)} Q(\sigma_{ab},T)d\sigma_{ab} = 0 \qquad (1)$$

Here, $Q(\sigma_{ab},T)d\sigma_{ab}$ is the local conductivity distribution, i.e., the volume fraction of the sample with a local (or intrinsic) conductivity between $\sigma_{ab}(T,H)$ and $\sigma_{ab}(T,H) + d\sigma_{ab}(T,H)$. [6,7] This local conductivity may be written as the sum of the normal conductivity plus the corrections due to the thermal fluctuations,

$$\sigma_{ab}(T,H) = \sigma_{abB}(T,H) + \Delta\sigma_{ab}(T,0) + \Delta\tilde{\sigma}_{ab}(T,H), \qquad (2)$$

where here $\Delta\tilde{\sigma}_{ab}(T,H) \equiv \sigma_{ab}(T,H) - \sigma_{ab}(T,0)$.

To approximate $Q(\sigma_{ab},T)$ in Eq. (1), we may note first that the basic effects of the stoichiometric inhomogeneities on $\sigma_{ab}^e(T,H)$, and therefore on $Q(\sigma_{ab},T)$, are due to the associated critical temperature inhomogeneities. For the corresponding distribution of T_{c0}'s, we will follow the MV procedure, which assumes a spatial Gaussian distribution characterized by the mean value of the critical temperature, $\bar{T}_{c0}$, and by the standard deviation $\Delta\bar{T}_{c0}$. Thus, the conductivity distribution may be written as

$$Q(\sigma_{ab}, T) d\sigma_{ab} = \frac{2}{\sqrt{\pi}\Delta\overline{T}_{c0}} \exp\left\{-\left(\frac{T_{c0} - \overline{T}_{c0}}{\Delta\overline{T}_{c0}}\right)^2\right\} dT_{c0} \tag{3}$$

As the exponential function in Eq. (3) is rapidly decreasing, to evaluate the integral in Eq. (1) we may change the integration limit to the interval $\overline{T}_{c0} \pm 2\Delta\overline{T}_{c0}$. Then, Eq. (1) becomes

$$\int_{\overline{T}_{c0} - 2\Delta\overline{T}_{c0}}^{\overline{T}_{c0} + 2\Delta\overline{T}_{c0}} \frac{\sigma_{ab}(T,H) - \sigma_{ab}^e(T,H)}{\sigma_{ab}(T,H) + 2\sigma_{ab}^e(T,H)} \frac{C}{\Delta\overline{T}_{c0}} \exp\left\{-\left(\frac{T_{c0} - \overline{T}_{c0}}{\Delta\overline{T}_{c0}}\right)\right\} dT_{c0} = 0 \tag{4}$$

where $C = 0.5448$ arises from the normalization conditions. This expression links, therefore, the intrinsic and the effective conductivities through $\Delta\overline{T}_{c0}$, the standard deviation of critical temperatures due to inhomogeneities. The corresponding *effective* in-plane paraconductivity and fluctuation induced in-plane magnetoconductivity may be defined by just using $\sigma_{ab}^e(\overline{\varepsilon}, H)$, calculated through Eq. (4), in the conventional definitions of the in-plane paraconductivity, $\Delta\sigma_{ab}(\varepsilon)$, and fluctuation induced magnetoconductivity, $\Delta\overline{\sigma}_{ab}(\varepsilon)$, i.e.,

$$\Delta\sigma_{ab}^e(\overline{\varepsilon}, 0) \equiv \upsilon_{ab}^e(\overline{\varepsilon}, 0) - \sigma_{abB}^e(\overline{\varepsilon}, 0), \tag{5}$$

and

$$\Delta\overline{\sigma}_{ab}^e(\overline{\varepsilon}, H) \equiv \sigma_{ab}^e(\overline{\varepsilon}, H) - \sigma_{abB}^e(\overline{\varepsilon}, 0), \tag{6}$$

where $\overline{\varepsilon} \equiv (T - \overline{T}_{c0})/\overline{T}_{c0}$ is the reduced temperature associated to the average transition temperature, $\overline{T}_{c0}$. Let us note here that, as already stressed in Ref. 6, the T_c-inhomogeneities will affect more strongly the observables with stronger intrinsic ε-dependencies. In the Bi-2212 crystals, where the thermal fluctuations of Cooper pairs around T_c are strongly two-dimensional (2D), the intrinsic critical exponents in the mean field region above T_{c0} are at around -3 and -1 for, respectively, $\Delta\overline{\sigma}_{ab}(T,H)$ (in the weak H-limit) and $\Delta\sigma_{ab}(\varepsilon)$.[7] So, we may expect that $\Delta\overline{\sigma}_{ab}(\varepsilon)$ is going to be much more affected by inhomogeneities than $\Delta\sigma_{ab}(\varepsilon)$.

As an example of application of the above approach, in Fig. 1(a) we present the in-plane paraconductivity (open circles) and of the fluctuation induced in-plane magnetoconductivity (open triangles and squares) measured by Pomar et al. [7] in a Bi-2212 crystal. The lines in this figure correspond to the intrinsic (without inhomogeneities) direct order parameter fluctuation effects calculated in a layered material with two layers per periodicity length and with the same Josephson coupling strength between adjacent layers, which is the case well suited for Bi-2212 crystals (see Refs. 7 and 9). The value of the in-plane superconducting coherence length amplitude

294

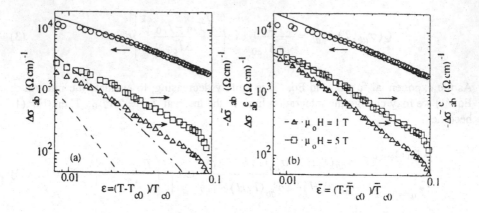

Figure 1. Reduced temperature dependence of the measured in-plane paraconductivity and of measured in-plane fluctuation induced magnetoconductivity of a Bi-2212 crystal for $\mu_0 H = 1$ T and 5 T. In (a) the lines correspond to the direct thermal fluctuation effects in an homogeneous crystal. In (b) the effects of the T_c-inhomogeneities on the theoretical expressions have been taken into account, and also the average reduced temperature is used. Figures from Ref. 7.

used here was $\xi_{ab}(0) \approx 1$ nm, which agrees with the one found by analyzing other fluctuation effects in this system (see Ref. 7). The results of Fig. 1 (a) show that whereas for the paraconductivity the agreement between the theory and the experimental data is excellent, there is a dramatic disagreement between the measured and the calculated fluctuation induced in-plane magnetoconductivity. Such a disagreement is confirmed by the results presented in Fig. 2 (a), where the measured excess conductivity is scaled as a function of $(T-T_c(H))(TH)^{-1/2}$, which is expected to hold for two-dimensional thermal fluctuations. [10] As it can be seen in this figure, there is no scaling below T_{c0}. These results may easily be understood at a qualitative level in terms of T_c-inhomogeneities, on the grounds of the comments presented at the end of the above paragraph.

The inhomogeneity effects on $\Delta\sigma_{ab}^e$ and $\Delta\tilde{\sigma}_{ab}^e$, have been estimated quantitatively in Ref. 7. For that, $\Delta\sigma_{ab}^e(\varepsilon,H)$, calculated through Eqs. (1) and (2) (and by using also the theoretical $\Delta\sigma_{ab}$ and $\Delta\tilde{\sigma}_{ab}$ for bilayered superconductors; see Ref. 9), was fitted to the experimental data with $\bar{T}_{c0}$ and $\Delta\bar{T}_{c0}$ as free parameters. As can be seen in Figs. 1 (b) and 2 (b), it is obtained an excellent and *simultaneous* agreement, the corresponding standard deviation for the critical temperature being $\Delta\bar{T}_{c0} = 0.6$ K. This value indicates that the stoichometric inhomogeneities, mainly of oxygen content, are quite small. But, in turn, these results of Ref. 7 clearly show that the presence of small T_c-inhomogeneities may dramatically affect the *measured* critical behaviour near T_c of the HTSC. These results also provide an alternative explanation in terms of T_c-inhomogeneities to the in-plane magnetoconductivity anomalies observed by other

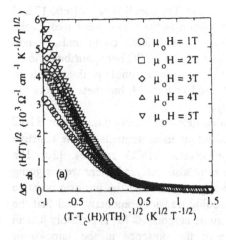
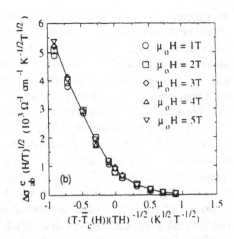

Figure 2. (a) Scaling of the excess conductivity for different magnetic fields measured in a Bi-2212 crystal. In (b) the effects of the T_c-inhomogeneities have been taken into account. The line through the 5 T data is a guide for the eyes. Figures from Ref. 7.

groups in Bi-2212 compounds and attributed by these authors to different *intrinsic* effects. [11]

2.2. SPLITTING OF THE RESISTIVE TRANSITION: INTRINSIC DOUBLE SUPERCONDUCTING TRANSITIONS VERSUS EXTRINSIC EFFECTS

The possibility of an intrinsic double superconducting transition in the HTSC was first open by some earlier heat capacity, C_p, measurements, which showed the presence in some samples of a double peak structure of the $C_p(T)$ curves near the average superconducting transition. [12] However, it was generally believed that these anomalies were probably an extrinsic effect, due for instance to the presence in these samples of stoichiometric inhomogeneities. The complexity of the HTSC chemistry, together with the strong influence of the oxygen content on their transition temperature, [13] made this simple explanation quite plausible. In fact, the extrinsic origin of the double transition was recently confirmed by heat capacity and magnetization measurements in $Y_1Ba_2Cu_3O_{7-\delta}$ (YBCO) samples with different oxygen contents. [14,15] However, in the last few years other groups, which have detected also double peak anomalies near T_c in different HTSC by measuring different observables (heat capacity, magnetic susceptibility and electrical resistivity), have claimed that these effects are intrinsic and that they are related to central, although different from one group to another, characteristics of these materials. [16-18] For instance, in Ref. 16 the double peak structure observed in $C_p(T)$ near T_c in the YBCO system is attributed to intrinsic T_c variations associated with the oxygen configuration around the CuO chains of these compounds. In contrast, in Refs. 17 and 18 it is proposed that the anomalous double transition effect is a manifestation of the possible unconventional symmetry of the superconducting order parameter of these materials. The conclusions of Refs. 17 and 18 about the existence of an intrinsic double transition in the HTSC are based on the observation of a splitting of the bulk resistive transition of different compounds, the non observation of these effects in other $\rho(T)$ measurements in HTSC being attributed to an insufficient experimental resolution. In fact, the presence of a double peak structure in the temperature derivative of $\rho(T)$ was first stressed in Ref. 19, but these authors did not conclude about the origin of such an anomaly.

To prove the existence of an intrinsic double superconducting transition in HTSC we have performed in our laboratory high resolution measurements of the resistive transitions in different single crystal and polycrystal YBCO samples. [20] The temperature derivatives of these resistivity data were also analyzed. Here we are going to summarize some of these results which strongly suggest that the intrinsic resistive transition of the HTSC does not present any double transition anomaly, and that the double peak structure observed in $d\rho(T)/dT$ by some authors[16-19] is probably just an extrinsic effect associated, in some cases, with the presence in the samples of stoichiometric inhomogeneities (mainly, small oxygen content inhomogeneities). In other cases, this double peak structure could just be an experimental artefact due, for instance, to the great sensitiveness of the temperature derivative of $\rho(T)$, near its maximum, to small electronic noise affecting the $\rho(T)$ data points.

An example of the in-plane (parallel to the CuO_2 layers) resistivity obtained in a YBCO single crystal (sample noted YS25) is presented in the Fig. 3. For clarity, in Figs. 3(a) only about 5% of the measured data points has been plotted. These values of $\rho_{ab}(T)$ and of $d\rho_{ab}(T)/dT$ in the normal region well above the transition, in the temperature region where $\rho_{ab}(T)$ is a linear function of T, are typical of excellent YBCO crystals, although somewhat twinned. [21,22] Another indication of the high quality of this sample is provided by the exceptional sharpness of its resistive transition, whose half width may be defined by[21]

$$\left(\frac{d\rho_{ab}(T)}{dT}\right)_{T_{CI}\pm\Delta T_{CI}^{\pm}} = \frac{1}{2}\left(\frac{d\rho_{ab}(T)}{dT}\right)_{T_{CI}} \tag{7}$$

where + or - correspond to, respectively, the upper and the lower half width and T_{cI} is the temperature where $d\rho_{ab}(T)/dT$ around the transition has its maximum. As can be seen in Figs. 3(b) and 3(c), the total width of $d\rho_{ab}(T)/dT$ around T_{cI} is, for this single crystal, less than 50 mK and 30 mK for, respectively, 3(b) and 3(c), the differences being associated with differences in the procedures used to obtain the temperature derivatives. In fact, these two figures also illustrate the crucial importance of the derivative procedure in analyzing with high temperature resolution the $\rho_{ab}(T)$ behaviour around T_c. In both cases, to obtain $d\rho_{ab}(T)/dT$ in each data point we use a polynomial determined by the best fit to its neighbour data points (together with the own data point). The use of a low degree polynomial or insufficient data points could introduce some spurious noise in $d\rho_{ab}(T)/dT$ which leads, in particular, to the onset of some kind of spurious structure around the $d\rho_{ab}(T)/dT$ maximum, the temperature region where the derivative is more sensitive to any irregularity. In Fig. 3(b), we have

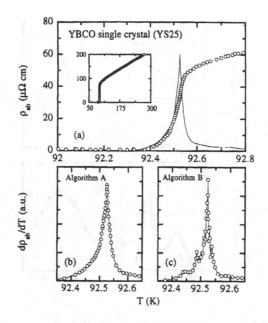

Fig. 3. (a) Temperature behaviour of the in-plane resistivity of one of the $Y_1Ba_2Cu_3O_{7-\delta}$ single crystals studied in this work (in this case, Ys7). In these two figures, for clarity only about 5% of measured data points are shown. (b) and (c) Temperature derivatives of the resistivity obtained by using the algorithms noted A and, respectively, B, and defined in the main text. Figure from Ref. 20.

used a degree-three polynomial, that for each temperature was determined by the four nearest data points to such a temperature. This procedure, called algorithm A, leads to a quite regular derivative, *without any double peak structure*. As the spacing between data points is of the order of 5 mK, the resolution in temperature relative of this derivative is better than 20 mK. So, with a resolution better than 20 mK, the results summarized in Figs. 3(a) to 3(b) show the absence of a double resistive transition in YBCO single crystals. This temperature resolution must be compared with the temperature shift between the two peaks reported by different authors, [14-19] which was 50 mK or more. Complementary, the results of Fig. 3(c), have been obtained by using a degree one polynomial (called here algorithm B) and only with the two nearest data points to each given temperature. The corresponding structure of peaks around the $d\rho_{ab}(T)/dT$ maximum is clearly a spurious effect associated with this inadequate derivative procedure. Let us stress here that these results were also confirmed by the analysis of our previous measurements of the resistivity in the a-direction, $\rho_a(T)$, not affected by the presence of the CuO chains, in two untwinned YBCO crystals. [20] However, in these measurements the temperature shift between the data points was of the order of 20 mK and, therefore, the temperature resolution of $d\rho_a/dT$ around T_{cI} was only of the order of 80 mK.

298

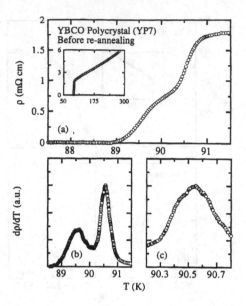

Fig. 4. (a) Temperature behaviour of the resistivity of one of the $Y_1Ba_2Cu_3O_{7-\delta}$ polycrystals studied in this work (in this case, sample YP7 before any re-annealing treatment). (b) Temperature derivative of the resistivity around the transition showing a two-peak structure typical of a two-phase sample. (c) Detail of $d\rho(T)/dT$ peak corresponding to the better oxygenated phase showing no sub-peak structure. Figure from Ref. 20.

As a complementary check of the possible existence of an anomalous resistive peak structure around T_{cI} in the YBCO system, in Figs. 4 and 5 we present an example of the results obtained in ceramic YBCO samples before and after re-annealing. This example correspond to the sample noted YP7. As may be seen in Figs. 4, before re-oxygenation the $\rho(T)$ behaviour of this sample shows the typical kink of a two-phase sample, with two different T_c's, their temperature shift being of the order of 1 K. This $\rho(T)$ behaviour is very similar to that of some of the YBCO samples studied in Ref. 16, and attributed by these authors to intrinsic effects associated with the oxygen arrangement around the CuO chains. However, such a double transition completely disappears after a new re-annealing, as shown in Fig. 5. This result clearly confirms that this double transition is just associated with the presence in the initially deficient oxygenated sample of small oxygen content inhomogeneities (less than 4%, the resolution of our x-ray analysis), at long length scales (i.e., at length scales much larger than the superconducting correlation length, even for temperatures relatively close to the transition) and uniformly distributed. A detail of the temperature derivative of the $\rho(T)$ part associated with the better oxygenated phase, with higher T_c, is presented in Fig. 4(c). We see here that within a temperature resolution of the order of 20 mK no sub-peak structure is observed. The absence of such an anomalous peak structure is also

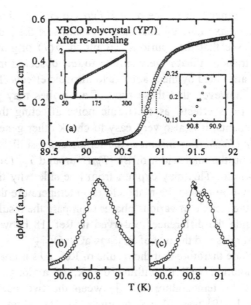

Fig. 5. (a) Temperature behaviour of the resistivity of the same polycrystalline sample as in Fig. 4 (sample YP7) but after a re-annealing treatment. In the scope in (a) we show an example of the defaults introduced artificially in the $\rho(T)$ curves to check the resolution of our temperature derivative procedure. This example consists in the shift in temperature (5 mK to lower temperatures) of a measured data point (open circle). The shifted data point is the solid circle. (b) and (c) Temperature derivatives of the resistivity obtained from the measured data points and, respectively, from the $\rho(T)$ curve resulting after the introduction of the default shown in the scope in (a). The solid circle here corresponds to the shifted point in the scope in (b). Figure from Ref. 20.

confirmed by the results presented in Fig. 5(b) for $d\rho(T)/dT$ of the re-annealed YP7 sample. We have also analyzed other ceramic YBCO samples, with different T_c's (i.e., with different oxygen content). These analyses show that the absence of an anomalous peak structure of $d\rho(T)/dT$ is a general behaviour of the YBCO system, independently of its oxygen content, at least until T_c's of the order of 80 K.

To check that the temperature derivative procedure introduced in Ref. 20 and based on the algorithm A, was able to detect the possible presence of small $\rho(T)$ anomalies around the transition, various types of artificial deformations were introduced in the experimental $\rho(T)$ curves. One of the deformations analyzed was the shift in temperature of just one of the $\rho(T)$ data points, mainly when it is located near T_{cI}. An example of such an artificial default, in this case introduced in the $\rho(T)$ curve of sample YP7 after re-annealing, may be seen in the inset of Fig. 5(a). The temperature shift between the measured data point (open circle) and the one artificially moved to the left (closed circle) is 5 mK. The resulting $d\rho/dT$ is presented in Fig. 5(c), which should be compared with the results of Fig. 5(b) for the non deformed $\rho(T)$ curve. This example clearly illustrates the dramatic influence on $d\rho(T)/dT$, mainly in the temperature region close to the derivative maximum, of small spurious shifts of the individual data points. In this example, we see that the spurious default in $\rho(T)$ originates a double peak structure around the transition that is very similar to that observed in Refs. 18 and 19 in YBCO samples, and attributed by these authors to intrinsic effects. These results of Ref. 20 strongly suggest, however, that these type of anomalies may probably be due to various types of spurious effects, as electronic noise affecting the resistivity or the temperature measurements. It is also very easy to check other general aspects of these spurious peaks of $d\rho(T)/dT$. For instance, the temperature shift between these peaks will directly depend on the total width of $d\rho(T)/dT$ around T_{cI} (and this shift will be less than this total width). This may explain from one side why the peaks associated with the $\rho(T)$ noise will in general be much closer in temperature than those associated with the presence in the sample of various phases (compare the results of Figs. 4 and 5), and from the other side the differences observed in Ref. 18 between the temperature splitting of YBCO samples and the one of $Bi_2Sr_2CaCu_2O_{8+x}$.

We may conclude therefore that the results of Ref. 20 summarized here show the absence, with a resolution well to within 20 mK, of a double peak structure of $d\rho(T)/dT$. Note that the temperature shift between the two peaks reported by the different authors [12,14-19] was 50 mK or more, i.e., larger than our experimental resolution. Complementarily, we have analyzed the influence on $d\rho(T)/dT$ of various types of extrinsic effects including the presence in the samples of various stoichiometric phases as well as the presence of spurious noise effects in the resistivity or in the temperature measurements. The present results strongly suggest that the double peak structure of the resistive transition observed by various groups [16-19] in different high temperature copper oxide superconductors is an extrinsic effect. Moreover, as already stressed in Ref.19 at present does not exist any firm theoretical link between the presence of a possible double transition in the HTSC and the symmetry of its superconducting order parameter. [23] So, these different results on $d\rho(T)/dT$ do not allow to draw any conclusion about the wave pairing state in HTSC. [24]

3. T_c-inhomogeneities non-uniformly distributed

It is now well established that the presence of T_c-inhomogeneities at long length scales but *non-uniformly* distributed may generate various types of striking anomalies in the transport properties around the superconducting transition in HTSC, including the so-called anomalous peaks of the in-plane resistivity[25] the thermoelectric power[26] and the magnetoconductivity. [27,28] In addition to their intrinsic interest, these results may also provide an alternative explanation, in terms of temperature independent current redistributions associated with non uniformly distributed T_c-inhomogeneities, of the anomalous resistivity and magnetoresistivity peaks observed above the average superconducting transition by different groups in other LTSC[29-31] and HTSC[32-35] compounds and that are being attributed to different, and in some cases not well settled, intrinsic effects. [29-35] In the case of the thermopower, it is also possible to explain the anomalous $S(T)$ peaks observed in YBCO samples in terms of oxygen content inhomogeneities. [26] In this case, this explanation does not need any particular distribution of the inhomogeneities but it needs the simultaneous presence of two possible consequences of these inhomogeneities [26]: differences between the T_c's of various sample domains and also differences between the sign of their corresponding thermopower. Both differences, in T_c and in the sign of S, may be a consequence of small oxygen content differences in almost full doped YBCO samples, where in fact these $S(T)$ anomalies have been observed. [26,36] This provides a simple explanation of the anomalous $S(T)$ peaks observed by different groups and attributed in some cases to sophisticated but not well settled intrinsic mechanisms. [36-38] In this Section, we will summarize some of our results on the influence of T_c-inhomogeneities on the in-plane magnetoconductivity in inhomogeneous HTSC crystals. We will also briefly comment on the thermopower anomalous peaks in YBCO samples with small oxygen content inhomogeneities.

3.1. ANOMALOUS PEAKS OF THE IN-PLANE MAGNETORESISTIVITY AROUND T_c

An example of the in-plane magnetoresistivity peaks observed in Ref. 28 in a Y-123 crystal (sample Y16 of Ref. 28) before re-oxygenation are presented in Figs. 6(a) and (b). These anomalous peaks lead to a negative and anisotropic (i.e., depending on the orientation of H with respect to the CuO_2 planes) in-plane magnetoresistivity excess which, for each magnetic field orientation, may be quantified through

$$\Delta\rho(T,H) \equiv \rho(T,H) - \rho(T,0) \qquad (8)$$

The data points in Figs. 7(a) and (b) correspond to the measured $\Delta\rho(T,H)$ for H perpendicular and, respectively, parallel to the ab planes. Only the $\Delta\rho(T,H)$ values corresponding to temperatures above the $\rho(T,0)$ peak have been represented. These important $\Delta\rho(T,H)$ values cannot be explained in terms of thermal fluctuations (see Ref. 3 and references therein). These results also show that, as may be expected from the $\rho(T,H)$ data, $\Delta\rho(T,H)$ is very anisotropic. In particular, for H parallel to the c direction the $\Delta\rho(H)_T$ saturation value is reached, at each temperature, for smaller field amplitudes than for H parallel to the ab planes. This behaviour of $\Delta\rho(T,H)$ is quite similar to that observed by other groups in other HTSC and LTSC compounds having $\rho(T,H)$ peaks near the superconducting transition and attributed by these authors to intrinsic effects[29,31-35] (compare, in particular, Figs. 7(a) and (b) with Fig. 2 of Ref. 34).

The experimental results for the Y16 sample after its reoxygenation are presented in Figs. 8(a) and (b) for H applied parallel and, respectively, perpendicularly to the CuO_2 planes [see inset in Figs. 6(a) and (b)]. In this new oxygen annealing the crystal was again placed in a boat within a tubular furnace with O_2 flowing. The furnace was then heated up to 600 °C (at a rate of 100 °C/h), kept at this temperature for

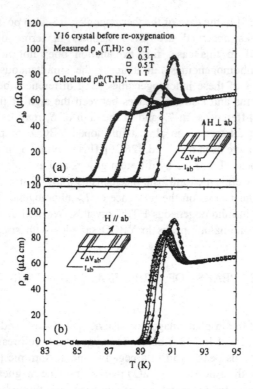

Fig. 6. In-plane magnetoresistivity versus temperature of the crystal Y16 before reoxygenation for different external magnetic fields with various amplitudes and orientations. In (a) the field was applied parallel to the crystallographic c-direction. In (b), it was applied parallel to the CuO_2 (ab) layers but still perpendicular to the injected current. The lines are the result of the simulations performed with the electrical resistor network represented in the inset of Fig. 8(b). Figures from Ref. 28.

2h, cooled to 400 °C in 1h and held at this temperature during four days. These latter processes were repeated three times. We see in Fig. 8(a) that the anomalous peak has completely disappeared from this $\rho_{ab}(T)$ curve. As, in addition, no structural changes were observed after these new annealings, these results confirm that the peak observed before is related to the presence in the sample of small (much less than 4% of the average oxygenation, the resolution of our x-ray diffraction measurements) oxygen-content inhomogeneities, that are strongly reduced by successive O_2 annealings. Let us note here that these results fully confirm the behaviour observed before in other YBCO crystals with resistivity peak anomalies [25] and, therefore, they provide new support to the explanation of the anomalous resistivity peak in terms of non uniformly distributed oxygen content inhomogeneities.

These anomalous $\rho_{ab}(T,H)$ peaks have been explained in terms of non-uniformly distributed T_c-inhomogeneities in Ref. 28. Here we will summarize some of these results. Let us first stress here that these inhomogeneities are going to be small (of the

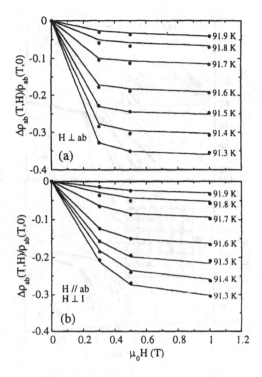

Fig. 7. Normalized magnetoresistivity excess of the crystal Y16 before reoxygenation at temperatures above but near the maximum of the anomalous $\rho(T,H)$ peak. (a) For H applied parallel to the c-direction. (b) For H applied parallel to the ab-planes and perpendicular to the injected current. The solid lines are the result of the simulation performed with the electrical resistor network represented in the inset of Fig. 8(b). Figures from Ref. 28.

order of two degrees or less, i.e., less than the 3% of the average T_c) and they may be due to small stoichiometric inhomogeneities (mainly of the oxygen content) extended over 10% or less of the sample volume. So, we are not able to directly determine these small local inhomogeneities. In addition, due to the oxygenation process, it is reasonable to expect that the best oxygenated parts of the crystal will be the edges of the sample domains. An example of an inhomogeneity distribution capable of generating the anomalous $\rho_{ab}(T,H)$ behaviour observed in the Y16 crystal is the one schematized in the inset of Fig. 8(a). In this figure, the shadowed domains at the upper edges of the crystal are better oxygenated and they have a higher T_c than the rest of the crystal. The dimensions of each high-T_c domain are $1/3L_x$, L_y, $1/7$ L_z, where L_x, L_y and L_z are the sample's dimensions. To study how this inhomogeneity distribution affects the measured $\rho_{ab}(T,H)$ and generates the anomalous peak, we simulate the measurement through an equivalent electrical network. The geometry of the inhomogeneity distribution and the contact arrangement allows us to reduce the equivalent electrical

304

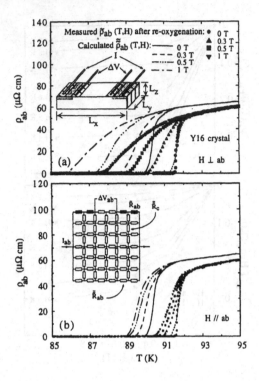

Fig. 8. In-plane magnetoresistivity versus temperature of the crystal Y16 after reoxygenation for different magnetic fields applied normally (a) and parallel (b) to the ab planes. The lines are the resistivities used in the electrical resistor network for the less oxygenated domains. Inset in (a): Schematic diagram of the T_c inhomogeneities of the crystal Y16. The shadowed parts correspond to the highest T_c domains and the dashed areas are the silver-coated electrical contacts. Inset in (b): Two dimensional electrical network for the sample schematized in (a). Figures from Ref. 28.

network, in principle three dimensional, to the bi-dimensional one represented in the inset of Fig. 8(b). The resistances noted as $\widetilde{\widetilde{R}}_{ab}(T,H)$ and $\widetilde{\widetilde{R}}_c(T,H)$ correspond to the less oxygenated domains (with lower T_c), while $\widetilde{R}_{ab}(T,H)$ corresponds to the domains with higher T_c. Each resistance in the network is related to the corresponding resistivity in the crystal by

$$\widetilde{R}_{ab}(T,H) = \frac{L_x(N+1)}{L_z L_y N} \widetilde{\rho}_{ab}(T,H),\qquad(9)$$

this relationship applying also to $\widetilde{\widetilde{R}}_{ab}(T,H)$ [with $\widetilde{\widetilde{\rho}}_{ab}(T,H)$], and by

$$\widetilde{R}_c(T,H) = \frac{L_z(N+1)}{L_x L_y N}\widetilde{\rho}_c(T,H) \quad .$$ (10)

In these equations, $\widetilde{\rho}_{ab}(T,H)$ and $\widetilde{\rho}_c(T,H)$ correspond to the less oxygenated domains (with lower T_c), whereas $\mathcal{P}_{ab}(T,H)$ corresponds to the domains with higher T_c, and $N\times N$ is the number of meshes of the network (6x6 in this case). For $\widetilde{\rho}_{ab}(T,H)$, we used the profiles also shown in Figs. 8(a) and (b), that are typical of non-fully oxygenated Y-123. The resistivity in the c-direction is assumed to be one hundred times the in-plane resistivity. We assume also that $\mathcal{P}_{ab}(T,H)$ for H parallel and perpendicular to the ab-planes may crudely be approximated by the resistivities measured after a new oxygen annealing, which are presented in Figs. 8 (a) and (b). However, to achieve the excellent agreement with the experimental results observed in Fig. 7, for $\mathcal{P}_{ab}(T,0)$ we have used the dotted curve represented in the same figure, that is slightly smoother than the experimental $\rho_{ab}(T,0)$. Such an excellent agreement is obtained in spite of the simplicity of our network, consisting only in 6x6 meshes and two different types of domains. [39]

    A first example of the results of these calculations are the solid lines in Figs. 6 and 7. As it can be seen, the agreement between the experimental data and the simulation is excellent, the anomalous behaviour of the magnetoresistivity and of the magnetoresistivity excess being reproduced at a quantitative level for both orientations of the external magnetic field. In Fig. 9(a), it is represented the current distribution in the network at $T = 95$ K, a temperature well above the superconducting transition. This case corresponds, therefore, to the trivial situation in which the different sample domains with different oxygen content have almost the same (normal) resistivity, so the current lines are parallel to the ab plane and uniformly distributed and no anomaly is then observed. In contrast, the electrical current distribution shown in Fig. 9(b) corresponds to $T = 91.2$ K, the temperature at which the maximum of the resistivity peak occurs. At this temperature, the domains with higher T_c are already superconducting and, therefore, the current density distribution is no longer uniform. There appears a higher current density in the top face of the crystal, where the voltage contacts are placed, giving rise to the anomalous voltage peak. Finally, in Fig. 9(c) it is represented the current distribution corresponding to $T = 91.2$ K in the presence of the external magnetic field of 1 T applied perpendicularly to the ab planes. The main effect of the magnetic field is to broaden the resistive transition, making the differences between the resistivities of the different domains much smaller than for H = 0. As a consequence, the current distribution is nearly uniform and the anomalous peak almost disappears from the effective (or measured) $\rho_{ab}(T,H)$ curves. Moreover, the broadening of the resistive transition is more pronounced for $H \perp$ ab than for $H /\!/$ ab, and this is because the field amplitude needed to completely quench the anomalous peak is bigger for the latter field orientation.

    As already noted in the introduction of this Section, anomalous magnetoresistivity peaks near T_c very similar to the ones described here have been observed in some thin films and granular samples of different LTSC. [29] The simplicity of the chemical structure of these compounds makes quite improbable the

306

presence of appreciable non-uniformly distributed compositional inhomogeneities. However, another source of T_c inhomogeneities could be related to the well known low-dimensionality effects, which appear when the superconducting coherence length, $\xi(T)$, becomes of the order of or bigger than one of the sample's dimensions (thickness in the case of thin films, or grain diameter in the case of granular samples) [4]. Due to the relatively large coherence length amplitude of the LTSC [$\xi(T = 0 \text{ K}) \approx 1000$ Å], these low dimensionality effects may easily be present in the LTSC films. In addition, the thin film edges are never completely sharp, but instead they have an irregular thinner shape, and so having a higher T_c than the rest of the film. When the temperature is well above any possible T_c in the film, the electrical current used to measure the resistivity must be uniformly distributed. But at temperatures in which the film edges are superconducting and the rest of the film still remains in the normal state, the current lines must concentrate in the edges. This current redistribution may give rise to an increment in the signal detected by a voltmeter connected at the film edges and, in some cases, to the anomalous resistivity peak. A condition for that is simply that the edge thickness should

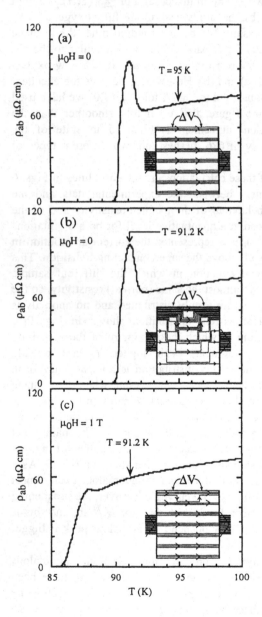

Fig. 9. Examples of the current redistributions originated in the electrical network corresponding to the crystal Y16 and represented in the inset of Fig. 8(b): (a) At $T = 95$ K, well above any superconducting transition in the sample. (b) At $T = 91.2$ K the temperature at which the anomalous peak has its maximum and the highest T_c domains become superconducting. (c) At $T = 91.2$ K, with a magnetic field of 1 T parallel to the c-crystallographic direction. These striking differences in both the current distributions and in the measured $\rho(T,H)$ are associated just with differences of the temperature and magnetic field dependence of $\rho(T,H)$ in each sample domain. Figures from Ref. 28.

be non uniform *along* the film. Otherwise, at the temperature at which the film edges are already superconducting a continuous superconducting path would connect both voltage terminals and no signal (and then no peak) would be detected. The anomaly so originated will be very dependent on the current used to perform the measurements. In fact, due to the smallness of the section of the highest T_c edges, it is very easy to reach the critical current density by increasing the applied current, making the effect to disappear. Some of the different experimental results of Ref. 29 may satisfactorily be explained on the grounds of these simple ideas based on the presence of low dimensionality induced T_c inhomogeneities, non uniformly distributed in the films. The anomalous $\rho(T,H)$ peaks observed in some granular LTSC samples[29] could also easily be explained in terms of low dimensionality effects, through the dependence of T_c with the grain diameter, if a non-uniform distribution of grains having different diameters exists in the sample.

3.2. NEGATIVE IN-PLANE LONGITUDINAL VOLTAGES AROUND T_c.

The results summarized in the precedent subsections, clearly suggest that the magnetoresistivity anomalies observed in inhomogeneous HTSC could strongly depend on the type of spatial distributions of the T_c-inhomogeneities. In other words, different locations of the non-uniformly distributed T_c-inhomogeneities could originate very different behaviours of $\rho(T,H)$ around the average T_c. To illustrate this conclusion, here we are going to summarize some of our measurements of the in-plane longitudinal magnetoresistivity of $Tl_2Ba_2Ca_2Cu_3O_{10}$ (Tl-2223) crystals with stoichiometric inhomogeneities. In some of these crystals we have observed that the in-plane longitudinal voltage measured in presence of magnetic fields applied perpendicularly to the ab-layers, $V(H)$, is negative just below T_c. This anomaly may be explained in terms of T_c-inhomogeneities non-uniformly distributed in the sample *surface*.[27]

An example of T_c-inhomogeneities non-uniformly distributed in the sample surface are schematized in Fig. 10(a). This example was studied in more detail in Ref. 27. Here again, the shadowed parts correspond to the domains of higher T_c, the corresponding three-dimensional electrical network being represented in Fig 10(b). The resulting resistivity and current redistributions are represented in Figs. 10(c) and (d). The striking result is the appearance, just below the average superconducting transition and in some parts of the sample surface, of counter-currents with sign opposite to that of the injected electrical current. This may lead to the appearance of a negative effective resistivity just below the transition as it is illustrated in Fig. 10(c). These effects could explain the magnetoresistivity anomalies that we have recently observed in some inhomogeneous Tl-2223 crystals.[27] Besides, these negative surface currents associated with T_c-inhomogeneities non-uniformly distributed in the sample surface, could maybe contribute to explain some of the anomalous negative behaviour of other longitudinal and transversal transport properties observed around T_c in LTSC[40] or HTSC[41] materials.

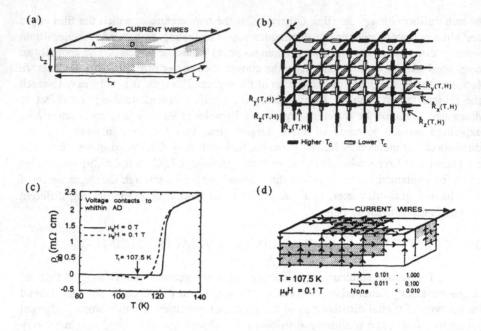

Figure 10. (a) Example of T_c-inhomogeneities non-uniformly distributed in the sample surface. (b) Corresponding three-dimensional resistor network. (c) Resulting in-plane effective resistivities for two applied magnetic fields. (d) Current density distribution at T_l and $\mu_0 H = 0.1$ T. Figure from Ref. 27.

3.3. ANOMALOUS PEAKS OF THE THERMOPOWER AROUND T_c: A BRIEF COMPARISON WITH THE ELECTRICAL RESISTIVITY PEAKS.

Since the measurements of Cabeza and coworkers, it is now well established that the *intrinsic* critical behaviour of the thermopower, $S(T)$, near T_c in copper-oxide superconductors is mainly driven by that of the electrical conductivity. In other words, the measurements in *homogeneous* HTSC show that their thermoelectric coefficient, $L(T)$, does not present any "sharp" critical divergence above the superconducting transition. [42] Instead, $L(T)$ in HTSC has around T_c the logarithmic divergence earlier predicted by Maki. [43] Complementarily, since the measurements of Mosqueira and coworkers[26] it is now also well established that the presence of small oxygen content inhomogeneities in $Y_1Ba_2Cu_3O_{7-\delta}$ (Y-123) samples, *uniformly or non-uniformly distributed*, may also deeply affect the behaviour of $S(T)$ around the average superconducting transition. In fact this provides a simple, and now widely accepted, explanation, in terms of the oxygen content inhomogeneities, of the $S(T)$ anomalous peaks observed by different groups in Y-123 samples. [36,37] Until the results of Ref. 26, these anomalous $S(T)$ peaks were attributed by various groups to different intrinsic, but not well settled, effects. [37,38]

An example corresponding to the inhomogeneities sketched in Fig. 11(b) and (c), of the influence of the relative amplitude of the inhomogeneity domain on the effective (measured) thermopower peak is represented in Fig. 12. This example corresponds to non-uniformly distributed T_c-inhomogeneities. The details may be seen in Ref. 26. In this reference it was also studied the quenching of the anomalous $S(T,H)$ peak by a magnetic field, an effect observed by various authors[36] and which has remained unexplained until the results of Ref. 26. The details of the generation of the $S(T)$ peaks by the presence of oxygen content inhomogeneities in Y-123 crystals may be seen in Ref. 26. However, it may be useful to briefly compare here how these inhomogeneities originate the anomalous $S(T)$ or the $\rho(T)$ peaks. The *independent* peaks that near T_c may present both observables in some $Y_1Ba_2Cu_3O_{7-\delta}$ samples may be understood in both cases in terms of oxygen content inhomogeneities at long length scales. However, in the case of $\rho(T)$, the anomalous peak may be explained by just taking into account only the associated T_c-inhomogeneities, which when they are non-uniformly distributed may lead near T_c to strong electrical current density inhomogeneities in the sample. [25] In fact, these

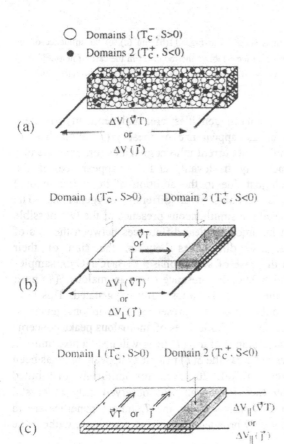

Fig. 11. (a) Schematic diagram of an example of uniformly distributed oxygen content inhomogeneities in polycrystalline samples. The better oxygenated domains (with $\delta \approx 0$) correspond to the smaller grains while the largest domains have a slight deficiency in oxygen content ($\delta \approx 0.1$). (b) and (c) Schematic diagrams of an example of non-uniformly distributed inhomogeneities in the oxygen content in single crystals for two different leads arrangements. The shadowed domains are well oxygenated ($\delta \approx 0$) while the white domains have $\delta \approx 0.1$. Figure from Ref. 26.

310

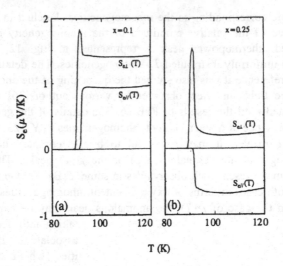

Fig. 12. An example, corresponding to situations sketched in Figs. 11(b) and (c), of the influence of the relative magnitude of the inhomogeneity domains on the effective thermopower in the case of non-uniformly distributed oxygen content inhomogeneities. In (a), x = 0.1, whereas in (b), x = 0.25. Figure from Ref. 26.

temperature-dependent current-density inhomogeneities are, independently of their origin, the only crucial ingredient for the appearance of these $\rho(T)$ anomalies. In contrast, in the case of $S(T)$ the possible heat current inhomogeneities (or, equivalently, the ∇T inhomogeneities) are practically irrelevant, and the appearance of the anomalous $S(T)$ peaks is, in general, just due to the addition of the different $S(T)$ associated with the various sample domains with different oxygenations. The appearance of a $S(T)$ peak will then need the simultaneous presence of the two possible consequences of the oxygen content inhomogeneities: differences between the T_c's of the different sample domains but also differences between the sign of their corresponding thermopowers. As in the case of single phase $Y_1Ba_2Cu_3O_{7-\delta}$ samples the sign change of $S(T)$ appears for $\delta \approx 0.06$, the presence of an anomalous $S(T)$ peak will be possible only if one part of the sample is almost fully oxygenated. This is in contrast with the $\rho(T)$ peak, that only needs the presence of T_c-inhomogeneities. Another difference to be stressed here between both types of anomalous peaks concerns the inhomogeneity distribution. The appearance of a $\rho(T)$ peak will need a non-uniform distribution of the T_c-inhomogeneities, whereas the $S(T)$ peak may appear, as it has been shown by Mosqueira and coworkers in Ref. 26, even for uniformly distributed inhomogeneities. This explains why the $\rho(T)$ peak was observed only in crystal samples and not in polycrystals: In these last samples the T_c-inhomogeneities are in general uniformly distributed and they only broaden the resistive transition, without the generation of a peak. [6]

4. Two examples of other inhomogeneity effects: How the crossing point of the magnetization and the full critical behaviour of the paraconductivity may be affected by the presence of inhomogeneities.

For completeness, we will summarize here two examples of other inhomogeneity effects. The first example will concern the thermal fluctuation effects of vortices below T_c in presence of stoichiometric and structural inhomogeneities. The second example will concern again the in-plane resistivity but this time very close to T_c and in presence of both types of T_c-inhomogeneities at long length scales: uniformly and non-uniformly distributed. In this last case both type of T_c-inhomogeneities are going to be analyzed separately, but they may simultaneously affect $\rho_{ab}(T)$ very close to T_c.

4.1. THE CROSSING POINT OF THE MAGNETIZATION IN HIGHLY ANISOTROPIC HTSC IN PRESENCE OF STRUCTURAL AND STOICHIOMETRIC INHOMOGENEITIES.

Although this review is centered on the effects of the T_c-inhomogeneities on the in-plane transport properties, it will be useful to note here that the presence in the samples of structural inhomogeneities, always at long length scales, may also deeply affect any observable and, in particular, the *measured* behaviour of $\rho(T,H)$ and of $S(T,H)$ around T_c. These structural inhomogeneity effects at long length scales, as those that exist in granular and ceramic HTSC, on $\rho(T,H)$ and on $S(T,H)$ around T_c have been earlier analyzed in our group, and a procedure to separate them from the intrinsic fluctuation effects was also proposed. [42,44,45] The interest of these anomalies of the structural inhomogeneity effects on $\rho(T,H)$ is enhanced by the fact that they are also directly related to the critical current densities in granular and ceramic HTSC. [46] However, these analyses are out of the scope of our present review. Some of the main aspects of these effects and of their interplay with the thermal fluctuations may be seen in Refs. 42, 44 and 45. Here, to illustrate the influence of these structural inhomogeneities al long length scales we are going to just summarize briefly some of our recent results on the in-plane magnetization, $M_{ab}(T,H)$, around T_c in highly anisotropic HTSC with small structural and stoichiometric inhomogeneities. Note also that until now we have reviewed in this paper some inhomogeneity effects on transport properties and its interplay with thermal fluctuations of Cooper pairs *above* T_c. In contrast, this new example concerns a static parameter and the interplay between stoichiometric and, mainly, structural inhomogeneities with thermal fluctuations of magnetic vortices *below* T_c.

The so-called "crossing point" of the excess magnetization versus temperature curves at a given magnetic field amplitude (with H applied perpendicularly to the CuO$_2$ layers), $\Delta M(T)_H$, probably provides one of the best and easiest scenarios to check the interplay between structural and stoichiometric inhomogeneities at long length scales and thermal fluctuation effects in HTSC. This crossing point of the $\Delta M(T)_H$ curves

312

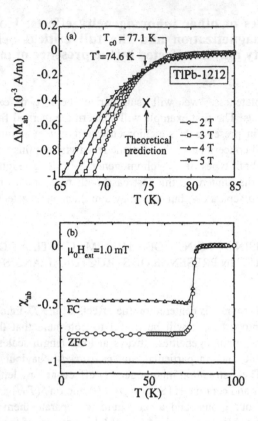

Fig. 13. (a) Measured in-plane excess magnetization vs. temperature, at different constant magnetic fields applied normally to the ab-planes, of a TlPb-1212 single crystal. (b) Corresponding in-plane field-cooled and zero-field-cooled susceptibilities. The main error source in these last observables is the demagnetization factor uncertainties.

was first observed experimentally by Kadowaki[47] and, independently, by Kes and coworkers. [48] In the case of highly anisotropic HTSC and for high magnetic field amplitudes (i.e., for $H \gtrsim (1/7)H_{c2}(T^*)$, where H_{c2} is the upper critical field at the crossing point temperature, T^*), these fluctuation effects are attributed to the creation and anihilation of quasi-two dimensional (pancake) vortices. [49] In this case, the Ginzburg-Landau model in the so-called lowest-Landau-level approximation (GL-LLL) predicts that the crossing point coordinates are related by[49]

$$-\Delta M_{ab}^* = \frac{k_B T^*}{\phi_0 s_e^v},$$

(11)

where k_B is the Boltzmann constant, s_e^v is an effective periodicity length which takes into account the possible multilayering effects on the vortex fluctuations and $\phi_0 = h/2e$ is the flux quantum (h is the Planck constant and e is the electron charge). The importance of Eq. (11) is enhanced by the fact that it relates directly the effective periodicity length, s_e^v, a microscopic parameter which in multilayered HTSC may depend on the Josephson and on the magnetic couplings between adjacent superconducting layers, to two directly measurable macroscopic observables, ΔM_{ab}^* and T^*, and that without any dependence on T_{c0}, the mean-field critical temperature, which is never directly accessible. This theoretical result has lead, therefore, to much experimental activity in the last years. [50] However, all the $\Delta M^*/T^*$ data published until now in polycrystalline or in single crystalline HTSC strongly disagree, in both the amplitude and the s dependence, with Eq. (11). In particular, in most of the experiments the measured $\Delta M_{ab}^*/T^*$ leads to an effective periodicity length, s_e^v, larger than s, in contradiction with the theoretical predictions. [49,51] An example of the disagreement between the measured crossing point coordinates and Eq. (11) may be seen in Fig. 13(a). This example correspond to a TlPb-1212 crystal. [52] The presence of strong stoichiometric inhomogeneities, which will appreciably reduce the superconducting fraction, has been discarded in most of the studied samples by independent measurements (x-ray and neutron diffraction, in particular). [50] Therefore, until now most of the authors propose that these $\Delta M_{ab}^*/T^*$ data are intrinsic and that the BLK and the TXBLS approaches do not explain, even at a qualitative level, the crossing points observed in highly anisotropic HTSC. [50]

It has been proposed recently, however, that the strong disagreement between the experimental data and the crossing point coordinates predicted by the theory of Tcšanovic and coworkers[49] could be resolved by taking into account all the possible non intrinsic effects on the magnetization. [52] These non intrinsic effects will be associated with structural and stoichiometric inhomogeneities, at different length scales and amplitudes, and not only with those due to the presence of strong stoichiometric inhomogeneities at long length scales (i.e., at length scales much larger than the superconducting coherence lengths, which are those easily observable with conventional x-ray and neutron diffraction techniques). This conclusion was strongly supported by simultaneous measurements of the crossing point in the high-magnetic-field limit [$H \lesssim H_{c2}(T^*)$; $H \gg H_0$] and of the field-cooled susceptibility (the so called Meissner fraction), χ_{ab}^{FC}, in different single crystals of various highly anisotropic HTSC families with different values of N and s. An example of the field-cooled (FC) and zero-field-cooled susceptibilities may be seen in Fig. 13(b). This example corresponds to the same sample than in Fig. 13(a). It may be easily concluded from Figs. 13(a) and (b) that the difference between the measured $\chi_{ab}^{FC}(T^*)$ (already corrected for demagnetization effects) and the susceptibility of an ideal superconductor is, within the experimental uncertainties, the same as the difference between the measured ΔM_{ab}^* the in-plane excess magnetization predicted by Eq. (11), with $s_e^v = s$. Similar measurements have been done in different HTSC with different values of N and s, in Ref. 52. These results demonstrate experimentally that in highly anisotropic HTSC crystals $\Delta M_{ab}^* / |\chi_{ab}^{FC}(T^*)|$ verifies, within the experimental uncertainties, Eq. (11), with $s_e^y = s$, independently of N. Complementarily, these results show that in spite of the fact that ΔM_{ab}^* and $\chi_{ab}^{FC}(T^*)$ are measured under very different magnetic field amplitudes [$H \lesssim H_{c2}(T^*)$ and, respectively, $H \leq H_{c1}(T^*)$, the lower critical magnetic field at T^*], the non intrinsic effects on both observables, associated with stoichiometric and structural inhomogeneities at different length scales, are the same within the experimental uncertainties. $\Delta M_{ab}^* / |\chi_{ab}^{FC}(T^*)|$ is, therefore, the intrinsic excess magnetization coordinate of the crossing point.

The above results on the crossing point versus the Meissner fraction, obtained on relatively good single crystals and for magnetic fields applied perpendicularly to the ab planes have been extended recently to granular samples in Ref. 53. In this paper, not only the inhomogeneities but also the random orientation effects have been separated from the intrinsic vortex fluctuation effects. Let us finally note here that the correction of the inhomogeneity effects on $\Delta M(T)_H$ through $\chi_{ab}^{FC}(T)$ is just an approximation which does not apply necessarily to all samples. In fact, we have experimentally observed the failure of such a correction for some of the very inhomogeneous samples studied, with quite low $|\chi_{ab}^{FC}(T^*)|$ values (let us say, with $|\chi_{ab}^{FC}(T^*)| \lesssim 0.3$).

4.2. FULL CRITICAL BEHAVIOUR OF THE PARACONDUCTIVITY VERSUS T_c-INHOMOGENEITIES UNIFORMLY AND NON-UNIFORMLY DISTRIBUTED.

Due to the high amplitude of their thermal fluctuations, it was earlier recognized that the HTSC could be excellent candidates to experimentally penetrate in the so-called full critical region. [3,54] In that region, the amplitudes of the thermal fluctuation effects are expected to be even bigger than the amplitudes of each observable in absence of fluctuations and, therefore, these effects cannot be any more considered as a small perturbation of the mean field like behaviour. Therefore, their corresponding critical exponents are expected to be different from those of the so-called mean-field-like region. For instance, in the case of the paraconductivity, the 3D-XY theory for full critical fluctuations predicts a critical exponent for the in-plane paraconductivity equal to -(2/3)(z-1), where z is the so-called dynamic critical exponent. [3,54,55] If, in addition, z=3/2, as predicted by the so-called E-model dynamics[56] (and which correspond to, for instance, the superfluid transition in the 4He liquid), then the critical exponent of the paraconductivity in the full critical region becomes -1/3, instead of -1/2 for the 3D mean-field region.

To our knowledge, the first experimental attempt to observe the full critical behaviour of the paraconductivity in HTSC was published in Refs. 56 and 57. In fact, an apparent critical exponent of -1/3 was observed for reduced temperatures below $\varepsilon \equiv (T - T_c)/T_c \lesssim 10^{-2}$, in excellent qualitative agreement with the estimated Ginzburg-Levanyuk reduced temperature, ε_{LG}, which corresponds to the limit (closer to T_c) of the mean field like region estimated in these compounds.[3] However, as the samples used in these experiments presented a quite wide resistive transition (the temperature width of $d\rho/dT$, ΔT_c, was of the order of 0.5 K or bigger), these results were not conclusive: as already stressed in these papers, this apparent full critical behaviour of the paraconductivity could as well be due to uniformly distributed T_c-inhomogeneities, associated for instance with relatively small oxygen content inhomogeneities. Moreover, a first analysis of the influence of the choice of T_c on the paraconductivity behaviour and of the uncertainties on the critical exponents associated with the uncertainties in the precise location of T_c, was already presented in these references, and later in Ref. 6 (see also the note in Ref. 58). In addition, the polycrystallinity of the samples used in these experiments prevented the use of the paraconductivity *amplitudes* as a further check of the intrinsic full critical and mean field like behaviours. So, it was concluded in these earlier papers that the measurements in samples having a relative resistive width, $\Delta T_c/T_c$, bigger than ε_{LG}, will not allow any quantitative conclusion on the full critical region: For these samples, the apparent full critical behaviour could just be associated with the extrinsic roundings due to T_c-inhomogeneities. [59]

Further attempts to observe experimentally the full critical behaviour of the paraconductivity were done by measuring the in-plane resistivity in apparently high quality single crystals, with very sharp resistive transitions, ΔT_c being of the order of 0.1 K. An earlier example of these attempts may be seen in Ref. 60, where a critical exponent of $-1/3$ was observed for $\varepsilon \leq 10^{-2}$. In addition, the corresponding absolute amplitude agrees with that expected by scaling the mean-field amplitude (obtained by taking into account the presence of two Josephson coupled CuO_2 layers per periodicity length) to the full critical region, and also such a behaviour is consistent with the in-plane fluctuation induced diamagnetism measured in the same untwinned crystals. Also, the temperature used as T_{c0}, T_{cl}, agrees at a quantitative level with the critical temperature estimated through $\Delta\chi_c$, the excess diamagnetism for H in the weak amplitude limit and applied parallel to the ab layers, that is not appreciably

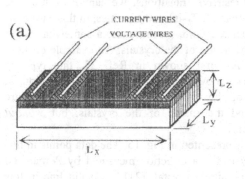

(a)

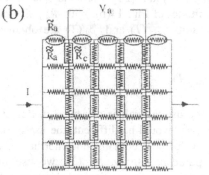

(b)

Fig. 14. (a) Schematic diagram of the sample YS3. For this crystal, the x direction in the figure corresponds to the crystallographic a direction. The shadowed parts correspond to the well-oxygenated domains and the dashed areas are the silver-coated electrical contacts. (b) Two-dimensional electrical-circuit model for the sample schematizaed in (a). Figure from reference 25.

316

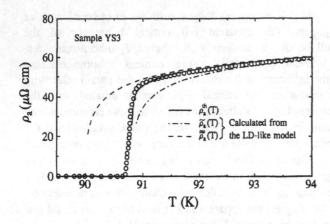

Fig. 15. Circles correspond to the measured resistivity in the a direction for sample YS3 around the transition. These data have been taken from Fig. 2(c) of Ref. 21. The dashed and dotted-dashed lines correspond to the resistivities calculated from the Lawrence-Doniach-like model by using different critical temperatures. The solid line is the result of the simulation performed with the electrical-circuit model of Fig. 14(b). Figure from reference 25.

fluctuations. [60,61] (In fact, probably the best possible estimation of the mean field critical temperature, never accessible directly, is through $\chi_c(T,H \to 0)$). Therefore, probably the results of Ref. 60 provide the more convincing, still at present, [59] experiment probing the full critical behaviour of the para-conductivity in any HTSC (see later).

However, as already stressed in that paper and later in Ref. 25, even in the highest quality samples, with very sharp resistive transitions, we cannot exclude the possible influence on the $\rho_a(T)$ behaviour of small T_c-inhomogeneities, in this case *non-uniformly distributed* in the crystals, associated with, for instance, a somewhat better oxygenation of the crystals surface than the inside of the crystals. An example of these inhomogeneities is shown in Fig. 14. As first shown in Ref. 25, this type of inhomogeneity may deform the $\rho_a(T)$ curve above but very close to the transition, at temperature distances to the transition of the order of the transition temperature difference, ΔT_c, between the surface and the inside of the crystals, but *without* broadening the measured resistive transition.

An example of such a deformation is presented in Fig. 15. The data points in this figure correspond to $\rho_a(T)$, the resistivity in the a direction measured by Pomar and coworkers in an untwinned $Y_1Ba_2Cu_3O_{7-\delta}$ single crystal. [21] The solid line in this figure was generated with the electrical circuit model of Fig. 14(b), with $\tilde{R}_a$, $\tilde{\tilde{R}}_a$ and $\tilde{\tilde{R}}_c$ given by Eqs. (9) and (10). In these equations, $\tilde{\rho}_a(T)$ and $\tilde{\tilde{\rho}}_a(T)$ are obtained through, [21]

$$1/\rho_a(T) = \Delta\sigma_a(T) + 1/\rho_{aB}(T), \tag{12}$$

where for $\rho_{aB}(T)$, the background resistivity, we use the values measured in Ref. 21. Although the background resistivity of each domain would be different, the use in our approximation of the same background resistivity for both domains does not have any qualitative influence in the results. For the paraconductivity, $\Delta\sigma_a(T)$, we use the Lawrence-Doniach (LD)-like expression, with the corresponding parameters also obtained, in the so-called mean-field region, in Ref. 21. The only difference between $\bar{\rho}_a(T)$ and $\tilde{\rho}_a(T)$ is, therefore, their critical temperatures that are, respectively, 90.8 K and 90 K. As can be seen in Fig. 15, the agreement between $\rho_a^{th}(T)$ and the measured data is excellent, even very close to the measured transition, in a temperature region that could be affected by (full) *critical* order parameter fluctuations (OPF). [21] When combined with the analysis made in Ref. 21, the present results show that the resistivity rounding very close to the transition (for reduced temperatures of $\varepsilon = 10^{-2}$ or less) could be explained by the presence of intrinsic (full) critical OPF or, alternatively, by mean-field-like OPF (as those calculated on the grounds of the LD-like approaches) plus small T_c inhomogeneities, associated with very small oxygen content inhomogeneities. However, let us stress here that $\rho_a(T)$ in the MFR (i.e., for $\varepsilon \gtrsim 10^{-2}$) will not be affected by these possible T_c inhomogeneities. Let us also stress here that the T_c-inhomogeneity distribution of Fig. 14(a) is just an example, and that other distributions also will produce a $\rho_a(T)$ deformation. This can be, for instance, the case of a central part of the crystal with lower T_c involved by a part (better oxygenated) with higher T_c.

The results summarized here just intend to show how crucial is to carefully check the possible extrinsic effects associated with structural and stoichiometric inhomogeneities in analyzing the critical behaviour of any observable very close to T_c in HTSC. In fact, as noted already in the Introduction, the complicate chemistry of these materials, together with the strong sensitiveness of their T_c to the stoichiometry appears, mainly in analyzing their critical behaviour very close to T_c, as the "counterpoint" to the high amplitude of their thermal fluctuations. Will these difficulties associated with inhomogeneities prevent any quantitative conclusion on the full critical region in HTSC? Although the results summarized here clearly show that very small inhomogeneities, almost undetectable by using conventional x-ray or neutron diffraction techniques, will suffice to strongly deform the behaviour of any observable very close to T_c, the answer is indeed not. In fact, systematic and reproducible data for different samples, for both *the critical exponent and the amplitude*, will strongly suggest an intrinsic behaviour. This was, in fact, the case of the results of Ref. 60: the data obtained in two different Y-123 crystals agree each other at a quantitative level, well to within the experimental uncertainties. As noted before, this provides, therefore, a quite convincing probe of the full critical behaviour, and also of the mean field like behaviour at bigger reduced critical temperatures!, of the paraconductivity in Y-123 crystals. Mainly in the case of the mean-field-like region, these conclusions were reinforced by independent paraconductivity measurements in other Y-123 untwinned crystals[62] and also by a recent comparison with other observables. [63,64] New reliable measurements of both the amplitude and the ε-behaviour of the paraconductivity very close to T_c ($\varepsilon < 10^{-2}$) in other high quality HTSC crystals will be, however, desirable.

5. Conclusions

The examples reviewed here show that the presence of small stoichiometric and structural inhomogeneities may deeply affect, even when they have characteristic

318

lengths much bigger than those of the superconductivity, the behaviour of any observable around T_C in HTSC. These extrinsic effects will arise simultaneously with the intrinsic thermal fluctuation effects. So, in analyzing the measurements of these last effects, it will be crucial to detect and to separate them from those associated with inhomogeneities. But, in addition, these inhomogeneity effects may concern other important fundamental and practical aspects of the HTSC. For instance, the results about $\rho(T,H)$ in presence of T_C-inhomogeneities clearly suggest that the local current redistributions that these inhomogeneities introduce may also deeply affect the critical current in practical HTSC. Moreover, there is considerable room for further work, in particular, to understand how the "intrinsic" inhomogeneous current redistribution associated with the Meissner effect, may affect many properties around $T_c(H)$, or to extend the present results to T_C-inhomogeneities with short characteristic lengths (when compared with the superconducting characteristic lengths).

6. Acknowledgements

This work was supported by the Spanish CICYT, under grants N° MAT 95-0279 and MAT 98-0371, the European Community Grant No. CHRX-CT93-0325, and by a grant from Union Fenosa, Spain, No.98-0666. CC acknowledges financial support from the Fundación Ramón Areces, Spain.

7. References

1. J.M. Kosterlitz and D. Thouless, in *Progress in Low Temperature Physics*, Edited by D.F. Brewer (North-Holland, Amsterdam, 1978) Vol. VIIB, p. 271.
2. J.G. Bednorz and K.A. Müller, Z. Phys. B Condensed Matter **64**, 189 (1986).
3. For a recent review of the thermal fluctuation effects of Cooper pairs above T_C in HTSC, see, e.g., F. Vidal and M.V. Ramallo, *in The Gap Symmetry and Fluctuations in High Temperature Superconductors*, Edited by J. Bok, G. Deutscher, D. Pavuna and A. Wolf (Plenum, London 1998), p. 477.
4. See, e.g., D. Shoenberg, *Superconductivity* (Cambridge University Press, Cambridge, 1962), p. 75. See also, L.R. Testardi, Phys. Lett. **35A**, 33 (1971).
5. See, e.g., B. Abeles, R.W. Cohen and G.W. Walker, Phys. Rev. Lett. **17**, 632 (1996); B. Abeles, R.W. Cohen and W.R. Stowell, Phys. Rev. Lett. **18**, 902 (1967).
6. J. Maza and F. Vidal, Phys. Rev. B **43**, 10560 (1991).
7. A. Pomar, M.V. Ramallo, J. Mosqueira, C. Torrón and F. Vidal, Phys. Rev. B **54**, 7470 (1996); J. Low Temp. Phys. **105**, 675 (1996).
8. See, e.g., R. Landauer, in *Electrical Transport and Optical Properties of Inhomogeneous Media*, Edited by J.C. Garland and D.B. Tanner (AIP, New York, 1978), p. 2.
9. M.V. Ramallo, A. Pomar and F. Vidal, Phys. Rev. B **54**, 4341 (1996).
10. S. Ullah and A.T. Dorsey, Phys. Rev. B **44**, 262 (1991).

11. See, e.g., C.M. Fu, W. Boon, Y.S. Wang, V.V. Moshchalkov and Y. Bruynseraede, Physica C **200**, 17 (1992); V. Calzona, M.R. Cimberle, C. Ferdeghini, G. Grasso, D.V. Livanov, D. Marre, M. Putti, A.S. Siri, G. Balestrino and E. Milani, Solid State Commun. **87**, 397 (1993); A.K. Prahdam, S.B. Ray, P.C. Chaddah, C. Chen, and B.M. Wanklyn, Phys. Rev. B **50**, 7180 (1994).

12. For earlier references on the anomalous peak structure of the specific heat around T_c in HTSC see, e.g., A. Junod, in *Physical Properties of High Temperature Superconductors II*, ed. D.M. Ginsberg (World Scientific, Singapore, 1990), p. 13.

13. See, e.g., C.H. Chen, in *Physical Properties of High Temperature Superconductors II*, ed. D.M. Ginsberg (World Scientific, Singapore, 1990), p. 261.

14. H. Claus, U. Gebhard, G. Linker, K. Röhberg. S. Riedling, J. Franz, T. Ishida, A. Erb, G. Müller-Vogt and H. Wühl, Physica C **200**, 271 (1992).

15. E. Janod, A. Junod, T. Graf, K.W. Wang, G. Triscone and J. Muller, Physica B **194-196**, 1939 (1994).

16. Y. Nakawaza, J. Takeya and M. Ishikawa, Physica C **225**, 71 (1994).

17. J.W. Loran, J.R. Cooper and K.A. Mirza, Supercond. Sci. Technol. **4**, S391 (1991).

18. R. Menegotto Costa, A.R. Jurelo, P. Rodrigues Jr., P. Pureur, J. Schaf, J.V. Kunzler, L. Ghuvelder, J.A. Campá and I. Rasines, Physica C **251**, 175 (1995).

19. J.A. Friedmann, J.P. Rice, J. Giapintzakis and D.M. Ginsberg, Phys. Rev. B **39**, 4258 (1989).

20. A. Pomar, S.R. Currás, J.A. Veira, and F. Vidal, Phys. Rev. B **53**, 8245 (1996).

21. A. Pomar, A. Díaz, M.V. Ramallo, C. Torrón, J.A. Veira and F. Vidal, Physica C **218**, 257 (1993).

22. See, e.g., J.P. Rice and D.M. Ginsberg, Phys. Rev. B **46**, 1206 (1992), and references therein.

23. Such a link was established by the authors of Refs. 6 and 7 just by analogy with that proposed by some authors for the heavy fermion superconductor UPt$_3$. See, e.g., R.A. Fisher, S. Kim, B.F. Woodfield, N.E. Phillips, L. Taifeller, K. Hasselbach, J. Flouquet, A.L. Giogi and J.L. Smith, Phys. Rev. Lett. **62**, 1411 (1989).

24. In contrast, the absence of *indirect* order parameter fluctuation effects on $\rho(T)$ above T_c may provide some indications about the wave pairing state in the HTSC. For the theory see, e.g., S.K. Yip, Phys. Rev. B **41**, 2012 (1990); J. Low Temp. Phys. **81**, 129 (1990). The first experimental evidence of the absence of indirect OPF effects on the paraconductivity in HTSC, suggesting then unconventional (non $^1s_0$) pair breaking wave pairing in these superconductors, was presented by J.A. Veira and F. Vidal, Phys. Rev. B **42**, 8748 (1990). For more recent developments on this subject see, e.g., A. Pomar. M.V. Ramallo, J. Maza and F. Vidal, Physica C **225**, 287 (1994).

25. J. Mosqueira, A. Pomar, A. Díaz, J.A. Veira, and F. Vidal, Physica C **225**, 34 (1994); J. Mosqueira, A. Pomar, J.A. Veira, J. Maza and F. Vidal, J. App. Phys. **76**, 1943 (1994).

26. J. Mosqueira, J.A. Veira and F. Vidal, Physica C **229**, 301 (1994); J. Mosqueira, J.A. Veira, J Maza, O. Cabeza and F. Vidal, Physica C **253**, 1 (1995).

27. Th. Siebold, C. Carballeira, J. Mosqueira, M.V. Ramallo and F. Vidal, Physica C **282-287**, 1181 (1997); J. Mosqueira, Th. Siebold, A. Pomar, A. Díaz, J.A. Veira, J. Maza and F. Vidal, Cryogenics **37**, 563 (1997). A negative voltage has also been observed by other groups in other HTSC. See, e.g., S. Aukkaravittayapum et al. Physica C **270**, 231 (1996); Y. Nishi et al. J. Matter Sci. Lett. **3**, 523 (1989).

28. J. Mosqueira, S.R. Currás, C. Carballeira, M.V. Ramallo, Th. Siebold, C. Torrón, J. Campá, I. Rasines and F. Vidal, Supercond. Sci. Techol. **11**, 1 (1998).

29. See, e. g., P. Lindqvist, A. Nordström and Ö. Rapp, Phys. Rev. Lett. **64**, 2941 (1990); P. Santhanam, C.C. Chi, S.J. Wind, M.J. Brady and J.J. Buchignano, ibid. **66**, 2254 (1991); E. Spahn and K. Keck, Solid State Commun. **78**, 69 (1991); Y. K. Kwong, K. Lin, P.J. Hakonen, M.S. Isaacson and J.M. Parpia, Phys. Rev.

320

B **44**, 462 (1991); A. Nordström and Ö. Rapp, Phys. Rev. B **45**, 12577 (1992); H. Vloeberghs, V. V. Moshchalkov, C. Van Haesendonk, R. Jonckheere and Y. Bruynseraede, Phys. Rev. Lett. **69**, 1268 (1992); A. W. Kleinsasser and A. Kastalsky, Phys. Rev. B **47**, 8361 (1993); S.G. Romanov, A.V. Fokin and K.Kh. Babamuratov, JETP Lett. **58**, 824 (1993); J.J. Kim, J. Kim, H.J. Shin, H.J. Lee, S. Lee, K.W. Park and E. Lee, J. Phys. Condens. Matter **6**, 7055 (1994); V. V. Moshchalkov, L. Gielen, G. Neuttiens, C. van Haesendonk and Y. Bruynseraede, Phys. Rev. B **49**, 15412 (1994); M. Park, M.S. Isaacson and J.M. Parpia, Phys. Rev. Lett. **75**, 3740 (1995); C. Strunk, V. Bruyndoncx, C. Van Haesendonk, V.V. Moshchalkov, Y. Bruynseraede, B. Burk, C.J. Chien and V. Chandrasekhar, Phys. Rev. B **53**, 11332 (1996); K. Yu. Arutyunov, Phys. Rev. B **53**, 12304 (1996); M. Park, M.S. Isaacson and J.M. Parpia, Phys. Rev. B **55**, 9067 (1997); B. Burk, C.-J. Chien, V. Chandrasekhar, C. Strunk, V. Bruyndoncx, C. Van Haesendonck, V.V. Moshchalkov, and Y. Bruynseraede, J. Appl. Phys. **83**, 1549 (1998); C. Strunk, V. Bruyndoncx, C. Van Haesendonck, V.V. Moshchalkov, Y. Bruynseraede, C.-J. Chien, B. Burk, and V. Chandrasekhar, Phys. Rev. B **57**, 10854 (1998). As stressed in the main text, many of the magnetoresistivity peak effects around T_C described in these papers and attributed by these authors to sophisticated intrinsic mechanisms, may be easily explained in terms of T_C inhomogeneities non-uniformly distributed in the samples. This last explanation was discarded by some of these authors due to the erroneous belief that these T_C inhomogeneities do not affect the magnetoresistivity measured with in-line electrical arrangements (see Ref. 25; see also the note in Ref. 30).

30. R. Vaglio, C. Attanasio, L. Maritato and A. Ruosi, Phys. Rev. B **47**, 15302 (1993). In that paper it was concluded that the $\rho(T)$ peaks observed near T_C in some low temperature superconductors by using a Van der Paw electrical arrangement (with the electrical leads in the sample corners) could be due to the presence in the samples of T_C inhomogeneities. However, it was erroneously suggested in that paper that in the case of an in-line electrical arrangement the T_C inhomogeneities could not produce a $\rho(T)$ peak. This last type of measurements were analyzed for the first time by Mosqueira and coworkers in Ref. 25.

31. C. Attanasio, L. Maritato and R. Vaglio in *Tunneling Phenomena in High and Low T_C Superconductors*, edited by A. de Chiara and M. Russo (World Scientific, Singapore, 1993).

32. A. Gerber, T. Grenet, M. Cyrot and B. Beille, Phys. Rev. Lett. **65**, 3201 (1990); L. Fabrega, M.A. Crusellas, J. Fontcuberta, X. Obrados, S. Piñol, C.J. van der Beck, P.H. Kes, T. Grenet and J. Beille, Physica C **185-189**, 1913 (1991); M. A. Crusellas, J. Fontcuberta and S. Piñol, Phys. Rev. B **46**, 14089 (1992); M.L. Trawick, S.M. Ammirata, C.D. Keener, S.E. Hebboul and J.C. Garland, J. of Low Temp. Phys. **105**, 1267 (1996).

33. S. Rubin, T. Schimpfke, B. Weitzel, C. Vossloh and H. Micklitz, Ann. Physik **1**, 492 (1992).

34. A.K. Pradham, S.J. Hazell, J.W. Hodby, C. Chen, A.J.S. Chowdury and B.M. Wanklyn, Solid State Commun. **88**, 723 (1993).

35. M.A. Crusellas, J. Fontcuberta and S. Piñol, Physica C **226**, 311 (1994).

36. H.J. Trodahl and A. Mawdsley, Phys. Rev. B **36**, 8881 (1987); W.N. Kang, K.C. Cho, Y.M. Kim and M.Y. Choi, Phys. Rev. B **39**, 2763 (1989); S. Yan, T. Chen, H. Zhang, J. Peng, Z. Shen, C. Wei, Q. Wen, K. Wu, L. Tong and H. Zhang, Modern Phys. Lett. B **2**, 1005 (1988); M.A. Howson, M.B. Salamon, T.A. Friedmann, S.E. Inderhees, J.P. Rice, D.M. Ginsberg and K.M. Ghiron, J. Phys.: Condens Matter **1**, 465 (1989); M.A. Howson, M.B. Salamon, T.A. Friedmann, J. P. Rice and D. Ginsberg, Phys. Rev. B **41**, 300 (1990); A.J. Lowe, S. Regan and M.A. Howson, Physica B **165-166**, 1369 (1990); Phys. Rev. B **44**, 9757 (1991); J. Phys.: Condens. Matter **4**, 8843 (1992); N.V. Zavaritskii, A.V. Samoilov and A.A. Yurgens, JETP Lett. **55**, 127 (1992); Y.N. Xiang, O.G. Shevchenko, and A.S. Panfilov, Sov. J. Low Temp. Phys. **18**, 916 (1992).

37. A.J. Lowe, S. Regan and M.A. Howson, Phys. Rev. B **47**, 15321 (1993); M.A. Howson, ibid. 15324 (1993); A.A.Varlamov and D.L. Livanov, Sov. Phys. JETP **71**, 325 (1990); A.V. Rapoport, Sov. Phys. Solid State **33**, 309 (1991).

38. For a more recent theoretical analyses of the thermal fluctuation effects on S(T) around T_C in HTSC see, A.A. Varlamov, G. Balestrino, E. Milani and D.V. Livanov (to be published).

39. Let us stress, however, that in this case (which corresponds to a typical non-uniformly distributed inhomogeneity) the behaviour of the calculated (th) ρ_{ab}^{th} does not depend on the number of meshes of the network, provided that the proportion and location of the different resistances is kept unchanged. This contrasts with the case of uniformly distributed T_C inhomogeneities for which with a small number of meshes it is not possible to represent adequately the inhomogeneity distribution. In this case, a small number of meshes could lead to the appearance of important *spurious* longitudinal and transversal voltages, which are just an artifact of an inadequate simulation. For instance, the calculations of the longitudinal and transversal voltages in superconductors with *uniformly* distributed inhomogeneities presented by R. Griessen and coworkers in Physica C **235-240**, 1371 (1994) may be affected by these spurious effects.

40. A negative Hall effect in a LTSC has been first observed by H. van Beelen et al., Physica **36**, 241 (1967), and by C.H. Weijsenfeld, Phys. Lett. **28A**, 362 (1968); a negative Ettinshausen effect in a LTSC has been first observed by F. Vidal, Phys. Rev B **8**, 1982 (1973).

41. See, e.g., S.J. Hagen, A.W. Smith, M. Rajeswari, J.L. Peng, Z.Y. Li, R.L. Greene, S.N. Mao, X.X. Xi, S. Bhattacharya, Q. Li, and C.J. Lobb, Phys. Rev. B **47**, 1064 (1993).

42. O. Cabeza, A. Pomar, A. Díaz, C. Torrón, J.A. Veira, J. Maza and F. Vidal, Phys. Rev. B **47**, 5332 (1993); A.J. López, J. Maza, Y.P. Yadava, F. Vidal, F. García Alvarado, E. Morán and M.A. Señaris Rodríguez, Supercond. Sci. Technol. **4**, S292 (1991).

43. K. Maki, J. Low. Temp. Phys. **14**, 419 (1974); Phys. Rev. B **43**, 1252 (1991).

44. J.A. Veira and F. Vidal, Physica C **159**, 468 (1989); C. Torrón, O. Cabeza, A. Díaz, J. Maza, A. Pomar, J.A. Veira and F. Vidal, J. of Alloys and Compounds **195**, 627 (1993).

45. O. Cabeza, G. Domarco, J.A. Veira, A. Pomar, C. Torrón, A. Díaz, J. Maza and F. Vidal, J. Alloys and Compounds **195**, 623 (1993); O. Cabeza, J. Maza, Y.P. Yadava, J.A. Veira, F. Vidal, M.T. Cascais, C. Cascales and I. Rasines, in *Properties and Applications of Perovskite-type Oxides*, Ed. L.G. Tejuca and J.L. Fierro (Marcel Dekker Inc. N.Y. 1992), p. 101.

46. See, e.g., A. Díaz, A. Pomar, G. Domarco, J. Maza and F. Vidal, App. Phys. Lett. **63**, 1684 (1993); Physica B **194-196**, 1933 (1994); A. Díaz, A. Pomar, G. Domarco, C. Torrón, J. Maza and F. Vidal, Physica C **215**, 105 (1993); J. App. Phys. **77**, 765 (1995).

47. K. Kadowaki, Physica C **185-189**, 2249 (1991).

48. P.H. Kes, C.J. van der Beck, M.P. Maley, M.E. McHenry, D.A. Huse, M.J.V. Menken and A.A. Menovsky, Phys. Rev. Lett. **67**, 2383 (1991).

49. Z. Tešanovic, L. Xing, L.N. Bulaevskii, Q. Li, and M. Suenaga, Phys. Rev. Lett. **69**, 3563 (1992).

50. Q. Li, M. Suenaga, T. Hikata and K. Sato, Phys. Rev. B **46**, 5857 (1992); Q. Li, K. Shibutani, M. Suenaga, I. Shigaki, R. Ogawa, ibid. **48**, 9877 (1993); Q. Li, M. Suenaga, L.N. Bulaevskii, T. Hikata, K. Sato, ibid. **48**, 13865 (1993); Q. Li, M. Suenaga, G.D. Gu, N. Koshizuka, ibid. **50**, 6489 (1994); J.R. Thompson, J.G. Ossandon, D.K. Christen, B.C. Chakoumakos, Yang Ren Sun, M. Paranthaman and J. Brynestad, ibid. **48**, 14031 (1993); Z. J. Huang, Y.Y. Xue, R.L. Meng, X.D. Qiu, Z.D. Hao, and C.W. Chu, Physica C **228**, 211 (1994); N. Kobayashi, K. Egawa, K. Miyoshi, H. Iwasaki, H. Ikeda, and R. Yoshizaki, ibid. **219**, 265 (1994); R. Jin, H.R. Ott, and A. Schilling, ibid. **228**, 401 (1994); G. Triscone, A.F. Khoder, C. Opagiste,

322

J.-Y. Genoud, T. Graf, E. Janod, T. Tsukamoto, M. Couach, A. Junod, and J. Muller, ibid. **224**, 263 (1994); A. Wahl, A. Maignan, C. Martin, V. Hardy, J. Provost, and Ch. Simon, Phys. Rev. B **51**, 9123 (1995); Y.Y. Xue, Y. Cao, Q. Xiong, F. Chen, and C.W. Chu, ibid. **53**, 2815 (1996); G. Villard, D. Pelloquin, A. Maignan, and A. Wahl, Physica C **278**, 11 (1997); Q. Li, M. Suenaga, T. Kimura, and K. Kishio, Phys. Rev. B **47**, 11384 (1993); B. Janossy, L. Fruchter. I.A. Campbell, J. Sanchez, I. Tanaka, and H. Kojima, Solid State Commun. **89**, 433 (1994); J.-Y. Genoud, G. Triscone, A. Junod, T. Tsukamoto, and J. Muller, Physica C **242**, 143 (1995); A. Junod, J.Y. Genoud, G. Triscone, and T. Schneider, Physica C **294**, 115 (1998).

51. See, e.g., L.N. Bulaevskii, M. Ledvig and V.G. Kogan, Phys. Rev. Lett. **68**, 3773 (1992); A.E. Koshelev, Phys. Rev. B **50**, 506 (1994).

52. J. Mosqueira, J.A. Campá, A. Maignan, I. Rasines, A. Revcolevschi, C. Torrón, J.A. Veira, F. Vidal, Europhys. Lett. **42**, 461 (1998); F. Vidal, C. Torrón, M.V. Ramallo, J. Mosqueira, *Superconducting and Related Oxides: Physics and Nanoengineering III*, Ed. D. Pavuna and I. Bozovic, (SPIE Publ., Bellingham, USA), p. 32.

53. J. Mosqueira, M.V. Ramallo, A. Revcolevschi, C. Torrón, F. Vidal, Phys. Rev. B **59** (Feb. 1999).

54. C.J. Lobb, Phys. Rev. B **36**, 3930 (1987).

55. See, e.g., P.C. Hohenberg and B.I. Halperin, Rev. Mod. Phys. **49**, 435 (1977).

56. F. Vidal, J.A. Veira, J. Maza, F. García-Alvarado, E. Morán, and M.A. Alario, J. Phys. C: Solid State Phys. **21**, L599-L606 (1988); J.A. Veira, J. Maza, F. Vidal, Phys. Lett. A **131**, 310 (1988); F. Vidal, J.A. Veira, J. Maza, J.J. Ponte, J. Amador, C. Cascales, M.T. Casais, I. Rasines, Physica C **156**, 165 (1988).

57. F. Vidal, J.A. Veira, J. Maza, J.J. Ponte, F. García-Alvarado, E. Morán, J. Amador, C. Cascales, A. Castro, M.T. Casais and I. Rasines, Physica C **156**, 807 (1988).

58. Let us stress that, obviously, the opposite procedure used since many years by some workers, which consist in the estimation of T_c by imposing a critical exponent in an almost arbitrary (in extent and location!) temperature region does not overcome at all these difficulties: If $\rho(T)$ is smoothly rounded by uniformly distributed T_c-inhomogeneities, it will be always possible to find successive and more or less extended ε-regions where the critical exponents take different values (to within almost zero and -3 or -4) which decrease when approaching the apparent T_c.

59. In spite of the earlier warnings, published in Refs. 26, 56 and 57, an appreciable number of papers were published since then by different groups (and still new papers are being published at present) which intend to conclude quantitatively on the paraconductivity full critical behaviour by just analyzing the temperature behaviour of the resistivity measured in different HTSC samples probably appreciably affected by uniformly (and maybe also by non uniformly) distributed T_c-inhomogeneities. See, e.g., Menegotto Costa, P. Pureur, L. Ghivelder, J.A. Campá, and I. Rasines, Phys. Rev. B **56**, 10836 (1997); S.H. Han, Yu. Eltsev and O. Rapp, Phys. Rev. B **57**, 7510 (1998). See also the note in Ref. 58.

60. A. Pomar, A. Díaz, M.V. Ramallo, C. Torrón, J.A. Veira, and F. Vidal, Physica C **218**, 257 (1993).

61. C. Torrón, A. Díaz, A. Pomar, J.A. Veira and F. Vidal, Phys. Rev. B **49**, 13143 (1994); M.V. Ramallo, C. Torrón and F. Vidal, Physica C **230**, 97 (1994).

62. W. Holm, Yu. Eltsev and Ö. Rapp, Phys. Rev. B **53**, 11992 (1995); J.T. Kim, N. Goldenfel, J. Giapintzakis and D. Ginsberg, Phys. Rev. B **56**, 118 (1997).

63. See, e.g., M.V. Ramallo and F. Vidal, Phys. Rev. **59** (Feb. 1999).

64. For an analysis of the influence of the T_c-inhomogeneities on the heat capacity measured very close to T_c in Y-123 crystals, see, F. Sharify, J. Giapintzakis, D.M. Ginsberg, D.J. van Harlingen, Physica C **161**, 555 (1989).

ELECTRON-BOSON EFFECTS IN THE INFRARED PROPERTIES OF METALS

What can we learn from the optical conductivity

S. V. SHULGA

Institute of Spectroscopy, RAS, Troitsk, 142092, Russia,
IFW-Dresden and the University of Bayreuth, Germany

Abstract. The interpretation of optical conductivity in the normal and superconducting states is considered in the frame of the standard Isotropic Single wide Band (ISB) model using the theory proposed by Nam (Nam S.B., *Phys. Rev.* **156** 470-486 (1967)). The *exact analytical* inversion of the normal state Nam equations is performed and applied for the recovery of the reliable information from the experimental data. The Allen formula is derived in the strong coupling approximation and used for the physically transparent interpretation of the FIR absorption. The phenomenological "generalised" Drude formula is obtained from the Nam theory in a high temperature approximation. It is shown, that the reconstruction of the shape of the spectral function $\alpha^2 F(\omega)$ from the normal state optical data at $T>T_c$ is not unique and the same data can be fitted by many spectral functions. This problem is considered in detail from different points of view. At the same time, using the exact analytical solution, one can get from the normal state data a useful piece of information, namely, the value of the coupling constant, the upper energy bound of the electron-boson interaction function, and the averaged boson frequency. Moreover, they are even visually accessible, if the optical data are presented in term of the frequency dependent optical mass and scattering rate. The superconducting state Nam formalism and related simplified theory are analysed from the user point of view. A novel *adaptive* method of the *exact numerical inversion* of the superconducting state Nam equations is presented. Since this approach uses the *first derivative* from the experimental curve, the signal/noise ratio problem is discussed in detail. It is shown that, the fine structure of the spectral function can be recovered from optical data in the case of s-wave pairing symmetry. In contrast, in the d-wave case the resulting image is approximately the convolution of the input electron-boson interaction function and the Gauss distribution $\exp(-\omega^2/\Delta_{opt}^2)$. A simplified visual accessibility (VA) procedure is proposed for "by eye" analyses

S.-L. Drechsler and T. Mishonov (eds.), High-Tc Superconductors and Related Materials, 323–360.

of the superconducting state optical reflectivity of the ISB metals with s and d pairing symmetry. The bosons, responsible for the superconductivity in $YBa_2Cu_3O_7$, exhibit a "phonon-like" spectral function with the upper frequency bound less than 500 cm^{-1} and the averaged boson frequency near 300 cm^{-1}.

1. Introductory remarks, reservations and basic equations

- The interpretation of the normal and superconducting state Far Infrared Region (FIR) properties of metals is still rather ambiguous. In this lecture I will restrict myself to the consideration of the of Isotropic Single wide Band (ISB) model based on the Nam formalism [1].

- From the optical conductivity $\sigma(\omega)$ one derives the so called transport spectral function $\alpha_{tr}^2 F(\omega)$. It has the same spectral structure, i.e. the same number of peaks with similar relative positions (see Fig.4 in [2] and Fig.1 in [3]) as a standard $\alpha^2 F(\omega)$ which enters the Eliashberg equations. But their amplitudes can be lower, and as a result the transport coupling constant λ_{tr} is less than the standard λ. In this lecture I will not distinct the $\alpha^2 F(\omega)$ and $\alpha_{tr}^2 F(\omega)$.

- If the electronic band is very narrow, the forward scattering could dominate. In this case the optical properties manifests themselves by the negligible coupling, while the tunnelling spectroscopy gives reasonable finite coupling strength. A similar problem arises when the quasiparticles from the flat or nested parts of the Fermi surface are mainly responsible for the superconductivity. Since they have small Fermi velocities, their influence in the transport response function which is proportional to v_F^2 is weak. Such a scenario was recently proposed for borocarbides [4].

- In the frame of the Eliashberg theory it is impossible to elucidate the nature of the bosons which are responsible for the superconductivity. In this context I will use the term boson, keeping for short the standard electron-phonon notation $\alpha^2 F(\omega)$.

- I will restrict myself to the consideration of the local (London) limit case, that is, to the normal skin effect scenario. The generalisation of results obtained to the nonlocal (Pippard) limit is possible but will not be considered here due to the lack of space.

- The power of the proposed methods is illustrated analysing available experimental data for a single-domain $YBa_2Cu_3O_{7-d}$ crystal [5]. Note, that to the best of my knowledge, the experimental HTSC superconducting state reflectivity was not been considered from the inverse problem viewpoint at all, despite the single early attempt [6].

- Theoretical analysis of any physical quantity, as optical conductivity, is performed in frame of some model which includes a theoretical formalism and some assumptions about the band structure, the anisotropy of the coupling function and so on. The considered here ISB model is the standard default. Nevertheless, in frame of its s-wave version the question: "What is the value of the gap? " has an unique answer, but for d-wave case the correct questions are either :"What are the values of the gaps? " or "What is the value of the maximum gap? ". In other words, the absolute values and the physical meaning of the quantities are model dependent.

- If the model is reach enough, one can calculate many physical properties using the same *set of input (material) parameters*. A solution of the inverse problem means the determination of these material parameters from the experimental data. A short list of the ISB model input parameters is given in section 1.1.

- As a rule, the inverse problem is ill-posed. The experimental data contain less information, than we would like to know. The reasons for are the temperature induced broadening and the finite signal to noise ratio S/N.

- In this lecture we will mainly be interested in the amount of the remaining information about the spectral function $\alpha^2 F(\omega)$, which can be obtained from the optical data. In spectroscopy the unit of information is a peak. A single peak needs four parameters for its description: the upper and the low frequency bounds Ω_{max} and Ω_{min}, the position of the maximum and its amplitude. An narrow symmetric peak within the broad band needs three parameters: the amplitude, the halfwidth, and the position. The sum of all peaks yields the upper and lower frequency bounds Ω_{max} and Ω_{min} of the spectral band. Since the Ω_{min} of the $\alpha^2 F(\omega)$ is usually equal to zero, the remaining three band parameters of $\alpha^2 F(\omega)$ can be determined even from the normal state data [7].

1.1. ISOTROPIC SINGLE BAND MODEL

The standard ISB model [8] is the most developed part of the modern theory of superconductivity. It describes *quantitatively* the renormalization of the physical properties of metals due to the electron-boson interaction. The input quantities of the ISB model are the density of states at the Fermi energy $N(0)$, the Fermi velocity v_F, the impurity scattering rate γ_{imp}, the Coulomb pseudopotential μ^*, and the spectral function $\alpha^2 F(\omega)$ of the electron-boson interaction. The scale of transport and optical properties is fixed by the plasma frequency $\hbar\omega_{pl} = \sqrt{4\pi e^2 v_F^2 N(0)/3}$. In this section I present only the final expressions of the Nam theory, more detailed information can be found in [1, 2, 9].

In order to calculate any physical property in the superconducting state one needs a solutions of Eliashberg equations (EE) [11]. In my opinion, the following EE representation is most suitable for numerical solution by iterations

$$
Im\tilde{\Delta}(\omega) = \frac{i\gamma_{imp}}{2}\frac{\tilde{\Delta}(\omega)}{\sqrt{\tilde{\omega}^2(\omega) - \tilde{\Delta}^2(\omega)}} + \frac{\pi}{2}\int dy\alpha^2 F(\omega - y)
$$

$$
\left[\coth\left(\frac{\omega - y}{2T}\right) - \tanh\left(\frac{y}{2T}\right)\right]Re\frac{\tilde{\Delta}(y)}{\sqrt{\tilde{\omega}^2(y) - \tilde{\Delta}^2(y)}} \quad (1)
$$

$$
Im\tilde{\omega}(\omega) = \frac{i\gamma_{imp}}{2}\frac{\tilde{\omega}(\omega)}{\sqrt{\tilde{\omega}^2(\omega) - \tilde{\Delta}(\omega)}} + \frac{\pi}{2}\int dy\alpha^2 F(\omega - y),
$$

$$
\left[\coth\left(\frac{\omega - y}{2T}\right) - \tanh\left(\frac{y}{2T}\right)\right]Re\frac{\tilde{\omega}(y)}{\sqrt{\tilde{\omega}^2(y) - \tilde{\Delta}^2(y)}}, \quad (2)
$$

where $\tilde{\Delta}(\omega)$ and $\tilde{\omega}(\omega)$ are the renormalized gap function and the renormalized frequency respectively, and γ_{imp} denotes the impurity scattering rate within the Born approximation. The real and the imaginary parts of the Eliashberg functions $\tilde{\Delta}(\omega)$ and $\tilde{\omega}(\omega)$ are connected by the Kramers-Kronig relations. Hence, they have the same Fourier images. This reasoning yields the fast solution procedure. The convolution type integrals (1-2) should be calculated by the Fast Fourier Transform (FFT) algorithm. The inverse *complex* Fourier transformations of the results obtained give *complex* values of $\tilde{\Delta}(\omega)$ and $\tilde{\omega}(\omega)$. If one would like to use this efficient FFT method also for d-wave pairing symmetry, the process of arithmetic-geometric mean [12] for the evaluation of the complex elliptic integrals (see below Eqs. (67,68) is strongly recommended. The density of states $ReN(\omega)$ and the density of pairs $ReD(\omega)$

$$
N(\omega) = \frac{\tilde{\omega}(\omega)}{\sqrt{\tilde{\omega}^2(\omega) - \tilde{\Delta}^2(\omega)}}, \quad (3)
$$

$$
D(\omega) = \frac{\tilde{\Delta}(\omega)}{\sqrt{\tilde{\omega}^2(\omega) - \tilde{\Delta}^2(\omega)}}, \quad (4)
$$

could be approximated by step functions. The convolution of any spectrum with the step function $sign(y^2 - \Delta_0^2)$ results in the shift by Δ_0 at $y > 0$

and by $-\Delta_0$ at $y < 0$ and its integration over y. It is important, that due to the presence of singularities in $N(\omega)$ and $D(\omega)$ at $\omega = \Delta_0$ the more complicated functions such as $\int d\omega' N(\omega') N(\omega - \omega')$ and $\int d\omega' D(\omega') D(\omega - \omega')$ still contain a jump (now at $\omega = 2\Delta_0$). Since the convolution is a quite general property of the Green functions approach, one can expect that the first derivatives from a spectrum (of different nature) over frequency will reproduce the input spectral function $\alpha^2 F(\omega)$ (may be with the appropriate phase distortion). When applied to the optical conductivity, this approach gives visual accessibility (VA) procedure discussed in the sec. 3.

In the normal state the set (2) is reduced to the following simple formula

$$Im\tilde{\omega}(\omega) = \frac{\gamma_{imp}}{2} + \frac{\pi}{2} \int dy \alpha^2 F(y) \left[\coth\left(\frac{y}{2T}\right) - \tanh\left(\frac{\omega - y}{2T}\right) \right], \quad (5)$$

which however looks unwieldy in comparison with the same formula presented in Matsubara formalism

$$\tilde{\omega}(i\omega_n) \equiv \tilde{\omega}_n = \tilde{\omega}_{n-1} + 2\pi T(1 + \lambda_n) \quad (6)$$

where

$$\lambda_k \equiv \alpha^2 F(i\nu_k) = 2 \int dz \alpha^2 F(z) z / (z^2 + \nu_k^2) \quad (7)$$

are kernels of the spectral function, $\omega_k = \pi T(2k + 1)$ and $\nu_k = 2\pi T k$ are fermion (ω_k) and boson (ν_k) Matsubara energies, respectively, and $\tilde{\omega}_0 = \gamma_{imp}/2 + \pi T(1 + \lambda_0)$.

The normal state ISB optical conductivity takes the following forms

$$\sigma(\omega, T) = \frac{\omega_{pl}^2}{8\pi i \omega} \int dy \frac{\tanh\left(\frac{y+\omega}{2T}\right) - \tanh\left(\frac{y}{2T}\right)}{\tilde{\omega}(\omega + y) - \tilde{\omega}^*(y),} \quad (8)$$

and

$$\sigma(i\omega_n, T) \equiv \sigma_n = \frac{\omega_{pl}^2}{4\pi n} \sum_{k=0}^{n-1} \frac{1}{\tilde{\omega}_k + \tilde{\omega}_{n-k-1}}, \quad (9)$$

in the real and imaginary axes techniques, respectively.

The tedious expression for the superconducting state ISB optical conductivity has the form [1, 9, 10]

$$\sigma(\omega) = \frac{\omega_{pl}^2}{16\pi\omega} \int dx \frac{g_{rr} \tanh(x/2T)}{\sqrt{\tilde{\Delta}^2(x) - \tilde{\omega}^2(x)} + \sqrt{\tilde{\Delta}^2(x+\omega) - \tilde{\omega}^2(x+\omega)}}$$

$$- \frac{g_{rr}^* \tanh[(x+\omega)/2T]}{\left(\sqrt{\tilde{\Delta}^2(x) - \tilde{\omega}^2(x)} + \sqrt{\tilde{\Delta}^2(x+\omega) - \tilde{\omega}^2(x+\omega)}\right)^*}$$

$$+ \frac{g_{ar} \{\tanh[(x+\omega)/2T] - \tanh(x/2T)\}}{\sqrt{\tilde{\Delta}^2(x) - \tilde{\omega}^2(x)} + \left(\sqrt{\tilde{\Delta}^2(x+\omega) - \tilde{\omega}^2(x+\omega)}\right)^*,} \quad (10)$$

where

$$
\begin{aligned}
g_{rr} &= 1 - N(x)N(x+\omega) - D(x)D(x+\omega), \\
g_{ra} &= 1 + N^*(x)N(x+\omega) + D^*(x)D(x+\omega)
\end{aligned}
\tag{11}
$$

are *coherent factors*.

2. Normal state optical conductivity

2.1. THE EXACT SOLUTION OF AN INVERSE PROBLEM

It is amusing, that the Eq. (9) is solvable for $\tilde{\omega}_k$ and Eq. (6) for λ_k. Hence, at least theoretically, an exact analytical solution exists. It takes three or four steps as shown below.

I) One has to perform the analytical continuation of the conductivity $\sigma(\omega)$ from the real energy axis, where we are living, to the poles of the Bose distribution function (Matsubara boson energies) $i\nu_k = 2\pi i T k$. This procedure is similar to the Kramers-Kronig analysis, which is nothing other than the analytical continuation from the real frequency axis to itself. For example, if one fits the data by a sum of Drude and Lorentz terms or by formula (33), one has simply to substitute the complex values $i\nu_k$ for the frequency. Note, that the non-metallic (IR active direct phonon contribution and interband transition) part of the dielectric permeability $\epsilon_{ph}(\omega)$ have to be subtracted from the total one $\epsilon(\omega)$ before starting of the analysis. In the genuine far infrared region one could use the real constant $\epsilon_\infty = \epsilon_{ph}(0)$, instead of $\epsilon_{ph}(\omega)$, but if the spectral range is wide, the subtraction of $\epsilon_{ph}(\omega)$ is not trivial.

II) The renormalized frequencies should be calculated from (9) as follows

$$
\tilde{\omega}_0 = \frac{\omega_{pl}^2}{8\pi\sigma_1},
\tag{12}
$$

$$
\tilde{\omega}_1 = \frac{\omega_{pl}^2}{4\pi\sigma_2} - \tilde{\omega}_0,
\tag{13}
$$

$$
\tilde{\omega}_{n-1} = \frac{\omega_{pl}^2}{2\pi A_n} - \tilde{\omega}_0, \text{ where } A_n = \sigma_n - \frac{\omega_{pl}^2}{4\pi n}\sum_{k=1}^{n-2}\frac{1}{\tilde{\omega}_k + \tilde{\omega}_{n-k-1}}.
\tag{14}
$$

III) Inverting Eq. (6), one could evaluate the values of the spectral function $\lambda_n = \alpha^2 F(\nu_n)$ we looked for

$$
\lambda_0 = \frac{\tilde{\omega}_0 - \gamma_{imp}/2}{\pi T} - 1
\tag{15}
$$

$$
\lambda_n = \frac{\tilde{\omega}_n - \tilde{\omega}_{n-1}}{2\pi T} - 1.
\tag{16}
$$

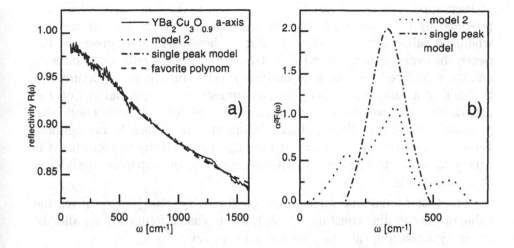

Figure 1. Panel (a) shows the frequency dependence of the normal state reflectivity (solid line, T=100 K, Schutzmann J. et al., *Phys. Rev.*, **B46**, 512 (1992)) in comparison with the model ones. The curves, shown by the dotted and the dotted-dashed lines, was calculated using "model 2" and "single peak" spectra (presented on the panel (b)). The dashed line is the least square fit of $YBa_2Cu_3O_{0.9}$ a-axis reflectivity by the favourite polynom Eq. 32 (with the fixed $\epsilon_\infty = 12$).

IV) Using, for example, Pade polynoms [14] one has to continue the electron-boson interaction function back to the real axis.

In practice, if the signal to noise ratio is not extremely large, the last step makes no sense due to the dramatic increase of the uncertainty in the λ_n with elevation of n. The point is that the small quantities λ_n Eq. (16) and A_n Eq. (14) are the difference of large quantities. Note, that the uncertainty in the λ_n values has to be small, not in comparison with λ_n itself, but with the difference between λ_n and its high energy asymptotic value

$$\tilde{\lambda}_n = \frac{2\lambda_0 < \Omega^2 >}{< \Omega^2 > +\nu_n^2} \tag{17}$$

where

$$< \Omega^2 >= \frac{1}{\lambda_0} \int_0^\infty d\Omega \ \Omega \alpha^2 F(\Omega) \tag{18}$$

If the average boson energy has the order of $2\pi T$, the last condition ($\delta\lambda_n \ll |\tilde{\lambda}_n - \lambda_n|$) may be violated even for n=3, since $\tilde{\lambda}_n$ and λ_n have similar values. For example, at T =100 K for the spectral function, shown by the dotted line in Fig. 1, $\tilde{\lambda}_n/\lambda_n$=1.37, 1.12, 1.06, 1.03 for n=1, 2, 3, 4 respectively. Three or four λ_n values only form a too small parameter set to recover correctly the shape of the spectral function. Actually, the possibility to obtain at least small amount of reliable information, is the additional advantage provided by the Matsubara solution in comparison with the real axis one where *all* values are unreliable [15]. Fig. 1 shows the good agreement between the experimental normal state 1-2-3 single crystal reflectivity data [5] and the calculated ones using for *various* spectral functions. Unfortunately, the lack of a unique solution promoted unreasonable speculations on the mechanisms in several novel superconducting systems in connection with proposed hand-made electron-boson interaction functions. In my opinion, twelve years after the discovery of the superconductivity in cuprates it is timely to return to a self-consistent description of the superconductivity in this compounds.

The first Matsubara value of the conductivity $\sigma(i\nu_1)(\equiv \sigma_1)$ gives the value of the coupling constant $\lambda(\equiv \lambda_0)$, if the plasma frequency ω_{pl} and the impurity scattering rate γ_{imp} are known *a priori*

$$\frac{1}{\sigma_1} = \frac{8\pi^2 T(1 + \lambda_0)}{\omega_{pl}^2} + \frac{4\pi\gamma_{imp}}{\omega_{pl}^2}. \tag{19}$$

The reflectivity $R(\omega)$ can also be continued to the imaginary frequency axis and included into the solution of the inverse problem. Following this way, one can derive from (19) the following formula

$$\lambda_0 = -\frac{\gamma_{imp}}{2\pi T} - 1 + \frac{\omega_p^2}{(2\pi T)^2(\coth^2(K_1/4) - \epsilon_\infty)}, \tag{20}$$

where ϵ_∞ is the low-frequency value of the non-metallic (phonon and interband transition) part of dielectric permeability, and K_m is the electromagnetic kernel

$$K(\nu_m) = \frac{2\nu_m}{\pi} \int_0^\infty dz \frac{\log|1 - R(z)|}{z^2 + \nu_m^2}. \tag{21}$$

The first λ_1 and the second λ_2 allow (for example) to estimate the average boson energy and *very approximately* the width of the spectral function. If there is the possibility to combine this incomplete information about $\alpha^2 F(\omega)$ with the results of other spectral measurements, for example, tunnelling or neutron data, one can qualify the transport spectral function. The so called "model 2" electron-boson interaction function [7] was recovered assuming, that the solution has to be resemble the phonon density of states, measured by inelasitc neutron spectroscopy.

2.2. OPTICAL MASS AND IMPURITY SCATTERING RATE

In ref.[7, 15] we shown, that it is useful to describe the optical conductivity in terms of the so called "extended" Drude formula

$$\sigma(\omega, T) = \frac{\omega_{pl}^2/4\pi}{W(\omega, T)} = \frac{\omega_p^2/4\pi}{\gamma_{opt}(\omega, T) - i\omega m_{opt}^*(\omega, T)/m_b}. \tag{22}$$

Here the optical relaxation $W(\omega)$, the optical mass $m_{opt}^*(\omega, T)/m_b$ and the optical scattering rate $\gamma_{opt}(\omega, T)$ as the complex, real and imaginary parts of an inverse normalised conductivity $\omega_{pl}^2/4\pi\sigma(\omega, T)$ have been introduced. This presentation is old and popular in spectroscopy, but the practical value of the Eq. (22) has not been exploited much. The optical relaxation will be compared with the effective mass $m_{eff}^*(\omega, T)/m_b$ and the effective scattering rate $\gamma_{eff}(\omega, T)$, which by definition are the real and imaginary parts of the renormalized frequency

$$\tilde{\omega}(\omega, T) = \frac{\omega m_{eff}^*(\omega, T)}{m_b} + \frac{i\gamma_{eff}(\omega, T)}{2}. \tag{23}$$

When the frequency dependencies of $\gamma_{opt}(\omega, T)$ and $m^*(\omega, T)/m_b$ are plotted, the coupling constant, the average boson energy and the upper boundary of spectral function are visually accessible (see Fig. 2). The appropriate visual accessibility rules can be derived from the properties of the renormalized frequency $\tilde{\omega}$ using the Allen formula Eq. (26) [2]. Since such useful expression was originally obtained in the weak coupling approximation ($\lambda \ll 1$) only, below I rederive it for the general case, including strong coupling.

For any k the values of λ_k are positive and are of the same order of magnitude, except may be λ_0. It means, that in the zero order the series $\tilde{\omega}_n$ defined by Eq. (6) is an arithmetical progression. Consequently, according to the Gauss rule for the arithmetical series the denominators $\tilde{\omega}_k + \tilde{\omega}_{n-k-1}$ in (9) at a given n do not depend on k, that is, they coincide. In view of Eq. (22), Eq. (9) takes the form

$$W_{n+1} \approx \tilde{\omega}_k + \tilde{\omega}_{n-k}. \tag{24}$$

Hence we can expand the denominator in (9)

$$\frac{4\pi\sigma_n}{\omega_{pl}^2} \equiv \frac{1}{W_n} = \frac{1}{n} \sum_{k=0}^{n-1} \frac{1}{W_n \left[1 + (\tilde{\omega}_k + \tilde{\omega}_{n-k-1} - W_n)/W_n\right]} \tag{25}$$

to the first order in power of $(\tilde{\omega}_k + \tilde{\omega}_{n-k-1} - W_n)/W_n$. After some simple algebra we arrive at the sought-for Allen formula

$$W_n = \frac{2}{n} \sum_{k=0}^{n-1} \tilde{\omega}_k. \tag{26}$$

The real axis versions of Eq. (26) at $T=0$ look similar

$$W(\omega) = \frac{1}{\omega} \int_0^\omega dz 2\tilde{\omega}(z), \tag{27}$$

$$\frac{\omega m^*_{opt}(\omega, T)}{m_b} = \frac{1}{\omega} \int_0^\omega dz \frac{2z m^*_{eff}(z, T)}{m_b}, \tag{28}$$

$$\gamma_{opt}(\omega) := \frac{1}{\omega} \int_0^\omega dz \gamma_{eff}(z). \tag{29}$$

and justify the following simple description of the interaction of light with the conducting subsystem of the normal state metal. When a photon with an energy ω has been absorbed, *two* excited *virtual* quasiparticles, an "electron" and a "hole" are created. If the first particle's frequency is ω', the second ones frequency should be $\omega - \omega'$. The exited "electron" and "hole" relax to the Fermi level according to the quasiparticle laws (see Eqs. 5,6). Since ω' can varies from 0 to ω, the optical relaxation $W(\omega)$ is the frequency-averaged $(\frac{1}{\omega} \int_0^\omega)$, double (electron+hole $\rightarrow$ factor 2) renormalized frequency $\tilde{\omega}$.

At $T = 0$ the ISB effective scattering rate $\gamma_{eff}(\omega) \equiv 2Im\tilde{\omega}(\omega)$

$$\gamma_{eff}(\omega) = \gamma_{imp} + 2\pi \int_0^\omega dy \alpha^2 F(y) \tag{30}$$

has a clear physical meaning. The exited quasiparticles with the energy ω can relax by the emission of virtual bosons with the energy varying the range from 0 to ω or by the impurity scattering.

At finite temperature the Allen formula

$$W(\omega) = \frac{1}{\omega} \int dz \frac{\tanh\left(\frac{\omega-z}{2T}\right) + \tanh\left(\frac{z}{2T}\right)}{2} 2\tilde{\omega}(z) \tag{31}$$

and the effective scattering rate (5) have exactly the same interpretation, since the only difference between them and Eqs. (27,27,29,30) is the presence of the Fermi distributions instead of step functions at $T=0$.

Figure 2 shows the frequency dependencies of the normal state optical and effective masses and scattering rates calculated using the model spectral function [7] with $\lambda=2$. The results are plotted for the low temperature $T=1$ K and for $T=100$ K which is close to T_c and has the order of $\Omega_{boson}/2\pi T$, where $\Omega_{boson}=330$ cm^{-1} is the averaged boson frequency. Both $m^*_{eff}(\omega)/m_b$ and $\gamma_{eff}(\omega)$ show the characteristic features corresponding the peculiarities of the phonon spectrum. When the frequency ω exceeds the upper energy bound of the boson spectrum Ω_{max}, the quasiparticle becomes "undressed". Its mass $m^*_{eff}(\omega)/m_b \approx 1$ and effective scattering rate $\gamma_{eff}(\omega) \approx const$, but of course, the absolute value of $\gamma_{eff}(\omega \to \infty)$ could be huge.

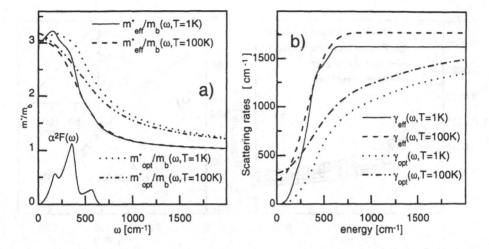

Figure 2. The frequency dependencies of the normal state optical and effective masses (a) and scattering rates (b) calculated using the "model 2" spectral function with $\lambda=2$ (shown in panel (a) for comparison). The quasiparticle effective mass $m^*_{eff}(\omega)/m_b$ and scattering rate $\gamma_{eff}(\omega)$ at $T=1$ K (solid lines, panels (a) and (b)) reproduce the fine structure of the spectral function, but at $T=100$ K (dashed lines) this function are smeared out due to averaging over Fermi distribution. At $\omega > \Omega_{max}$ the quasiparticles become "undressed", $m^*_{eff}(\omega)/m_b \approx 1$ and $\gamma_{eff}(\omega) \approx const$. The optical mass $m^*_{opt}(\omega)/m_b$ and the scattering rate $\gamma_{opt}(\omega)$ even at $T=1$ K (dotted lines) do not reproduce the structure of the spectral function due to the energy averaging, all the more at high temperature ones ($T=100$ K, dashed-dotted lines).

The optical mass $m^*_{opt}(\omega)/m_b$ and scattering rate $\gamma_{opt}(\omega)$ have no features at $\omega \approx \Omega_{max}$ due to the energy averaging (27,28,29,31). One can only *calculate* the approximate value of Ω_{max} as the frequency where the $m^*_{opt}(\omega)/m_b - 1$ and $\gamma_{opt}(\omega)$ reach the halves of their maximum magnitudes. Fig. 2 shows, that this my rule can not be applied even to the calculated frequency dependence of the optical scattering rate $\gamma_{opt}(\omega)$, the more to the experimental curve, where *the high energy* behaviour is distorted by the interband transition contribution to the dielectric permeability. In contrast the frequency dependence of the optical mass $m^*_{opt}(\omega)/m_b$ is suitable for the application of the simple criterion

$$m^*(\omega = \Omega_{max})/m_b - 1 \approx 0.5(m^*(\omega = 0)/m_b - 1). \qquad (32)$$

Moreover its *low energy* value $m^*_{opt}(\omega = 0)/m_b \approx 1 + \lambda_{tr}$ itself contains a useful piece of information. Fig. 2a shows, that the approximate values of the transport coupling constant $\lambda_{tr} \approx (m^*_{opt}(0)/m_b - 1)$ and the upper frequency bound of the electron-boson coupling function $[m^*_{opt}(\Omega_{max})/m_b - 1 \approx 0.5(m^*_{opt}(0)/m_b - 1)]$ are visually accessible. If the spectral function is

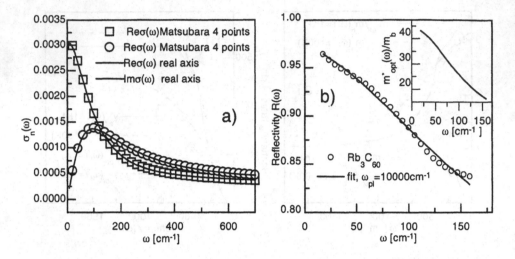

Figure 3. Panel (a) shows the frequency dependencies of the real and imaginary parts of the calculated normal state optical conductivity using the real axis formalism (solid lines) and Matsubara technique (squares - Re$\sigma(\omega)$, circles - Im$\sigma(\omega)$). The analytical continuation from the imaginary axis to the real one was made by the Pade polynom with the degree of polynomial n=4. Calculations was made for T=100 K using "model 2" spectral function. Panel (b) shows the frequency dependence of the normal state reflectivity R(ω) of Rb$_3$C$_{60}$ measured at T=40 K by Degiorgi *et al.*, *Phys. Rev. Lett.* **69**, 2987 (1992) (circles) in comparison with the least square fit by the favourite polynom. Since four parameters were too much, the value of the plasma frequency ω_{pl}=10000 cm^{-1} was fixed. The obtained frequency dependence of the optical mass is shown in the inset.

wide and its peak position (or averaged frequency) coincide roughly with the $\Omega_{max}/2$, the optical relaxation W(ω) is smooth and the elucidation of Ω_{boson} at high temperature is impossible. When an exotic superconductor with the narrow, δ-function like spectral function will be discovered, following this way one can easily construct the recipe for the evaluation of the strength and the position of the single peak.

2.3. MY FAVOURITE POLYNOM

The Pade polynom analytical continuation [14] is a capricious and not completely correct procedure. Nevertheless, if the temperature is high enough in comparison with the energy under consideration it works splendidly. The numerical experiment has given to my great surprise the following (and to some extent obvious) result (see Fig. 3). The good agreement between the real axis calculations and the Matsubara ones takes place every time when the approximation using 1000 σ_n points gives the same result as the *four* points Pade approximation.

This remarkable Pade polynom with only four parameters $\omega_{pl}^2/4\pi$, A, B and C

$$\sigma(\omega) = \frac{\omega_{pl}^2/4\pi}{A - i\omega + \frac{B\omega}{\omega + iC}} \tag{33}$$

fits well the majority of the experimental data at $T \geq T_c$ and itself is the demonstration of the result obtained in section 2.1: the few first Matsubara values σ_k are responsible for the frequency dependence of $\sigma(\omega)$.

The Pade polynom (33) is a formula of merit. At first, it has the correct analytical properties. Its two poles are located in the lower half-plane of the complex energy. The high ($\omega \to \infty$) and low ($\omega \to 0$) energy asymptotic behaviour are reasonable [23]. At second, it is ideal for the least square fits of experimental $R(\omega)$ [24], as well as for simple analytical estimations. Having the exact solution (see sec.2.1) we have to substitute the complex values $i\nu_k$ for energy, perform easily the analytical continuation to the imaginary axis and analyse the values of σ_i. In the way we obtain the interpretation of A, B, C. Another possibility is a fast and easy analysis of the data in terms of the optical mass $m_{opt}^*(\omega, T)/m_b$ and the optical scattering rate $\gamma_{opt}(\omega, T)$

$$\frac{m_{opt}^*(\omega, T)}{m_b} - 1 = \frac{BC}{\omega^2 + C^2} \tag{34}$$

$$\gamma_{opt}(\omega, T) = A + \frac{B\omega^2}{\omega^2 + C^2}. \tag{35}$$

At third, if the low-frequency reflectivity $R(\omega)$ measured at high temperature can not be fitted by Eq. (33) or the obtained value of λ_0 contradicts the generally accepted values, it might indicate, that the material under consideration is *not* a standard Isotropic Single wide Band (ISB) metal and its properties have to be analysed if terms of *another*, unconventional model. In order to illustrate the utility of the this approach, let us consider another superconducting compound with relatively high transition temperature Rb_3C_{60} [16, 18, 17]. Since the ratio of the experimental reflectivities in the superconducting and normal states $R_s/R_n \approx 1$ above 100 cm^{-1} [16], in frame of the ISB model it means, that the high energy bound of the electron-boson interaction function $\Omega_{max} < 100$ cm^{-1}. The low energy part of the normal state reflectivity $R(\omega)$, measured at T=40 K, was fitted (with the adopted fixed ω_{pl}=10000 cm^{-1}) by (33) (see Fig. 3b), $m_{opt}^*(\omega, T)/m_b$ is shown in the inset. One can see, that the optical mass at small ω is huge itself and gives $\lambda > 100$. This values points to heavy fermion or polaron models, rather than to the generally accepted weak or intermediate coupling scenario. On the other hand, the analysis of the normal state specific heat data gives an extremely small value of the coupling constant. As a result we concluded that Rb_3C_{60} is not a ISB metal and its properties should be treated in terms of a more complicated model which takes into account the self-energy and conductivity vertex corrections, the small value of the bandwidth $\Delta E \approx 100$ meV, possible influence of the electron-electron correlation, the strong anisotropy of the coupling function and the Fermi velocity, and etc.

And the last but not least, if the data can be fitted by (33), that is by the formula with four parameters, it is impossible to recover from this *four* parameters the detailed shape of the electron-boson interaction function $\alpha^2 F(\omega)$.

In conclusion, it is naturally to define the amount of the available information by the number of poles and zeros of the function which fits well the experimental curve or the number of the determinate Matsubara values.

2.4. TEMPERATURE DEPENDENCE OF THE OPTICAL MASS AND RELAXATION RATE AND ITS TWO-BAND ERSATZ

In frame of the ISB model we assume, that all quasiparticles have the same Fermi velocity, but in real metals there is some anisotropy of v_F. If the impurity scattering rate γ_{imp} is small in comparison with the complete optical scattering rate γ_{opt} and the electron-boson coupling is isotropic, we can use the average value $< v_F^2 >$ and keep the ISB model. On the other hand, if the impurity scattering, given by the mean free path l, dominates, the quasiparticles from different parts of the Fermi surface will have different scattering rates $\gamma_{imp,i} = v_{F,i}/l$. Let us accept for simplicity, that we can divide the Fermi surface between two parts with the approximately uniform properties. The optical conductivity of this two-band system is the sum of two Drude terms

$$\sigma(\omega) = \frac{\omega_{pl,1}^2/4\pi}{\gamma_1 - i\omega} + \frac{\omega_{pl,2}^2/4\pi}{\gamma_2 - i\omega}. \tag{36}$$

One can see, that Eq. (36) and Eq. (33) are equivalent and four parameters $\omega_{pl,1}^2, \omega_{pl,2}^2$, γ_1, γ_2 in (36) connected with ω_{pl}^2, A, B, C in (33) as follows

$$
\begin{aligned}
\omega_{pl}^2 &= \omega_{pl,1}^2 + \omega_{pl,2}^2 \\
C &= \frac{\omega_{pl,1}^2 \gamma_2 + \omega_{pl,2}^2 \gamma_1}{\omega_{pl,1}^2 + \omega_{pl,2}^2} \\
A &= \frac{\gamma_1 \gamma_2 (\omega_{pl,1}^2 + \omega_{pl,2}^2)}{\omega_{pl,1}^2 \gamma_2 + \omega_{pl,2}^2 \gamma_1} \\
B &= \gamma_1 + \gamma_2 - A - C
\end{aligned}
\tag{37}
$$

It means, that even if there is no electron-boson interaction in the two-band system, the frequency dependence of its optical conductivity (36) will

337

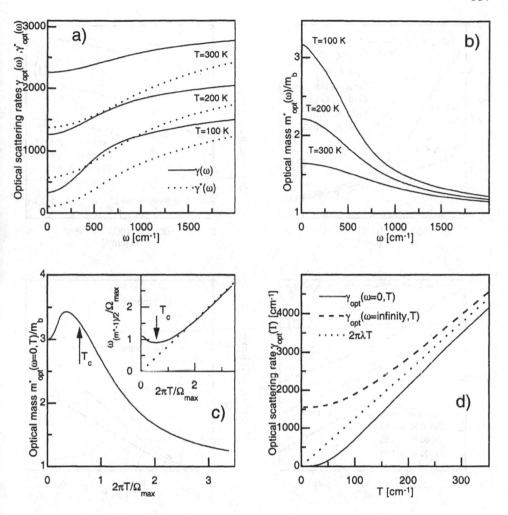

Figure 4. Panels (a) and (b) show the frequency dependence of the optical mass and the optical scattering rate for T=100K, 200 K, 300 K. Panels (c) and (d) present the corresponding temperature dependence of $m^*_{opt}(T)/m_b$ and $\gamma_{opt}(T)$ for $\omega - 0$ and $\omega = \infty$. All curves (solid lines) was calculated using model 2 spectral function. For comparison the normalised optical scattering rates $\gamma^*(\omega, T) = \gamma_{opt}(\omega, T)/m^*_{opt}(\omega, T)/m_b$ are plotted by dotted lines in panel (a). The temperature dependence of the quantity $\omega_{(m^*-1)/2}$ (see text) is shown by solid line on inset (c) together with the phenomenological line $\omega = 5T$ (dotted line). The common for $\gamma_{opt}(\omega = 0, T)$ and $\gamma_{opt}(\omega = \infty, T)$ asymptote $\gamma = 2\pi\lambda T$ is shown by dotted line in panel (d).

reproduce a ISB model one (8) with some frequency dependent "mass" and "scattering rate". The multiband ersatz of the electron-boson effects could be revealed by the investigation of the temperature dependence of

338

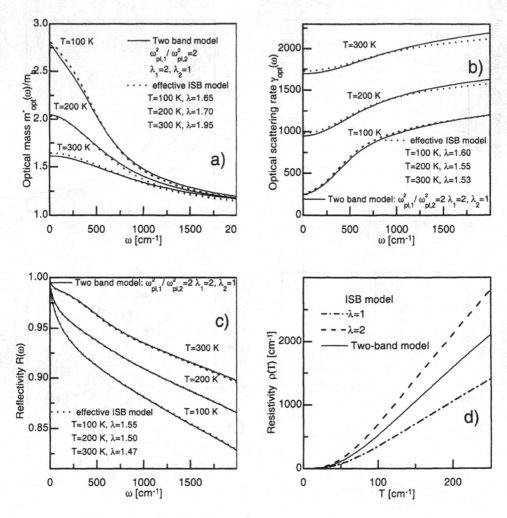

Figure 5. The frequency and temperature dependencies of the two-band optical mass, scattering rate, reflectivity and DC resistivity are shown by solid lines in panels a, b, c, d respectively. The corresponding ISB fitted counterparts are plotted by the dotted lines.

$m^*_{opt}(\omega, T)/m_b$ and $\gamma_{opt}(\omega, T)$. It originates mainly from the presence of the Bose factor in Eqs.(5) ($\equiv$ sampling in (6)) and to a less degree from the Fermi distributions in Eqs. (5-9).

In subsection 2.2 the low ($T \to 0$) temperature behaviour of $m^*_{opt}(\omega)/m_b$ and $\gamma_{opt}(\omega)$ was discussed in detail. We found, that at a given quasipartical frequency ω in accord with the frequency conservation principle only virtual bosons with $\Omega < \omega$ can contribute to the relaxation.

At high temperature the bosons lose their individuality. At $T \to \infty$ the expression (5) can be substantially simplified, since the Bose factor $coth(z/2T) \approx 2T/z$ dominates in comparison with the Fermi one $tanh[(\omega - z)/2T]$. Keeping the leading term $2T/z$ we arrive at

$$Im\tilde{\omega}(\omega) \equiv \gamma_{eff}(\omega) \approx \frac{\gamma_{imp}}{2} + \pi T \int \frac{dy\alpha^2 F(y)}{y} = \frac{\gamma_{imp} + 2\pi\lambda T}{2}. \tag{38}$$

In Matsubara technique the same result can be obtained even more easily. At $T \to \infty$ all kernels of the spectral function λ_k, defined by Eq. (7), vanish as $1/(<\Omega_{boson}^2> + 4\pi^2 k^2 T^2)$ for all k, except k=0. Therefore, the Eq. (6) for the renormalized frequency becomes trivial

$$\tilde{\omega}_n = \omega_n + \gamma_{imp}/2 + \pi\lambda T. \tag{39}$$

If one substitutes the obtained $\tilde{\omega}(\omega)$ in Eq. (8) or $\tilde{\omega}(i\omega_n)$ into Eq. (9) after some algebra one arrives at the Drude type expression for the optical conductivity

$$\sigma(\omega) = \frac{\omega_{pl}^2}{4\pi} \frac{1}{\gamma_{imp} + 2\pi\lambda T - i\omega}. \tag{40}$$

Note, that the relation $\gamma_{opt} = 2\gamma_{eff} \equiv -2Im\Sigma$ is correct, if and only if the self energy $\Sigma(\omega) = \omega - \tilde{\omega}(\omega)$ does not depend on frequency. Otherwise, the relation

$$\gamma_{opt}(\omega) = -2Im\Sigma(\omega) \tag{41}$$

does not satisfy the Kramers-Kronig (K.-K.) relations, that is, it violates the causality. The proof is very simple. Due to the charge conservation law, when the photon has been absorbed, *two* quasiparticles, "the electron" and "the hole" come into being. The optical conductivity is a bosonic function defined by its values at ν_k, while the quasiparticle self energy $\Sigma(\omega)$ is a fermionic function defined by the values at ω_k. As a result, the right and left hand sides of Eq. (41) have different statistics, what is very strange. In frame of the ISB model considered here the proof is a little bit more complicated. The optical conductivity itself and its inverse $1/\sigma(\omega)$ are response functions. It means, that they have no poles on the upper half-plane of the complex frequency or, in other words, their real and imaginary parts are connected by Kramers-Kronig relations. The same is valid for the self energy. From Eq. (41) after K.-K. transformation one obtains

$$m_{opt}^*(\omega) - 1 = 2(m_{eff}^*(\omega) - 1). \tag{42}$$

The optical and effective masses have the same asymptotic value $1+\lambda$ at $\omega, T \to 0$. Due to the presence of the factor of 2 in (42) it would mean, that 1=2.

Let us turn to the temperature dependence of the optical properties of ISB metals, which are shown in Fig. 4. All curves were calculated using model 2 spectral function and $\gamma_{imp} = 0$. Panels 4a presents the frequency dependence of the optical scattering rate $\gamma_{opt}(\omega, T)$ (solid lines) for T=100 K, 200 K, 300K. The corresponding optical masses are shown in panel 4b. One can see, that the absolute value of $\gamma_{opt}(\omega, T)$ increases with T, but the frequency dependent component decreases. It leads to diminishing of $m^*_{opt}(\omega, T)/m_b$ when T is rising. The temperature dependence of the minimum ($\omega = 0$,solid line) and maximum ($\omega = \infty$,solid line) values of $\gamma_{opt}(\omega)$ are shown in panel 4d in comparison with their common asymptote $\gamma = 2\pi\lambda T$ (dotted line). The corresponding function $m^*_{opt}(\omega = 0, T)/m_b$, named sometimes $1 + \lambda_{opt}(T)$ is presented in panel 5c. The position of maximum of the small hump at low T is approximately equal to $\Omega_{min}/5$, where Ω_{min} is the frequency position of the lowest boson peak in $\alpha^2 F(\omega)$. At the same temperature the DC resistivity $\rho(T) = 4\pi\gamma_{opt}(\omega = 0, T)/\omega^2_{pl2}$ starts the linear rise. In subsection 2.2 the simple rule for determination of the *upper* frequency bound of the spectral function Ω_{max} was proposed. At $\omega \approx \Omega_{max}$ the optical mass without unity $(m^*_{opt}(\Omega_{max}, T)/m_b - 1)$ is approximately the half of its value at $\omega = 0$. The temperature dependence of the quantity $\omega_{(m^*-1)/2}$ defined by

$$2[m^*_{opt}(\omega = \omega_{(m^*-1)/2}, T)/m_b - 1)] = m^*_{opt}(\omega = 0, T)/m_b - 1 \qquad (43)$$

is shown by solid line on inset in panel 4c in comparison with the phenomenological line $\omega = 5T$. One can see, that

$$\omega_{(m^*-1)/2} \approx max(5T, \Omega_{max}) \qquad (44)$$

The normalised optical scattering $\gamma^*(\omega) = \gamma_{opt}(\omega)/m^*_{opt}(\omega, T)/m_b$ is shown in Fig(4a) (dotted lines). In our opinion [7] this quantity is far useless in comparison with $\gamma_{opt}(\omega)$ and $m^*_{opt}(\omega, T)/m_b$. With the exception of the low frequency region where its value is constant, $\gamma^*(\omega)$ is a quasilinear function $\propto \omega$ up to $max(4\Omega_{max}, 2\pi\lambda T)$. Even the simple analysis by the favourite polynom Eq. (33) allows to determine the four parameters, while $\gamma^*(\omega)$ depends on two. It is interesting, that the Gilbert transformation from $\gamma^*(\omega)$ is not equal to 0. In this connection we note that the use of the BSC normal state conductivity expression

$$\sigma(\omega) = \frac{\omega^{*2}_{pl}}{4\pi(\gamma^* - i\omega)}, \qquad (45)$$

is correct, only if $\gamma^* = const$. Here $\omega^{*2}_{pl} = \omega^2_{pl}/(1 + \lambda)$ is the frequency-independent renormalised plasma frequency.

Note, that according to the Allen formula (31), the optical relaxation is approximately *linear* with respect to the spectral function $\alpha^2 F(\omega)$. It means, that the conversion from the low temperature scenario to the asymptotic high temperature one takes place for each boson with frequency Ω_i separately at $T \approx \Omega_i/5$, that is, when Ω_i becomes of the order of the *first* Matsubara energy $\nu_1 = 2\pi T$.

Since the isotropic single wide band model is the subject of the present lecture, discussing the properties of two-band systems we have to pay special attention to possible artefact results. Let us start our consideration from the temperature independent two-Drude terms simulator (36). Since it is equivalent to the favourite polynom (33), we can "extract the information about spectral function" including the temperature dependence of the coupling constant $\lambda(T)$. It is known, that if the material parameters are temperature independent, the optical scattering rate $\gamma_{opt}(\omega, T)$ is a monotonically increasing function. Since the simulator's $\gamma_{opt}(\omega)$ does not changes, when T is rising, one may conclude, that the coupling "constant" $\lambda(T)$ is a monotonically decreasing function with the high temperature asymptote $1/T$. On the other hand, the analysis of the optical mass, which is usually a monotonically decreasing function, gives the opposite result: $\lambda(T)$ dramatically increases when temperature is growing. The study of the reflectivity $R(\omega, T)$ gives the third fitted function $\lambda(T)$. The reasonable way to resolve this conflict is the replacement of the model. If one restricts oneself to the consideration of the scattering rate only, the contradiction can be overlooked. The fact is that the strong temperature dependence of material parameters exists, for example, in heavy fermion systems. Another interesting scenario was realised in fullerides, where the DC resistivity (measured at the constant pressure) changes with temperature as T^2 and the same DC resistivity (measured at the constant volume) follows conventional linear law $\rho \propto T$, as it should be, since the theory assumes the V=const condition.

The second example is the clean limit of anisotropic systems. As usually, one can divide the Fermi surface between two (for simplicity) parts with the approximately uniform properties. This disjoined representation of the Fermi surface is known as the multiband model. We accept also, that the intra- and interband interactions are governed by the electron-boson spectral functions having the same "model 2" shape, but different coupling constants λ_{ij}. The scales of transport and optical properties of the bands are fixed by the plasma frequencies $\hbar\omega_{pl,i}^2 = 4\pi e^2 v_{F,i}^2 N_i(0)/3$. In the normal state is possible to diagonalise the equations by the substitution $\lambda_1 = \lambda_{11} + \lambda_{12}$ and $\lambda_2 = \lambda_{21} + \lambda_{22}$. The resulting two-band optical conductivity is the sum of two ISB terms (8). For the model calculations the arbitrary values $\lambda_1 = 2$, $\lambda_2 = 1$ and $\omega_{pl,1}^2/\omega_{pl,2}^2 = 2$ was chosen.

342

The frequency and temperature dependencies of the two-band optical mass, scattering rate, reflectivity and DC resistivity are shown by solid lines in Fig. (5)a, b, c, d respectively. The *by-eye-analysis* can not reveal the difference between clean limit two-band optical properties and the standard ISB ones. Fortunately, the ISB model can pose several possibilities for a quantitative study. I simply fit the two-band curves, shown on panels (5)a, b, c, by the appropriate ISB ones (dotted lines), calculated using the same "model 2" spectral function. The coupling constant λ was used as fitting parameter. One can see, that the temperature dependencies of $m^*_{opt}(\omega, T)/m_b$, $\gamma_{opt}(\omega, T)$ and $R(\omega, T)$ of the anisotropic system are yet weaker, that it should be according to the ISB rules. The fitted functions $\lambda(T)$ grow or drop, when T is rising, depending on which quantity has been chosen for the fit.

For the few metals, where the mutual anisotropy of $\lambda(\mathbf{k})$ and of $v_F(\mathbf{k})$ was investigated, the following regularity was found. The quasiparticles with smaller $v_F(\mathbf{k})$ have larger $\lambda(\mathbf{k})$. Usually this information is unavailable. One can estimate the anisotropy, if one compares the material parameters obtained from the optical spectra with the ones revealed by the analysis of thermodynamic, or point-contact tunnelling data. The scale of the optical conductivity is fixed by $\omega^2_{pl} \propto v^2_F N(0)$. The normal state specific heat data are proportional to $(1 + \lambda) N(0)$. The point-contact tunnelling spectra are weighted by the factor v_F. As a result, if one treats optical spectra of anisotropic metals in terms of the ISB the model, one reveals the electron-boson coupling function $\alpha^2_{opt} F(\omega)$, which distinguish itself by smaller value of the coupling function λ_{opt} in comparison with lambdas, obtained by other physical methods. This difference is usually assigned to the kinetic coefficient (1-cos) [2] or, in another language, to the vertex corrections to the quasipartical Green function[1]. It is interesting, that if the system under consideration is completely isotropic, and $\alpha^2 F(\omega, \mathbf{k}, \mathbf{k}')$ do not depend on $\mathbf{k}$ and $\mathbf{k}'$, the vertex corrections vanish as soon as considered in this subsection usual anisotropy effects. What kind of corrections are more important in each particular case, first order vertex ones or discussed here zero order distortions caused by the interplay of $v_f(\mathbf{k})$ and $\lambda(\mathbf{k})$, it is the question *for* the gallop band structure calculations.

In conclusion, the temperature dependence of the optical properties of ISB metals really exists and can be well quantitatively described. It originates mainly from the presence of the Bose factor in Eqs.(5) ($\equiv$ sampling in (6)) and to a lesser extent from the presence of the Fermi distributions in Eqs.(5-9). The principal deviation from the predicted frequency and temperature behaviour manifests that the chosen ISB model is not complete and has to be replaced.

2.5. OPTICAL MASS AND SCATTERING RATE IN MID-INFRARED: ARTEFACTS GENERATED BY OVERLOOKING OF INTERBAND TRANSITIONS AND UNDERESTIMATION OF ϵ_∞.

The formal textbook relation between the conductivity $\sigma(\omega)$ and permeability $\epsilon(\omega)$

$$\epsilon(\omega) = 1 + \frac{4\pi i \sigma(\omega)}{\omega} \qquad (46)$$

is not suitable for a practical use, since the entering Eq. (46) $\sigma(\omega)$ reflects the total response of all excitations.

The simplest reasonable decomposition of (46), valid over the spectral region up to frequencies just above plasma edge, is

$$\epsilon(\omega) = \epsilon_\infty + \epsilon_{inter}(\omega) + \epsilon_{phonon}(\omega) + \frac{4\pi i \sigma(\omega)}{\omega}, \qquad (47)$$

where ϵ_{inter} and ϵ_{phonon} are the contributions of interband transition(s) and direct IR active phonons centred below plasma edge. The response of high energy excitations and wide interbands, overlapping the plasma edge is presented by the complex constant ϵ_∞. $\sigma(\omega)$ is *the quasiparticle* conductivity, defined by Eqs.(8,10).

The direct phonon band contribution will not be considered, since it is usually weak and plays some role only in the selected compounds containing light elements and having small value of ω_{pl}, for example, in doped fullerenes.

The used here model interband contribution

$$\epsilon_{inter}(\omega) = \frac{\omega_{pl,inter}^2}{\Omega_0^2 - \omega^2 - i\omega\gamma_{inter}} \qquad (48)$$

is the simple Lorentzian with $\omega_{pl,inter}=16000$ cm^{-1}, $\Omega_0=4000$ cm^{-1}, and $\gamma_{inter}=6000$ cm^{-1}. The overlooking of the interband contribution was simulated as follows. The sum of two terms given by Eqs. (48) and (8) was treated as the quasiparticle response in terms of the optical mass and scattering rate

$$\gamma_{opt}(\omega,T) - i\omega m_{opt}^*(\omega,T)/m_b = \frac{\omega_{pl}^2 + \omega_{pl,inter}^2}{4\pi\sigma(\omega) - i\omega\epsilon_{inter}(\omega)}. \qquad (49)$$

For simplicity I set $\omega_{pl} = \omega_{pl,inter}$ and following to

The results are shown in Figs. 6a, 6b and 6c by solid lines. The "true" masses and scattering rates

$$\gamma_{opt}(\omega,T) - i\omega m_{opt}^*(\omega,T)/m_b = \frac{\omega_{pl}^2}{4\pi\sigma(\omega)}. \qquad (50)$$

344

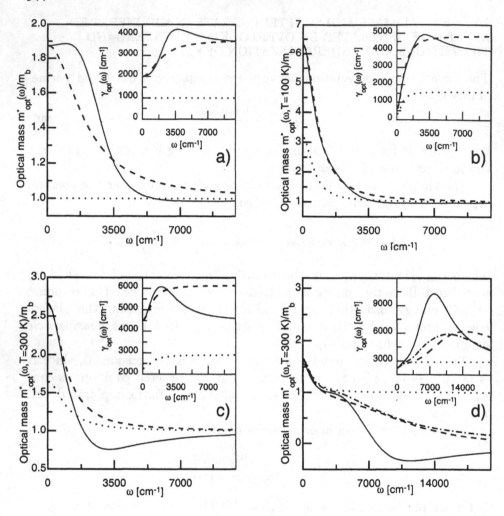

Figure 6. The artefact frequency dependencies of the optical masses and scattering rates (insets) generated by the overlooking of interband transition (panels a, b, c) and the underestimation of ϵ_∞ (panel d) are shown by solid lines. The true, unpertubed frequency dependencies of $m^*_{opt}(\omega)/m_b$ and $\gamma_{opt}(\omega)$ are plotted by dotted lines. The dashed lines in panels a, b, c are the *chi-by-eye* fitted curves (by Eq. 32). For parameters of the model interband transition and trial ISB quasiparticle optical conductivities see text.

and the *chi-by-eye [19]* fit by the formula (33) are plotted in Figs. 6a, 6b and 6c by dotted and dashed lines respectively.

Note, that due to the frequency averaging, the quasipartical optical conductivity (8,10) is a slowly varying function in the mid-infrared region. In contrast, the interband permeability near transition edge (here, near the center of the Lorentzian) can changes rapidly. The second important feature of $\epsilon_{inter}(\omega)$ is its very weak temperature dependence.

The mixture of the standard Drude term ($\gamma=1000$ cm^{-1}) with the Lorentzian (48) is shown in panel 6a. This simple artefact can be easily recovered, since it is temperature independent and the optical mass changes too quickly.

The second and third examples are the half-by-half blends of the "model 2" ISB conductivities (T=100 K, panel 6b; T=300 K, panel 6c) with the same $\epsilon_{inter}(\omega)$ Eq. (48). It is more complicated case, since the nonmonotonic behaviour could be screened by forthcoming transitions. In this case detailed quantitative analysis of both the frequency and the temperature dependencies is needed. One has to pay the special attention to the temperature dependencies of $m_{opt}^*(\omega \to 0, T)/m_b$, $\omega_{(m^*-1)/2}$ and $\gamma_{opt}(\omega \approx 2\pi\lambda T, T)$.

If the interband transitions are wide spectral bands without unqualified features, their contribution to the permeability is usually included in ϵ_∞. The effects caused by the underestimation of its real part was considered by us in [7]. Here we study the problems produced by the use of Eq. 46 and by the underestimation of the imaginary part of ϵ_∞. Since the last quantity can be approximated by the simple constant only in the limited frequency range, for the sake of the analytical correct the model function was chosen in the form

$$Im(\delta\epsilon_\infty(\omega)) = \left[\tanh\left(\frac{\omega - 2000}{5000}\right) + \tanh\left(\frac{50000 - \omega}{30000}\right)\right] \quad \text{at } \omega > 0,$$
$$Im(\delta\epsilon_\infty(-\omega)) = -Im(\delta\epsilon_\infty(\omega)),$$
$$Im(\delta\epsilon_\infty(0)) = 0,$$
$$Re(\delta\epsilon_\infty(\omega)) = 2.5 - \frac{2\omega}{\pi}\int_0^\infty \frac{dx\, Im(\delta\epsilon_\infty(x))}{x^2 - \omega^2}. \tag{51}$$

In the mid infrared $(\delta\epsilon_\infty(\omega)) \approx 4 + 2i$.

The model frequency dependencies of the optical masses and scattering rates

$$\gamma_{opt}(\omega, T) - i\omega m_{opt}^*(\omega, T)/m_b = \frac{\omega_{pl}^2}{4\pi\sigma(\omega) - i\omega\tilde{\delta}(\epsilon_\infty)}. \tag{52}$$

are shown in panel 6d for $\tilde{\delta}(\epsilon_\infty) = \delta(\epsilon_\infty(\omega)) - 1$ (solid lines), $\tilde{\delta}(\epsilon_\infty) = \delta(\epsilon_\infty(\omega)) - 4$ (dashed-dotted lines), and $\tilde{\delta}(\epsilon_\infty) = 2i$ (dashed lines). The unperturbed $\gamma_{opt}(\omega, T)$ and $m_{opt}^*(\omega, T)/m_b$ ("model 2" spectral function, $\lambda=2$, T=300 K) are plotted by dotted lines for comparison. As usually, the masses do not lie, and manifest that the "unusual" frequency dependence of the scattering rate is no more than the artefact of the chosen interpretation procedure.

The early band structure calculations [27] predicted, that the interband contribution into the in-plane HTSC permeability is a real constant between 10 and 15 at $\omega \to 0$, and becomes complex for $\omega > 0.4eV$. In the vicinity of the plasma edge its value is 4+2i. This information was used by us for the successful description of the *normal state* optical properties of HTSC materials both in FIR and MIR spectral regions [7, 15].

In conclusion, the overlooked interband transitions produces the "unusual" frequency dependence of the optical relaxation, which can be nevertheless analysed properly within the ISB model.

2.6. WHY IS THE FIRST DERIVATIVE FROM THE EXPERIMENTAL CURVE BETTER THAN THE SECOND ONE?

Let us rewrite the valid at T=0 Eqs.(29,30) in the differential form

$$\alpha^2 F(\omega) = \frac{1}{2\pi} \frac{d\gamma_{eff}(\omega)}{d\omega} \tag{53}$$

$$\alpha^2 F(\omega) = \frac{1}{2\pi} \frac{d^2(\omega\gamma_{opt}(\omega))}{d\omega^2}. \tag{54}$$

At a glance both Eqs. (53) and (54) look nice and simple. But if one attempts to apply them to an experimental data, the distinction becomes evident. The point is that the experimental data are known only approximately. A finite value of the signal to noise ratio (S/N) is the reason why the Eq. (54) can not be applied to the FIR data. Typical value of the reflectivity S/N do not exceed 50, the multi-reflection techniques [3] allow to reach S/N=2000. Let us adopt for simplicity, that $f(\omega) = \omega\gamma_{opt}(\omega)$ is sampled in equidistant points ω_i, $i = 1, M$ and all values $f(\omega_i)$ have the same errors

$$\frac{f(\omega_i)}{\delta f(\omega_i)} = \left(\frac{S}{N}\right)_f. \tag{55}$$

Note, that in the normal state at $T = 0$ (when the application of (53) and (54) is correct) both $f(\omega) = \omega\gamma_{opt}(\omega)$ and its first derivative $f'(\omega)$ are monotonically increasing functions, if a metal under consideration is the ISB one.

The difference $f(\omega_{i+1}) - f(\omega_i)$ has the order of $f(\omega_i)/M$. Its error is the $\sqrt{\delta f^2(\omega_{i+1}) + \delta f^2(\omega_i)} \approx \sqrt{2}\delta f(\omega_i)$. As a result the first derivative $f'(\omega)$ has the signal to noise ratio

$$\left(\frac{S}{N}\right)_{f'} = \frac{f'(\omega_i)}{\delta f'(\omega_i)} = \frac{f(\omega_{i+1}) - f(\omega_i)}{\sqrt{\delta f^2(\omega_{i+1}) + \delta f^2(\omega_i)}} \approx \frac{1}{\sqrt{2}M} \left(\frac{S}{N}\right)_f, \tag{56}$$

reduced by the factor $K_1 = \sqrt{2}M$ in comparison with the original (S/N) in $f(\omega)$. Similar the second derivative $f''(\omega)$ has the signal to noise ratio

$$\left(\frac{S}{N}\right)_{f''} = \frac{f''(\omega_i)}{\delta f''(\omega_i)} = \frac{(f'(\omega_{i+1}) - f'(\omega_i))'}{\sqrt{\delta f'^2(\omega_{i+1}) + \delta f'^2(\omega_i)}} \approx \frac{1}{2M^2}\left(\frac{S}{N}\right)_f, \tag{57}$$

reduced by the factor $K_2 = 2M^2$.

In subsection 2.3 the four parameters formula (33) was promoted as the universal fitting tool. $M = 4$ means $K_2 = 32$. In another words, the normal state $\gamma_{opt}(\omega)$ has to have the original signal to noise ratio (S/N)=100, in order to define *four* points in its second derivative $\gamma''_{opt}(\omega)$ with accuracy of 30%.

As for K_1, this quantity grows reasonably. Of course, one can not measure the effective scattering rate $\gamma_{eff}(\omega)$ and use the formula (53), nevertheless the title question is worth-while. The point is that the first derivatives from the superconducting state optical scattering rate $\gamma'^s_{opt}(\omega)$ and the reflectivity $R'(\omega)$ reproduce the input spectral function $\alpha^2 F(\omega)$[2, 3, 20, 21]. The well structured spectral function of lead obtained from the analysis of the first derivative of $A_s(\omega) - A_n(\omega)$ looks convincing (at least for me) due to the reported value of S/N=6000.

Analysing Eq. (54) we arrive at the same conclusions as discussed above. The normal state optical conductivity provides us with restricted information similarly as other physical methods. The only way to extract the all required material parameters values is the analysis of different measurements in frame of the *same* model.

3. Superconducting state optical conductivity

3.1. THE INVERSION PROCEDURE

3.1.1. *Adaptivity, what does it mean?*
Let us consider from a general point of view the problems, which one has to take in mind, when one try to reconstruct a spectral function $\alpha^2 F(\omega)$ from experimental reflectivity data. This analysis should be performed in frame of a model which is based on an appropriate formalism. Thus, it means, that the results obtained are model dependent. The ideal, but rare situation occur, when the exact analytical solution of the inverse problem is available

$$\alpha^2 F(\omega) = f(R(\omega), \omega) \tag{58}$$

where f is a known function, which could depend on $\alpha^2 F(\omega)$. In the last case the numerical solution can be found by iterations.

But what should we do, if such a function f is not available by reason of the complexity of the direct formalism. In this case we shell use the numerical *adaptive strategy*, based on the substitution of equations of the direct formalism by the *adaptive* formula(s), suitable for the numerical solution of ill-posed problems. Since it is a strategy, let us consider how does the adaptivity work.

Generally speaking, one can use *any* adaptive formula. For example, one can consider the $\alpha^2 F(\omega)$ as an input function and the reflectivity as an output function for some formalism. Then one could write

$$R(\omega) = \int dz K(\omega - z)\alpha^2 F(z), \tag{59}$$

or in the Fourier space

$$\mathcal{F}(R(x)) = \mathcal{F}(K(x))\mathcal{F}(\alpha^2 F(x)), \tag{60}$$

where $\mathcal{F}(f(x))$ is the Fourier image of the function f and $K(x)$ is an for the time being unknown kernel. $K(x)$ could be determined as follows.

- At first, one has to substitute a *trial* spectral function $\alpha^2_{trial}F(\omega)$ into the equations of the main formalism (Eliashberg equations + Nam-Rainer formula) and obtains a *trial* reflectivity $R_{trial}(\omega)$.
- Next, one should substitute $\alpha^2_{trial}F(\omega)$ and the *calculated* $R_{trial}(\omega)$ into the adaptive formula (60) and solve Eq. (60) for $K_1(\omega)$. The subscript 1 means "the first iteration".
- Finally, one has to substitute the *now known* kernel $K_1(\omega)$ and the *experimental* $R_{exp}(\omega)$ into the same adaptive formula (60), and solve it for $\alpha_1^2 F(\omega)$.

So, on each iteration we use the model formalism ones and the adaptive formula twice. At first, one have to calculate $R_{k-1}(\omega)$ using $\alpha^2_{k-1}F(\omega)$, as input function. At second, one should find the adaptive function(s)

$$\mathcal{F}(K_k(x)) = \frac{\mathcal{F}(R_{k-1}(x))}{\mathcal{F}(\alpha^2_{k-1}F(x))}, \tag{61}$$

and finally calculate the spectral function

$$\mathcal{F}(\alpha_k^2 F(x)) = \frac{\mathcal{F}(R_{exp}(x))}{\mathcal{F}(K_k(x))}. \tag{62}$$

The presented "trick" is quite general. One could replace analogously, for example, $R(\omega)$ in (59-62) by the tunnelling density of states $N(\omega)$ or the ARPES spectral function $A(\omega)$, and obtain adaptive methods for the tunnelling and electronic spectroscopies respectively, which is expected can work, with any formalism.

As mentioned above, the experimental data are known only approximately, therefore the inverse problem is ill-posed. Up to now the solution of

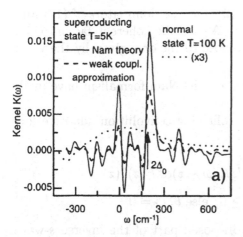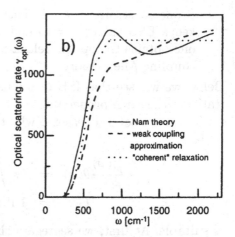

Figure 7. Panel (a) shows the frequency dependence of the superconducting state kernels $K(\omega)$ (solid and dashed lines) in comparison with the normal state one (dotted line). The source superconducting state optical scattering rates, calculated in the strong (solid line), and weak (dashed line) coupling approximations are shown in panel (b). The convolution of the input spectral function with the master peak (="coherent" relaxation) is drown in panel (b) by the dotted line.

ill-conditioned problems is the art more than the routine procedure. Since one can use any adaptive formula(s), it will be pragmatic for him/her/AI to choose it from the short (alas!) list of well investigated by numerical mathematics equations (see, for example, sec. 18.4-18.7 in [19] and references within).

On this understanding the discussed above convolution adaptive formula is one of the best possible selections, since the Fourier transformation is linear and unitary.

3.1.2. *The inversion method for the s-wave ISB reflectivity.*

The inverse problem for the superconducting state optical absorptivity of Pb was successfully solved by B. Farnworth and T. Timusk [3] within the weak coupling ISB model, based on the theory proposed by P.B. Allen [2]. The difference between weak coupling theoretical calculations and experimental data [20] was analysed in detail in [2]. It was pointed, that:

- The first derivative from the superconducting state optical scattering rate contains the component, proportional to the input spectral function $\alpha^2 F(\omega)$ due to the influence of coherent factors (11).
- The weight $\omega/2\Delta_0$ of this component is strongly underestimated.

- If one substitutes the solutions of (2) into the dominators of (10) and keeps BSC values $\tilde{\omega} = \omega$ and $\tilde{\Delta} = \Delta_0$ for the coherent factors (11), one obtains the optical relaxation, close to the one, given by the weak coupling Allen theory.

Below we will see that ISB model based on the Nam formalism have qualitatively the same properties.

There are several reasons, why the following convolution adaptive formula

$$\frac{\gamma_{opt}^s(\omega)}{d\omega}\theta(\omega) = 2\pi \int dz K(\omega - z)\alpha^2 F(z)\theta(z) \tag{63}$$

$$\theta(x) = 1 \text{ if } x \geq 0, \text{ else } \theta(x) = 0$$

is suitable. At first, we segregate the ill-posed part of the inverse s-wave problem (taking of the first derivative) into the separate step. At second, the low and high frequency values of $\gamma_{opt}^{\prime s}(\omega)$ coincide (=0), it is important for the FFT algorithm. At third, the obtained kernel $K(\omega)$ reminds us the instrumental resolution function of a badly adjusted spectrometer, allowing to profit by the applied spectroscopy experience. If one thinks in terms of the transfer functions, Laplace transformation adaptive formula could be more suitable.

The equation (63) was solved with respect to $K(\omega)$ using optical scattering rates *calculated* in strong (solid line, Fig. 7b) and weak (dashed line, Fig. 7b) coupling approximations and input "model 2" $\alpha^2 F(\omega)$. The results are plotted by the same type lines in Fig. 7a. No doubt, all features of the kernel are important for the recovery of the precise peak amplitudes and energy positions. Nevertheless, the general shape and main physics of $\gamma_{opt}^s(\omega)$ are defined by the *master* peak, located just above $\omega = 2\Delta_0$. The arrow, pointing in Fig. 7a the frequency position of $2\Delta_0$, is located within this peak. The master peak income is illustrated by dotted line in Fig. 7b. It was calculated as follows.

- All collateral maxima and minima of the shown by solid line $K(\omega)$ were rejected and artificial kernel $K_{coh}(\omega)$ was set to zero anywhere besides the master peak frequency region, where it coincides with $K(\omega)$.
- $K_{coh}(\omega)$ was convoluted with the input spectral function $\alpha^2 F(\omega)$ and integrated over ω
- Since the discussed phenomenon is governed by the coherence factors, the resulting "scattering rate" (dotted line, Fig. 7b) is labelled as *coherent relaxation*.
- The simple frequency dependence of the coherent relaxation, reminding us the properties of the quasiparticle effective scattering rate $\gamma_{eff}(\omega)$, allow to enunciate the Visual Accessibility procedure, described in forthcoming section.

The normal state kernel (multiplied on 3) is shown by the dotted line on Fig. 7a for comparison. The pessimistic conclusion, made in the previous chapter, can be couched in current terms: the (induced by the frequency and temperature averaging) width of $K(\omega)$ is too large to resolve the fine structure of the spectral function.

The domination of the master peak makes the choice of the trial $\alpha^2 F(\omega)$ on the first iteration easy. It can be the normalised $\gamma_{opt}^{',s}(\omega)$ or $R'_s(\omega)$. Note, that the master peak is asymmetric. The final frequency resolution is defined by the small value of its left half-width.

The self-consistent determination of the value of the superconducting gap is theoretically possible in frame of adaptive approach, but strongly not recommended. Following this way one transforms the weakly nonlinear task to the strongly nonlinear problem. For the case of ill-conditioned equations it is the wrong way. The value of Δ_0 has to be known a $priori$. The related issue is the difference between $\alpha^2 F(\omega)$ and $\alpha_{tr}^2 F(\omega)$. Substituted to the Eliashberg equations transport spectral function $\alpha_{tr}^2 F(\omega)$ lands us in difficulty, since the obtained value of the gap can be lower than the known Δ_0. In my opinion, for the sake of simplicity in this case one might sacrifice the unknown (from the first principle calculations) Coulomb pseudopotential and use μ^* as a simple fitting parameter.

In conclusion, the *adaptive method of the exact numerical inversion* of the s-wave superconducting state Nam equations is presented. In my opinion, one day the FIR spectroscopy will provide not an alternative to the tunnelling, but a powerful independent tool for the determination of $\alpha^2 F(\omega)$ in s-wave superconductors.

3.2. RULES OF VISUAL ACCESSIBILITY PROCEDURE

The frequency dependence of the superconducting state ISB reflectivity $R_s(\omega)$ in the local limit has the distinctive forms. This features are determined by the physical properties of the optical scattering rate $\gamma_{opt}^s(\omega)$. In this section we consider our simplified approach (Visual Accessibility) which allow us to analyse "by eye" the experimental reflectivity curve and get qualitatively the same results as the sophisticated method, described in the previous section, able to obtain quantitatively.

3.2.1. *Optical scattering rate and reflectivity*
As far as the physics is determined by the $\gamma_{opt}(\omega)$, but the measured quantity is $R(\omega)$, let us derive a simple relation between them. In view of a high value of the permeability $\epsilon(\omega) \gg 1$ one can expand the formula

$$R(\omega) = \left| \frac{1 - \sqrt{\epsilon(\omega)}}{1 + \sqrt{\epsilon(\omega)}} \right|^2 \qquad (64)$$

in powers of $1/\sqrt{\epsilon(\omega)}$. In view of Eq. (22), substituting $1 + \lambda$ instead of m_{opt}^*/m_b, one arrives the following approximate relation

$$R(\omega) \approx 1 - \frac{2\gamma_{opt}(\omega)}{\omega_{pl}\sqrt{1+\lambda}} \tag{65}$$

valid at $\omega \leq \Omega_{max} + 2\Delta$ both in the normal ($\Delta=0$) and superconducting ($\Delta=\Delta_0$) states, except for small frequencies $\omega < \gamma_{opt}(0)/(1 + \lambda)$, where Hagen-Rubens approximation is suitable.

Fig. 8a shows the frequency dependence of the superconducting state reflectivity $R_s(\omega)$ (T=10 K, solid line) in comparison with the normal state reflectivity $R_n(\omega)$ (T=100K, dashed line). Both curves have been calculated using the "model 2" spectral function. The frequency dependencies of the optical scattering rates $\gamma_{opt}^s(\omega)$ and $\gamma_{opt}^n(\omega)$ are shown in the inset. One observes that $R(\omega)$ looks like the mirror image of the appropriate $\gamma_{opt}(\omega)$, reproducing all important features.

3.2.2. The visual accessibility criteria for s-wave superconductors

Let us summarise our knowledge about the s-wave superconducting state optical relaxation rate and the local (London) limit reflectivity (see Fig. 8). The following features are visually accessible.

- The doubled gap value $2\Delta_0$ coincides with the absorption edge. If the $\omega < 2\Delta_0$, the reflectivity R(ω)=1 and $\gamma_{opt}^s(\omega)$=0.
- The impurity scattering gives a sharp decrease of the reflectivity just above $2\Delta_0$. The magnitude of the "jump" is proportional to γ_{imp}. $\gamma_{opt}^s(\omega)$ jumps approaching γ_{imp} from above.
- In superconducting state the coherent relaxation dominates. Hence, the first derivatives from $\gamma_{opt}^{\prime s}(\omega)$ and from R(ω) reproduce the shape of the spectral function (see Fig. 8b).
- Since the frequency dependence of the coherent relaxation rate grows faster, than the frequency averaged normal state one, and is fulfilled its maximum value at the frequency $\omega = 2\Delta_0 + \Omega_{max}$, at this point the ratio of the $\gamma_{opt}^s(\omega)/\gamma_{opt}^n(\omega)$ has a maximum, while the reflectivity ratio $R_s(\omega)/R_n(\omega)$ exhibits a minimum.

3.3. VISUAL ACCESSIBILITY AND SEPARABLE D-WAVE MODEL

In this section we consider briefly, how the procedure described in section 3.2.2 should be corrected in order to make the main features of the d-wave ISB reflectivity visually accessible.

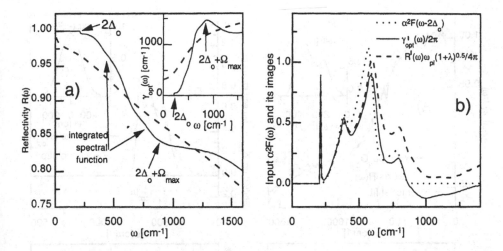

Figure 8. Panel (a) shows the frequency dependence of the superconducting state reflectivity $R_s(\omega)$ (T=10 K, solid line) in comparison with the normal state reflectivity $R_n(\omega)$ (T=100K, dashed line). Both curves were calculated using "model 2" spectral function. The frequency dependencies of the optical scattering rates $\gamma_{opt}^s(\omega)$ and $\gamma_{opt}^n(\omega)$ are shown in the inset. Panel (b) shows the frequency dependencies of the first derivatives from the optical scattering rate $\gamma_{opt}'^s(\omega)/2\pi$ and from the normalised reflectivity $R'(\omega)\omega_{pl}(1+\lambda_0)^{0.5}/4\pi$ in comparison with the shifted on $2\Delta_0$ input spectral function $\alpha^2 F(\omega - 2\Delta)$. One can see, that due to domination of the coherent relaxation in the frequency region between $2\Delta_0$ and $2\Delta_0 + \Omega_{max}$, $R'(\omega)\omega_{pl}(1+\lambda_0)^{0.5}/4\pi$ and $\gamma_{opt}'^s(\omega)/2\pi$ reproduces well the peaks of the spectral function. The important features are indicated in panel (a). One can see, that the absorption onset at $2\Delta_0$, the frequency dependence of the coherent relaxation between $2\Delta_0$ and $2\Delta_0 + \Omega_{max}$, and its upper energy bound at $2\Delta_0 + \Omega_{max}$ are visually accessible. The term "integrated spectral function" means, that the first derivative over ω from this part of the spectrum reproduces $\alpha^2 F(\omega - 2\Delta_0)$.

3.3.1. *Separable d-wave model*

Formally speaking, the frequency dependencies of the isotropic and the anisotropic parts of the spectral function are not obliged to coincide. At the same time, if the shapes differ substantially, the density of states (67) looks strange. So, for the sake of tunnelling, the model spectral function was chosen in the simple form

$$\alpha^2 F(\omega, \varphi) = \alpha_0^2 F(\omega)(1 + 2g\cos(2\varphi)\cos(\varphi')) \tag{66}$$

If one substitutes the function (66) into the $s + id$ ISB Eliashberg equations, one finds that at $g > 1$ the d-wave component entirely dominates. Hence, instead of the sophisticated $s + id$ approach, we can use for our analysis the pure d-wave model described in detail in [31]. For the sake of simplicity we restrict ourselves here to the consideration of the clean limit $\gamma_{imp} = 0$

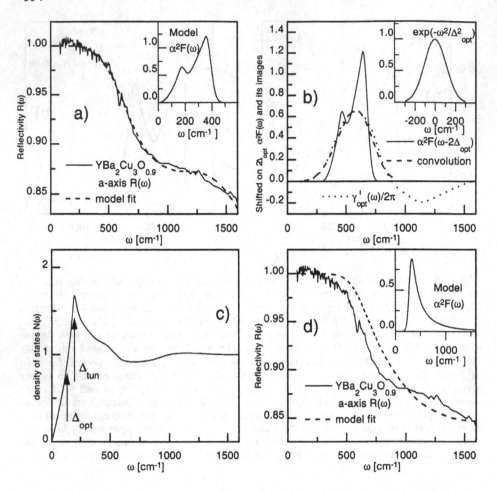

Figure 9. Panels (a) shows the frequency dependency of the reflectivity $R_s(\omega)$ (dashed lines) calculated within the separable d-wave model in comparison with the experimental data (solid line, Schutzmann *et al Phys. Rev.* **B46** 512 (1992)). The spectral functions $\alpha_0 F(\omega)$ are shown in the insets. The coupling constants were chosen to obtain the value of the tunnelling gap Δ_{tun}=200 cm^{-1}. The values of the plasma frequencies ω_{pl} and the low-frequency values of the non-metallic part of dielectric permeability ϵ_∞ were obtained by the least-square fit. Panel (b) shows the frequency dependence of the first derivative from the calculated optical scattering rate $\gamma'_{opt}(\omega)$ (dotted line) in comparison with the shifted on $2\Delta_{opt}$ input spectral function $\alpha_0^2 F(\omega - 2\Delta_{opt})$(solid line). The convolution of the $\alpha_0^2 F(\omega - 2\Delta_{opt})$ with the Gauss distribution function (shown in inset (b)) is plotted by the dashed line. Panel (c) shows the frequency dependence of the calculated tunnelling density of states $N(\omega)$, the frequency positions of the tunnelling gap Δ_{tun} and the optical gap Δ_{opt}. Panel (d) shows the frequency dependencies of the reflectivity $R_s(\omega)$ (dashed lines), calculated using caudate spectral function (inset), in comparison with the experimental data.

and set the Coulomb pseudopotential $\mu^*(\varphi, \varphi')$ to zero. The value of the parameter g was chosen little a bit larger than unity, since the normal state properties (defined in frame of *this* model by $\alpha_0^2 F(\omega)$) manifest the strong renormalization due to the phonon-electron interaction.

It is interesting, that it is possible after minor rehash to use the available s-wave computational programs for the strong coupling d-wave calculations. The clean limit version of the separable d-wave model keeps the convolution structure of the set (2). One has to substitute the angle averaged density of states

$$\left\langle \frac{\tilde\omega(y)}{\sqrt{\tilde\omega(y)^2 - \tilde\Delta^2(y)}} \right\rangle_\varphi = \frac{1}{\pi} \int_0^\pi \frac{\tilde\omega(y)d\varphi}{\sqrt{\tilde\omega^2(y) - 2\tilde\Delta^2(y)\cos^2(2\varphi)}} \tag{67}$$

instead of $N(\omega)$ (3) in (2), and the angle averaged density of pairs

$$\left\langle \frac{\tilde\Delta(y)}{\sqrt{\tilde\omega(y)^2 - \tilde\Delta^2(y)}} \right\rangle_\varphi = \frac{1}{\pi} \int_0^\pi \frac{2\tilde\Delta(y)\cos^2(2\varphi)d\varphi}{\sqrt{\tilde\omega^2(y) - 2\tilde\Delta^2(y)\cos^2(2\varphi)}} \tag{68}$$

instead of $D(\omega)$ (4) in (2). Similarly, the conductivity can be calculated using the Nam formula (10). For this purpose one should substitute $\Delta_d(x) = \sqrt{2}\Delta(x)\cos(2\varphi)$ instead of the $\Delta_s(\omega)$ into the element of integration in (10) and averages the obtained $\sigma(\omega, \varphi)$ over the angle φ.

Fig. 9a shows the model reflectivity $R_s(\omega)$, calculated in the frame of the separable d-wave model (dashed line) in comparison with the experimental data [5] (solid line). The spectral function $\alpha_0 F(\omega)$ is shown in the inset in Fig. 9(a), its coupling constant λ was chosen to reproduce the value of the *tunnelling* gap Δ_{tun}=200 cm^{-1} derived from the tunnelling spectra [32]. Note, that it was the same "model 2" [7] curve, used above, but without any high energy peak. The value of the plasma frequency ω_{pl} and low frequency value of the non-metallic part of dielectric permeability ϵ_∞ were obtained by the least-square fit.

Fig. 9b shows the first derivative from the d-wave optical scattering rate $\gamma'_{opt}(\omega)$ (dotted line) in comparison with the shifted on $2\Delta_{opt}$ input spectral function $\alpha_0^2 F(\omega - 2\Delta_{opt})$ (solid line). Here we define the *optical gap* $\Delta_{opt} \approx \Delta_{tun}/\sqrt{2}$ as the physical quantity, with defines the shift of the spectral function $\alpha_0^2 F(\omega)$ in FIR and tunnelling spectroscopy. The distinction between Δ_{opt} and Δ_{tun} arises due to the angle averaging $(< ... >_\varphi)$ both in the Eliashberg equations and in the conductivity calculation. The tunnelling gap Δ_{tun} (see Fig. 9c) is the maximum value of the d-wave gap

$$\Delta^d(\varphi) = \sqrt{2}\Delta\cos(2\varphi) \tag{69}$$

and the optical one is its mean square angle average. Note, that the dip in the density of states $N(\omega)$ (see Fig. 9c) arises at $\Delta_{opt} + \Omega_{max}$ and the electron-boson region starts in $N(\omega)$ at $\Delta_{opt} < \Delta_{tun}$. It means, that the low-energy excitations (if they contribute to the spectral function) will degrade the sharp peak in the tunnelling density of states.

If one substitutes $\gamma'_{opt}(\omega)$ in Eq. 63, one finds that coherent relaxation still dominates, but the main peak is substantially wider and can be approximated by a Gaussian function $Qexp(-\omega^2/\Delta^2_{opt})$ ((Fig. 9b, inset). The function

$$C(\omega) = \alpha_0^2 F(\omega - 2\Delta_{opt}) \otimes \frac{1}{\sqrt{\pi}\Delta_{opt}} \exp\left(-\frac{\omega^2}{\Delta^2_{opt}}\right) \qquad (70)$$

is plotted by the dashed line in the Fig. 9b and agrees well with the $\gamma'_{opt}(\omega)/2\pi$. Here $\otimes$ means a convolution. Note, that the coincidence of the prefactor Q with the $1/\sqrt{\pi}\Delta_{opt}$ is come-by-chance.

In my opinion, the additional ill-posed step, the *real* deconvolution of the Gaussian can not be performed using the data with reasonable value of (S/N), since the spectral function in these compounds have half-width of the order of Δ_{opt}. On the other side, if one (by chance) will find that $\gamma'_{opt}(\omega)$ or $R'(\omega)$ are structured, it means, that the ISB separable d-wave model is not suitable for HTSC and have to be replaced.

Strictly speaking, the d-wave reflectivity is less than unity at any frequency. At the same time, there is an onset of the absorption at some ω_Δ, where $R(\omega)$ becomes visually different from unity. Since the convolution $C(\omega)$ is wider than the input spectral function $\alpha_0^2 F(\omega)$, the position of the onset $\omega_\Delta < 2\Delta_{opt}$. For the model spectral function shown in the inset in Fig. 9a $\omega_\Delta \approx 1.4\Delta_{opt} \approx \Delta_{tun}$. Generally speaking, the selected boson modes could be responsible for the d-wave pairing. In this case the position of the absorption onset will be between $\Omega_{min} + \Delta_{opt}$ and $\Omega_{min} + \Delta_{tun}$, where Ω_{min} is the low frequency bound of the electron-boson interaction function. Similarly, the point where the coherent relaxation rate reaches its maximum, will be shifted by Δ_{opt} towards the high frequencies in comparison with its position in the s-wave superconductors. As a result, the reflectivity ratio $R_s(\omega)/R_n(\omega)$ exhibits a minimum and the scattering rate ratio $\gamma^s_{opt}(\omega)/\gamma^n_{opt}(\omega)$ has a maximum approximately at $\Omega_{max} + 2\Delta_{tun}$.

3.3.2. The visual accessibility criteria for d-wave superconductors

In ISB s-wave metals at the same frequency $\omega = \Delta_0$ the density of states exhibits three important features. At first, in this point $N(\omega)$ has a maximum. At second, since the averaged over $\mathbf{k}$ value of Δ_0 coincides in this case with itself, the density of states at $\omega = \Delta_0$ changes most rapidly. At third, for $\omega < \Delta_0$, the value of $N(\omega)$ is equal to zero. In d-wave superconductors this structural feature has a different energy position. One could denote

- the position of the maximum in the density of states Δ_{tun} as the *tunnelling* gap;
- the value of the shift $2\Delta_{opt}$ of the image of the spectral function in the first derivative from the optical relaxation rate $\gamma'_{opt}(\omega)$ as the *optical* gap ,
- the frequency ω_Δ where $R_s(\omega)$ becomes visually distinct from unity as the onset of the absorption or as the *absorption* gap.

The approximate relations between the gaps are $\Delta_{tun} \approx \omega_\Delta \approx \sqrt{2}\Delta_{opt}$. Similar to the description of the visually accessible features made in subsection 3.2.2 for a s-wave metals, we can write the same criteria for the separable d-wave model. Note, that in comparison with the s-wave case, all presented below itemised statements are valid approximately only.

- The single gap value Δ_{tun} coincides with the absorption edge. If the $\omega < \Delta_{tun}$, the reflectivity $R(\omega) \approx 1$ and $\gamma^s_{opt}(\omega) \approx 0$.
- The impurity scattering does not give the sharp decreasing of the reflectivity just above the Δ_{tun}. Instead, it decreases the gap itself.
- The coherent electron-boson scattering dominates in the superconducting state. The first derivatives from the $\gamma'^s_{opt}(\omega)$ and from the $R(\omega)$ reproduce the spectral function *convoluted* with the Gauss distribution function $exp(-\omega^2/\Delta^2_{opt})$ (see Fig. 9b).
- Since the frequency dependence of the coherent relaxation rate grows faster than the normal state one, and reaches its maximum value when the frequency $\omega \approx 2\Delta_{tun} + \Omega_{max}$, at this point the ratio of the $\gamma^s_{opt}(\omega)/\gamma^n_{opt}(\omega)$ will exhibit a maximum and reflectivities ratio $R_s(\omega)/R_n(\omega)$ will exhibit a minimum.

Finally, let us make the simple acceptance test for the presented above visual accessibility procedure. The test spectral function has a long history. At that old time, when people did not distinguish between the *effective* scattering rate $\gamma_{eff}(\omega)$ and the *optical* one $\gamma_{opt}(\omega)$, Collins et al [22] interpreted the high energy asymptotic behaviour of the $\gamma_{opt}(\omega)$ (see Fig. 2) in terms of its step-like effective counterpart $\gamma_{eff}(\omega)$ (5, 30). As the result, they arrived at the spectral function having a long tail up to the very high frequency (see Fig. 9d, inset). Nowadays, similar spectral functions occasionally arise in different models, since this shape suggests the spin fluctuations spectrum reported by neutron spectroscopy.

The calculations was performed in the same fashion as for model spectral function shown in Fig. 9a. The coupling constant was chosen to reproduce the value of the *tunnelling* gap $\Delta_{tun}=200$ cm^{-1}. The values of the plasma frequency ω_{pl} and low-frequency value of the non-metallic part of the dielectric permeability ϵ_∞ were obtained by the least-square fit. The peak position had the value 330 cm^{-1} which was close to the one in the original paper [22]. I choose the power law $1/\omega^{2.5}$ for the tail and the high energy cut-off 1600 cm^{-1}.

Let us compare the calculated curve and the experimental one. Since the low frequency bound in the model spectrum $\Omega_{min} \approx 200$ cm^{-1}, the absorption edge takes place at $\Omega_{min} + \Delta_{tun}$ in accord with the results discussed in [26]. At second, in the region between 400 cm^{-1} and 800 cm^{-1}, where the reflectivity "integrates" the spectral function, there is a systematic shift on 140 cm^{-1}. For the compensation of this shift one has to decrease the value of tunnelling gap by a. factor of two The most important discrepancy is connected with the upper frequency bound of the single-particle relaxation, that is, with the Ω_{max}. The reflectivity ratio R_s/R_n has a minimum at 1400 cm^{-1} in comparison with the 900 cm^{-1} in the experimental one.

In conclusion, the separable d-wave model reproduce all important features of the traditional ISB s-wave one with the single exception: the resulting images of the spectral function are the convolution of the input electron-boson interaction function and the Gauss distribution $\exp(-\omega^2/\Delta^2)$.

4. Acknowledgements

I express my deep gratitude to S.-L. Drechsler, O. V. Dolgov, E. G. Maksimov, A. Golubov, D. Rainer, M. Kulic, N. Yu. Boldyrev, V. M. Burlakov, and R. S. Gonnelli for numerous discussions, fruitful collaboration, and generous support of this work. I am deeply grateful to K.D. Schotte, K.F. Renk, E.A. Vinogradov, V.N. Agranovich, J. Fink, and V.L. Ginzburg for their stimulate decisive support during my work.

The author acknowledges gratefully for financial support of SFB 311 and 463.

References

1. Nam, S.B. (1967) Theory of electromagnetic properties of superconducting and normal systems, *Phys. Rev.* **156**, 470-493 ·
2. Allen, P.B. (1971) Electron-phonon effects in the infrared properties of metals, *Phys. Rev.* **B3**, 305-320
3. Farnworth, B. and Timusk, T. (1976) Phonon density of states of superconducting lead, *Phys. Rev.* **B14**, 5119-5120
4. Shulga, S.V., Drechsler, S.-L., Fuchs, G., Müller, K.-H., Winzer K., Heinecke, M., and Krug K. (1998) Upper critical field peculiarities of superconducting YNi$_2$B$_2$C and LuNi$_2$B$_2$C *Phys. Rev. Lett.*,**80** , 1730-1733
5. Schutzmann, J., Gorshunov, B., Renk, K.F., Munzel, J., Zibold, A., Geserich, H.P., Erb, A. and Muller-Vogt G. (1992) Far-infrared hopping conductivity in the CuO chains of a single-domain YBa$_2$Cu$_3$O$_{7-d}$ crystal, *Phys. Rev.* **B46**, 512-515
6. Sulewski, P.E., Sievers, A.J., Buhrman, R.A., Tarascon, J.M., Greene, L.H. and Curtin, W.A. (1987) Far-infrared measurement of $\alpha^2(\omega)F(\omega)$ in superconducting

359

$La_{1.84}Sr_{0.16}CuO_{4-y}$, *Phys. Rev.* **B35**, 5330-5333

7. Shulga, S.V., Dolgov, O.V., and Maksimov, E.G. (1991) Electronic states and optical spectra of HTSC with electron-phonon coupling, *Physica* **C178**, 266-274

8. Carbotte, J.P. (1990) Properties of boson-exchange superconductors, *Rev. Mod. Phys.* **62**, 1027-1158

9. Lee, W., Rainer, D. and Zimmerman, W. (1989) Holstein effect in the far-infrared conductivity of high T_c superconductors, *Physica* **C159**, 535-544

10. Dolgov, O.V., Golubov, A.A., and Shulga, S.V. (1990) Far infrared properties of high T_c superconductors, *Physics Letters* **A147**, 317-322

11. Allen, P.B. and Mitrovic, B. (1982) Theory of superconducting T_c, in *Solid State Physics* **37**, edited by Ehrenreich H., Zeitz F., and Turnbull, D., Academic publ., N.Y., 1-92

12. Abramowitz, M. and Stegun, I.A. (1970) *Handbook of Mathematical functions*, Dover publications, Inc., N.Y, 589-598

13. Marsiglio, F., Startseva, T. and Carbotte, J.P. (1998) Inversion of K_3C_60 reflectance data, *Phys. Lett.* **A245**, 172-176

14. Vidberg, H.J. and Sirene, J. (1977) Solving the Eliashberg equations by means of N-point Pade approximants, *J. Low Temp. Phys.* **29**, 179-192

15. Dolgov, O.V. and Shulga, S.V. (1995) Analysis of intermediate boson spectra from FIR data for HTSC and heavy fermion systems, *J. of Superconductivity* **8**, 611-612

16. Degiorgi, L., Wachter, P., Gruner, G., Huang, S.-M., Wiley, J. and Kaner, R.B. (1992) Optical response of the superconducting state of K_3C_{60} and Rb_3C_{60}, *Phys. Rev. Lett* **69**, 2987-2990

17. Degiorgi, L., Nicol, E.J., Klein, O., Gruner, G., Wachter, P., Huang, S.-M., Wiley, J. and Kaner, R.B. (1994) Optical properties of the alkali-metal-doped superconducting fullerenes: K_3C_{60} and Rb_3C_{60}, *Phys. Rev.* **B49**, 2787-2990

18. Mazin, I.I., Dolgov, O.V., Golubov, A. and Shulga, S.V. (1993) Strong-coupling effects in alkali-metal-doped C_{60}, *Phys. Rev.* **B47**, 538-541

19. Press, V.H., Teukolsky, S.A., Vetterling, W.T.,and Flannery, B.P. (1992) *Numerical recipes*, sec. 18.4-18.7, Cambridge uni. press, N.Y, 795-817

20. Joyce, R.R. and Richards, P.L. (1970) Phonon contribution to the far-infrared absorptivity of superconducting and normal lead, *Phys. Rev. Lett.* **24**, 1007-1011

21. Farnworth, B. and Timusk, T. (1974) Far-infrared measurement of the phonon density of states of superconducting lead, *Phys. Rev.* **B10**, 2799-2802

22. Collins, R.T., Schlesinger, Z., Holtzberg, F., Chaudhary, P. and Field, C. (1989) Reflectivity and conductivity of $YBa_2Cu_3O_7$, *Phys. Rev.* **B39**, 6571-6574

23. Sulewski, P.E., Sievers, A.J., Maple, M.B., Torikachvili, M.S., Smith, J.L. and Fisk, Z. (1988) Far-infrared absorptivity of UPt_3, *Phys. Rev.* **B38**, 5338-5352

24. Burlakov, V.M., Shulga, S.V., Keller, J. and Renk, K.F. (1992) Evidence for a Fano-type interaction of infrared active phonons with electronic excitations in $Tl_2Ba_2Ca_2Cu_3O_{10}$, *Physica* **C190**, 304-308

25. Basov, D.N, Liang, R., Dabrowski, B., Bonn, D.A., Hardy, W.N. and Timusk, T. (1996) Pseudogap and charge dynamics in CuO_2 planes in YBaCO, *Phys. Rev. Lett.* **77**, 4090-4093

26. Kamaras, K., Herr, S.L., Porter, C.D., Tache, N., Tanner, D.B., Etemad, S., Venkatesan, T., Chase, E., Inam, A., Wu, X.D., Hegde, M.S. and Dutta, B. (1991) In a clean high-T_c superconductor you do not see the gap, *Phys. Rev. Lett.* **64**, 84-87

27. Maksimov, E. G., Rashkeev, S. N., Savrasov, S. Yu., Uspenskii, Yu. A. (1989) Microscopic studies of the optical spectra of $YBa_2Cu_3O_7$ *Phys. Rev. L* **63**, 1880-1883

28. Dervenagas, P., Bullock, M., Zarestky, J., Canfield, P., Cho, B.K., Harmon, B., Goldman, A.I. and Stassis, C. (1995) Soft phonons in superconducting $LuNi_2B_2C$, *Phys. Rev.* **B52**, R9839-R9842

29. Kawano, H., Yoshizawa, H., Takeya, H., Kadowaki K. (1997) New phonon peak in superconducting state of YNi_2B_2C *Physica* **C282-287** 1055-1056

360

30. Kawano, H., Yoshizawa, H., Takeya, H., Kadowaki K.(1996) Anomalous photon scattering below Tc in YNi_2B_2C *Phys. Rev. Lett.* **77** 4628-4631

31. Schachinger, E., Carbotte, J.P. and Marsiglio, F. Effect of suppression of the inelastic scattering rate on the penetration depth and conductivity in a $d_{x^2-y^2}$ superconductor, *Phys. Rev.* **B56**, 2738-2750

32. Zhao, G.L., Browne, D. A. and Callaway, J. (1995) Quasiparticle spectrum in superconducting $YBa_2Cu_3O_7$, *Phys. Rev.* **B52**, 16217-16222

OPTICAL STUDY OF YBCO THIN FILMS: SPECTRAL ANOMALES OF LOW TEMPERATURE DYNAMICS OF STRONGLY CORRELATED CHARGES

I.YA.FUGOL, V.N.SAMOVAROV AND M.YU.LIBIN
B.Verkin Inst. for Low Temperature Physics & Engineering,
National Academy of Sciences, 310164 Kharkov, Ukraine
e-mail: fugol@ilt.kharkov.ua

Based on the low-temperature optical experiments (absorption and reflection) in visible and middle infrared frequency regions with epitaxial films of copper oxide materials like a $Y_1Ba_2Cu_3O_{6+x}$, we considered the anomalous spectral effects in HTSC having no analogy with conventional BCS superconductors: (1) the optical response to superconducting transition at T_c; (2) the spectral weight redistribution induced by doping and temperature; (3) the drastic enhancement of low-temperature photodoping; (4) the long spin-structure relaxation via temperature variations seen in the optical spectra. The thorough analysis of the results obtained is fully compatible with the concept of two-component system of light and heavy carriers (holes), being in dynamical coexistence. The dynamical coexistence of the intraband carriers occurs on the background of strongly correlation interrelation of the carriers with the interband (charge transfer) excitations.

1. Introduction

In recent years the optical spectroscopy of HTSC at frequencies much higher than the superconducting gap, $\hbar\omega \gg \Delta$, has been found to be informative for understanding the nature of high-T_c-superconductivity, in particular, for studying the anomaly of the normal-state properties of HTSC. With regards to room temperature (RT) optical experiments, the following distinct features of the HTSC-spectra, that are originally associated with the unusual electron structure, deserve mention: (i) the pronounced optical effect of strongly correlated electrons (holes) under doping, manifesting itself in spectral weight redistribution between the high-energy (interband) region of spectra and the low-energy (intraband) one; (ii) the declination of the optical conductivity from the Drude-like behavior (expected for free carri-

S.-L. Drechsler and T. Mishonov (eds.), High-T_c Superconductors and Related Materials, 361–374.
© 2001 *Kluwer Academic Publishers. Printed in the Netherlands.*

362

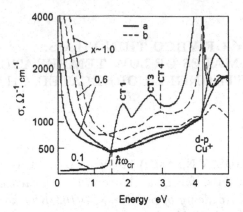

Figure 1. Optical conductivity of untwinned YBCO crystals for various composition x at RT [2]: – along *a*-axis; – – along *b*-axis.

ers) and the appearance of an additional band in the middle infrared region; (iii) the inherent potentiality of photodoping, which results in pumping of carriers under irradiation with light quanta at the frequencies of interband charge transfer trasitions above the optical gap $\hbar\omega > \hbar\omega_g \cong 1.6$ eV; (iv) the optical relaxation of the reflection coefficient in the course of aging and the oxygen ordering of HTSC samples after cooling from high temperature ($T \cong 450$ K) to RT.

The mentioned important results at RT are presented, for example, in [1-6]. All the above features were measured in optical range from 10^{-2} eV to 5 eV with different dopping regimes of $Y_1Ba_2Cu_3O_{6+x}$: the dielectric phase, $x < 0.3$; the strange metal phase at underdoping, $0.35 < x < 0.8$; the metal at optimal doping, $x \cong 0.9$; the overdoped metal with depressed superconductivity, $x > 0.95$. Because there are two different cuprate structures in the ab geometry of $Y_1Ba_2Cu_3O_{6+x}$ - CuO_2 (active plane) and CuO_x (chain structure along the *b*-axis), it is essential that the contributions from these planes to absorption and reflection should be separated in experiments with light polarization $E \parallel ab$. In this respect most intriguing are the data of polarization spectrum measurements on untwinned single crystals of YBaCuO for *a*- and *b*- axes [2, 6]. For example, the optical conductivity spectrum of $Y_1Ba_2Cu_3O_{6+x}$ for different doping at RT is shown in Fig. 1 [2]. Below energy of the charge transfer optical gap, $E_g \cong 1.6$ eV, separating the lower valent and the upper Hubbard bands there are intraband transitions. Above this energy, $E > E_g$, one can observe interband absorption bands with charge transfer from O^{2-} to Cu^{2+} - (CT), that are most prominent at low doping. Still higher in energy, at $E = 4.1$ eV there is a band which belongs to the transition (d-p) in the local center of Cu^+ in CuO_x. Fig. 1 clearly shows the evolution of optical conductivity spectra with doping from the dielectric to a superconducting state. The correlation effect in the spectra with doping is revealed in the form of integral spectral weight redistribution from the interband charge transfer (CT) transitions in the visible (VIS) region to the intraband transitions in the middle infrared one (MIR). As seen from Fig. 1 there is an crossover point ω_{cr} at which $\sigma_1(\omega)$ is constant with doping. In should be mentioned that in general the existence of the doping-independent crossover point ω_{cr} suggests that the optical gap in $Y_1Ba_2Cu_3O_{6+x}$ is not generated by the charge density wave (the spectrum of optical excitations associated with the charge density wave shifts dramatically to a red region as a whole on doping). The optical spectra of YBaCuO in the vicinity of intraband transitions have a Drude peak, centered at $\omega = 0$ with width of $1/T$ and additional spectral weight in the MIR region (maximum at 0.6 eV). It is the occurrence of additional absorption in MIR that results into difference the frequency dependence from the law $\sigma(\omega) \sim \omega^{-2}$ expected for the ordinary Fermi liquids. And that entailed, in one time, the need for development of new concepts of HTSC electronic spectrum (for instance, of "marginal" and "nesting" Fermi-liquid [7]. It should be emphasized that in BCS-superconductors no radically new optical effect were observed at low temperature (LT) as compared to RT in the region $\hbar\omega \gg \Delta$, besides, no optical response to the emergence of the SC-state was detected. In the case of HTSC the LT dynamics of charges displays new effects which are signaled on precursory processes to SC-ity.

This paper presents the experimental results on some nonordinary spectral properties of the normal phase of $Y_1Ba_2Cu_3O_{6+x}$ in VIS and MIR, revealed at different regimes of doping and temperature variations within the LT range below $T = 200$ K down to 20 K. In the main these are the data of investigations under the Program of the National Academy of Sciences of Ukraine. All optical measurements were carried out on the thin epitaxial films $\ell \cong 2000 \div 3000$ Å in thickness oriented along the ab-planes. The films were grown in German scientific centers (University of Erlangen; Munchen Technische University). The films were tested for their electrical, magnetic and structure characteristics. Some details of the LT optical measurements were described [8].

At present it can be said with confidence that the hole subsystem dynamics in HTSC materials is largely determined by the Fermi-liquid effects and the strong correlation interaction between intra- and interband transitions. However, up to now the correlation redistribution along spectra has been studied only for the doping effect at RT and has not been considered for other processes of charge dynamics. According to the data in [8] and the results presented here, the integral redistribution of spectral weight in HTSC also occurs in experiments with temperature variations, in photoinduced spectra, in experiments with provocation of spin-structure

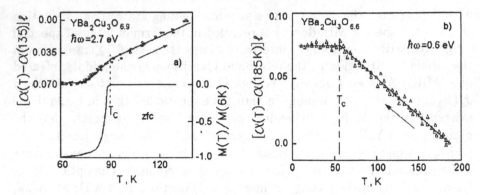

Figure 2. Temperature dependences of the differential absorption coefficient for VIS (a) and MIR (b) with different x (in (a) shown are magnetic measurements of the film for zfc regime also).

instability and so on. Therefore, the LT-spectral data should be analyzed from the general viewpoint. We would like to stress that all LT-features in the MIR and VIS spectra can be naturally examined in terms of the Mott-Hubbard model of 2D strongly correlated electrons with taking into account the intense spin fluctuations at LT.

2. Optical response to SC-transition at T_c and temperature dependences of spectral functions

One of the unexpected spectroscopic results in copper oxides was the finding of sensitivity to the onset of SC-ity for the transmitted electromagnetic field of optical frequencies $\hbar\omega \gg \Delta$. The optical response in absorption and reflection to T_c was detected both in VIS and MIR. It was shown that the response arose under optical transitions belonging to the conducting plane of CuO_2 [9-12]. The subsequent experiments confirmed the occurrence of this peculiar effect in YBaCuO and other HTSC in a wide range of doping. The T-dependence of the differential absorption coefficient at fixed frequency $\Delta\alpha_\omega(T) \cdot \ell = [\alpha_\omega(T) - \alpha_\omega(T_0)]\ell$ for the CT-maximum at 2.6 eV is shown in Fig. 2 *a* and that for the differential absorption coefficient for the MIR maximum at 0.6 eV is shown in Fig. 2 *b*. As it can be seen the T_c-point coincide closely with the kink of the functions $\alpha_\omega(T)$ and $R_\omega(T)$. Note that the optical response to the SC-ing transition in HTSC was also detected in luminescence spectra on oxygen F-centers at frequencies near 2-3 eV [13].

The temperature behavior of the optical spectra in YBaCuO were found to be uncommon. Below T_c one can observe a continuos freezing of the T-dependences. Particularly striking is the T-behavior of $\alpha_\omega(T)$ and $R_\omega(T)$ for the normal state at $T > T_c$: for the regimes of underdoping and opti-

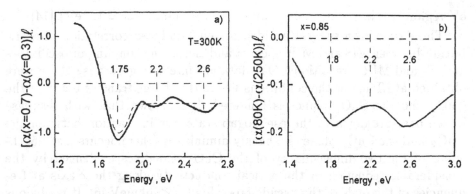

Figure 3. The differential absorption spectra YBCO in the CT-interband region of frequencies: a) - on doping from x=0.3 to x=0.7 at RT; b) - on cooling from $T = 250$K to $T = 80$K for x=0.85.

cal doping the T-dependences of the absorption and reflection coefficients have different sign in the VIS and MIR region: both $\alpha_\omega(T)$ and $R_\omega(T)$ are increasing functions of T in VIS and they are decreasing ones of T in MIR (Fig. 2 a, b). However, in the overdoping regime ($x = 0.95$) the CT-transitions in VIS undergo fundamental changes with temperature and the VIS spectral function (T) measured decreases with temperature like the MIR one [8]. It is like that, the correlation spectral weight redistribution is terminated in the overdoping regime where YBaCuO becomes an uncorrelated metal.

Take notice that the dependences $\alpha_\omega(T)$ and $R_\omega(T)$ have no selective character in frequency and the T-changes touch on the whole wide region in the VIS and MIR spectra. The differential absorption spectra $\Delta\alpha(\omega)\ell = \left(\alpha^f(\omega) - \alpha^i(\omega)\right)\ell$ (where f, i - mark the final and initial conditions of measurements) are shown in Fig. 3 a for doping and in Fig. 3 b for cooling from 250 to 80 K (for comparison). In both cases one can observe a decrease in the integral spectral weight of the CT-absorption in favour of the MIR-absorption (but the scale of the T-effect is much smaller than that of the doping one). Another remarkable common feature of the spectral evolution under doping and T-variations is the existence of the crossover point ω_{cr} positioning near E_g and dividing the VIS interband and the MIR intraband transitions: $\omega_{cr} \cong 1.55$ eV at doping and $\omega_{cr} \cong 1.4$ eV at T-variations.

The above data on sensitivity of spectral functions to T_c and on their different signs of the temperature dependences above and below the crossover point are in agreement with the results in [14]. The ratio of reflection coefficients for the superconducting and the normal phases near T_c was measured by applying the temperature modulation of HTSC samples. The response of the reflection spectra to the SC transition is prominent for the $1.0 - 2.5$

366

eV region within which there is an crossover point ($\hbar\omega \cong 1.4$ eV) [14].

The data shown suggest an idea of the T-induced correlation redistribution between VIS and MIR-spectra and the strong coupling of oscillators at CT- and MIR- transitions. Our data are interesting because these were obtained at LT under the conditions where the charge transfer between the CuO_2 and the CuO_x planes is hindered. It is known that with decrease in temperature down to the pseudogap state, the interaction between the CuO_2 and the CuO_x planes is sharply diminished (there occurs a strengthening of the two-dimensinality of the CuO_2 plane), as evidenced by the considerable reduction in the optical conductivity along the c-axis at frequencies of the scale of the pseudogap value $\hbar\omega < 40$ meV [6]. It is obvious that this process reflects the freezing of charge transfer between the different planes of the YBCO cell.

So, there exists an optical response of all the spectral functions at frequencies hundreds times higher than the superconducting gap which is opened under transition of HTSC systems to a superconducting state. It may be assumed that this nonconventional effect in HTSC is responsible for by a high ratio $(\Delta/\varepsilon_F)^2 \cong 10^{-2}$ which determines the effect scale and is hundreds times more than that in conventional BCS materials. The optical response mechanism was considered theoretically from different viewpoints in [15,16] and still remains debatable. The sensitivity of spectral functions to T_c occurs against the background of temperature-induced correlation mixture low- and high-frequency excitations.

3. Photodoping at Low Temperature

One of the important questions of the optical spectroscopy of HTSC is the question of photoirradiation critical dose, D_o, and threshold spectral energy, $\hbar\omega_{th}$, which determine permissible doses and energies of photoirradiation without persistent photo-induced effects in the lattice structure and in the current characteristics. It is known that a HTSC-material is sensitive to exposure dose with VIS light quanta in the region of CT-transition (with the threshold spectral energy $\hbar\omega_{th} > \hbar\omega_g$) and with the doses $D > D_o \cong 10^{19-20}$ ph/cm^2. Namely, using these conditions of photoirradiation at RT it becomes possible to observe an enhancement of T_c in HTSC-underdoped samples after photopumping [5,17]. The accepted explanation of the photoinduced conductivity is based on the assumption that the photoexcitation at RT causes the ordering of oxygen vacancies in the CuO_x chains that in ones turn leads to a subsequent transfer of electrons between the CuO_2 and the CuO_x planes and the emergence of excess holes in CuO_2 [17].

At the same time in our works it is found that the threshold dose of

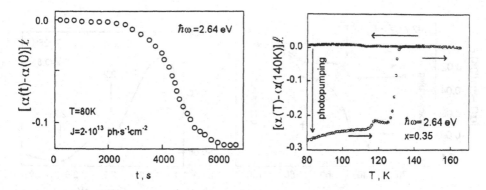

Figure 4. Dose dependence of the absorption of YBCO with x=0.35 under irradiation and at measurements with $\hbar\omega = 2.64$ eV. At doses $D = J \cdot t > 10^{17}$ ph·cm^{-2} the sharp decrease of VIS-absorption occurs due to photoinduced insulator-metal phase transition.

Figure 5. Temperature dependences of YBCO with x=0.35 before and after LT-photopumping with $\hbar\omega = 2.64$ eV, $D = 10^{17}$ ph·cm^{-2}.

photoinduced effects become less by about two orders of magnitude if photopumping is carried out at low temperatures, $D_o \cong 10^{17}$ phot/cm^2 [9]. The progressive increase of photodoping and the LT photoinduced phase transition of $Y_1Ba_2Cu_3O_{6.35}$ from dielectric to metal can be traced by the dose dependence of the absorption. In Fig. 4 shown are the changes in the absorption coefficient as compared to the initial moment of irradiation with a light of $J = 2 \cdot 10^{13}$ ph·s^{-1}·sm^{-2} at 80 K for $Y_1Ba_2Cu_3O_{6+x}$ (x=0.35). As is evident, there occurs a sharp passage of the dependence $\Delta\alpha_\omega(t)$ to saturation at $D = J \cdot t > D_0 \cong 10^{17}$ ph·cm^{-2}, suggesting the insulator-metal phase transition. In Fig. 5 the process of YBaCuO metallization under LT photopumping is seen from the temperature dependence of the absorption. The upper curve (cooling) corresponds to the dielectric state of YBaCuO with x=0.35 prior photopumping. As is evident, in this case the behavior of $\alpha(T)$ is temperature independent what is typical of the dielectric phase. After exposure of the sample to the light flux of $D = 10^{17}$ ph·cm^{-2} at 80 K and with subsequent increase in temperature, one can observe a T-dependence of absorption typical for HTSC metals (compare with Fig. 2 a). At $T = 135$ K a sharp reduction in the photopumped holes occurs due to their recombination with electrons so that the sample reverts to the initial insulator state. The temperature of the reduction in the photopumped carriers is dependent on doping of samples.

The effect of photodoping is also observed when pumping immediately into the superconducting phase. The experimental results for a two-phase sample with $T_{c_1} = 50$ K and $T_{c_2} = 75$ K are illustrated in Fig. 6. After the LT photopumping the sample is seen to become more transparent in

368

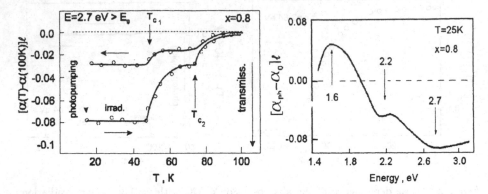

Figure 6. Temperature dependences YBCO with x=0.8 before (the upper curve) and after (the lower curve) LT-photopumping into SC-phase with $\hbar\omega = 2.7$ eV, $D \geq 10^{17}$ ph·cm^{-2}.

Figure 7. The differential absorption spectra of YBCO with x=0.8 in the CT-interband region of frequencies after LT photopumping.

VIS and with subsequent heating, it returns to the initial state. The sharp decrease in the nonequilibrium concentration of photopumped holes occurrs at T_c. We assume that the superconducting state is favourable to retention of photopumped holes. As is seen from Fig. 7, photopumping results in the correlation spectral weight redistribution between VIS and MIR, i.e. it acts like chemical doping. We can see a decrease in the integral absorption coefficient for $Y_1Ba_2Cu_3O_{6.8}$ after LT photopumping due to generation of nonequilibrium extra holes.

Our understanding of the process of persistent photodoping at LT is distinguished from the accepted ones at RT and is as follows. We think that the photoinduced holes appear in the valence band as a result of the charge-transfer excitations in the CuO_2 plane with $\hbar\omega > E_g = 1.6$ eV: $Cu^{2+}O^{2-} \stackrel{\hbar\omega}{\Longrightarrow} Cu^+O^-$. Photopumping will be persistent if some obstacles to a reversible process exist. We suggest that at LT-photodoping a new process arises which hampers the reversible recombination. This new process is considered in [18]. It is based on the concept of appropriate generation of low-temperature heavy quasiparticles - fluctuons. The physical concept of fluctuons was first introduced by I.Lifshitz in the physics of solids, and their phenomenological model was developed in [19]. In HTSC systems a hole fluctuon in the active CuO_2 plane is easily generated through hole trapping by spin fluctuations inherent in the pseudogap state at $T < T^*$. In the 2D case particularly deep fluctuon wells are generated with reduction in temperature, the well depth increasing as T^{-2}. It is the hole trapping by spin fluctuations that results in the retention and accumulation of photoinduced holes.

4. LT relaxation of YBCO samples seen by optics

Yet another low-temperature peculiarity in the behavior of spectral functions of HTSC at $T < T^*$ was observed with varying direction and velocity of T-scanning of the samples, namely, hysteresis effects in absorption and reflection. The hysteresis loop under cooling and heating of YBaCuO was mentioned in our former work for VIS absorption [12] and then subsequently the same effect was observed in measurements of MIR absorption and reflection. It is important to stress that the hysteresis loop in the MIR spectra is temperature inverse with that of the VIS spectral measurements. This means that the absorption (reflection) coefficient undergoes changes in different directions in VIS and MIR during relaxation. Generally speaking, the occurrence of hysteresis effects is associated with the existence of relaxation processes within a hysteresis loop temperature range.

The differential spectra of VIS-absorption of a sample with $x = 0.4$ (in the vicinity of dielectric-metal transition) measured for two relaxation times ($t_1 = 5700$ s and $t_2 = 7200$ s) after cooling from 220 to 88 K are plotted in Fig. 8. It can be seen that the absorption decreases through the whole VIS region i.e. the integral bleaching of the C'1-transitions occurs. The charateristic time of LT-relaxation could be estimated as $\tau \cong (1 \div 2) \cdot 10^3$ s. Time modifications in the absorption and reflection spectra are also observed in the MIR region after rapid cooling down to a certain temperature. These modifications occur with increasing both the integral MIR absorption and the MIR reflection, i.e. during relaxation the MIR and VIS optical characteristics undergo changes in different directions just as it occurs under chemical or photo-doping. At the same time, in the near UV region for $\hbar\omega = 4.1$ eV in which there are optical transitions (3d-4p) on local centers of Cu^+, one cannot observe relaxation effects in absorption at LT, i.e. the concentration of monovalent Cu^+ ions in the chain structure of CuO_x remains unchanged after temperature variations in the $T < T^*$ region (Fig.9).

In Fig. 10 presented are the relaxation curves for the underdoped metal film of $Y_1Ba_2Cu_3O_{6+x}$ with $x = 0.7$, $T_c = 70$ K. The data for different temperatures were obtained in sequential measurements upon cooling from 250 K. Differential absorption $\Delta\alpha\ell = \left(\alpha^f - \alpha^i\right)\ell$ vs T is plotted relative to the initial moment of temperature stabilization (T_i) of $\alpha^i(t = 0)$. After each previous case where the curve $\Delta\alpha(t)\ell$ reached saturation at a certain temperature, the temperature was quickly decreased to a next level and relaxation was measured once again and so on. On this slow cooling with reaching the equilibrium state (saturation) one can see that relaxation is absent below $T < T_c$.

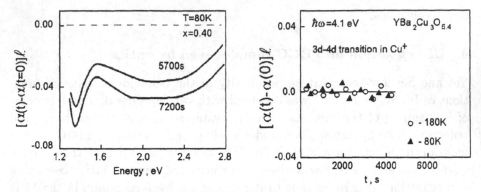

Figure 8. The differential spectra of YBCO with x=0.4 in CT-interband region of frequencies at different time of relaxation after fast cooling.

Figure 9. Time dependence of the differential asorption coefficient at $\hbar\omega = 4.1$ eV after fast cooling of YBCO from 220K up to 180K ($\bigcirc$) and up to 80K ($\triangle$)

It seems to be likely that the T-induced time instability revealed in VIS optical absorption of CT-transitions, which belong to the CuO_2 plane, reflects the inherent electron (spin)-structure relaxation in the active CuO_2 plane at LT-variations. In the course of this relaxation the spectral weight redistribution from the VIS to the MIR region occurs. If we accept this assumption two questions arise immediately: 1) what is the nature of the electron-structure relaxation at low temperatures? 2) what is the reason why the LT-relaxation is not so clearly observed in experiments on the d.c. resistivity?

Let us remember that the T-induced structure relaxation resulting in ortho-ordering and variations in the spectral functions of HTSC samples after fast cooling from high temperature ($400^\circ C$) down to RT are described in a series of works [20, 3]. The commonly accepted model of RT-relaxation of a order parameter is based on the mechanism of diffusion rearrangment of oxygen vacancies making the copper-oxygen chain (Cu-O-Cu) along the b-axis in CuO_x longer and following the charge transfer between the CuO_x and CuO_2 planes. One of the arguments in favour of this model was the evidence for the variations in the Cu^+ concentration (the variations observed in the optical reflection of the (3d-4d) transitions of the Cu^+ ions at 4.1 eV vs time) [3].

In contrast to the RT-relaxation of the spectral functions, the LT-relaxation of the CT-interband and MIR-intraband spectral weights is not accompanied by variations in the Cu^+ concentration (Fig. 9). This result is against the electron-structure rearrangement in CuO_x, where the Cu^+ ions reside, and the evidences that the LT-relaxation occurs exactly in the CuO_2 plane without any charge transfer between the planes (the interplane

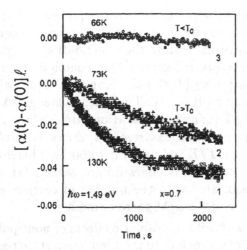

Figure 10. Time dependences of the differential absorption coefficient at $\hbar\omega = 1.49$eV in process of relaxation after cooling to different temperatures $T > T_c = 70$K and $T < T_c$: 1 - $T = 130$K; 2 - $T = 73$K; 3 - $T = 66$K.

charge transfer is likely to be hindered at LT). We assume that the most reliable explanation of the LT-relaxation of the spectral functions in VIS and MIR is the temperature-induced spin-structure rearrangement in the CuO_2 plane due to the formation of hole-spin fluctuons at LT. Strong influence of the system prehistory and its cooling rate on the LT state and long relaxation times in underdoped HTSC suggest that these are similar to irreversible effects in nonergodic systems with competing interactions. From this viewpoint, the transition to the pseudogap state at T^* may be considered as a transition to a nonergodic system where ordering regions alternate with disordering ones. In the fluctuon model the ordering occurs within a region round a hole trapped by spins. Note also that the concept of holes trapped in spin bag [21] as well as the concept of holes trapped in "string" potential created by AF-background [22], or the model of stripes, that is the subject of wide speculation [23] are not dramatically different from our pattern of hole-spin fluctuons.

Thus, we think it quite reasonable to associate the LT-instability of optical spectra and the effective photodoping at low temperatures with two processes - the hole trapping by spin fluctuations at low temperature and the induced spectral weight redistribution from VIS to MIR (due to strong correlation).

In parallel with the fluctuon domains where the trapped heavy holes reside, there exists a current network for light holes. The spectral weight of intraband transitions is distributed between the contributions from light and heavy carriers. Note that in the MIR region dominant is the contribution from heavy holes while in the far low-frequency region ($\omega \to 0$) the contribution from light holes is prevalent. The measurements of HTSC sample resistivity ($\omega \to 0$) reveal mainly the contribution of light carriers not related to the spin-structure relaxation. D.c. conductivity $\sigma(0)$ is defined by plasma frequency $\omega_p^2 \sim n_L(T,t)$ and damping $\Gamma(T,t)$ (where n_L is the number of light carriers): $\sigma(0) \sim n/\Gamma$. The relative changes $\Delta\sigma/\sigma$ in the experiments with varying T (or due to relaxation) are $\Delta\sigma/\sigma = \Delta n/n - \Delta\Gamma/\Gamma$. As is evident from the optical measurements, $\Delta n/n \sim \Delta\alpha/\alpha \cong 5\%$. At the same time $\Delta\Gamma/\Gamma \sim \Delta T/T$ and amount to 50 %. Therefore, unlike the optical measurements, the resistive ones do not sense relaxation and temperature variations associated with the number of carriers and shows the T-variations in damping of the light carriers mainly.

So, the optical data reveal the dramatic effects of nonequilibrium inherent in HTSC after their fast cooling to the pseudogap state region below T^*. The effects are most pronounced for underdoped samples and much weaker for the optimal doping regime. Generally speaking, the high nonequilibrium is resulted from the competitive interactions occurred in the HTSC system which bring two types of carriers (holes) with different spin surrounding into the dynamic coexistence: light (coherent) holes and heavy holes trapped in spin bag. As a result, the optical conductivity is determined by two contributions - a coherent part at the expense of Drude-like light carriers and a noncoherent one at the expense of heavy hole fluctuons. The optical relaxation observed arises from the self-consistent hole trapping by spin fluctuations and the subsequent spin ordering inside a fluctuon. In this case the ordering time is supposed to be $\tau \cong exp(a\sqrt{N})$ where $N \gg 1$, the number of spins in a fluctuon, may be rather high.

5. Conclusion

After the discovery of high-T_c superconductivity a great quantity of works were concentrated on the problem of adequate description of the behavior of electrons in anisotropic cuprates. For the analysis of the above optical results we used the consequences of the theoretical concepts developed in a series of theoretical works as a basis. These works are directly or indirectly connected with a broader Hubbard-like model of strongly correlated 2D electrons submerged in spin media: [22, 24-27]. One of the important general results of these theories is that the behavior of holes in the AF-background is responsible for by the competition between the superexchange energy lost near by a hole and its kinetic energy. Thus, the Hubbard-like model evidences that there is a dynamical coexistence of mobile (coherent) holes and heavy holes trapped in a spin bag. The impressive calculations of the temperature evolution of DOS in terms of the Hubbard model are carried out by [27]. Away from half-filling of LHB a sharp resonance peak appears near the chemical potential. As the doping is increased the resonance width also increases and it starts to merge with a lower Hubbard band. For the overdoping regime, both low-energy peaks are indistinguishable, implying that the system has become an uncorrelated usual metal. The narrow peak in DOS evolves like the MIR band in the optical spectra. The narrow peak appears to be strongly temperature dependent, thus it is pronouced only in the region of low temperature and within a regime from underdoping to optical doping.

Recent low-temperature experimental and theoretical works have made it clear that in HTSC materials there appears a state below $T < T^*$ which is precursory to superconductivity. The temperature T^* coincides with the temperature of the pseudogap opening with spin singlet ordering. This LT precursory region turns out to be optically active in the absorption and reflection spectra. Using the basic consequences of the theory of correlated electrons and the optical experimental findings, we propose the following principal ansatze:

1. The consistent concept of strongly correlated hole excitations interacting with spin fluctuations in the CuO_2-plane is attractive for understanding and interpreting the anomalous optical effects in HTSC at LT having no analogue with conventional BCS-superconductors (the sensitivity to T_c, the T-induced spectral weight redistribution, the LT-relaxation of optical functions, the LT-enhanced photodoping).

2. A nontrivial consequence of the correlation model in the intermediate regime, is the coexistence of light (Drude-like) and heavy (hole-spin-fluctuons) holes, the latter being generated in the LT region below the temperature of spin ordering (opening spin gap). Thus, the high-T_c- superconductive scenario is realized in dynamical two-component systems of carriers.

3. Subsystems of light and heavy carriers are in dynamical equilibrium upon doping or T-variations. If the doping of HTSC is increased the localization effects are weakened, while if the temperature is decreased, the localization effects become stronger and the density of heavy trapped holes increases. Thus, the temperature serves as fine tuning for high-T_c-superconductivity.

6. Acknowledgements

We are pleased to express our deep gratitude to Prof. G.Saemann-Ischenko and his colleaques for their useful scientific cooperation and preparation of HTSC samples for our optical measurements. We are grateful to Prof. H.Kinder and Dr. P.Berberich for providing the samples for the MIR-measurements. We thank Dr. S.Shulga for useful discussion and Dr. S.-

L.Drecshler for his interest in our work.

References

1. S.Uchida, I.Ido, H.Takagi, T.Arima, Y.Tokura, S.Tojima, *Phys. Rev.* **B43**, 7942 (1991).
2. S.L.Cooper, D.Reznik, A.Kotz, M.A.Karlow, R.Liu, M.V.Klein, W.C. Lee, *Phys. Rev.* **B47**, 8233 (1993).
3. J.Kircher, M.Cardona, A.Zibold, K.Widder, H.P.Geserich, J.Giapintzanis, D.M.Ginsberg, B.W.Veal, A.P.Paulikas, *Phys. Rev.* **B48**, 9684 (1993).
4. G.Yu, C.H.Lee, D.Mihailovic, A.J.Heeger, C.Fincher, N.Herron, E.M.McCarron, *Phys. Rev.* **B48**, 7545 (1993).
5. V.Kudinov, A.Kirilyuk, N.Kreines, R.Laiha, E.Lahderanta *Phys. Lett.* **A151**, 358 (1990).
6. A.V.Puchkov, D.N.Basov, T.J.Timusk, *J.Phys. Condeus. Matter* **8**, 10049 (1996).
7. C.M.Varma, *Int. J. Mod. Phys.* **B3**, 2083 (1989).
8. I.Fugol, V.Samovarov "Low Temperature Optical Spectroscopy of YBCO Thin Films" in "Studies of HTSC (Advances in Research and Applications)" **v. 22** Narlikar, ed. Nova Science Commack, N.Y. (1997).
9. I.Fugol, G.Saemann-Ischenko, V.Samovarov, Yu.Rybalko, V.Zhuravlev, Y.Strobel, B.Holzapfel, P.Berberich, *Solid. State Commun.* **80**, 201 (1991).
10. H.L.Dewing, E.K.H.Salje, *Supercond. Sci. Technol.* **5**, 50 (1992); *J. Solid State Chem.* **100**, 363 (1992).
11. C.H.Ruscher, M.Gotte, B.Schmidt, C.Quitmann, G.Guntherodt, *Physica* **C204**, 30 (1992).
12. I.Fugol, V.Samovarov, A.Ratner, V.Zhuravlev, G.Saemann-Ischenko, *Solid State Commun.* **86**, 385 (1993); *Physica* **C216**, 391 (1993).
13. I.Fugol, C.Politis, A.Ratner, V.Samovarov, V.Zhuravlev, *J. Lumines* **62**, 291 (1994).
14. H.J.Holcomb, C.L.Perry, J.P.Collman, W.A.Little, *Phys. Rev.* **B53**, 6734.
15. J.E.Hirsch, *Physica* **C199**, 305 (1992); **201**, 347 (1992).
16. A.S.Alexandrov, A.M.Bratkovshy, N.F.Mott, E.K.H.Salje, *Physica* **C215**, 359 (1993).
17. E.Osquiquil, M.Maenhoudt, B.Wayts, *Phys. Rev.* **B49**, 3675 (1994).
18. L.Kukuschkin, V.Samovarov, M.Libin, I.Fugol, *Low Temp. Phys.* **22**, 290 (1996).
19. M.Krivoglaz, *Uspechy Fis. Nauk* **111**, 617 (1973).
20. H.Shaked, J.Jorgensen, B.Hunter, R.Hittermen, A.Paulikas, B.Veal, *Phys. Rev.* **B51**, 547 (1995).
21. J.R.Schrieffer, X.-G.Wen, S.-C.Zhang, *Phys. Rev.* **B39**, 11663 (1989).
22. E.Dagotto, *Rev. Mod. Phys.* **66**, 763 (1994).
23. B.Levi, *Physics Today*, June, 19 (1998).
24. B.Stojkovic, D.Pines, *Phys. Rev.*, **B55**, 8575 (1997).
25. S.M.Quinlarr, P.J.Hirschfeld, D.J.Scalapino, *Phys. Rev.* **B53**, 8575 (1996).
26. A.Sherman, M.Schreiber, *Phys. Rev.* **B50**, 12887 (1994).
27. Th.Pruschke, M.Jarrell, J.K.Freerecks, *Adv. in Phys.* **44**, 187 (1995).

INTRODUCTION TO D-WAVE SUPERCONDUCTIVITY

H. WON
Department of Physics and IRC, Hallym University
Chunchon 200-702, South Korea

AND

K. MAKI AND E. PUCHKARYOV
Department of Physics and Astronomy,
University of Southern California,
Los Angeles, CA 90089-0484, USA

Abstract. After a brief survey on experiments supporting d-wave supercon-ductivity in the hole-doped high-T_c cuprates, we present the weak-coupling theory for d-wave superconductors. Remarkable effects of impurity scatter-ing in d-wave superconductors are also reviewed.

1. Introduction

Perhaps the most gratifying event after the discovery of high-T_c cuprate su-perconductors by Bednorz and Müller[1] is that d-wave symmetry is finally established in the hole-doped high-T_c cuprates [2, 3]. However, it is worth recalling that the superconductivity of the electron-doped superconductor $Nd_{2-x}Ce_xCuO_4$ appears to be of s-wave type[4, 5]. In the following we shall first recall briefly how the d-wave symmetry is established in the hole-doped high-T_c cuprate superconductors. The order parameter characterizing the d-wave symmetry is given by

$$\Delta(\vec{k}) = \Delta \cos(2\phi) \tag{1}$$

where ϕ is the angle the vector $\vec{k}$ (within the a-b plane) makes from the a-axis. The angular dependence of $\Delta(\vec{k})$ is shown in Fig. 1. In contrast to a variety of s-wave order parameters there are two remarkable characteris-tics. First, when ϕ crosses $\pi/4$, $3\pi/4$, $5\pi/4$, and $7\pi/4$, $\Delta(\vec{k})$ changes sign. So

375

S.-L. Drechsler and T. Mishonov (eds.), High-T$_C$ Superconductors and Related Materials, 375–386.
© *2001 Kluwer Academic Publishers. Printed in the Netherlands.*

moving around $\phi = 0$ to 2π, $\Delta(\vec{k})$ changes sign, 4 times. Second, associated with these crossings $\Delta(\vec{k})$ vanishes, $\Delta(\vec{k}) = 0$ in these 4 points. Or if we take the cylindrical Fermi surface, there are 4 nodal lines running parallel to k_z in $\Delta(\vec{k})$. Therefore, the experiments can test these two fundamental aspects. First, the presence of nodes in $\Delta(\vec{k})$ is seen by ARPES [6, 7], the T-linear dependence of the superfluid density[8], and the T^2 dependence of the low temperature part of the specific heat [9, 10], the electronic Raman scattering [11] and the quasi-particle density of states [12, 13] and the thermal conductivity in a magnetic field within the a-b plane [14, 15].

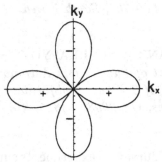

Fig.1 $\Delta(\vec{k})$ in the k_x-k_y plane is shown. It has four lobes extended in (1,0), (0,1), (-1,0), and (0,-1) directions.

Second, the phase sensitive experiments have been done using the Josephson junction between YBCO and a classic s-wave superconductor like Pb [16, 17] and Nb [18]. Also an ingenious tricrystal geometry formed by 3 pieces of crystals grown together gives rise to 1/2 of the flux quantum which is detected by a sensitive SQUID magnetometer. In this way Tsuei, Kirtley and their collaborators were able to test d-wave symmetry in YBCO, Tl2201, Bi2212, GdBCO, and LSCO [2, 19]. Perhaps these two sets of experiments appear to be more than adequate to confirm the presence of the d-wave symmetry.

However, the impurity effect (e.g. Zn-substitution of Cu in the Cu-O$_2$ plane) will provide very unique insight. In order to rule out any possible s-wave models (so called extended s-wave model) we have to just study the quasi-particle density of states in the presence of impurities. In superconducting order parameter with substantial s-wave component the energy gap will open up immediately [20] which has not been seen. We shall discuss the impurity effect later.

Perhaps you may ask what it all means. We believe that this means the superconductivity in the hole-doped high-T_c cuprates is not due to the electron-phonon interaction but due to the Coulomb interaction between electrons. In other words, the high-T_c cuprates is characterized by the Coulomb dominance or more popularly the strong electron correlation. In the celebrated Anderson dogmas I-III [21], Anderson postulated looking at the phase diagram of the high-T_c cuprates;

1) The principal action is taking place within the Cu-O$_2$ plane in all high-T_c cuprates. 2)The dominant interaction is the Coulomb interaction between holes. 3) In order to describe these features the simplest model will be the one band Hubbard model. We believe these three dogmas are valid, though we need a further elaboration on 3). It appears to us the Hubbard model has still a few problems but its descendent the t-J model should be more appropriate. First of all, the energy gap observed in the antiferramagnetic phase is not Hubbard gap but the charge transfer gap. Second, in the limit of the dilute hole concentration, the midgap states develop which is nicely interpreted by Zhang and Rice[22]. Therefore the basic Hamiltonian should describe these low energy states. Surprisingly it appears that the effective t-J model with $J/t \simeq 0.3$ provides a good starting model [23, 24, 25, 26]. We will not go into the remaining dogmas IV-VI since they are simply logically unsound and unacceptable. There is no reason to doubt the Fermi liquid theory in 2D system [27]

So what can we get from the dogmas I-III? I believe the simplest model will be the antiparamagnon model proposed by a few people [20, 29, 30] in the dawn of the d-wave model. In particular, the presence of the antiparamagnons is established by NMR and later by inelastic neutron scattering experiments [31]. But perhaps we shall not elaborate on this point.

2. Weak-coupling model

If we assume that the d-wave superconductor is rather similar to the s-wave superconductor, except that the interaction potential favors the d-wave symmetry due to its Coulombian origin, it is very natural to take the weak-coupling model á la BCS (Bardeen, Cooper, and Schrieffer [32]). Then the interaction potential is given by [33]

$$N_0 V(\vec{k}, \vec{k}') = 2\lambda \cos(2\phi) \cos(2\phi') - \mu \tag{2}$$

where the first term gives an attractive interaction while the second term the on-site Coulomb repulsion. Here N_0 is the quasi-particle density of states in the normal state on the Fermi surface. Also for a homogenous state (i.e. $\Delta(\vec{r}) = $ const.) the Coulomb term plays no role though in the inhomogenous situation like in the vortex state the Coulomb term has the secondary effect [34]. Here we shall summarize the result for the homogenous situation [33].

First, the gap equation is given by

$$1 = 2\lambda \int_0^{E_c} dE \langle \mathrm{Re}(\frac{\cos^2(2\phi)}{\sqrt{E^2 - \Delta^2 \cos^2(2\phi)}}) \rangle \tanh(\frac{E}{2T}) \tag{3}$$

where E_c is the cut-off energy and $< \ldots >$ means average over ϕ. Eq. (3) is rewritten as

$$-\ln(\frac{\Delta(T)}{\Delta_0}) = \frac{8}{\pi} \int_0^\infty dx \Phi(x)(1 + e^{\beta\Delta x})^{-1} \tag{4}$$

where

$$\Phi(x) = \langle \frac{\cos^2(2\phi)}{\sqrt{x^2 - \cos^2(2\phi)}} \rangle = \begin{cases} K(x) - E(x) & \text{for } x \leq 1 \\ x(E(1/x) - K(1/x)) & \text{for } x > 1 \end{cases} \tag{5}$$

and $K(x)$ and $E(x)$ are the complete elliptic integrals. Here we assumed $E_c/\Delta_0 \gg 1$ and $\Delta_0 = \Delta(0)$.

In particular, we have

$$T_c = 1.136 E_c e^{-\frac{1}{\lambda}}, \qquad \Delta_0 = 2.14 T_c \tag{6}$$

The first relation in Eq. (6) is identical to the one in an s-wave superconductor but Δ_0 is about 20% larger than the one in s-wave superconductor. Eq. (4) has the asymptotic forms

$$\Delta(T)/\Delta_0 = \begin{cases} 1 - 3\zeta(3)(\frac{T}{\Delta_0})^3 + \frac{135}{8}\zeta(5)(\frac{T}{\Delta_0})^5 & \text{for } T \ll T_c \\ \frac{2\pi T_c}{\Delta_0}(\frac{8}{21\zeta(3)})^{1/2}(1 - \frac{T}{T_c})^{1/2} & \text{for } T \simeq T_c \end{cases} \tag{7}$$

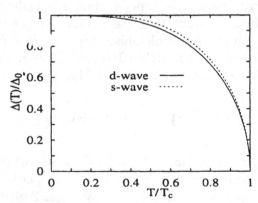

Fig.2 The order parameter $\Delta(T)$ is shown as a function of the temperature for d-wave and s-wave superconductors.

Making use of $\Delta(T)$, we can calculate the thermodynamic potential, specific heat, and the superfluid density $\rho_s(T)$ and the out-of-plane super-fluid density $\rho_{sc}(T)$. We will not go into these details, but we show $\Delta(T)$, $C_s(T)/\gamma_N$, $\rho_s(T)$, $\rho_{sc}(T)$ in Figs. 2, 3, and 4.

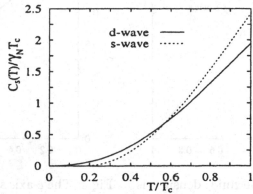

Fig.3 The specific heat is shown as a function of temperature for d-wave and s-wave superconductors.

Perhaps the low temperature asymptotic forms of some of these functions are useful.

$$C_s(T) \simeq \frac{27\zeta(3)}{\pi^2} \gamma_N \frac{T^2}{\Delta_0}, \qquad \text{for } T \ll T_c \tag{8}$$

$$\rho_s(T) = (\frac{\lambda(0)}{\lambda(T)})^2 = 1 - \chi_s/\chi_N$$

$$= 1 - 2(\ln 2)(\frac{T}{\Delta}) - \frac{9}{4}\zeta(3)(\frac{T}{\Delta})^3 - \frac{2025}{64}\zeta(5)(\frac{T}{\Delta})^5, \text{for } T \ll T_c \tag{9}$$

$$\rho_{sc}(T) = 1 - \frac{\pi^2}{6}(\frac{T}{\Delta})^2 - 3\zeta(3)(\frac{T}{\Delta})^3 - \frac{7\pi^4}{120}(\frac{T}{\Delta})^4, \quad \text{for } T \ll T_c \tag{10}$$

where $\rho_{sc}(T) = (\frac{\lambda_c(0)}{\lambda_c(T)})^2$. If we assume that the Fermi surface is basically cylinder and the quasi-particle is the Fermi liquid as expleained in the beginning, the c-axis superfluid denisty is given by the Josephson current across the barrier between conducting layers. There is no question of coherence or incoherence. The quasi-particle moves coherently as in usual tunneling junction. So we calculate $\rho_{sc}(T)$ by making use of the Ambegaokar-Baratoff formula [35] extended for the d-wave superconductor.

$$\rho_{sc}(T) = \frac{\pi^2 T}{2\Delta_0} \sum_n \langle \frac{\Delta^2 \cos^2 2\phi}{(\omega_n^2 + \Delta^2 \cos^2 2\phi)} \rangle$$

$$= \frac{\pi}{2\Delta_0} < \Delta \cos 2\phi \tanh(\frac{\Delta \cos 2\phi}{2T})) \tag{11}$$

380

Very recently $\rho_{sc}(T)$ YBCO is determined very precisely in [36]. $\rho_{sc}(T)$ in YBCO decreases like T^2 consistent with Eq.(10).

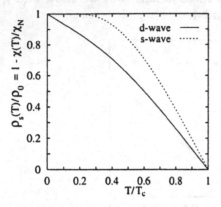

Fig.4 The superfluid density is shown as a function of temperature for d-wave and s-wave superconductors.

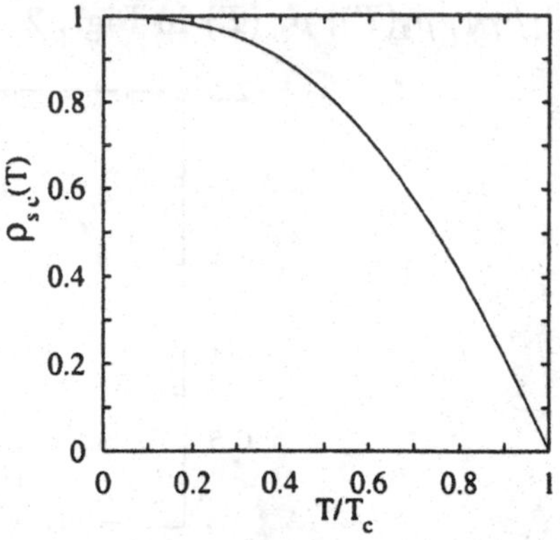

Fig.5 The c-axis superfluid density is shown as a function of temperature.

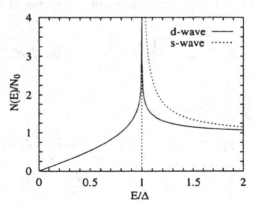

Fig.6 The density of states is shown as a function of E/Δ for $d$wave and s-wave superconductors.

Finally, the quasi-particle density of states is given by

$$N(E)/N_0 = \langle \mathrm{Re}\frac{E}{\sqrt{E^2 - \Delta^2 \cos^2(2\phi)}}\rangle$$
$$= = \begin{cases} \frac{2}{\pi} x K(x) & \text{for } x \leq 1 \\ \frac{2}{\pi} K(\frac{1}{x}) & \text{for } x > 1 \end{cases} \quad (12)$$

where $x = E/\Delta$, which is shown in Fig. 6.

As already mentioned briefly in the introduction, these predictions are, if not quantitatively, established semi-quantitatively by a variety of experiments.

3. Impurity Scattering

It is now well established that Zn-substitution of Cu in the Cu-O$_2$ plane of the hole-doped high-T_c cuprate produces dramatic effects. First of all, the superconducting transition temperature is rapidly suppressed. Second, the residual density of states (i.e. the quasi-particle density states at $E = 0$) increases very rapidly. These features are very well described in a model with impurity scattering in the unitarity limit [37, 38, 39, 40, 41, 42]. Here we shall summarize the result in [41, 42]. At the same time we contrast the result for a weak-scatter (i.e. the impurity scattering in the Born limit) and the one for a strong scatter (the unitarity limit) [43]. We suspect that Ni-substitution of Cu will provide a good chance to test the Born limit. Unfortunately, there are still very few experiments available for Ni-substitution.

The quasi-particle Green function in the presence of impurity scattering is given by

$$G^{-1}(\vec{k}, \omega) = \tilde{\omega} - \xi\rho_3 - \Delta(\vec{k})\rho_1 \tag{13}$$

where ρ_1 and ρ_3 are the Pauli spin matrices operating in the Nambu space, $\tilde{\omega}$ is the renormalized frequency determined from

$$\tilde{\omega} = \omega + i\Gamma \frac{\langle \frac{\tilde{\omega}}{\sqrt{\tilde{\omega}^2 - \Delta^2 \cos^2 2\phi}} \rangle}{\cot^2 \delta + (\langle \frac{\tilde{\omega}}{\sqrt{\tilde{\omega}^2 - \Delta^2 \cos^2 2\phi}} \rangle)^2} \tag{14}$$

where $\Gamma = n_i/\pi N_0$, n_i is the impurity concentration and N_0 is the electron density of states in the normal state on the Fermi surface per spin and δ is the scattering phase shift at $E = 0$. For a small δ (i.e. in the Born limit) Eq(13) reduces to

$$\tilde{\omega} = \omega + i\Gamma \sin^2 \delta \langle \frac{\tilde{\omega}}{\sqrt{\tilde{\omega}^2 - \Delta^2 \cos^2 2\phi}} \rangle \tag{15}$$

while for $\delta = \frac{\pi}{2}$ (the unitarity limit)

$$\tilde{\omega} = \omega + i\Gamma \langle \frac{\tilde{\omega}}{\sqrt{\tilde{\omega}^2 - \Delta^2 \cos^2 2\phi}} \rangle^{-1} \tag{16}$$

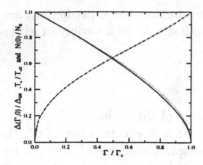

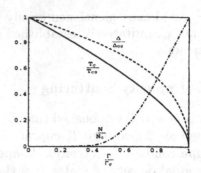

Fig.7 $\Delta(\Gamma, 0)/\Delta_{00}(\ldots)$, $T_c/T_{c0}(--$), and the residual density of states (-·-·-·), in the unitarity limit, are shown as functions of Γ/Γ_c where $\Gamma_c = 0.88T_{c0}$

Fig.8 $\Delta(\Gamma, 0)/\Delta_{00}$, T_c/T_{c0}, the residual density of states, in the Born limit, are shown as functions of Γ/Γ_c.

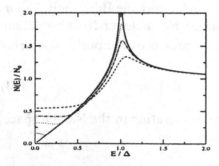

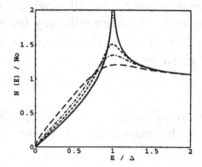

Fig.9 The quasi particle density of states, in the unitarity limit, for $\Gamma/\Delta = 0(—)$, $0.01(\ldots)$, $0.05(---)$, $0.1(-·-·-·)$, and $0.2(---)$ is shown as a function of E/Δ.

Fig.10 The quasi particle density of states, in the Born limit, is shown as as a function of E/Δ with $\Gamma/\Delta = 0, 0.01, 0.05, 0.1$, and 0.2 with the peak going down correspondingly.

We shall not go into the details of the results but summarize some salient features. In Fig. 7 and 8, we show T_c/T_{c0}, $\Delta(\Gamma, 0)/\Delta(0, 0)$, and $N(0)/N_0$ vs. Γ/Γ_c for the unitarity limit and the Born limit respectively where $\Gamma_c = (\frac{\pi}{2\gamma})T_{c0}$. Here T_{c0} and $\Delta(0, 0) = \Delta_{00}$ are the transition temperature and the superconducting order parameter at $T = 0$ K in the absence of impurities. We note that T_c/T_{c0} follows the Abrikosov-Gor'kov formula [44] in the both limits, while Γ-dependence of $\Delta(\Gamma, 0)$ and $N(0)$ are quite different. Perhaps these differences are more clearly seen from the quasi-particle density of states shown in Fig. 9 and 10, respectively. In the Born limit $N(0)/N_0 = 0$ practically until $\Gamma/\Gamma_c \geq 0.5$. In Fig. 11 and 12 are shown the superfluid density $\rho_s(\Gamma, 0)$ for the unitarity limit and the Born limit and Josephson current $J(\Gamma, 0)$ for the same limit respectively. As explained in 2, $J(\Gamma, 0) =$

$\rho_{sc}(\Gamma, 0)$ where $\rho_{sc}(\Gamma, 0)$ is the superfluid density parallel to the c-axis.

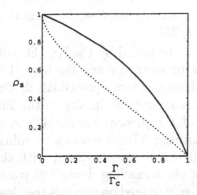

Fig.11 The superfluid density is shown as a function of Γ/Γ_c for the unitarity limit (...) and the Born limit (—).

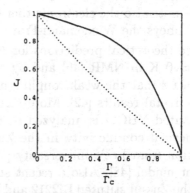

Fig.12 The Josephson current density is shown as a function of Γ/Γ_c for the unitarity limit (...) and the Born limit (—)

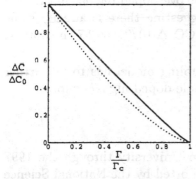

Fig.13 The jump in the in the specific heat is shown as a function of Γ/Γ_c for the unitarity limit (...) and the Born limit (—).

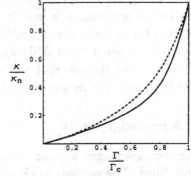

Fig.14 The low temperature electronic thermal conductivity κ normalized by the one in the normal state, κ_n is shown as a function of Γ/Γ_c for the unitarity limit (...) and the Born limit (—).

Finally, in Fig. 13 we show the jump in the specific heat for both limit and in Fig. 14 the thermal conductivity normalized by the one in the normal state $\kappa_n = \frac{2\pi^2}{3}\frac{n}{m\Gamma}T$ for both limit. The jump in the specific heat is given by

$$\frac{\Delta C}{\Delta C_0} = 14\zeta(3)\frac{T_c}{T_{c0}}(1-\rho\psi^{(1)}(\frac{1}{2}+\rho))^2[-\psi^{(2)}(\frac{1}{2}+\rho)\pm\frac{2}{9}\rho\psi^{(3)}(\frac{1}{2}+\rho)]^{-1} \quad (17)$$

where $\pm$ sign in the denominator refers to the Born limit and the unitarity limit respectively and $\rho = \Gamma/2\pi T_c$ and $\psi^n(z)$ are the poly-gamma functions.

We note also the electronic thermal conductivity for $T \ll T_c$ is linear in T and obeys the Wiedemann-Franz law [42, 45].

The theoretical predictions for $N(0)/N_0$ are tested by the Knight shift at $T = 0$ K in NMR [46] and by the low temperature specific heat [10]. It appears that the weak-coupling model describes semi-quantitatively the experimental results [42]. More recently the superfluid density in the Zn-substituted YBCO is analyzed in terms of the present model [47]. Also the thermal conductivity in the Zn-substituted YBCO exhibits a subtle deviation from Lee's universality [39], which is again consistent with the present model [48]. Also a recent study of the tunneling density of states from the Zn-substituted Bi2212 and the Ni-substituted one reveals the clear difference between 2 systems [49] again consistent with Ref.[43].

In summary, we present the prediction of the weak-coupling model for d-wave superconductors. Of course we don't expect the perfect agreement with experiment. Nevertheless, we discovered that the weak-coupling model works very well for LSCO in particular in the overdoped region. In the case of YBCO the agreement is within 30% suggesting there is already some strong coupling effect. For example for YBCO $\Delta(0)/T_c = 2.7$ in contrast to the weak-coupling ratio $\Delta(0)/T_c = 2.14$.

Therefore, we believe that the weak-coupling model has to be further studied to clarify the many aspects of the hole-doped high-T_c cuprates.

4. Acknowledgements

H. Won acknowledges support from Hallym University through the 1997 research fund. The present work is also supported by the National Science Foundation under grant number DMR95-31720.

References

1. Bednorz, J. G. and Müller, K. A. (1986) Z. Phys. B **64**, 189.
2. Tsuei, C. C. and Kirtley, J. R. (1997) Physica C **282-287**, 4.
3. Van Harlingen, D. J. (1997) Physica C **282-287**, 128.
4. Wu, P.H., Mao J., Mao, S.N., Peng, J.L., Xi, X.X., Venkatesan, T., Greene, R.L. and Anlage, S.M. (1993) Phys. Rev. Lett. **70**, 85-88; Anlage, S.M. Wu, D.H., Mao, J., Mao, S.N., Xi, X.X., Venkatesan, T., Peng, J.L., Greene, R.L., Phys. Rev. B **50**, 523.
5. Stadlober, B., Krug, G., Nemetshek, R. Hackl, R., Cobb, J.T., and Markert, J.T. (1995) Phys. Rev. Lett. **74**, 4911.
6. Shen, Z.-X., Dessau, D.S., Wells, B.O., King, D.M., Spicer, W.E., Arko, A.J., Marshall, D., Lombardo, L.W., Kapitulnik, A., Dickinson, P. Doniach, S. DiCarlo, J., Loesser, A.G., and Park, C.H. (1993) Phys. Rev. Lett. **70**, 1553.
7. Ding, H., Campuzano, J.C., Gofron, K., Gu, C., Liu, R., Veal, B.B., Jennings, G. (1994) Phys. Rev. B **50**, 1333.

8. Hardy, W.H., Bonn, D.A., Morgan, D.C., Liang, R., and Zhang, K. (1993) Phys. Rev. Lett. **70**, 3999.
9. Moler, K.A., Baar, D.J. Urbadr, J.S., Liang, R., Hardy, W.N., and Kapitulnik, A. (1994) Phys. Rev. Lett. **73**, 2744.
10. Momono, N., Ido, M. Nakano, T., Oda, M., Okajima, Y., and Yamaya, K. (1994) Physica C **233**, 395; Momono, N. and Ido, M. (1996) Physica C **264**, 311.
11. Devereaux, T.P., Einzel, D., Stadlober, B., Hackl, R., Leach, D.H. and Neumeier, J.J. (1994) Phys. Rev. Lett. **72**, 396.
12. Renner, Ch. and Fischer, Ø. (1995) Phys. Rev. B **51**, 9203.
13. Oda, M., Mañabe, C., and Ido, M. (1996) Phys. Rev. B **53**, 2253.
14. Yu, F., Salamon, M.B., Leggett, A.J., Lee, W.C., and Ginsberg, D.M. (1995) Phys. Rev. Lett. **77**, 3056.
15. Aubin, H., Behnia, K., and Ribault, M., Gagnon, R., and Taillefer, L. (1997) Phys. Rev. Lett. **78**, 2624.
16. Wollman, D.A., Van Harlingen, D.J., Lee, W.C., Ginsberg, D.M., and Leggett, A.J. (1993) Phys. Rev. Lett. **71**, 2134.
17. Wollman, D.A., Van Harlingen, D.J., Giapintzakis, J., Ginsberg, D.M. (1995) Phys. Rev. Lett. **74**, 797.
18. Brawner, D.A. and Ott, H.R. (1994) Phys. Rev. B **50**, 6530; (1996) Phys. Rev. B **53**, 8249.
19. Tsuei, C.C., Kirtley, J.R., Chi, C.C., Yu-Jahnes, L.S., Gupta, A., Shaw, T., Sun, J.Z., and Ketchen, M.B. (1994) Phys. Rev. Lett. **73**, 593-596; Kirtley, J.R., Tsuei, C.C., Sun, J.Z., Chi, C.C., Yu-Jahnes, L.S., Gupta, A., Rupp, M. and Ketchen, M.B. (1996) Nature **373**, 225.
20. Borkowski, L.S. and Hirschfeld, P.J. (1994) Phys. Rev. B **49**, 15404.
21. Anderson, P.W. (1997) "The theory of superconductivity in the high-T_c cuprates," (Princeton Univ. Press).
22. Zhang, F.C. and Rice, T.M. (1988) Phys. Rev. B **37**, 3754.
23. Eskes, H. and Sawatzky, G.A. (1988) Phys. Rev. Lett. **61**, 1415.
24. Matsukawa, H. and Fukuyama, H. (1989) J. Phys. Soc. Jpn. **58**, 2845; (1990) ibid **59**, 3687.
25. Hybersten, M.S., Stechel, F.B., Schluter, M. and Jennison, D.R. (1990) Phys. Rev. B **41**, 11068.
26. Tohyama, T. and Maekawa S. (1991) Physica C **185-186**, 1575.
27. Shankar, R. (1994) Rev. Mod. Phys. **66**, 129; Metzner, W. (1994) Physica B **197**, 457.
28. Bickers, N.E., Scalapino, D.J. and Scaletter, R.T. (1987) 687.
29. Monthoux, P., Balatsky, A.V., and Pines, D. (1991) Phys. Rev. Lett. **67**, 3448.
30. Moriya, T., Takahashi, Y. and Ueda, K. (1990) J. Phys. Soc. Jpn. **59**, 2905.
31. Rossat-Mignod, J., Regnault, L.P., Bourges, P., Burlet, P., Vettier, C. and Henry, J.Y. (1993) in "Selected Topics in Superconductivity" edited by Gupta, L.C. and Multani, M.S. (World Scientific, Sigapore) pp.265-346.
32. Bardeen, J., Cooper, L.N. and Schrieffer, J.R. (1957) Phys. Rev. **108**, 1175.
33. Won, H. and Maki, K. (1994) Phys. Rev. B **49**, 1397.
34. Won, H. and Maki, K. (1995) Europhys. Lett. **30**, 421; (1996) Phys. Rev. B **53**, 5927.
35. Ambegaokar, V. and Baratoff, A. (1963) Phys. Rev. Lett. **10**, 486; erratum **11** 104.
36. Hosseini, A., Kamel, S., Bonn, D.A., Liang, R., and Hardy, W. H. (1998) Phys. Rev. Lett. **81**, 1298.
37. Hotta, T. (1993) J. Phys. Soc. Jpn. **62**, 274.
38. Hirschfeld, P.J. and Goldenfeld, N. (1993) Phys. Rev. B **48**, 4219; Hirschfeld, P.J., Putikka, W.O. and Scalapino, D.J. (1993) Phys. Rev. Lett. **71**, 3705.
39. Lee, P.A. (1993) Phys. Rev. Lett. **71**, 1887.
40. Kim, H., Preosti, G. and Muziker, P. (1994) Phys. Rev. B **49**, 3544.
41. Sun, Y. and Maki, K. (1995) Phys. Rev. B **51**, 6059.

42. Sun, Y. and Maki, K. (1995) Europhys. Lett. **32**, 335.
43. Puchkaryov, E. and Maki, K. (1998) Europhys. J. B. **4** 191.
44. Abrikosov, A.A. and Gor'kov, L.P. (1961) Soviet Phys. JETP **12**, 1243.
45. Graf, M.J., Yip, S.K., Sauls, J.A. and Rainer, D. (1996) Phys. Rev. B **53**, 15147.
46. Ishida, K., Kitaoka, Y., Ogata, N., Kamino, T., Asayama, K., Cooper, J.R. and Athanassopoulou (1993) J. Phys. Soc. Jpn. **62**, 2803.
47. Bernhard, C. Tallon, J.L., Bucci, C., De Renzi, R., Guidi, G., Williams, G.V.M. and Niedermayer, Ch. (1996) Phys. Rev. Lett. **71**, 2304.
48. Taillefer, L., Lussier, B., Gagnon, R., Behnia, K. and Aubin, H. (1997) Phys. Rev. Lett. **79**, 483.
49. Hancotte, H., Deltour, R., Davydov, D.N., Jansen, A.G.M and Wyder, P. (1997) Phys. Rev. B **55**, R3410.

NONLOCAL ELECTRODYNAMICS AT SUPERCONDUCTING SURFACES

R. BLOSSEY
Laboratorium Voor Vaste Stoffysica en Magnetisme
Katholieke Universiteit Leuven
Celestijnenlaan 200D
B-3001 Leuven, Belgium

The following four lectures discuss properties of interfaces of superconductors with vacuum, metals and other superconductors, first from the viewpoint of superconductivity (Lecture 1 & 2) and then from the physics of phase transitions at surfaces or interfaces (Lecture 3 & 4).

1. Lecture: Nonlocal Electrodynamics of Superconductors

1.1. PIPPARD'S PHENOMENOLOGICAL APPROACH

The two basic *macroscopic* properties of a superconductor, its perfect DC-conductivity and the expulsion of a magnetic field (Meissner effect), are governed by the London equations (Landau, 1980-2; Tinkham, 1996)[1]

$$\mathbf{E} = \partial_t [\Lambda \mathbf{j}_s] \ , \tag{1}$$

$$\mathbf{B} = -c \nabla \times [\Lambda \mathbf{j}_s] \ , \tag{2}$$

where $\Lambda \equiv m/(n_s e^2)$ is a phenomenological parameter which contains the superconducting electron density n_s. According to eq.(1), an electric field *accelerates* superconducting electrons rather than helps them against the metallic resistance, as described by Ohm's law. From eq.(2), in combination with Maxwell's equation $\nabla \times \mathbf{B} = (4\pi/c)\mathbf{j}_s$, follows the equation

$$\nabla^2 \mathbf{B} = \lambda_L^{-2} \mathbf{B} \ . \tag{3}$$

Its solutions demonstrate that the magnetic field $\mathbf{B}$ decays *exponentially* from the surface of a superconductor into its bulk on a scale set by the

---

[1] This text mostly uses *cgs*-units, following in notation Landau-Lifshitz' book on Statistical Mechanics.

S.-L. Drechsler and T. Mishonov (eds.), High-T$_c$ Superconductors and Related Materials, 387–412.
© *2001 Kluwer Academic Publishers. Printed in the Netherlands.*

London penetration depth

$$\lambda_L = \sqrt{\frac{mc^2}{4\pi n_s e^2}}, \tag{4}$$

obviously a materials constant.

In Table I, the values of λ_L are compared to experimentally measured values of the penetration depth, λ (de Gennes, 1989). The discrepancy

TABLE 1. Penetration depths and coherence lengths of 'soft' superconductors (in Å)

	λ_L	λ	ξ_0	λ_{nl}
Al	157	≈ 500	1.6×10^4	530
Sn	355	510	2.3×10^3	560
Pb	370	390	830	480

between λ_L and λ is apparent. A possible source for the discrepancy may be that the local relationship between the supercurrent and vector potential, $\mathbf{j}_s \sim \mathbf{A}$, implicit in eqs.(1) and (2), is invalid for 'soft' superconductors. These are elemental metals which according to the London theory have a very short penetration depth (or, equivalently, a low value of the Ginzburg-Landau parameter $\kappa \equiv \lambda_L/\xi$, where ξ is the correlation length).

In order to cure the discrepancy between λ and λ_L, Pippard invoked an analogy between the behavior of superconductors and that of normal metals at high frequencies (Pippard, 1953). In the anomalous skin effect, Ohm's law has to be replaced by a *nonlocal* relation, in which the range of the electric field controlling the current is smeared out over the electronic mean free path l. Pippard thus introduced a similarly generalized current-vector potential relation

$$\mathbf{j}_s(\mathbf{r}) = -\frac{3c}{16\pi^2\lambda_L^2\xi_0} \int d\mathbf{r}' \frac{(\mathbf{r} - \mathbf{r}')[(\mathbf{r} - \mathbf{r}') \cdot \mathbf{A}(\mathbf{r}')]}{|\mathbf{r} - \mathbf{r}'|^4} \exp(-|\mathbf{r} - \mathbf{r}'|/\xi_0) \tag{5}$$

where ξ_0 is the so-called *coherence length*. It measures the 'size' of the superconducting wave function, and can be estimated by an elementary argument (Tinkham, 1996). The momentum range Δp of electrons participating in the superconducting state for $T \approx T_c$ is roughly given by $\Delta p \approx kT_c/v_F$, where T_c is the transition temperature, and v_F the Fermi velocity. By the uncertainty relation, Δp is associated with a spatial range

$\Delta x \geq \hbar/\Delta p \approx \hbar v_F/kT_c$. Thus, the relevant length for the coherence of the superconducting state is

$$\xi_0 \equiv a \frac{\hbar v_F}{kT_c} \tag{6}$$

with some numerical constant a. Pippard could fit the experimental data on Al and Sn with a single parameter, $a = 0.15$. The spectacular success of Pippard's "guess" became apparent, when it was shown that the nonlocal kernel calculated from BCS-theory quite closely resembles eq.(5), and that $a_{BCS} = 0.18$ (Mattis, Bardeen, 1958; Tinkham, 1996).

1.2. BCS THEORY

We now attempt to understand the origin of the nonlocality expressed in eq.(5) from the more microscopic viewpoint of BCS theory. The latter is an *effective* rather than a fundamental theory: it assumes a free electron gas with a *weak short-range attraction*, but does not explain the origin of this interaction. Its action functional reads (Kree, 1988)[2,3]

$$S[\bar{\psi}, \psi] = \int_{1;s} \bar{\psi}_s \left[\partial_\tau - \epsilon(i\nabla) \right] \psi_s + g \int_1 \bar{\psi}_\uparrow \bar{\psi}_\downarrow \psi_\downarrow \psi_\uparrow \equiv S_0 + S_{int}. \tag{7}$$

The first contribution in eq.(7) represents the kinetic energy of electrons with spin $\uparrow$ or $\downarrow$. The second is the attractive attraction with coupling strength g. Microscopically, it is caused by the electron-phonon interaction; the expression eq.(7) takes this into account in the simplest possible way.

Note that eq.(7) makes use of a very concise notation. $\bar{\psi}$, ψ are anticommuting (fermionic) Grassmann fields, with the overbar denoting complex conjugation. The advantage of these objects is that they allow the use of functional or path integral methods (Itzykson, Zuber, 1980). Second, $\int_{1;s} \equiv \int_0^\beta d\tau \int d^3x \sum_{\uparrow,\downarrow}$, where the integral $\int_0^\beta d\tau$ is the Matsubara finite temperature integral with $\beta = 1/T$. We will not go into the details of the properties of Grassmann fields here, as they will be eliminated from our calculations immediately (the interested reader should consult, e.g., (Itzykson, Zuber, 1980; Berezin, 1966)). However, we will soon come back to the use of the Matsubara technique.

We want to calculate the partition function

$$Z = \int \mathcal{D}[\bar{\psi}, \psi] \exp S \tag{8}$$

---

[2]In sections 1.2-1.4 $\hbar = 1$.

[3]The methods used here are not the most general, but suffice for the present purpose. A conceptually advanced approach is provided by the Eilenberger equations (Eilenberger, 1968). The author is indebted to D. Rainer for a discussion of this point.

which is, due to the nonlinearity of S, not immediately possible. We first use the Hubbard-Stratonovich transform (Itzykson, Zuber, 1980)

$$\exp\left[g\int_1 \bar{\psi}_\uparrow \bar{\psi}_\downarrow \psi_\downarrow \psi_\uparrow\right] = \int \bar{\mathcal{D}}[\Delta, \Delta^*] \exp\left[-\int_1 \frac{|\Delta|^2}{g} - \Delta^* \psi_\downarrow \psi_\uparrow - \Delta \bar{\psi}_\uparrow \bar{\psi}_\downarrow\right].$$

$$(9)$$

The use of this expression allows to reduce the argument of the exponential in eq.(8) to a quadratic form, which can then be completed by a G(r)aussian integration. The result is

$$Z = \int \mathcal{D}[\Delta^*, \Delta] \exp\left[-\int_1 \frac{|\Delta|^2}{g} + Tr\ln \mathcal{G}^{-1}(\Delta^*, \Delta)\right],\qquad (10)$$

where the relation $detA = exp\, Tr\ln A$ was used (Itzykson, Zuber, 1980). In eq.(10),

$$\mathcal{G}^{-1} = \begin{bmatrix} \partial_\tau - \epsilon(i\nabla) & \Delta \\ \Delta^* & \partial_\tau + \epsilon(-i\nabla) \end{bmatrix}.\qquad (11)$$

By adding a source term $S_h \equiv \int_1 (h\Delta^* + h^*\Delta)$ to the effective functional S_Δ (the exponential in eq.(10)), the saddle point contributions to Z can be found by a varying this functional with respect to Δ, Δ^*. For $\Delta = \Delta_0 = const.$, one finds in the limit $h \to 0$ the *gap equation*

$$-gT\sum_{\omega_n} \int \frac{d^3q}{(2\pi)^3} \frac{1}{[\omega_n^2 + \epsilon^2 + |\Delta_0|^2]} = 1\qquad (12)$$

from which the exponential dependence of Δ_0 on the coupling strength is obtained (de Gennes, 1989).

Intermezzo: Calculation of Matsubara frequency sums. In the following we will frequently encounter integrals of the type of eq.(12) so that it is useful to demonstrate how they can be evaluated. This is done in an elegant way by complex integration methods (Kree, 1988). Consider the contour integral

$$I \equiv -\oint_C \frac{dz}{2\pi} f(z) h(z)\qquad (13)$$

where $f(z) = [e^{\beta z} + 1]^{-1}$, $\beta = 1/T$. The poles of $f(z)$ lie on the imaginary axis at $z_n = (2n+1)i\pi T$, and their residue is

$$Res f(z)|_{z=z_n} = -T.\qquad (14)$$

Assuming that the poles of $h(z)$ are in the complex plane, as shown in Figure 1, one has from eq.(13)

$$I = iT \sum_{z_n} h(z_n) . \tag{15}$$

If $h(z)$ is regular with $z \cdot h \cdot f \rightarrow 0$ for $|z| \rightarrow \infty$, one can deform the

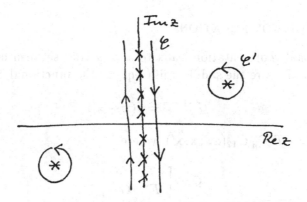

Figure 1. Pole structure of f, h and corresponding integration contours C and C'.

integration path C into C', and hence has

$$\sum_{z_n} h(z_n) = \frac{1}{T} \oint_{C'} \frac{dz}{2\pi i} f(z) h(z) . \tag{16}$$

Now we come back to our example, eq.(12). With the help of the above formulae one has

$$I = \sum_{z_n} \frac{1}{(z_n + \epsilon)(z_n - \epsilon)} = \frac{1}{T} \oint \frac{dz}{2\pi i} \frac{f(z)}{(z + \epsilon)(z - \epsilon)} \tag{17}$$

so that

$$I = -\frac{1}{2\epsilon T} [f(-\epsilon) - f(\epsilon)] = \frac{1}{2\epsilon T} tanh(\beta \epsilon / 2) . \tag{18}$$

The gap equation now reads:

$$\frac{g}{2} \int_q \frac{tanh(\beta E(q)/2)}{E(q)} = 1 \tag{19}$$

with $E(q) = \sqrt{\epsilon^2 + \Delta_0^2}$.

The nonanalytic, and hence nonperturbative, dependence of the gap function on the coupling strength g can now be determined. For that purpose, it is helpful to change from the integral over momentum space to an energy integral. The dominant momentum space region $|\mathbf{q} - \mathbf{q}_F| < q_0$ corresponds to the energy shell $|\epsilon| < \epsilon_0$. Approximating the density of states by $\mathcal{N}(\epsilon) \approx \mathcal{N}(0)$ in the integration range, the gap equation reduces to a one-dimensional integral. At $T = 0$, $tanh(1/T...) \to 1$, and the integral yields

$$|\Delta(T = 0)| \approx \exp -(1/g\mathcal{N}(0)). \tag{20}$$

1.3. THE GOR'KOV EQUATIONS

The functional representation leads to a very concise form of the Gor'kov equations, which are the saddle-point eqs. to the functional S_Δ,

$$\mathcal{G}^{-1} \cdot \mathcal{G} = \delta(\mathbf{x} - \mathbf{x}') \tag{21}$$

with $\Delta(\mathbf{x}) = g T \sum_n G_{12}(\omega_n; \mathbf{x}, \mathbf{x}')$, where

$$\mathcal{G} = \begin{bmatrix} G & F \\ F^* & G^* \end{bmatrix}. \tag{22}$$

We now want to solve these equations. Explicitly they read (assuming constant $\Delta = \Delta^*$):

$$\left[i\omega_n - \frac{\nabla^2}{2m} + \mu \right] G + \Delta F^* = \delta \tag{23}$$

$$\left[-i\omega_n - \frac{\nabla^2}{2m} + \mu \right] F^* - \Delta G = 0. \tag{24}$$

These equations are linear and hence can be solved by Fourier transform with the result

$$G = -\frac{i\omega_n + \eta_p}{\omega_n^2 + \eta_p^2 + \Delta^2}, \quad F^* = \frac{\Delta}{\omega_n^2 + \eta_p^2 + \Delta^2}, \tag{25}$$

where $\eta_p = p^2/2m - \mu$.

1.4. GINZBURG-LANDAU EQUATIONS

In order to obtain the familiar Ginzburg-Landau eqs. of the theory of superconductivity one has to assume that Δ, Δ^* are small so that one can expand (Gor'kov, 1959)

$$\mathcal{G}^{-1} = \begin{pmatrix} G_0^{-1} & 0 \\ 0 & (G_0^*)^{-1} \end{pmatrix} + \begin{pmatrix} 0 & \Delta^* \\ \Delta & 0 \end{pmatrix} = \mathcal{G}_0^{-1} + \hat{\Delta} \tag{26}$$

and hence has

$$G = G_0 \left[(1 - \hat{\Delta})G_0 + ...\right], \tag{27}$$

which is a series familiar from quantum mechanical perturbation theory, only that here matrix Green functions are involved. The saddle point eqs. now acquire the form

$$0 = \int d^3x' L(x, x')\Delta(x') + \tag{28}$$

$$\int d^3x' d^3x'' d^3x''' M(x, x', x'', x''')\Delta(x')\Delta^*(x'')\Delta(x''') + \mathcal{O}(\Delta^5)$$

where the kernels L, M result from a combination of Green functions, e.g.

$$L(x, x') = L(x - x') = -T \sum_{\omega_n} G_0(\omega_n; |x - x'|)G_0(\omega_n; |x - x'|) + \frac{1}{g}, \tag{29}$$

which can be evaluated in a similar way to eq.(12).

To finally obtain the GL-eqs, two steps still have to be taken: *i)* one additionally has to assume *small spatial variations* of the gap functions to allow for an expansion of the kernels in momentum q, *ii)* the inclusion of the magnetic field. The latter is usually done in a semi-classical approximation to the Green function (Gor'kov, 1959)

$$G_0(\omega_n; x, x', \mathbf{A}) \approx e^{i\phi(x,x')}G_0(\omega_n, x - x', \mathbf{A} = 0) \tag{30}$$

where

$$\phi(x, x') \approx \frac{e^*}{c}\mathbf{A}(x) \cdot (x - x') \tag{31}$$

so that

$$L(x, x'; \mathbf{A}) = e^{\frac{2ie^*}{c}\mathbf{A}(x)\cdot(x-x')} L(x - x'; \mathbf{A} = 0) \tag{32}$$

and an expansion in q results in the gauge-invariant derivative $(1/i)\nabla - (e^*/c)\mathbf{A}$.

Thus, we now see how nonlocality arises in BCS/GL-theory: if one assumes small Δ, but does *not* perform an expansion in q, the saddle-point equations are polynomial in the gap function, but each term carries a nonlocal integral kernel. If one does *not* assume that Δ, Δ' (and the c.c.) themselves are small, but rather

$$\left(\Delta(x) - V^{-1}\int \Delta(x)d^3x\right) \ll V^{-1}\int \Delta(x)d^3x \tag{33}$$

(and Δ^*, respectively), it is possible to derive a nonlocal generalization of the GL-equations in the form of an even polynomial in $|\Delta(x)|^2 \cdot |\Delta(x')|^2$

containing nonlocal kernels and a kinetic term of the form (Schattke, 1966)

$$\frac{1}{2m}\int d^3x \int d^3x' e^{-i\chi(x)}\widehat{Q}(x-x')e^{i\chi(x')} \times$$

$$\times \left[\left(\frac{1}{i}\nabla - \frac{e^*}{c}\mathbf{A}\right)\Delta(x)\right]\left[\left(-\frac{1}{i}\nabla' - \frac{e^*}{c}\mathbf{A}'\right)\Delta^*(x')\right]. \quad (34)$$

where χ is a phase factor. The full free energy functional from the expansion in eq.(33) has a very complicated form, and is clearly not easy to use in analytic calculations.

For special geometries, such as thin films, it is possible to simplify these expressions considerably, as the spatial dependence of the gap function can often be neglected (Wu, Lei, 1965; Thompson, Baratoff, 1968). In this case, the saddle-point equations in Ginzburg-Landau form read, with $v \propto \Delta$

$$\alpha + \beta|\psi|^2 + \int d^3x \int d^3x' \mathbf{A}(x)\widehat{Q}(x-x')\mathbf{A}(x') = 0 \quad (35)$$

and

$$j(x) = -|\psi|^2 \int d^3x' \widehat{Q}(x-x')\,\mathbf{A}(x'), \quad (36)$$

where the parameters α, β are determined by the behaviour of the kernels for small momenta. Thus, the phenomenological expression introduced by Pippard enters in suitably generalized Ginzburg-Landau equations.

2. Lecture: The nonlocal kernel; penetration depths

2.1. THE NONLOCAL KERNEL

We now want to calculate the nonlocal kernel Q explicitly. In slightly more general form it is given by eq.(36)

$$j_i(\mathbf{x}) \equiv -\int d\mathbf{x}' Q_{ik}(\mathbf{x}-\mathbf{x}'))A_k(\mathbf{x}') \quad (37)$$

Q_{ik} is a matrix kernel. Note that (for the moment) boldface is used to characterize the arguments as three component vectors.

As the current density is the first variation of the free energy with respect to $\mathbf{A}$, the kernel Q is the second variation, and therefore

$$Q_{ik}(\mathbf{x}-\mathbf{x}') = Q_{ki}(\mathbf{x}'-\mathbf{x}) \quad (38)$$

or, in momentum space, $Q_{ik}(\mathbf{k}) = Q_{ki}(-\mathbf{k})$. A further property of Q follows from the invariance of the current under the gauge transformation $\mathbf{A}(\mathbf{k}) \rightarrow$

$\mathbf{A}(\mathbf{k}) + i\mathbf{k}\,\chi(\mathbf{k})$. This invariance holds only if $Q_{ik}k_k = 0$, i.e. when the current is orthogonal to $\mathbf{k}$. If the crystal lattice is cubic,

$$Q_{ik}(\mathbf{k}) = \left(\delta_{ik} - \frac{k_i k_k}{k^2}\right) Q(\mathbf{k}). \tag{39}$$

In the transverse gauge $\nabla \cdot \mathbf{A} = 0 = \mathbf{k} \cdot \mathbf{A}$, so that $\mathbf{j}(\mathbf{k}) = -Q(\mathbf{k})\mathbf{A}(\mathbf{k})$, and only the scalar function $Q(\mathbf{k})$ needs to be determined.

To compute $Q(\mathbf{k})$, we now turn to the Gor'kov eqs., (21). The inverse Green function with field reads in explicit form:

$$\tag{40}$$

$$\mathcal{G}^{-1} = \begin{bmatrix} i\omega_n - \frac{1}{2m}\left[\frac{1}{i}\nabla - \frac{e^*}{c}\mathbf{A}\right]^2 + \mu & \Delta \\ \Delta^* & -i\omega_n - \frac{1}{2m}\left[\frac{1}{i}\nabla + \frac{e^*}{c}\mathbf{A}\right]^2 + \mu \end{bmatrix}$$

We now expand $G = G^0 + G^1$, and similarly F, where the index '0' indicates the solution to the Gor'kov equations without field we had found earlier. One finds for constant Δ up to first order in G^1 and $\mathbf{A}$ the equation

$$\left(i\omega_n + \frac{1}{2m}\nabla^2 + \mu\right) G^1 + \Delta F^{*1} = \frac{ie}{mc}\mathbf{A} \cdot \nabla G^{*0} \tag{41}$$

and analogous equations for F^{*1} and G^1. Here the gauge $\nabla \cdot \mathbf{A} = 0$ has been used, since otherwise derivative terms of the vector potential arise. Expressing $\mathbf{A}$ in Fourier components and defining $G^1 = g(\omega_n; x - x') \exp(ik(x - x')/2)$ (and again similarly for F^{*1} with a function f), the amplitude functions g, f fulfill a set of two algebraic equations.

In terms of the Green functions, the current $\mathbf{j}$ generally reads

$$\mathbf{j}(x) = \frac{ie^*}{m}\left[(\nabla' - \nabla)G(\tau, x; \tau', x')\right]_{x=x',\tau'=\tau+0} - \frac{(e^*)^2 n_s}{mc}\mathbf{A}(x). \tag{42}$$

Expressed in terms of the function g this is

$$\mathbf{j}_s(\mathbf{k}) = \frac{2eT}{m}\sum_{n=-\infty}^{\infty}\int\frac{d^3p}{(2\pi)^3}\left[\mathbf{p} \cdot g(\omega_n; \mathbf{p})\right] - \frac{n_s e^2}{mc}\mathbf{A}. \tag{43}$$

Inserting g, the final expression for Q is found (returning to cgs-units)

$$Q(\mathbf{k}) = \frac{3\pi T n_s e^2}{4mc}\sum_{n=-\infty}^{\infty}\int_{-1}^{+1}dx\frac{\Delta^2(1-x^2)}{[\omega_n^2 + \Delta^2 + (\hbar k v_F x/2)^2][\omega_n^2 + \Delta^2]^{1/2}}, \tag{44}$$

with $\omega_n = (2n+1)\pi T$. Eq.(44) is obtained after simplifying the momentum-space integration by averaging over the **p**-values orthogonal to **k**, which is possible since **j** and **A** are transverse, and by picking the dominant contribution of the integral in the vicinity of the Fermi surface (Landau, 1980-2). As a result, we find that $Q \to const.$ for $k \to 0$, while it decays as $Q \sim 1/k$ for $k \to \infty$ (see Figure 2).

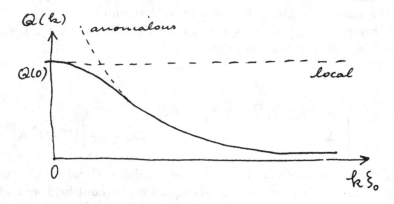

Figure 2. The nonlocal kernel in momentum space.

In a similar fashion also the *frequency* dependence of the electromagnetic kernel can be calculated (Abrikosov, Gor'kov, Khalatnikov, 1959). In general, $Q(\omega)$ is complex, with the imaginary part being reponsible for absorption. For small frequencies, $\omega \ll v_F k$, momentum dependences can be neglected. It is then found that

$$Re\,(Q(\omega) - Q(0)) \sim \omega^2 \quad , \quad Im\,Q(\omega) \sim -\omega \qquad (45)$$

where the proportionality factors differ in various regimes depending on temperature, the size of the gap Δ_0 and frequency. In some cases, they also bring in additional frequency dependencies (e.g., if the usually small 'electronic term' is retained aside to the 'pair term' $\sim \omega$ in the imaginary part of the kernel.)

2.2. PENETRATION DEPTHS

We will now use eq. (44) to calculate the penetration depth in a superconductor. Assuming the one-dimensional configuration in Figure 3, the current-vector potential relationship reads (Landau, 1980-2)

$$A''(x) = -\frac{4\pi}{c}j(x) \quad , \quad x > 0. \qquad (46)$$

Next, the boundary condition at the surface needs to be specified. The simplest case is given by *specular reflection* (Tinkham, 1996), for which

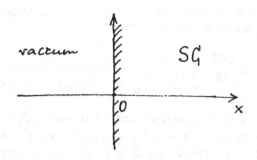

Figure 3. Geometry for the calculation of penetration depths.

$A(x) = A(-x)$ and the derivative of A has a jump discontinuity at the surface, $A'(+0) - A'(-0) = 2H$ with the applied field H. Integrating eq.(46) with this boundary condition, one obtains in k-space

$$A(k) = -\frac{2H}{k^2 + (4\pi/c)Q(k)}.$$ (47)

Defining the penetration depth by $\lambda \equiv \int_0^\infty dx\, B/H = -A(0)/H$ one has

$$\lambda = \frac{1}{\pi} \int_{-\infty}^{+\infty} dk (k^2 + (4\pi/c)Q(k))^{-1}.$$ (48)

For $k \to 0$, $\lambda = \lambda_L$, while in the *extreme anomalous limit* for $k \to \infty$, where $Q(k) \sim 1/k$, the penetration depth reads

$$\lambda \sim (\lambda_L^2 \xi_0)^{1/3}$$ (49)

which confirms the profound influence of the coherence length on the penetration depth.

We close this section by a remark on the penetration depth for a d-wave (HT) superconductor. This is a problem of considerable interest, as the penetration depth might be a sensitive probe of the nature of the order parameter in high T_c materials (Scalapino, 1995). The local theory predicts for an s-wave superconductor the temperature dependence $\lambda(T) - \lambda(0) \sim \exp -\Delta/kT$, while a local d-wave superconductor obeys $\lambda(T) - \lambda(0) \sim T$ because of the nodes in the gap. The latter temperature dependence was found in experiments with various high-T_c materials. At very low T, however, the linear law was found to cross over to a higher power law which may have different origins (e.g., impurities). For the penetration depth along the c-axis, i.e. orthogonal to the CuO-planes, nonlocality may be another source as pointed out recently (Kosztin, Leggett, 1997). In the d-wave case, the nonlocal kernel reads

$$Q(k) = \frac{2\pi T}{\lambda_L^2} \sum_{n=-\infty}^{\infty} \left\langle \hat{p}_{\parallel}^2 \frac{\Delta_p^2}{\sqrt{\omega_n^2 + \Delta_p^2(\omega_n^2 + \Delta_p^2 + \alpha^2)}} \right\rangle \qquad (50)$$

where $\Delta_p \equiv \Delta(\hat{p})$ is the anisotropic order parameter, $\hat{p}_{\parallel}$ is the projection of $\hat{p}$ on the boundary, $< .. >$ stands for averaging over the $2D$-Fermi surface, and $\alpha = (kv_F/2)\hat{k} \cdot \hat{p}$. The vector $\hat{k}$ is a unit vector perpendicular to the boundary in the direction of the applied magnetic field. If $\hat{k} \cdot \hat{p} = 0$, $\alpha = 0$, and the local expression holds. At $T = 0$, the limiting expression for $k \to 0$ is $Q \propto 1 - \pi^2\sqrt{2}k\xi_0/16$, while for $k \to \infty$

$$Q(k) \propto \frac{2}{3} \frac{\ln(k\xi_0)}{(k\xi_0)^2} \qquad (51)$$

where $\xi_0 = \hbar v_F/\Delta_0$, where $\Delta_0 \equiv max\{\Delta(\hat{p})\}$. The d-wave nonlocal kernel decays faster than the kernel of the s-wave superconductor. The resulting correction to the penetration depth is therefore less than 1 %, unobservable in experiment.

3. Lecture: Interface potential approach to 'wetting' in type-I superconductors

3.1. INTRODUCTION

In this and the following lecture we will encounter "nonlocality in superconductors" in the context of interfacial phase transitions. As J.O. Indekeu and J.M.J. van Leeuwen have shown, "wetting"-type phase transitions can occur in superconductors in a very similar fashion as in fluids (Indekeu, van Leeuwen, 1995). Consider the situation sketched in Figure 4. If the three interfacial tensions (surface free energies per unit area) between the wall W, the superconducting sheath SC and the normal region N fulfill the inequality

$$\gamma_{W,N} \leq \gamma_{W,SC} + \gamma_{SC,N} \qquad (52)$$

the wall will be "dry", i.e. neighbored by a normal region. If the equal holds in eq.(52), a macroscopic superconducting sheath will adjoin the wall. The transition from the first situation to the second is a "wetting" transition, and is very similar to the one encountered in liquids: the analogy becomes apparent if one just replaces W by S (*solid*), SC by L (*liquid*) and N by V (*vapor*). Eq.(52) is a simple consequence of the classic Young equation for the equilibrium of a liquid droplet on a substrate. (For more detail on the wetting transition in superconductors, see the lectures by J.O. Indekeu.)

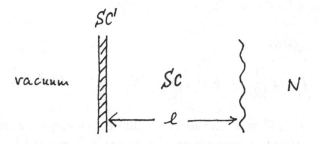

Figure 4. Geometry for wetting in superconductors.

We now define the surface free energy

$$\tilde{\gamma} \equiv \gamma_{W,SC} + \gamma_{SC,N} + V(\ell) \tag{53}$$

where the function $V(\ell)$ is called an *effective interface potential* of a superconducting sheath of thickness ℓ. If we suppose for a moment that $V(\ell) = 0$ for $\ell = \ell_0, \ell_\infty$, then according to the equality in eq.(52), $\tilde{\gamma} = \gamma_{W,N}$ for two different values of ℓ. In this case a *first-order transition* between a thin and a thick layer is encountered.

3.2. CALCULATION OF $V(\ell)$

The starting point for the calculation of $V(\ell)$ for wetting in superconductors is the Ginzburg-Landau functional used by Indekeu and van Leeuwen (Indekeu, van Leeuwen, 1995),

$$
\begin{aligned}
\gamma_{GL}[\psi, \mathbf{A}] \;=\; & \frac{\hbar^2}{2mb}|\psi(0)|^2 + \int d^2y \int_0^\infty dx \left[\frac{(\nabla \times \mathbf{A} - \mathbf{H})^2}{8\pi} \right. \\
& \left. + \;\; \alpha|\psi|^2 + \frac{\beta}{2}|\psi|^4 + \frac{1}{2m}|(\frac{\hbar}{i}\nabla - \frac{e^*}{c}\mathbf{A})\psi|^2 \right]
\end{aligned}
\tag{54}
$$

The first term in eq.(54) is the surface free energy of a superconducting wall. Its presence is a necessary requirement for the existence of a wetting transition in superconductors (Indekeu, van Leeuwen, 1995). The parameter b is the extrapolation length which governs the decay of the superconducting order parameter from the wall into the bulk. Supposing translational invariance in the directions parallel to the wall, ψ can be chosen as real, and $\mathbf{A}(x) = (0, A(x), 0)$ with $\mathbf{H} = He_z$. Then, with a number of suitable rescalings one finally obtains $\gamma \equiv \tilde{\gamma}\beta/(\alpha^2\xi)L^2$ as

$$\gamma[\phi, a] = \frac{\xi}{b}\phi^2(0) + \tag{55}$$
$$+ \int_0^\infty dx \left[\frac{1}{2}(1 - \phi^2)^2 + \dot\phi^2 + (H_0^2 - \frac{1}{2}) + \kappa(\kappa a^2\phi^2 + \kappa\dot a^2 - \dot a H_0)\right].$$

where $\kappa \equiv \lambda_L/\xi$, $\xi^2 \equiv \hbar^2/2m|\alpha|$. In the limit $\kappa \to 0$, γ loses its dependence on a, $\dot a$. The resulting decoupling of a and ϕ is illustrated in Figure 5. On the scale of the correlation length ξ, the field does not penetrate into the superconducting region anymore.

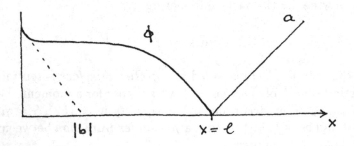

Figure 5.   Decoupling of ϕ and a.

The decoupling of ϕ and a has nice technical consequences, which we will now exploit (Blossey, Indekeu, 1996). To calculate the surface free energy by minimization, γ only needs to be varied with respect to ϕ only, leading to the equations

$$\delta\gamma/\delta\phi = 0 \tag{56}$$

$$\dot\phi(0) = \frac{\xi}{b}\phi(0). \tag{57}$$

The first integral of eq.(56) reads $\dot\phi^2 = \frac{1}{2}(1 - \phi^2)^2 + \delta$, where δ is an integration constant. Figure 6 shows the *phase portrait* $-\dot\phi(\phi)$ for eqs.(56,57). The boundary, or initial, value $\phi_1 \equiv \phi(0)$ is determined with the help of the boundary condition. The value of ϕ_1 at the transition follows from an equal-areas rule indicated by the shaded areas in Figure 6. This condition fixes the value of b, and hence of $\phi_1(\delta = 0)$. Note the jump discontinuity of $\dot\phi(\phi)$ at the origin: the profile $\phi(x)$ has a discontinuous first derivative at $x = \ell$ (see Figure 5).

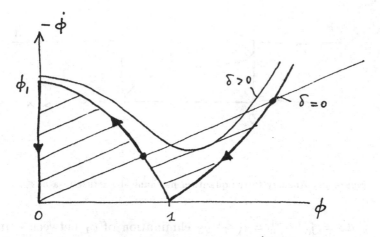

Figure 6. Phase portrait $-\dot\phi(\phi)$.

For the computation of $V(\ell)$ it is instructive to first consider the case of small ℓ. Small ℓ means small ϕ, so that one can use the quadratic part of the functional eq.(55)

$$\hat\gamma_{qu}[\phi] = -\phi(0)^2 + \int_0^\Lambda dy[t\phi^2 + \dot\phi^2 + H^2] \tag{58}$$

(where again some rescalings have been made: $H \equiv H_0|b|/\xi$, $\Lambda \equiv \ell\xi/|b|$, $|t| \equiv b^2/\xi^2$, $\sigma \equiv sgn(b)\xi/|b|$.) The variational eq. to $\hat\gamma_{qu}$ reads

$$\ddot\phi = t\phi, \tag{59}$$

which is accompanied by the boundary conditions $\dot\phi(0) = -\phi(0)$ and $\phi(\Lambda) = 0$. Eq.(59) for $t < 0$ $(T > T_c)$ has sines, and for $t > 0$ $(T < T_c)$ hyperbolic sines as solutions. The case $t = 0$ is degenerate: ϕ is linear in this case. By invoking the boundary conditions one obtains transcendental equations for the solution manifold. The determination of the interface potential for small ℓ thus becomes quite similar to the determination of the ground state energy of a quantum particle in a box, albeit for curious asymmetric boundary conditions, see Figure 7.

Finally, the determination of the free energy of the solutions shows that for $\ell < \ell_0$, the solution $\phi = 0$ is minimal, where ℓ_0 is given by (Blossey, Indekeu, 1996)

$$\ell_0/\xi = \begin{cases} tan^{-1}(|b|/\xi), & T \leq T_c \\ tanh^{-1}(|b|/\xi), & T \geq T_c \end{cases} \tag{60}$$

The interface potential is *linear* for small values of ℓ, $V(\ell) = H_0^2\ell/8\pi$. For $\ell > \ell_0$, a numerical calculation is required to obtain $V(\ell)$ from

$$V(\ell) = \gamma[\phi]|_{\phi(\ell)=0} \tag{61}$$

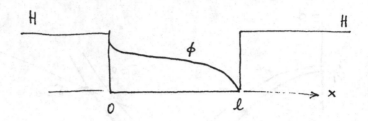

Figure 7. Analogy to the quantum mechanical particle-in-a-box.

and $\ell = \int_0^\ell dx = \int_0^{\phi_1} \dot\phi^{-1} = \ell(\phi_1)$ by elimination of ϕ_1 between ℓ and V. The nonextremal profiles of $\phi(x)$ are traced by variation of δ, as is indicated in Figure 6.

We conclude this section with a brief sketch of the analytic calculation of the asymptotic behaviour of $V(\ell)$ for $\ell \to \infty$. Using the saddle-point eqs., the expression for $\ell(\delta)$ reads

$$\ell = \int_0^{\phi_1(\delta)} d\phi[(\phi^2 - 1)^2/2 + \delta]^{-1/2}. \tag{62}$$

This integral has a singularity at $\phi = 1$ for $\delta \to 0$ which controls the behaviour of V for diverging ℓ. To determine this singularity, consider

$$\ell_{sing} \equiv \int_{1-\epsilon}^{1+\epsilon} d\phi[2(\phi - 1)^2 + \delta]^{-1/2} \tag{63}$$

where we put $(\phi^2 - 1) = (\phi - 1)(\phi + 1) \approx 2(\phi - 1)$ for $\phi \to 1$. Introducing $(\phi - 1)/\sqrt\delta = x$, we have

$$\ell_{sing} = \int_{-\epsilon/\sqrt\delta}^{\epsilon/\sqrt\delta} dx(2x^2 + 1)^{-1/2} = \frac{1}{\sqrt2} \ln\left\{ \frac{(1 + \delta/(2\epsilon^2))^{1/2} + 1}{(1 + \delta/(2\epsilon^2))^{1/2} - 1} \right\} \tag{64}$$

$$=_{\delta \to 0} -\frac{1}{\sqrt2} \ln\delta + const.$$

An analogous calculation for V according to eq.(61) yields $V(\delta) = c_0 + c_1\delta$ (with c_1, c_2 as constants) so that we finally obtain the result (Blossey, Indekeu, 1996)

$$V(\ell) \propto \exp{-\sqrt2\ell/\xi}. \tag{65}$$

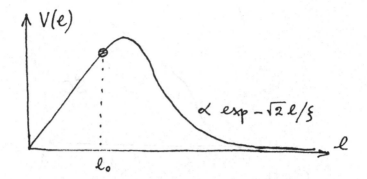

Figure 8. The effective interface potential $V(\ell)$ for superconductors with $\kappa = 0$.

The full effective interface potential is sketched in Figure 8. The double-well structure is visible, albeit with a *boundary minimum* at $\ell = 0$. The value of ℓ_0, beyond which $\phi \neq 0$, is indicated as a 'quantum dot'.

In liquids, $V(\ell)$ decays for large ℓ as (de Gennes, 1981)

$$V(\ell) \sim \frac{A}{\ell^2} \tag{66}$$

or as

$$V(\ell) \sim \frac{A^*}{\ell^3} \tag{67}$$

which is caused by microscopic long ranged dispersion (van der Waals) forces. Eq.(66) typically holds for film thicknesses $\ell < 800$ Å. For $\ell > 800$ Å, retardation effects cause a crossover to eq.(67). The origin of the dispersion forces are the fluctuations of the electromagnetic field; this will occupy us in Lecture 4. Here, we only note that eqs. (66), (67) are hallmarks of *nonlocal* behaviour: the presence of long range forces with their algebraic decay shows that the interactions in the system are *nonadditive*.

The section is closed by a remark on a possible use of the effective interface potential $V(\ell)$. E.g., the shape distortion of an SC/N-interface meeting the free surface at a finite (contact) angle θ can be calculated from the effective interface model (Blossey, Indekeu, 1996)

$$\mathcal{H}[\ell] = \int d^2y \left[(\gamma_{SC/N}/2)(\nabla\ell)^2 + V(\ell) \right]. \tag{68}$$

The result is shown in Figure 9. The applied field H is directed out of the paper plane. The interface has a characteristic macroscopic S-shape on the scale of the correlation length ξ and ends in a parabolic foot. This is caused by the boundary minimum structure of $V(\ell)$ at $\ell = 0$.

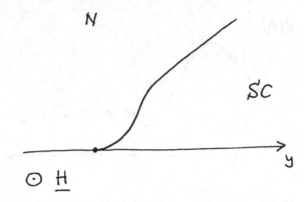

Figure 9.   Distortion of the SC/N-interface near the wall, in the partial wetting regime $(T < T_w)$.

3.3. ANALOGY WITH TYPE-II SUPERCONDUCTORS

We conclude this lecture by pointing out an interesting analogy of the wetting transition in superconductors with vortex depinning in type-II materials. If, in eq.(54) (or rather its scaled version) the formal limit $\kappa \gg 1$ is taken, γ_{GL} changes into the (somehow complementary) free energy functional (de Gennes, 1989; Tinkham, 1996)

$$\gamma_{GL}(\kappa \gg 1) = \frac{1}{8\pi} \int d^3x [\mathbf{B}^2 + \lambda_L^2 (\nabla \times \mathbf{B})^2] \tag{69}$$

where $\mathbf{A} = -\lambda_L^2 (\nabla \times \mathbf{B})$ has been used. Assuming a rotationally symmetric field configuration around a vortex located at $r = 0$, the variation of γ_{GL} with respect to $\mathbf{B}$ yields the vortex energy $\tau \equiv (\Phi_0/4\pi\lambda_L)^2 \ln \kappa$ where $\Phi_0 = \int \mathbf{A} \cdot ds$ is the flux enclosed in the vortex. A calculation of the vortex-vortex interaction energy is possible by the superposition of two vortex fields. It yields the repulsive interaction (Tinkham, 1996)

$$F_{12} = \frac{\Phi_0}{4\pi} H_1 = (\Phi_0/2\pi\lambda_L)^2 K_0(x_{12}/\lambda_L)/2 \sim e^{-|x_{12}|/\lambda_L}/\sqrt{|x_{12}|} \tag{70}$$

where K_0 is the Hankel function of order zero.

If one now, based on these results, considers a vortex at a distance x_L from a vacuum-superconductor interface, a vortex interaction potential $V(x_L)$ can be obtained by considering the interaction between the vortex and its image located at $x = -x_L$. The result, again obtained from superposition, is (de Gennes, 1989)

$$F_{12}(x_L) = \frac{\Phi_0}{4\pi} \left[H \exp{-x_L/\lambda_L} - \frac{1}{2} H_1(x_L) + \Delta H \right] \tag{71}$$

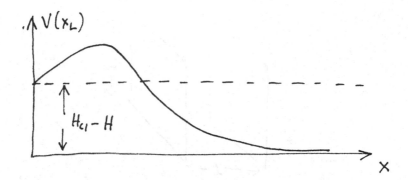

Figure 10. The vortex interaction potential $V(x_L)$ for superconductors with $\kappa \gg 1$.

where $\Delta H = H_{c1} - H$ with the critical field H_{c1}. Thus, at $H = H_{c1}$ (the coexistence condition for wetting, but now in the type-II regime) $V(x_L) = 0$ and $V(x_L = \infty) = 0$. Figure 10 indeed bears a striking similarity to the effective interface potential (Kolomeisky, Levanjuk, 1990). Note that the vortex potential decays on the scale of λ_L.

4. Lecture: Long-range forces in Superconductors.

4.1. INTRODUCTION

In Lecture 3 we have seen that, for $\kappa \to 0$, the effective interface potential $V(\ell)$ decays exponentially for $\ell \to \infty$: $V \propto \exp -\sqrt{2}\ell/\xi$. In fluids, this kind of exponential decay is usually synonymous with forces of *short range*. However, since $\xi \approx 1000$ Å, the exponential decay of the interface potential easily extends into a regime where in fluids retardation effects have already become important (see eq.(67)). Do similar long range, i.e. *algebraic*, contributions exist in superconductors? This lecture addresses this question, which was originally posed by de Gennes (de Gennes, 1995). He had pointed out before (de Gennes, 1981) that the presence of long-range forces profoundly affects the interfacial profiles, which acquire van der Waals tails dominating the exponential decay of purely short-range forces.

4.2. CASIMIR EFFECT

The origin of dispersion forces which lead to algebraically decaying contributions to the interface potential are the fluctuations of the electromagnetic field. This is most transparently illustrated by the simplest case, which bears the name of H.G.B. Casimir (Casimir, 1948). He considered two ideally conducting plates of side length L at a distance ℓ enclosing a vacuum, see Figure 11.

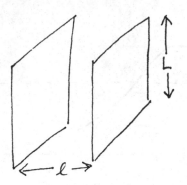

Figure 11. Plate geometry for the Casimir effect.

The electric field is oriented perpendicular to the plates, the magnetic field parallel. The modes of the electric field confined between the plates give rise to a force between the plates. Dimensional analysis shows that the force per unit surface must be proportional to

$$F \sim \frac{\hbar c}{\ell^4} . \tag{72}$$

What is the magnitude of F, and what is its sign? To answer this question, the free energy per unit surface

$$\mathcal{E} \equiv \frac{E - E_0}{L^2} \tag{73}$$

has to be determined, where $E - E_0$ is the energy difference between the confined and the free system (of identical volume). The calculation, e.g. found in (Itzykson, Zuber, 1980), yields $\mathcal{E} = (\pi/120)(\hbar c/\ell^3)$ so that the force between the plates is given by

$$F = -\frac{\pi \hbar c}{240 \ell^4} . \tag{74}$$

The force is attractive.

For a more realistic system like the solid/liquid/vapor system studied in wetting of liquids, or the superconducting/normal systems studied here, the dielectric properties of the materials have to be taken into account, which are governed by the dielectric functions $\epsilon(\omega, q)$. The mode spectrum of the electromagnetic fields is thus considerably more complex compared to that of a simple 'vacuum' (indicated in Figure 12).

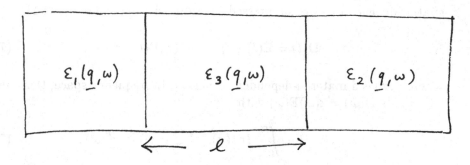

Figure 12. Plate geometry for material systems.

A more general theory than Casimir's is thus needed to answer this question. This theory was developed in the 1950's and 1960's with the methods of quantum electrodynamics, predominantly by Russian physicists (Dzyaloshinskii, Lifschitz, Pitaevskii, 1961; Barash, Ginzburg, 1975). Although this theory meanwhile has become textbook material (Landau, 1980-2), a derivation of these theories is technically involved and cannot be achieved here. The pragmatic philosophy we will therefore adopt is to simply apply the expressions from DLP-theory to metals and superconductors. The general expression for the force acting on two interfaces between three media of dielectric functions $\epsilon_i(\omega)$, $i = 1, 2, 3$ as obtained in the DLP-theory of dispersion forces is given by

$$F = \frac{\hbar}{2\pi^2 c^2} \int_0^\infty d\zeta \zeta^3 \int_1^\infty dp p^2 \epsilon_3(i\zeta)^{3/2} [A_1(i\zeta, p) + A_2(i\zeta, p)] \qquad (75)$$

where, with $i = 1, 2$, $A_i(i\zeta, p) = [a_i(i\zeta, p) \exp[2p\zeta \ell \sqrt{\epsilon_3}/c] - 1]^{-1}$. Here, $a_1 = \prod_i [(s_i + p)/(s_i - p)]$, and $a_2 = \prod_i [(s_i + p\epsilon_i/\epsilon_3)/(s_i - p\epsilon_i/\epsilon_3)]$, with $s_i = [\epsilon_i/\epsilon_3 - 1 + p^2]^{1/2}$.

As we will see, for metals and superconductors this formula will only allow us to discuss the case of retarded forces.

4.3. DIELECTRIC FUNCTIONS; SURFACE IMPEDANCE

Having 'found' the expression which generalizes the Casimir force for real materials, we are left with the task to determine the dielectric function for metals and superconductors. In this subsection we will first recall some basic properties of dielectric functions. In material electrodynamics, the relevant field is not the electric field but the *displacement* (in the following, spatial dependences are suppressed for simplicity) (Landau, 1980-1)

$$D(t) = E(t) + \int_0^\infty d\tau f(\tau) E(t - \tau) \tag{76}$$

where $f(\tau)$ is a material-dependent function. In frequency space, this equation reads $D(\omega) = \epsilon(\omega) E(\omega)$ with

$$\epsilon(\omega) = 1 + \int_0^\infty d\tau f(\tau) e^{i\omega\tau} \equiv \epsilon'(\omega) + i\epsilon''(\omega). \tag{77}$$

From this definition of the *complex dielectric function*, $\epsilon(-\omega) = \epsilon^*(\omega)$, so that ϵ' is an even, while ϵ'' is an odd function of frequency. Consequently, $\epsilon'(\omega) = \epsilon_0 + \mathcal{O}(\omega^2)$, $\epsilon''(\omega) = \mathcal{O}(\omega)$, where ϵ_0 is the static dielectric constant.

In metals, the situation is slightly more complicated. For $\omega \to 0$, the dielectric function diverges, so that $\epsilon_0 = \infty$. For small values of ω, in a range where the metal can be characterized by its conductivity σ,

$$\epsilon(\omega) = \frac{4\pi i\sigma}{\omega} \tag{78}$$

while for high frequencies, UV and beyond, $\epsilon(\omega) = 1 - \omega_p^2/\omega^2$ where $\omega_p = (4\pi N e^2/m)^{1/2}$ is the plasma frequency of electrons of number density N.

The dielectric function of a superconductor differs from that of a metal. It can be deduced from the field penetration depth. As ϵ also diverges in the static limit as for a metal, the penetration of an electromagnetic field into the superconductor fulfills $\lambda \sim c/(\omega\sqrt{\epsilon}) \ll \lambda_{vac} = c/\omega$. Therefore one can treat the field as a plane wave. The boundary conditions for the tangential field components are

$$E_t = \hat{\zeta}[H_t \times n] \tag{79}$$

where $\hat{\zeta} = \sqrt{\mu/\epsilon}$ is the surface impedance (in the following, we set $\mu = 1$). For the superconductor, one finds (Landau, 1980-1)

$$\hat{\zeta} = -\frac{i\omega}{c}\lambda. \tag{80}$$

Eq.(80) expresses a relation between λ and $\hat{\zeta}$, and hence ϵ, so that one may wonder how this result relates to the frequency dependence of the nonlocal BCS-kernel Q. Eq.(80) then generalizes to

$$\hat{\zeta} \propto -\frac{i\omega}{c}[Q(\omega)]^{-1/3} \tag{81}$$

where $Q(\omega)$ is the frequency dependent kernel (Abrikosov, Gor'kov, 1959; Abrikosov. Gor'kov, Khalatnikov, 1959; Miller, 1960).

4.4. LONG RANGE FORCES

We finally turn to the computation of eq.(75) for the SC/N systems, and consider first the case where $i = 1, 2$ are normal metals, and $i = 3$ is vacuum. Then, eq.(74) is indeed recovered with $\epsilon_{1,2} = \infty$, and $\epsilon_3 = 1$. If $\epsilon_3 \to \infty$, i.e. we replace the vacuum in the gap between the plates by a metal, $F \to 0$. Thus, in order to get a finite result, the differences in the frequency behaviour for finite frequencies need to be considered. Considering high frequency behaviour with $\epsilon(i\zeta) \sim 1/\zeta^2$, the dependence on ζ drops out of the exponential in eq.(75) altogether, leaving only exponential contributions (Dzyaloshinskii, Lifschitz, Pitaevskii, 1961). Thus, we have to go to smaller frequencies and resort to eq.(78), and obtain

$$F = -0.34 \cdot 10^{-3} \frac{\hbar c^2}{\sigma_3 \ell^5} \tag{82}$$

i.e., an attractive force that decays more slowly than the retarded forces in liquids.

Turning to the case of superconductors, as $\epsilon_3 \sim 1/\zeta^2$ from eq.(80), the same problem as in metals arises: there are only exponential contributions. Again one has to look into this problem in more detail and invoke the results from BCS theory. For small frequencies, the imaginary part of the kernel Q dominates, and thus leads to a behaviour of the dielectric function as $\epsilon(i\zeta) \sim \zeta^{-4/3}$. Performing the calculation in an analogous fashion as in the case of metals, one now finds

$$F = -\frac{\bar{A}}{\ell^6} \tag{83}$$

with $\bar{A} = \hbar c^3 \alpha^{9/2}$, where α is a function of material properties (Abrikosov, Gor'kov, 1959; Abrikosov, Gor'kov, Khalatnikov, 1959). Note that in this calculation, as for metals, $\epsilon_{1,2}$ were both taken as *infinite*, so that the resulting force between the SC/N interfaces is attractive.

It is interesting to see that for a superconductor enclosed between two metals or other superconductors, the dominant contribution decays even *faster* than for a metal (and hence is faster than that of retarded forces in liquids). The forces given in eq.(83) act on lengths $\ell > \xi$, while for $\ell < \xi$, exponentially decaying forces exist. It is in this regime where long range forces can exist that decay more rapidly than eq.(83).

In order to obtain those possible contributions, however, the dielectric functions need to be considered in frequency ranges in which spatial dispersion effects are non-negligible (Prange, 1963). For this purpose, DLP-theory needs to be extended to spatial dispersion effects (Barash, Ginzburg, 1975; Kats, 1978).

4.5. VAN DER WAALS FORCES BETWEEN VORTICES

We conclude this lecture by another brief look into HTSC-materials. As we have seen in Lecture 3, vortices in classic superconductors repel or attract each other with exponential forces. Do long range, or van der Waals, forces exist between vortices?

The answer to this question is yes, as was recently demonstrated (Blatter, 1996). In the layered HTSC materials, the vortices have the structure of a stack of pancakes, where each pancake is confined to one layer of the material. Furthermore, the vortices are big and polarizable, and, most importantly, being confined to a two-dimensional world, interact with a logarithmic law (Brandt, Mints, Snapiro, 1996). The interaction energy of a pancake dipole in a layer of thickness d is given by $F_{12} = -2d[2(\mathbf{u}_1 \cdot \hat{\mathbf{n}})(\mathbf{u}_2 \cdot \hat{\mathbf{n}}) - \mathbf{u}_1 \cdot \mathbf{u}_2]/R^2$, where R is the distance between the vortices, the $\mathbf{u}_i$ are the vortex displacements, and $\hat{\mathbf{n}}$ is a unit vector along the axis between the vortices. A displacement of vortex 1 by $\mathbf{u}_1$ causes a displacement $\mathbf{u}_2$ of vortex 2. The response of the vortices to displacements is limited by the elastic forces exerted on them by the other flux lines, which are proportional to $d\mathbf{u}_2/\lambda^2 \ln(\lambda/d)$. A balance of these forces leads to an attractive potential (Mukherji, Nattermann, 1997)

$$V_{vdW} \sim -\frac{\lambda}{R^4} \frac{<\mathbf{u}^2>}{\ln(\lambda/d)} \tag{84}$$

where the averaged squared displacements $<\mathbf{u}^2>$ well depend on d, but not on R. The cause for these displacements can both be thermal or disorder fluctuations (Mukherji, Nattermann, 1997).

Acknowledgement. The author thanks Joseph Indekeu for many enjoyable discussions about wetting (and other topics). R.B. is a Research Fellow with the KU Leuven supported under VIS/97/01. Part of this research was done at the Heinrich-Heine-Universität Düsseldorf under support of the SFB 237 "Unordnung und Große Fluktuationen".

References

Abrikosov, A.A., Gor'kov, L.P. (1959) On the theory of superconducting alloys I. the electrodynamics of alloys at absolute zero, *Sov. Phys. JETP* **35**, 1090-1098

Abrikosov, A.A., Gor'kov, L.P., Khalatnikov, I.M. (1959) A superconductor in a high frequency field, *Sov. Phys. JETP* **35**, 182-189

Barash, Yu. S., Ginzburg, V.L. (1975) Electromagnetic fluctuations in matter and molecular (van-der-Waals) forces between them, *Sov. Phys.-Usp.* **18**, 305-322

Berezin, F.A. (1966) *The Method of Second Quantization*, Academic Press, New York

Blatter, G., Geshkenbein, V. (1996) Van der Waals Attraction of Vortices in Anisotropic and Layered Superconductors, *Phys. Rev. Lett.* **77**, 4958-4961

Blossey R., Indekeu, J.O. (1996) Interface-potential approach to surface states in type-I superconductors, *Phys. Rev.* **B 53**, 8599-8603

Brandt, E.H., Mints, R.G., Snapiro, I.B. (1996) Long-Range Fluctuation-Induced Attraction of Vortices to the Surface in Layered Superconductors, *Phys. Rev. Lett.* **76**, 827-831

Casimir, H.G.B. (1948) *Proc. Acad. Sci. Amst.* **60**, 793

de Gennes, P.G. (1989) *Superconductivity of Metals and Alloys*, Addison-Wesley, Reading

de Gennes, P.G. (1981) Some effects of long range forces on interfacial phenomena, *J. Physique Lett.* **42**, L377-L379

de Gennes, P.G., (1995) private letter to J.O. Indekeu

Dzyaloshinskii, I.E., Lifshitz, E.M., Pitaevskii, L.P. (1961) The General Theory of Van der Waals Forces, *Advances in Physics* **10** 165-209

Eilenberger, G. (1968) Transformation of Gorkov's Equation for Type II Superconductors into Transport-Like Equations *Z. Physik* **214**, 195-213

Gor'kov, L.P. (1959) Microscopic derivation of the Ginzburg-Landau equations in the theory of superconductivity, *Sov. Phys. JETP* **36**, 1364-1367

Indekeu, J.O., van Leeuwen, J.M.J (1995) Interface Delocalization Transition in Type-I Superconductors, *Phys. Rev. Lett.* **75**, 1618-1621

Itzykson, C., Zuber, J.-B. (1980) *Quantum Field Theory*, McGrawHill, New York

Kats, E.I. (1978) Influence of nonlocality effects on van der Waals interaction, *Sov. Phys. JETP* **46**, 109-113

Kolomeisky E.B., Levanjuk, A.P. (1990) Wetting-like phase transition of a vortex near a boundary of superconductor, *Solid State Comm.* **73**, 223-224

Kosztin, I., Leggett, A.J. (1997) Nonlocal effects on the Magnetic Penetration Depth in d-wave Superconductors, *Phys. Rev. Lett.* **79**, 135-138

Kree, R. (1988) *Theorie der Supraleitung*, Lectures at the University of Düsseldorf

Landau, L.D., Lifschitz, E.M., (1980) *Elektrodynamik der Kontinua*, 3. Auflage, Akademie-Verlag, Berlin

Landau, L.D., Lifschitz, E.M., (1980) *Statistische Physik II* Akademie-Verlag, Berlin

Mattis, D.C., Bardeen, J. (1958) Theory of the Anomalous Skin Effect in Normal and

Superconducting Metals, *Phys. Rev.* 111, 412-417

Miller, P.B. (1960) Surface Impedance of Superconductors, *Phys. Rev.* 118, 928-934

Mukherji, S. Nattermann, T. (1997) Steric Repulsion and van der Waals attraction between Flux Lines in Disordered High T_c superconductors, *Phys. Rev. Lett.* 79, 139-142

Pippard, A.B. (1953) *Proc. Roy. Soc.* (London) A216, 547

Prange, R.E. (1963) Dielectric Constant of a Superconductor, *Phys. Rev.* 129, 2495-2503

Scalapino, D.J. (1995) The case for $d_{x^2-y^2}$ pairing in the cuprate superconductors *Phys. Rep.* 250, 329-365

Schattke, W. (1966) A nonlocal theory for the superconductor in a magnetic field, *Phys. Lett.* 20, 245-247

Thompson R., Baratoff, A. (1968) Superconducting Thin Film in a Magnetic Field - Theory of Nonlocal and Nonlinear Effects. I. Specular Reflection, *Phys. Rev.* 167, 361-381

Tinkham, M. (1996) *Introduction to Superconductivity* 2nd Ed., McGraw-Hill, Inc., New York

Wu Hang-sheng, Lei Hsiao-lin (1965) A nonlocal extension of the Ginzburg-Landau equations in the theory of superconductivity, *Acta Physica Sinica* 21, 1355-1369

FERMI LIQUID SUPERCONDUCTIVITY

Concepts, Equations, Applications

M. ESCHRIG, J.A. SAULS

Department of Physics & Astronomy, Northwestern University, Evanston, IL 60208, USA

AND

H. BURKHARDT, D. RAINER

Physikalisches Institut, Universität Bayreuth, D-95440 Bayreuth, Germany

1. Introduction

The theory of Fermi liquid superconductivity combines two important theories for correlated electrons in metals, Landau's theory of Fermi liquids and the BCS theory of superconductivity. In a series of papers published in 1956-58 Landau [31] argued that a strongly interacting system of Fermions can form a "Fermi-liquid state" in which the physical properties at low temperatures and low energies are dominated by fermionic excitations called *quasiparticles*. These excitations are composite states of elementary Fermions that have the same charge and spin as the non-interacting Fermions, and can be labeled by their momentum $\mathbf{p}$ near a Fermi surface (defined by Fermi momenta $\mathbf{p}_f$). Landau further argued that an ensemble of quasiparticles is described by a classical distribution function in phase space, $f(\mathbf{p}, \mathbf{R}; t)$, and that the low-energy properties of such a system are governed by a classical transport equation, which we refer to as the *Boltzmann-Landau transport equation*. A significant feature of Landau's theory is that quasiparticles are well defined excitations at low energy, yet their interactions are generally large and can never be neglected. These interactions lead to internal forces acting on the quasiparticles, damping of quasiparticles, and give rise to many of the unique signatures of strongly correlated Fermi liquids. The quasiparticle interactions are parametrized by phenomenological interaction functions that determine the interaction energy, internal forces between quasiparticles, and damping terms.

The modern theory of superconductivity started in 1957 with the publica-

S.-L. Drechsler and T. Mishonov (eds.), High-T_C Superconductors and Related Materials, 413–446.

tion by Bardeen, Cooper and Schrieffer [8] on the *Theory of Superconductivity*, wildly known as the *BCS theory*. Theorists on both the 'east' and 'west' established within a few years basically a complete "standard theory of superconductivity" which was finally comprehensively reviewed by leading western experts in *Superconductivity* edited by Parks in 1969 [45].

At about the time Parks' books were edited two papers by Eilenberger [19] and Larkin & Ovchinnikov [32] were published, which demonstrated, independently, that the complete standard theory of (equilibrium) superconductivity can be formulated in terms of a *quasiclassical* transport equation. Somewhat later this result was generalized to non-equilibrium conditions by Eliashberg [21] and Larkin & Ovchinnikov [33]. We consider this theory the generalization of Landau's theory of normal Fermi liquids to the superconducting states of metals or superfluid states of liquid ^{3}He. It combines Landau's semiclassical transport equations for quasiparticles with the concepts of pairing and particle-hole coherence that are the basis of the Bardeen, Cooper and Schrieffer theory. We will call this theory alternatively the "*quasiclassical theory*", as it was coined by Larkin and Ovchinnikov, or the "*Fermi liquid theory of superconductivity*". Several limits of the quasiclassical theory were known before its general formulation was established by Eilenberger, Larkin, Ovchinnikov and Eliashberg. De Gennes [17] had shown that equilibrium superconducting phenomena for T near T_c could be described in terms of classical correlation functions, which may be calculated from a Boltzmann equation [39]. In 1964 Andreev [4] developed a set of equations (Andreev equations) which are equivalent to the clean limit equations of the quasiclassical theory. In the mid 60's Leggett [37] discussed in a series of papers the effects of Landau's interactions on the response functions for a superconductor. Geilikman [24] as well as Bardeen, Rickayzen and Tewordt [9] introduced in the late fifties a semiclassical transport equation, which corresponds to the long-wavelength, low-frequency limit of the quasiclassical dynamical equations. The linear response theory obtained by ξ-integrating the Kubo response function [1] is also equivalent to the linearized quasiclassical transport equation [49]. These early theories are predecessors of the complete quasiclassical theory which provides a full description of superconducting phenomena ranging from inhomogeneous equilibrium states to superconducting phenomena far from equilibrium. The theory is valid at all temperatures and excitation fields of interest, and it covers clean and dirty systems as well as metals with strong electron-phonon or electron-electron interactions.

In section 2 we introduce the quasiclassical propagators, quasiclassical self-energies, as well as the set of quasiclassical equations. We briefly discuss their foundations and interpretation. The quasiclassical propagators are the generalization of Landau's distribution functions to the superconducting state, and the quasiclassical self-energies describe the effects of quasiparticle interactions, quasiparticle-phonon interactions and quasiparticle-impurity scattering. For further details of the quasiclassical theory and its derivation we refer to various review articles [35, 18, 50, 47] and references therein. In sections 3 and 4 we present two recent applications of the quasiclassical theory. Section 3 discusses the Josephson effect and charge transport across a junction of differently oriented d-wave superconductors, while section 4 presents calculations of the electromagnetic response of a two-dimensional "pancake" vortex in a layered superconductor.

2. Quasiclassical theory

2.1. QUASICLASSICAL TRANSPORT EQUATIONS

Derivations of the quasiclassical equations were given by Eilenberger, Larkin, Ovchinnikov and Eliashberg in their original papers [19, 32, 21, 33], by Shelankov [52], and in several review articles [35, 18, 50, 48]. All derivations start from a formulation of the theory of superconductivity in terms of Green's functions ($\check{G}$), self-energies ($\check{\Sigma}$), and Dyson's equation,

$$\check{G} = \check{G}_0 + \check{G}_0 \check{\Sigma} \check{G} \ . \tag{1}$$

We use here the notation of Larkin & Ovchinnikov [35] who introduced Nambu-Keldysh matrix Green's functions and self-energies (indicated by a háček accent), whose matrix structure comprises Nambu's particle-hole index and Keldysh's doubled-time index [29]. Nambu's particle-hole matrix structure is essential for BCS superconductivity since off-diagonal terms in the particle-hole index indicate particle-hole coherence due to pairing. Keldysh's doubled-time index, on the other hand, is a very convenient tool for describing many-body systems out of equilibrium. Spin dependent phenomena require an additional matrix index for spin $\uparrow$ and $\downarrow$. The Green's functions depend on two positions ($\mathbf{x}_1, \mathbf{x}_2$) and two times ($t_1, t_2$) or, alternatively, on an average position, $\mathbf{R} = (\mathbf{x}_1 + \mathbf{x}_2)/2$, average time $t = (t_1 + t_2)/2$ and, after Fourier transforming in $\mathbf{x}_1 - \mathbf{x}_2$ and $t_1 - t_2$, on a momentum $\mathbf{p}$ and energy ϵ.[1] All the physical information of interest is contained in the Green's functions, whose calculation requires i) an evaluation of the self-energies and ii) the solution of Dyson's equation. The Fermi-liquid theory of superconductivity provides a scheme for calculating self-energies and Green's functions consistently by an expansion in the small parameters T_c/E_f, $1/k_f\xi_0$, $1/k_f\ell$, ω/E_f, q/k_f, ω_D/E_f, where E_f and k_f are Fermi energy and momentum, T_c and ξ_0 the superconducting transition temperature and coherence length, ℓ the electron mean free path, ω_D the Debye

---

[1] We set in this article $\hbar = k_B = 1$. The charge of an electron is $e < 0$.

frequency, and ω, q are typical frequencies and wave-vectors of external perturbations, such as electromagnetic fields, ultra-sound or temperature variations. We follow [50] and assign to these dimensionless expansion parameters the order of magnitude "*small*". To leading order in *small* the full Green's functions, $\check{G}(\mathbf{p}, \mathbf{R}; \epsilon, t)$, can be replaced by the "ξ-integrated" Green's functions (quasiclassical propagators), $\check{g}(\mathbf{p}_f, \mathbf{R}; \epsilon, t)$, and the full self-energies, $\check{\Sigma}(\mathbf{p}, \mathbf{R}; \epsilon, t)$, by the quasiclassical self-energies, $\check{\sigma}(\mathbf{p}_f, \mathbf{R}; \epsilon, t)$. The quasiclassical propagator describes the state at position $\mathbf{R}$ and time t, of quasiparticles with energy ϵ (measured from the Fermi energy) and momenta $\mathbf{p}$ near the point $\mathbf{p}_f$ on the Fermi surface. The reduction to the Fermi surface ($\mathbf{p} \to \mathbf{p}_f$) is an essential step. It establishes the bridge between full quantum theory and quasiclassical theory.

The dynamical equations for the quasiclassical propagators are obtained from Dyson's equation for the full Green's functions (1), and one finds (see, e.g., the review [47] and references therein)

$$[\epsilon\hat{\tau}_3 - \check{\sigma} - \check{v}, \check{g}]_\otimes + i\mathbf{v}_f \cdot \boldsymbol{\nabla}\check{g} = 0 , \tag{2}$$

$$\check{g} \otimes \check{g} = -\pi^2\check{1} , \tag{3}$$

where the $\otimes$-product is defined by

$$\check{A} \otimes \check{B}(\epsilon, t) = e^{\frac{i}{2}(\partial_\epsilon^A \partial_t^B - \partial_\epsilon^B \partial_t^A)} \check{A}(\epsilon, t)\check{B}(\epsilon, t) , \tag{4}$$

and the commutator is given by

$$[\check{A}, \check{B}]_\otimes = \check{A} \otimes \check{B} - \check{B} \otimes \check{A} . \tag{5}$$

Eq. (2) turns in the normal-state limit into Landau's classical transport equation for quasiparticles. Hence, one should consider eq. (2), which has the form of a transport equation for matrices, as a generalization of Landau's transport equation to the superconducting state. This interpretation becomes more transparent if one drops the Keldysh-matrix notation, and writes down the equations for the three components $\hat{g}^{R,A,K}$ (advanced, retarded and Keldysh-type) of the Keldysh matrix propagator separately.

$$\left[\epsilon\hat{\tau}_3 - \hat{v}(\mathbf{p}_f, \mathbf{R}; t) - \hat{\sigma}^{R,A}(\mathbf{p}_f, \mathbf{R}; \epsilon, t), \hat{g}^{R,A}(\mathbf{p}_f, \mathbf{R}; \epsilon, t)\right]_\circ$$
$$+i\mathbf{v}_f \cdot \boldsymbol{\nabla}\hat{g}^{R,A}(\mathbf{p}_f, \mathbf{R}; \epsilon, t) = 0 , \tag{6}$$

$$\left(\epsilon\hat{\tau}_3 - \hat{v}(\mathbf{p}_f, \mathbf{R}; t) - \hat{\sigma}^R(\mathbf{p}_f, \mathbf{R}; \epsilon, t)\right) \circ \hat{g}^K(\mathbf{p}_f, \mathbf{R}; \epsilon, t) \tag{7}$$
$$-\hat{g}^K(\mathbf{p}_f, \mathbf{R}; \epsilon, t) \circ \left(\epsilon\hat{\tau}_3 - \hat{v}(\mathbf{p}_f, \mathbf{R}; t) - \hat{\sigma}^A(\mathbf{p}_f, \mathbf{R}; \epsilon, t)\right)$$
$$-\hat{\sigma}^K(\mathbf{p}_f, \mathbf{R}; \epsilon, t) \circ \hat{g}^A(\mathbf{p}_f, \mathbf{R}; \epsilon, t) + \hat{g}^R(\mathbf{p}_{,f} \mathbf{R}; \epsilon, t) \circ \hat{\sigma}^K(\mathbf{p}_f, \mathbf{R}; \epsilon, t)$$
$$+i\mathbf{v}_f \cdot \boldsymbol{\nabla}\hat{g}^K(\mathbf{p}_f, \mathbf{R}; \epsilon, t) = 0 .$$

The ○-product stands here the following operation in the energy-time variables (the superscripts a, b refer to derivatives of $\hat{a}$ and $\hat{b}$ respectively).

$$[\hat{a} \circ \hat{b}](\mathbf{p}_f, \mathbf{R}; \epsilon, t) = e^{\frac{i}{2}\left(\partial_\epsilon^a \partial_t^b - \partial_t^a \partial_\epsilon^b\right)} \hat{a}(\mathbf{p}_f, \mathbf{R}; \epsilon, t)\hat{b}(\mathbf{p}_f, \mathbf{R}; \epsilon, t) , \qquad (8)$$

and the commutator $[\hat{a}, \hat{b}]_\circ$ stands for $\hat{a} \circ \hat{b} - \hat{b} \circ \hat{a}$. An important additional set of equations are the normalization conditions

$$\hat{g}^R(\mathbf{p}_f, \mathbf{R}; \epsilon, t) \circ \hat{g}^K(\mathbf{p}_f, \mathbf{R}; \epsilon, t) + \hat{g}^K(\mathbf{p}_f, \mathbf{R}; \epsilon, t) \circ \hat{g}^A(\mathbf{p}_f, \mathbf{R}; \epsilon, t) = 0, \quad (9)$$

$$\hat{g}^{R,A}(\mathbf{p}_f, \mathbf{R}; \epsilon, t) \circ \hat{g}^{R,A}(\mathbf{p}_f, \mathbf{R}; \epsilon, t) = -\pi^2 \hat{1}. \qquad (10)$$

The normalization condition was first derived by Eilenberger [19] for superconductors in equilibrium. An alternative, more physical derivation was given by Shelankov [52]. The quasiclassical transport equations (6,7) supplemented by the normalization conditions, Eqs. (9-10), are the fundamental equations of the Fermi-liquid theory of superconductivity. The various steps and simplifications done to transform Dyson's equations into transport equations are in accordance with a systematic expansion to leading orders in small.

The matrix structure of the quasiclassical propagators describes the quantum-mechanical internal degrees of freedom of electrons and holes. The internal degrees of freedom are the spin (s=1/2) and the particle-hole degree of freedom. The latter is of fundamental importance for superconductivity. In the normal state one has an incoherent mixture of particle and hole excitations, whereas the superconducting state is characterized by the existence of quantum coherence between particles and holes. This coherence is the origin of persistent currents, Josephson effects, Andreev scattering, flux quantization, and other non-classical superconducting effects. The quasiclassical propagators, in particular the combination $\hat{g}^K - (\hat{g}^R - \hat{g}^A)$, are intimately related to the quantum-mechanical density matrices which describe the quantum-statistical state of the internal degrees of freedom. Non-vanishing off-diagonal elements in the particle-hole density matrix indicate superconductivity, and the onset of non-vanishing off-diagonal elements marks the superconducting transition. One reason for the increased complexity of the transport equations in the superconducting state (3 coupled transport equations for the 3 matrix distribution functions $\hat{g}^{R,A,K}$) in comparison with the normal state of the Fermi liquid (1 transport equation for a scalar distribution function) is the fact that the quasiparticle states in the normal state are inert to the perturbations and to changes in the occupation of quasiparticle states. Hence, the only dynamical degrees of freedom are here the occupation probabilities of a quasiparticle state. On the other hand, quasiparticle states of energy $\epsilon \lesssim \Delta$ are coherent mixtures

418

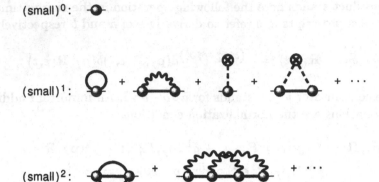

(small)^0:

(small)^1:

(small)^2:

Figure 1. The figure shows the leading order self-energy diagrams that contribute to the Fermi liquid theory of superconductivity. The diagram in the first row represents an effective potential which affects the shape of the quasiparticle Fermi surface and the mass of a quasiparticle. The diagrams in the second row are of 1st order in the expansion parameter and represent Landau's quasiparticle interactions and the quasiparticle pairing interaction (first diagram), the Migdal-Eliashberg self-energy which leads to mass enhancement and damping of quasiparticles due to their coupling to phonons (second diagram), quasiparticle-impurity scattering in leading order in $1/k_f\ell$ (the third and fourth diagram are representatives of an infinite series of diagrams whose sum is the T-matrix for quasiparticle scattering at an impurity, multiplied by the impurity concentration). The diagrams in the third row describe quasiparticle-quasiparticle collisions (first diagram) and a small correction to the quasiparticle-phonon interaction term of Migdal and Eliashberg.

of particle and hole states, and react sensitively to external as well as internal forces in the superconducting state. Thus the quasiclassical transport equations describe the coupled dynamics of the quasiparticle states and their occupation probability (distribution functions). It is only in limiting cases possible to decouple to some degree the dynamics of states and occupation. An important such case, which admits using scalar distribution functions, are low frequency ($\omega \ll \Delta$) phenomena in superconductors, as first discussed by Betbeder-Matibet and Nozieres [13] (see also [50]).

To conclude the section on the general quasiclassical theory we discuss the quasiclassical self-energies. The set of leading order self-energies arranged according to their power in the expansion parameter *small* is shown in Fig. 2.1. The shaded spheres with connections to phonon lines (wiggly), quasiparticle lines (smooth) or impurity lines (dashed) represent interaction vertices describing quasiparticle-quasiparticle interactions, quasiparticle-phonon interactions in Migdal-Eliashberg approximation [43, 20], and quasiparticle-impurity scattering. The full lines represent quasiparticle propagators (smooth) and phonon propagators (wiggly). The diagrams shown in Fig. 2.1 comprise the interaction processes taken into account in the standard theory of superconductivity. We note that the interaction vertices are in leading order inert, i.e. independent of temperature, not affected by perturbations or changes in quasiparticle occupations and, in particular, not affected by the superconducting transition. These vertices are phenomenological parameters in the Fermi liquid theory of superconductivity which must be taken from experiment.

3. The Effect of Interface Roughness on the Josephson Current

In this section we present an application of the quasiclassical theory to Josephson tunneling in d-wave superconductors. Tunneling experiments in superconductors probe, in general, the quasiparticle states and the pairing amplitude at the tunneling contact. Such experiments give valuable information on the quasiparticle density of states and the symmetry of the order parameter. For superconductors with a single isotropic gap parameter the superconducting state is in most cases not distorted by the tunneling contact, and one measures basically bulk properties of the superconductor. A typical example is the Josephson current of an $S-I-S$ tunnel junction. The Josephson current for traditional s-wave superconductors is well described by the universal formula of Ambegaokar and Baratoff [2],

$$I_J(\psi, T) = I_c(T)\sin(\psi) = \frac{\pi}{2|e|R_N}\Delta(T)\tanh\left(\frac{\Delta(T)}{2k_BT}\right)\sin(\psi) , \qquad (11)$$

which holds for isotropic BCS superconductors and a weakly transparent, non-magnetic tunneling barrier of arbitrary degree of roughness. It describes the dependence of the Josephson current on the temperature T, and the phase difference ψ across the junction. This current-phase relation depends only on two parameters, the bulk energy gap $\Delta(T)$ which characterizes the superconductors, and the normal state resistance R_N which characterizes the barrier.

The universality of the Ambegaokar-Baratoff relation is lost for junctions involving anisotropic superconductors, in particular superconductors with strong anisotropies and sign changes of the gap function on the Fermi surface. The Josephson current depends in these cases on the orientation of the crystals with respect to the tunneling barrier, and on the quality of the barrier. Ideal barriers, which conserve parallel momentum, will give, in general, a different Josephson current than rough barriers of the same resistance R_N. The origin of these non-universal effects is scattering at the tunneling barrier, which may lead to a depletion of the order parameter in its vicinity [3], to new quasiparticle states bound to the barrier

[14, 26, 41, 15, 54], and eventually to spontaneous breaking of time reversal symmetry ([41, 53, 23] and references therein). All these special features, which reflect the anisotropy of the gap function, react sensitively to barrier roughness which smears out the anisotropy, broadens the bound state spectrum, changes the order parameter near the barrier, and thus influences the Josephson currents and the tunneling spectra.

The quasiclassical theory of superconductivity is particularly well suited for studying the effects of roughness of tunneling barriers, surfaces, or interfaces. A barrier between two superconducting electrodes is modeled in the quasiclassical theory by a thin (atomic size) interface which may reflect electrons with a certain probability or transmit them across the interface. Reflection and transmission may be ideal or to some degree diffuse. We focus here on the effects of interface roughness on the Josephson critical current. In the following we present our model for rough interfaces which combines Zaitsev's model [55] for ideal (no roughness) interfaces with Ovchinnikov's model [44] for a rough surface. We then discuss briefly the interface resistance in the normal state and present analytical results for the Josephson current across a weakly transparent interface. Finally we present our numerical results for d-wave pairing which demonstrate the strong dependence of the Josephson current of junctions with unconventional superconductors on the junction quality, and discuss, in particular, the Ambegaokar-Baratoff relation for the Josephson current of weakly coupled junctions of d-wave superconductors.

3.1. MODEL FOR A ROUGH BARRIER

Our quasiclassical model of rough barriers combines two models known from the literature, i.e., the ideal interface first discussed by A.V.Zaitsev [55] and the "rough layer" introduced by Yu.N. Ovchinnikov [44]. The ideal part of the interface determines in this model the reflectivity and transmittivity of the interface, whereas the rough layers lead to some degree of randomness in the directions of the reflected and transmitted quasiparticles. Zaitsev's boundary condition relates the propagators $\check{g}^{l,r}(\mathbf{p}_f, \mathbf{R}_I; \epsilon, t)$ on any set of four trajectories with the same momentum parallel to the interface (see Fig.1). These trajectories are characterized by the four momenta, $\mathbf{p}^l_{f\,in}$, $\mathbf{p}^l_{f\,out}$, $\mathbf{p}^r_{f\,in}$, $\mathbf{p}^r_{f\,out}$, at the left (superscript l) and right (superscript r) sides of a point $\mathbf{R}_I$ on the interface (see Fig.1). Zaitsev's relations are

$$\check{d}^l + \check{d}^r = 0 \ , \tag{12}$$

$$\check{d}^r (\check{s}^r)^2 = i\pi \frac{1 - \mathcal{R}}{1 + \mathcal{R}} \left[\check{s}^l, \check{s}^r \left(1 - \frac{i}{2\pi} \check{d}^r \right) \right] \ , \tag{13}$$

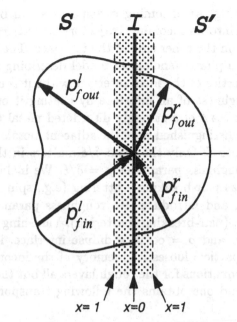

Figure 2. Sketch of our interface model. The ideal interface (I) separates two different superconductors (S,S'), whose Fermi surfaces (curved lines) are shown on the right and left sides. Parallel momentum and energy are conserved in a transmission and reflection process which fixes, as shown, the four momenta, $p^{l,r}_{f\,in,out}$. The interface is coated by rough layers (dotted area).

where $\mathcal{R} = \mathcal{R}(\mathbf{p}^l_{f\,in}) = \mathcal{R}(\mathbf{p}^r_{f\,in})$ is the reflection parameter, and $\check{d}^s$, $\check{s}^s$ ($s = l, r$) are defined as the difference and sum of propagators of quasiparticles on incoming trajectories before reflection (incoming velocity, momentum $\mathbf{p}^s_{f\,in}$) and on outgoing trajectories after reflection (outgoing velocity, momentum $\mathbf{p}^s_{f\,out}$),

$$\check{d}^s(\mathbf{p}^s_{f\,in}, \mathbf{R}_I; \epsilon, t) = \check{g}^s(\mathbf{p}^s_{f\,out}, \mathbf{R}_I; \epsilon, t) - \check{g}^s(\mathbf{p}^s_{f\,in}, \mathbf{R}_I; \epsilon, t) , \quad (14)$$

$$\check{s}^s(\mathbf{p}^s_{f\,in}, \mathbf{R}_I; \epsilon, t) = \check{g}^s(\mathbf{p}^s_{f\,out}, \mathbf{R}_I; \epsilon, t) + \check{g}^s(\mathbf{p}^s_{f\,in}, \mathbf{R}_I; \epsilon, t) . \quad (15)$$

We follow here the notation of A.I. Larkin & Yu.N. Ovchinnikov [35] introduced in section 1, and use Nambu-Keldysh matrix propagators, indicated by a "háček". Parallel momentum is conserved at an ideal interface, which fixes, together with energy conservation, the kinematics of interface scattering as shown in Fig.1. In the normal state a quasiparticle moving towards the point $\mathbf{R}_I$ at the interface with a momentum $\mathbf{p}^s_{f\,in}$ can either be reflected with probability $\mathcal{R}(\mathbf{p}^s_{f\,in})$ or transmitted with probability $\mathcal{T}(\mathbf{p}^s_{f\,in}) = 1 - \mathcal{R}(\mathbf{p}^s_{f\,in})$ to the outgoing trajectory on the other side of the interface. Two more channels open in the superconducting state because of Andreev scattering. An incoming quasiparticle can be "retroreflected" (velocity reversal) into its incoming trajectory or transmitted into the incoming trajectory on the other side of the interface. The function $\mathcal{R}(\mathbf{p}^s_{f\,in})$ is considered here a phenomenological model describing the reflection and transmission properties of the ideal interface without roughness.

We model the roughness of an interface by coating it on both sides by a thin layer of thickness δ of a strongly disordered metal (see Fig.1). These "rough layers" are distinguished from the adjacent metals only by their very short mean free path ℓ. Only the ratio δ/ℓ matters in the limit δ, $\ell \to 0$, and defines the roughness parameter $\rho = \delta/\ell$. We include regular elastic scattering as well as pair-breaking scattering (e.g., spin-flip scattering, inelastic scattering), and introduce two roughness parameters, ρ_0 (regular scattering) and ρ_{in} (pair-breaking scattering). Vanishing ρ's correspond to a perfect interface, and $\rho = \infty$ to a diffuse interface, i.e., a reflected or transmitted quasiparticle looses its memory of the incoming direction. In the quasiclassical equations for the rough layers all but the scattering terms can be dropped, and one obtains the following transport equation in the rough layer.

$$- \left[\frac{\rho_{in}}{2\pi} \langle \check{g}_3^{l,r} \rangle_{\pm} + \frac{\rho_0}{2\pi} \langle \check{g}^{l,r} \rangle_{\pm}, \check{g}^{l,r} \right] + i \overline{v}_{\perp}^{l,r} \partial_x \check{g}^{l,r} = 0 \ , \qquad (16)$$

where the superscripts l and r distinguish the metals on the left and right sides of the interface, $\check{g}_3$ is the Nambu-Keldysh matrix obtained from $\check{g}$ by keeping only the $\hat{\tau}_3$-components of its Nambu submatrices, x is the spatial coordinate perpendicular to the interface, measured in units of δ, and the dimensionless velocity $\overline{v}_{\perp}^{l,r}$ is the perpendicular component of the quasiparticle velocity normalized by an averaged Fermi velocity, $\overline{\mathbf{v}}^{l,r} = \mathbf{v}_f^{l,r}/\sqrt{\langle |\mathbf{v}_f^{l,r}|^2 \rangle}$. The left and right rough layers are located between $x = -1$ and $x = +1$, and are separated by Zaitsev's ideal interface at $x = 0$ (see Fig.1). In general, the quasiclassical propagators in the rough layers depend on $\mathbf{p}_f$, $\mathbf{R}_I$, ϵ, t, and the spatial variable x which specifies the position in the infinitesimally thin rough layer. We use Ovchinnikov's model [44] for the scattering processes in the rough layers. In this model scattering preserves the sign of $\overline{v}_{\perp}$; particles moving towards the interface are not scattered into outgoing directions and vice versa. This "conservation of direction" is indicated in (16) by the indices $\pm$ in the scattering terms, which stand for averaging over the momenta corresponding to $\overline{v}_{\perp} > 0$ and $\overline{v}_{\perp} < 0$, respectively. We assume equal scattering probability into all states compatible with Ovchinnikov's conservation of direction. A reversal of the velocity may only happen at Zaitsev's interface which separates the two rough layers.

In order to obtain the current-phase relationship of the junction we solve the quasiclassical transport equations in both superconductors and in the rough layers. The solution in the left superconductor has to match continuously the solution at $x = -1$ in the left rough layer and, equivalently, the solution in the right superconductor has to match the solution in the right rough layer at $x = 1$. At $x = 0$ the solutions in the rough layers are matched via Zaitsev's conditions (12, 13). The order parameter has to be determined self-consistently. To get a finite current we fix the phase difference of the order parameter across the junction,

$$\psi(\mathbf{R}_I) = \psi^l(\mathbf{R}_I) - \psi^r(\mathbf{R}_I) , \tag{17}$$

where the phases ψ^l and ψ^r at the interface point $\mathbf{R}_I$ are determined by

$$\Delta^{l,r}(\mathbf{p}_f^{l,r}, \mathbf{R}_I) = \tilde{\Delta}^{l,r}(\mathbf{p}_f^{l,r}) \exp(\psi^{l,r}(\mathbf{R}_I)) . \tag{18}$$

Here, $\tilde{\Delta}^{l,r}(\mathbf{p}_f^{l,r})$ are convenient reference order parameters on the left and right sides. The reference order parameter can be taken real in the cases discussed in this paper. Note that the phase difference is measured directly at the interface at point $\mathbf{R}_I$. The current density is obtained by standard formulas of the quasiclassical theory, i.e.,

$$\mathbf{j}(\mathbf{R}, t) = \int \frac{d\epsilon}{4\pi i} \int \frac{d^2 \mathbf{p}_f}{(2\pi)^3 \mid \mathbf{v}_f \mid} e \mathbf{v}_f \, tr(\hat{\tau}_3 \hat{g}^K(\mathbf{p}_f, \mathbf{R}; \epsilon, t)) \tag{19}$$

in the Keldysh formulation, and

$$\mathbf{j}(\mathbf{R}, t) = T \sum_{\epsilon_n} \int \frac{d^2 \mathbf{p}_f}{(2\pi)^3 \mid \mathbf{v}_f \mid} e \mathbf{v}_f \, tr(\hat{\tau}_3 \hat{g}^M(\mathbf{p}_f, \mathbf{R}; \epsilon_n)) \tag{20}$$

in the Matsubara formulation. The critical Josephson current is obtained as the maximum supercurrent across the junction in equilibrium. The Fermi velocity, $\mathbf{v}_f(\mathbf{p}_f)$, is a function of the momentum. This function has the symmetry of the lattice, and will be understood here as a material parameter, to be taken from theoretical models or from experiment, if available.

This completes our brief review of the quasiclassical equations for a rough interface.

3.2. NORMAL STATE RESISTANCE

In this section we calculate the normal state resistance of our model interface, and show that interface roughness does not affect the interface resistance. This is a unique feature of Ovchinnikov's model for a rough layer which is very convenient for studying the effects of interface roughness at fixed interface resistance.

In the normal state the retarded and advanced propagators $\hat{g}^R$ and $\hat{g}^A$ are trivial, and given by $\hat{g}^{R,A} = \mp i\pi\hat{\tau}_3$, and the Keldysh propagator can be parameterized in terms of a distribution function $\Phi(\mathbf{p}_f, \mathbf{R}; \epsilon, t)$,

$$\hat{g}^K(\mathbf{p}_f, \mathbf{R}; \epsilon, t) = 4\pi i \begin{pmatrix} \Phi(\mathbf{p}_f, \mathbf{R}; \epsilon, t) & 0 \\ 0 & \Phi(-\mathbf{p}_f, \mathbf{R}; -\epsilon, t) \end{pmatrix} . \tag{21}$$

Given Φ one can calculate the current density from

$$\mathbf{j}(\mathbf{R}, t) = 2 \int d\epsilon \int \frac{d^2 \mathbf{p}_f}{(2\pi)^3 |\mathbf{v}_f|} e\mathbf{v}_f \Phi(\mathbf{p}_f, \mathbf{R}; \epsilon, t) . \tag{22}$$

The distribution function in the rough layers can be calculated by solving the Landau-Boltzmann transport equation,

$$\overline{\mathbf{v}}_\perp^{l,r} \partial_x \Phi^{l,r} + \rho_{tot} \left(\Phi^{l,r} - \langle \Phi^{l,r} \rangle_\pm \right) = 0 , \tag{23}$$

which follows by taking the normal state limit of the general quasiclassical transport equation (16). In this normal state limit the scattering rates for elastic and inelastic scattering can be added up to a total scattering rate $\rho_{tot} = \rho_0 + \rho_{in}$. Zaitsev's boundary conditions (12, 13) at an ideal interface turn in the normal state into the following classical boundary conditions for the distribution function at the left and right sides of the boundary at $x = 0$,

$$\Phi^l(\mathbf{p}^l_{f\,out}) = \mathcal{R}(\mathbf{p}^l_{f\,in}) \, \Phi^l(\mathbf{p}^l_{f\,in}) + \mathcal{T}(\mathbf{p}^r_{f\,in}) \, \Phi^r(\mathbf{p}^r_{f\,in}) , \tag{24}$$

$$\Phi^r(\mathbf{p}^r_{f\,out}) = \mathcal{R}(\mathbf{p}^r_{f\,in}) \, \Phi^r(\mathbf{p}^r_{f\,in}) + \mathcal{T}(\mathbf{p}^l_{f\,in}) \, \Phi^l(\mathbf{p}^l_{f\,in}) . \tag{25}$$

In addition, the solutions of (23) in the two rough layers have to be matched at $x = \pm 1$ to the physical distribution functions on the left and right sides of the interface.

In order to calculate the interface resistance we apply a voltage V to our junction, solve the transport equation (23) subject to the proper matching conditions at $x = 0$ and $x = \pm 1$, and calculate the current from (22). We assume that the junction is formed by very good conductors separated by a weakly transparent interface, such that the total resistance of the junction is dominated by interface scattering, and the voltage drop is to a good approximation localized at the interface. This leads to a difference eV of the electrochemical potentials in the two conductors, which drives the current. It is assumed that both conductors are in thermal equilibrium far away from the interface. This fixes the incoming parts of the distribution function on the left and right sides of the rough interface,

$$\Phi^l(\mathbf{p}^l_{f\,in}, \mathbf{R}; \epsilon) = f(\epsilon), \quad \Phi^r(\mathbf{p}^r_{f\,in}, \mathbf{R}; \epsilon) = f(\epsilon + eV) , \tag{26}$$

where f is the Fermi function at temperature T. The outgoing excitations are not in equilibrium, and their distribution function must in general be calculated from the transport equation (23). This calculation can be skipped in Ovchinnikov's model. In this model the number of incoming and outgoing excitations are conserved separately, and the incoming part of the equilibrium distribution function on the left (right) side of the interface are solutions of the transport equation (23) at $x < 0$ ($x > 0$). This pins down the total current in terms of the incoming parts of the thermal distribution functions far away from the interface. The distribution functions at $x = -0$ ($x = +0$) are given by the sum of the incoming equilibrium distribution function on the left (right) side, the reflected distribution function, and the transmitted distribution function from the other side. On the right side of the interface we have

$$\Phi^r(\mathbf{p}^r_{f\,in}, x = +0; \epsilon) = f(\epsilon + eV) , \qquad (27)$$

$$\Phi^r(\mathbf{p}^r_{f\,out}, x = +0; \epsilon) = \mathcal{R}(\mathbf{p}^r_{f\,in})f(\epsilon + eV) + \mathcal{T}(\mathbf{p}^l_{f\,in})f(\epsilon) , \qquad (28)$$

and the equivalent formulas hold on the left side.

The current density is constant, and can be calculated at any convenient point of the junction. We pick $x = +0$, and obtain by inserting the distribution functions (27,28) into (??)

$$j_\perp = 2e \int d\epsilon \left(\int \frac{d^2 \mathbf{p}^r_{f\,out}}{(2\pi)^3 \mid \mathbf{v}^r_f \mid} \mathbf{v}^r_{f\perp}[\mathcal{R}(\mathbf{p}^r_{f\,in})f(\epsilon + eV) + \mathcal{T}(\mathbf{p}^l_{f\,in})f(\epsilon)] \right.$$

$$\left. + \int \frac{d^2 \mathbf{p}^r_{f\,in}}{(2\pi)^3 \mid \mathbf{v}^r_f \mid} \mathbf{v}^r_{f\perp} f(\epsilon + eV) \right) . \qquad (29)$$

We now use the relations $\mathcal{R}(\mathbf{p}^r_{f\,in}) = 1 - \mathcal{T}(\mathbf{p}^r_{f\,in})$, $\mathcal{T}(\mathbf{p}^l_{f\,in}) = \mathcal{T}(\mathbf{p}^r_{f\,in})$, and

$$\frac{d^2 \mathbf{p}^r_{f\,in}}{(2\pi)^3 \mid \mathbf{v}^r_f(\mathbf{p}^r_{f\,in}) \mid} \mathbf{v}^r_{f\perp}(\mathbf{p}^r_{f\,in}) = -\frac{d^2 \mathbf{p}^r_{f\,out}}{(2\pi)^3 \mid \mathbf{v}^r_f(\mathbf{p}^r_{f\,out}) \mid} \mathbf{v}^r_{f\perp}(\mathbf{p}^r_{f\,out}) , \qquad (30)$$

and obtain

$$j_\perp = \frac{1}{R_N A} V = \left(2e^2 \int \frac{d^2 \mathbf{p}^r_{f\,out}}{(2\pi)^3 \mid \mathbf{v}^r_f \mid} \mathbf{v}^r_{f\perp} \mathcal{T}(\mathbf{p}^l_{f\,in}) \right) V . \qquad (31)$$

Equation (30) is a geometric relation which follows from conservation of parallel momentum. For clarity we write down in (30) explicitly the momentum dependence of the velocity. The factor in brackets in (31) is the interface conductance per area, $1/(R_N A)$, where A is the area of the interface. The conductance is obviously independent of the scattering rate ρ_{tot}, i.e., independent of the roughness of the interface.

3.3. WEAKLY TRANSPARENT INTERFACE

Zaitsev's boundary conditions (12, 13) at $x = 0$ can be simplified substantially for a weakly transparent interface ($\mathcal{T} \ll 1$). We expand the Matsubara propagator $\hat{g}^M$ to first order in the transmission amplitude, $\mathcal{T}$: $\hat{g}^M = \hat{g}^M_{(0)} + \hat{g}^M_{(1)} + O(\mathcal{T}^2)$. The solution $\hat{g}^M_{(0)}$ corresponds to the equilibrium solution of a non-transparent interface. Both sides are decoupled in this case, and Zaitsev's boundary conditions turn into $\hat{d}^{M\,l}_{(0)} = \hat{d}^{M\,r}_{(0)} = 0$. Hence, the 0th order propagators on the left and right sides of the interface describe superconductors bounded by a rough, non-transparent surface. The first order correction to the difference of the incoming and reflected propagator, $\hat{d}^M_{(1)} = \hat{g}^M_{(1)}(\mathbf{p}^{out}_f, x = 0; \epsilon_n) - \hat{g}^M_{(1)}(\mathbf{p}^{in}_f, x = 0; \epsilon_n)$, can be obtained directly from (13) in terms of the 0th order propagators, i.e. without having to solve the quasiclassical transport equation for $\hat{g}^M_{(1)}$. Expansion of (13) to first order in the transparency $\mathcal{T}$ [6] leads to

$$\hat{d}^{M\,r}_{(1)} = -\frac{i}{8\pi}\mathcal{T}\left[\hat{s}^{M\,l}_{(0)}, \hat{s}^{M\,r}_{(0)}\right]. \tag{32}$$

Equation (30) implies that the current density (20) can be written in terms of the difference, $\hat{d}^M$, of outgoing and incoming propagators alone, and does not depend on the sum. Hence, one can calculate from Eq. (32) the Josephson current in terms of the 0th order quantities $s^{M\,l}_{(0)}$ and $s^{M\,r}_{(0)}$. The straight forward calculation leads to the following formula for the current-phase relation.

$$j_J(\psi, T) = \frac{2e}{\pi} k_B T \sum_{\epsilon_n} \int \frac{d^2 \mathbf{p}^r_{f\,out}}{(2\pi)^3 \mid \mathbf{v}^r_f \mid} v^r_{f\perp} \mathcal{T}(\mathbf{p}^r_f)$$

$$\times \Big([f^{M\,l}_{1(0)}(\mathbf{p}^l_f; \epsilon_n) f^{M\,r}_{1(0)}(\mathbf{p}^r_f; \epsilon_n) + f^{M\,l}_{2(0)}(\mathbf{p}^l_f; \epsilon_n) f^{M\,r}_{2(0)}(\mathbf{p}^r_f; \epsilon_n)] \sin(\psi)$$

$$+ [f^{M\,l}_{1(0)}(\mathbf{p}^l_f; \epsilon_n) f^{M\,r}_{2(0)}(\mathbf{p}^r_f; \epsilon_n) - f^{M\,l}_{2(0)}(\mathbf{p}^l_f; \epsilon_n) f^{M\,r}_{1(0)}(\mathbf{p}^r_f; \epsilon_n)] \cos(\psi) \Big) ,\tag{33}$$

where f^M_1 and f^M_2 are the off-diagonal components of the propagator $\hat{g}^M(\mathbf{p}_f, x = 0; \epsilon_n)$, defined via $\hat{g}^M = f^M_1 \hat{\tau}_1 + f^M_2 \hat{\tau}_2 + g^M \hat{\tau}_3$. In the case of real reference order parameters $\tilde{\Delta}^{l,r}$ (this is a possible choice for an s-wave or d-wave superconductor but, e.g., not for d+is) the term proportional $\cos(\psi)$ vanishes, and we get the standard sinusoidal behavior of the Josephson current j_J [12].

We note that the Josephson current vanishes in the limit of strong inelastic roughness, $\rho_{in} \to \infty$. The reason is that the off-diagonal components of $\hat{g}^M$ decay to zero in the dirty layer. Hence, one has $\hat{g}^M = -i\pi sgn(\epsilon_n)\hat{\tau}_3$ at $x = 0$, and the current is zero.

3.4. CALCULATION OF THE JOSEPHSON CURRENT

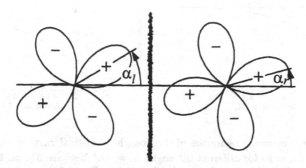

Figure 3. The figure introduces the tilt angles, α_l and α_r, which describe the orientations of the d-wave order parameters on the left and right sides of the interface.

We are interested here in the dc Josephson current of a junction of two d-wave superconductors, and study first the case of a weakly transparent interface ($\mathcal{T} \ll 1$) introduced in Sect.3.3. The results are then compared with numerical calculations for a finite $\mathcal{T}$.

We present selected numerical results for the Josephson current in $S - I - S'$ junctions of layered, tetragonal d-wave superconductors. Of special interest are the effects of changing the orientation of the two crystalline super-conductors [40], and we consider for this reason junctions with interfaces parallel to the c-direction, and the crystals rotated in the a-b plane by the tilt angles α^l and α^r (see Fig.2). In order to calculate the current we fix the phase difference ψ across the interface, solve the quasiclassical equations, and obtain the order parameter $\Delta(\mathbf{p}_f, \mathbf{R})$, and the Matsubara propagator $\hat{g}^M(\mathbf{p}_f, \mathbf{R}; \epsilon_n)$. The supercurrent across the interface can then be calculated from (20).

We neglect the coupling in c-direction, which seems a good first approxima-tion for layered cuprate superconductors, and model the conduction elec-trons of our d-wave superconductors by a cylindrical Fermi surface with radius p_f, and an isotropic Fermi velocity $\mathbf{v}_f$ in the a-b plane. The pairing interaction $V(\mathbf{p}_f, \mathbf{p}_f')$ is taken as purely d-wave, i.e.,

$$V(\mathbf{p}_f, \mathbf{p}_f') = 2V_0 \cos(2\phi) \cos(2\phi') . \qquad (34)$$

The angles ϕ and ϕ' are the polar angles of the momenta $\mathbf{p}_f$ and $\mathbf{p}_f'$ in the a-b planes. The critical temperature is determined by the dimen-sionless coupling constant V_0 and a cut-off energy ϵ_c via the BCS relation $T_c = 1.13\epsilon_c e^{-1/V_0}$. In this simple model the superconductors are isotropic in the a-b planes, except for the pairing interaction, which leads to an order

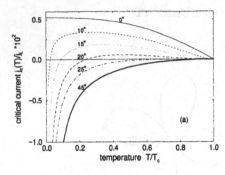
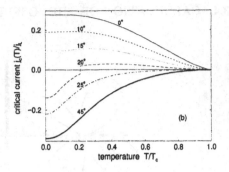

Figure 4. Temperature dependence of the Josephson critical current $j_c(T)$ of an ideal interface ($\rho_0 = \rho_{in} = 0$) for different tilt angles $\alpha^l = -\alpha^r$ between $0°$ and $45°$. (a) Tunnel junctions ($\mathcal{T}_0 = 0.01$), (b) strongly coupled junctions ($\mathcal{T}_0 = 0.50$).

parameter, $\Delta(\mathbf{p}_f) = \Delta \cos(2\phi)$, with gap zeros and a sign change in (1,1) direction.

Our interface is modeled by a reflection probability $\mathcal{R}(\theta)$, and a roughness parameter ρ_0. The pair-breaking scattering rate ρ_{in} is zero in this section. We follow [30] and choose the following one-parameter model for the dependence of $\mathcal{R}$ on the angle of incidence θ,

$$\mathcal{R}(\theta) = \frac{\mathcal{R}_0}{\mathcal{R}_0 + (1 - \mathcal{R}_0)\cos^2(\theta)} . \tag{35}$$

It interpolates smoothly between a reflection probability $\mathcal{R}_0$ at perpendicular incidence and total reflection ($\mathcal{R} = 1$) at glancing incidence. All results presented in this section were obtained by numerical calculations, which involve a self-consistent calculation of the order parameter in the superconducting electrodes and of the scattering self-energy in the two rough layers.

In Figs.3a,b we show the temperature dependence of the Josephson critical current for d-wave superconductors, and ideally smooth ($\rho_0 = 0$) interfaces. We compare the cases of a very weak transparency ($\mathcal{T}_0 = 1 - \mathcal{R}_0 = 0.01$) and a rather large transparency ($\mathcal{T}_0 = 0.50$). The critical currents are calculated for a representative set of tilt angles, $\alpha^l = -\alpha^r = 0°-45°$. The critical current is normalized in all figures by $j_L = |e|v_f N_f \Delta_0$ which is the Landau critical current at $T = 0$ of a clean s-wave superconductor with the same T_c as the d-wave superconductor. Our results for d-wave superconductors differ significantly from the Ambegaokar-Baratoff curve for s-wave superconductors. One finds [7]:

a) A strong dependence on the orientation of the crystals.

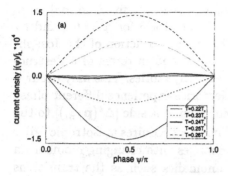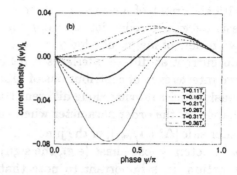

Figure 5. Current-phase relationship $j_J(\psi)$ of an ideal interface ($\rho_0 = \rho_{in} = 0$) at different temperatures. (a) $T_0 = 0.01$, (b) $T_0 = 0.50$. The tilt angles are fixed at $\alpha^l = -\alpha^r = 20°$.

b) The temperature dependence shows an anomalous positive curvature near T_c. This is a consequence of the increasing width of the region of a depleted order parameter with increasing coherence length $\xi(T)$.

c) For intermediate tilt angles the junction characteristics switches with increasing temperature from having its maximum current at phase difference $\psi = \pi$ (π-type) to a maximum at $\psi = 0$ (0-type).

d) This leads for junctions with a weak transparency to a non-monotonic temperature dependence of the critical current (see also [54]). It decreases with increasing T at low T, reaches a minimum, and increases again before it decreases to 0 when approaching T_c.

e) Junctions with a weak transparency show an anomalous enhancement of the critical current at low T. This is a consequence of the existence of zero-energy bound states at the interface [26].

f) Strongly coupled junctions show a similar behavior, but the anomalies are reduced in this case.

The current-phase relation for the junctions of Figs.3a,b is shown in Figs.4a,b for the specific tilt angles, $\alpha^l = -\alpha^r = 20°$, where the anomaly in $j_c(T)$ is most pronounced. The temperatures are chosen below, above and at the characteristic temperature at which the junction jumps from 0-type to π-type. The figure demonstrates that the anomalies in the critical current are accompanied by equally significant anomalies in the current-phase relation. At the jump we find a nearly sinusoidal behavior of the critical current with a doubled period, i.e., $j_J(\psi) \sim \sin(2\psi)$. This is a consequence of the finite transparency of the interface and cannot be understood within the framework of equation (33), which is valid only to linear order in the transparency $\mathcal{T}$.

The influence of interface roughness on the Josephson critical current is shown in Figs.5a,b. Roughness reduces strongly the magnitude of the

430

critical current of anisotropic superconductors, even at fixed normal state resistance. The reduction is in our cases by a factor 5 for $\alpha^l = 0°$, and by two orders of magnitude for $\alpha^l = 45°$. The strong reduction of the Josephson current at diffuse interfaces can be understood in terms of a destructive interference of contributions of different trajectories. This may occur if quasiparticles moving along different trajectories experience different phase changes of the order parameter when going from one side $[\Delta^s(\mathbf{p}_{f\,in})]$ to the other side $[\Delta^{s'}(\mathbf{p}_{f\,out})]$ of the junction. This effect requires anisotropic order parameters whose phase (e.g., the sign) changes with changing momentum direction. It is important to note that anomalies such as the transitions from a 0-junction to a π-junction, and the enhancement at low temperatures are sensitive to interface roughness, and have disappeared for $\rho_0 = 2$. At $\rho_0 = 2$ we have already reached to a good approximation the rough limit ($\rho_0 = \infty$) of our model. The results for $\rho_0 > 2$ are essentially unchanged compared to $\rho_0 = 2$, except for $\alpha^l = -\alpha^r = 45°$ where j_J vanishes for $\rho_0 = \infty$.

Figs.6a,b finally demonstrate the role of the distorted order parameter near the interface. They show the calculated critical currents for a non-selfconsistent, constant order parameter. These approximate results exhibit the same qualitative features as Figs.3a,5a, but show significantly different magnitudes and T-dependences at temperatures above $\approx 0.5\,T_c$.

3.5. AMBEGAOKAR-BARATOFF RELATION

[2] derived a universal relation between the critical Josephson current across a junction of weakly coupled isotropic BCS superconductors, the energy gap and the junction resistance in the normal state (11). The maximum

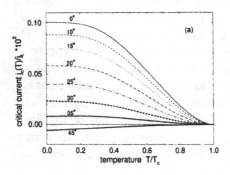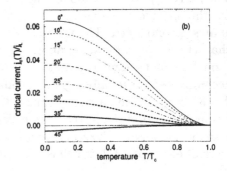

Figure 6. The same as in Figs.3a,b but for a rough interface ($\rho_0 = 2, \rho_{in} = 0$).

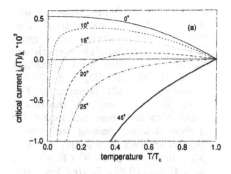
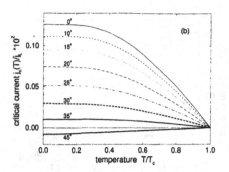

Figure 7. The same as in Figs.3a,5a, but with an artificially suppressed depletion of the order parameter near the interface. A comparison with Figs.3a,5a shows the effect of the depletion layers.

Josephson current (Ambegaokar-Baratoff limit) is obtained at $T = 0$:

$$I_c R_N = \frac{\pi \Delta_0}{2 |e|} , \qquad (36)$$

This relation no longer holds for anisotropic superconductors. The $I_c R_N$-product is non-universal and depends, in general, on the orientation of the order parameter and the quality of the interface.

The critical Josephson current, $I_c = A j_c$, is given in section 5 in terms of the Landau critical current

$$j_L = |e| \, v_f N_f \Delta_0 . \qquad (37)$$

In addition one obtains from (31) and the reflection probability (35) in the limit of weak transparency the following interface resistance in the normal state.

$$R_N = \frac{3\pi}{4 e^2 A N_f v_f T_0} . \qquad (38)$$

By combining (37) and (38) one obtains

$$R_N I_c = \frac{\pi \Delta_0}{2 |e|} \times \frac{3}{2 T_0} \frac{j_c}{j_L} \qquad (39)$$

We factorized the right side of (39) into the Ambegaokar-Baratoff result and a correction term which is 1 for isotropic BCS superconductors and determines, in general, the deviation from the Ambegaokar-Baratoff relation. The correction factor is written in terms of the ratio j_c/j_L given in figures 3, 5, 6. One can infer from these results that the correction factor

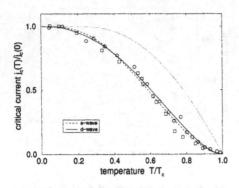

Figure 8. Temperature dependence of the critical current density of a symmetric single grain boundary junction. The experimental data are taken from Mannhart et al. (1988) (10° ($\square$) and 15° (o) $YBa_2Cu_3O_{7-\delta}$ tilt boundary). The solid (dashed) line is a theoretical curve obtained from our model for d-wave (s-wave) pairing with interface parameters $\mathcal{T}_0 = 0.001$, $\rho_0 = 0.5$, $\rho_{in} = 0$, $\alpha^l = -\alpha^r = 5°$ ($\mathcal{T}_0 = 0.001$, $\rho_0 = 0$, $\rho_{in} = 0.34$). Dotted line: Ambegaokar-Baratoff formula [Eq. (1)].

may be negative and larger than 1 for *ideal* junctions with d-wave superconductors (see Fig.3a), and is much smaller than 1 for *rough* junctions with d-wave superconductors (see Fig.5a). In addition, the correction factor depends strongly on the orientation of the order parameter (see Figs.3, 5). In Fig.7 we compare temperature dependent critical current measurements on symmetric single grain boundary junctions by J. Mannhart et al. [40] (10° and 15° YBCO tilt boundaries) with our theoretical calculations. Our calculations have shown that the standard Fermi-liquid theory of superconductivity in correlated, anisotropic metals predicts characteristic differences for d-wave and s-wave superconductors in the temperature dependence of the Josephson current and the current-phase relation. These differences are to some degree washed out by interface roughness, such that the best way to observe these effects would be experiments on ideally clean grain boundary junctions with a weak transparency, and the possibility of continuously varying the tilt angles.

4. Electromagnetic Response of a Pancake Vortex in Layered Superconductors

In oder to demonstrate the capacities of the Fermi liquid theory of superconductivity we apply in this section the quasiclassical theory of superconductivity to a non-trivial dynamical problem. We calculate the response of the currents in the core of a vortex to an alternating electric field. An electric field has two principal effects on a superconductor. It changes the supercurrents by accelerating the superfluid condensate, and it generates dissipation by exciting "normal" quasiparticles. This two-fluid picture of a *condensate* and *normal excitations* is clearly reflected in the optical conductivity of standard s-wave superconductors in the Meissner state [42]. The conductivity has a superfluid part,

$$\sigma_s(\omega) = \frac{e^2 n_s}{m} \left(\delta(\omega) + iP\frac{1}{\omega} \right) , \qquad (40)$$

and a dissipative normal part. At $T = 0$ dissipation starts at the threshold frequency for creating quasiparticles, $\omega = 2\Delta$. At finite temperature the coupling to thermally excited quasiparticles leads to dissipation at all frequencies. The two-fluid picture no longer holds for type-II superconductors in the vortex state. The response to an electric field consists of contributions from gapless[2] vortex core excitations [16], and contributions from excitations outside the cores, where the quasiparticle spectrum shows the bulk gap. For high-κ superconductors the electric response at low frequencies of the region outside the core can be described very well by the London equations [38], i.e. by two-fluid electrodynamics. The response of the core is more complex. In the traditional model of a "normal core" (Bardeen and Stephen [10]) the conductivity in the core is that of the normal state, $\sigma_{core} = \sigma_n$. The Bardeen-Stephen model [10] is a plausible approximation for dirty superconductors with a mean free path (ℓ) much shorter than the size of the core ($\ell \ll \xi_0$). In this limit the vortex core excitations of Caroli et al. [16] may be considered as a continuum of normal excitations.

In 1969 Bardeen et al. [11] published a detailed discussion of the bound states in the core of a vortex, and argued that the bound states contribute significantly to the circulating supercurrents in the vortex core. This effect was confirmed by recent self-consistent calculations of free and pinned vortices in clean and medium dirty superconductors [48]. The authors show that all currents in the core (circulating currents as well as superimposed transport currents) are carried predominantly by the bound states. This means that the model of a normal core needs to be modified for relatively clean ($\ell \gtrsim \xi_0$) superconductors. Furthermore the response of the bound states is expected to show dissipative as well as superfluid features.

4.1. QUASICLASSICAL THEORY OF THE ELECTROMAGNETIC RESPONSE

A convenient formulation of the quasiclassical theory for our purposes is in terms of the quasiclassical Nambu-Keldysh propagator $\check{g}(\mathbf{p}_f, \mathbf{R}; \epsilon, t)$, which is a 4×4-matrix in Nambu-Keldysh space, and a function of position $\mathbf{R}$, time t, energy ϵ, and Fermi momentum $\mathbf{p}_f$. We consider a superconductor with random atomic size impurities in a static magnetic field, and an externally applied $a.c.$ electric field, $\mathbf{E} = -\frac{1}{c}\partial_t \delta\mathbf{A}$. In a compact notation the transport

---

[2]We ignore the "mini-gap" of size Δ^2/E_f predicted by Caroli et al. [16].

equation for this system and the normalization condition read

$$[(\epsilon + \frac{e}{c}\mathbf{v}_f \cdot \mathbf{A})\check{\tau}_3 - \check{\Delta}_{mf} - \check{\sigma}_i - \delta\check{v}, \check{g}]_\otimes + i\mathbf{v}_f \cdot \boldsymbol{\nabla}\check{g} = 0 \ , \tag{41}$$

$$\check{g} \otimes \check{g} = -\pi^2 \check{1} \ , \tag{42}$$

where $\mathbf{A}(\mathbf{R})$ is the vector potential of the static magnetic field, $\mathbf{B} = \boldsymbol{\nabla} \times \mathbf{A}$, $\check{\Delta}_{mf}(\mathbf{p}_f, \mathbf{R}; t)$ the mean-field order parameter matrix, and $\check{\sigma}_i(\mathbf{p}_f, \mathbf{R}; \epsilon, t)$ is the impurity self-energy. The perturbation $\delta\check{v}(\mathbf{p}_f, \mathbf{R}; t)$ includes the external electric field and the field of the charge fluctuations, $\delta\rho(\mathbf{R}; t)$, induced by the external field. For convenience we describe the external electric field by a vector potential $\delta\mathbf{A}(\mathbf{R}; t)$ and the induced electric field by the electro-chemical potential $\delta\varphi(\mathbf{R}; t)$. Hence in the Nambu-Keldysh matrix notation the perturbation has the form,

$$\delta\check{v} = -\frac{e}{c}\mathbf{v}_f \cdot \delta\mathbf{A}(\mathbf{R}; t)\check{\tau}_3 + e\delta\varphi(\mathbf{R}; t)\check{1} \ , \tag{43}$$

and is assumed to be sufficiently small so that it can be treated in linear response theory.

Equations (41) and (42) must be supplemented by self-consistency equations for the order parameter and the impurity self-energy. We use the weak-coupling gap equations,

$$\hat{\Delta}_{mf}^{R,A}(\mathbf{p}_f, \mathbf{R}; t) = \int_{-\epsilon_c}^{+\epsilon_c} \frac{d\epsilon}{4\pi i} \langle V(\mathbf{p}_f, \mathbf{p}_f')\hat{f}^K(\mathbf{p}_f', \mathbf{R}; \epsilon, t)\rangle \ , \tag{44}$$

$$\hat{\Delta}_{mf}^K(\mathbf{p}_f, \mathbf{R}; t) = 0 \ , \tag{45}$$

and the impurity self-energy in Born approximation with isotropic scattering,

$$\check{\sigma}_i(\mathbf{R}; \epsilon, t) = \frac{1}{2\pi\tau} \langle \check{g}(\mathbf{p}_f', \mathbf{R}; \epsilon, t)\rangle \ , \tag{46}$$

where $\hat{f}^K$ is the off-diagonal part of the 2×2 Nambu matrix $\hat{g}^K$, and the Fermi surface average is defined by

$$\langle \dots \rangle = \frac{1}{N_f} \int \frac{d^2\mathbf{p}_f'}{(2\pi)^3 |\mathbf{v}_f'|} \dots \ . \tag{47}$$

The materials parameters that enter the self-consistency equations are the pairing interaction, $V(\mathbf{p}_f, \mathbf{p}_f')$, the impurity scattering lifetime, τ, in addition to the Fermi surface data $\mathbf{p}_f$ (Fermi surface), $\mathbf{v}_f$ (Fermi velocity), and $N_f = \int \frac{d^2\mathbf{p}_f}{(2\pi)^3 |\mathbf{v}_f|}$.

In the linear response approximation one splits the propagator and the self-energies into an unperturbed part and a term of first order in the perturbation,

$$\breve{g} = \breve{g}_0 + \delta\breve{g}, \quad \breve{\Delta}_{mf} = \breve{\Delta}_{mf0} + \delta\breve{\Delta}_{mf}, \quad \breve{\sigma}_i = \breve{\sigma}_0 + \delta\breve{\sigma}_i , \qquad (48)$$

and expands the transport equation and normalization condition through first order. In 0th order we obtain

$$[(\epsilon + \frac{e}{c}\mathbf{v}_f \cdot \mathbf{A})\breve{\tau}_3 - \breve{\Delta}_{mf0} - \breve{\sigma}_0, \breve{g}_0]_\otimes + i\mathbf{v}_f \cdot \nabla \breve{g}_0 = 0 , \qquad (49)$$

$$\breve{g}_0 \otimes \breve{g}_0 = -\pi^2 \breve{1} , \qquad (50)$$

and in 1st order

$$[(\epsilon + \frac{e}{c}\mathbf{v}_f \cdot \mathbf{A})\breve{\tau}_3 - \breve{\Delta}_{mf0} - \breve{\sigma}_0, \delta\breve{g}]_\otimes + i\mathbf{v}_f \cdot \nabla \delta\breve{g} = [\delta\breve{\Delta}_{mf} + \delta\breve{\sigma}_i + \delta\breve{v}, \breve{g}_0]_\otimes ,$$
$$(51)$$

$$\breve{g}_0 \otimes \delta\breve{g} + \delta\breve{g} \otimes \breve{g}_0 = 0 . \qquad (52)$$

In order to close this system of equations one has to supplement the transport and normalization equations with the self-consistency equations of 0th and 1st order:

$$\hat{\Delta}_{mf0}^{R,A}(\mathbf{p}_f, \mathbf{R}) = \int_{-\epsilon_c}^{+\epsilon_c} \frac{d\epsilon}{4\pi i} \langle V(\mathbf{p}_f, \mathbf{p}_f')\hat{f}_0^K(\mathbf{p}_f', \mathbf{R}; \epsilon)\rangle, \quad \hat{\Delta}_{mf0}^K = 0 , \qquad (53)$$

$$\delta\hat{\Delta}_{mf}^{R,A}(\mathbf{p}_f, \mathbf{R}; t) = \int_{-\epsilon_c}^{+\epsilon_c} \frac{d\epsilon}{4\pi i} \langle V(\mathbf{p}_f, \mathbf{p}_f')\delta\hat{f}^K(\mathbf{p}_f', \mathbf{R}; \epsilon, t)\rangle, \quad \delta\hat{\Delta}_{mf}^K = 0 ,$$
$$(54)$$

and

$$\breve{\sigma}_0(\mathbf{R}; \epsilon) = \frac{1}{2\pi\tau} \langle \breve{g}_0(\mathbf{p}_f', \mathbf{R}; \epsilon)\rangle , \qquad (55)$$

$$\delta\breve{\sigma}_i(\mathbf{R}; \epsilon, t) = \frac{1}{2\pi\tau} \langle \delta\breve{g}(\mathbf{p}_f', \mathbf{R}; \epsilon, t)\rangle . \qquad (56)$$

Finally, the electro-chemical potential, $\delta\varphi$, is determined by the condition of local charge neutrality [25, 5]. This condition follows from the expansion of charge density to leading order in the quasiclassical expansion parameters. One obtains $\delta\rho(\mathbf{R}; t) = 0$, i.e.

$$-2e^2 N_f \delta\varphi(\mathbf{R}; t) + eN_f \int \frac{d\epsilon}{4\pi i} \langle \mathrm{Tr}\delta\hat{g}^K(\mathbf{p}_f', \mathbf{R}; \epsilon, t)\rangle = 0 . \qquad (57)$$

The self-consistency equations (53)-(56) are of vital importance in the context of this paper. Equations (54) and (56) for the response of the quasiclassical self-energies are equivalent to *vertex corrections* in the Green's function response theory. They guarantee that the quasiclassical theory does not violate fundamental conservation laws. In particular, (55) and (56) imply charge conservation in scattering processes, whereas (53) and (54) imply charge conservation in a particle-hole conversion process. Any charge which is lost or gained in a particle-hole conversion process is balanced by the corresponding gain or loss of condensate charge. It is the coupled quasiparticle dynamics and collective condensate dynamics which conserves charge in superconductors. Neglect of the dynamics of either component, or use of a non-conserving approximation for the coupling of quasiparticles and collective degrees of freedom leads to unphysical results. Condition (57) is a consequence of the long-range of the Coulomb repulsion. The Coulomb energy of a charged region of size ξ_0^3 and typical charge density $eN_f\Delta$ is $\sim e^2 N_f^2 \Delta^2 \xi_0^5$, which should be compared with the condensation energy $\sim N_f \Delta^2 \xi_0^3$. Thus, the cost in Coulomb energy is a factor $(E_f/\Delta)^2$ larger than the condensation energy. This leads to a strong suppression of charge fluctuations, and the condition of local charge neutrality holds to very good accuracy for superconducting phenomena.

Equations (49)-(57) constitute a complete set of equations for calculating the electromagnetic response of a vortex. The structure of a vortex in equilibrium is obtained from (49), (50), (53) and (55), and the linear response of the vortex to the perturbation $\delta\mathbf{A}(\mathbf{R};t)$ follows from (51), (52), (54), (56) and (57). The currents induced by $\delta\mathbf{A}(\mathbf{R};t)$ can then be calculated directly from the Keldysh propagator $\delta\hat{g}^K$ via

$$\delta\mathbf{j}(\mathbf{R};t) = eN_f \int \frac{d\epsilon}{4\pi i} \langle \mathbf{v}_f(\mathbf{p}_f')\mathrm{Tr}\left(\hat{\tau}_3 \delta\hat{g}^K(\mathbf{p}_f', \mathbf{R}; \epsilon, t)\right) \rangle . \tag{58}$$

4.2. STRUCTURE AND SPECTRUM OF A PANCAKE VORTEX

We consider an isolated pancake vortex in a strongly anisotropic, layered superconductor, and model the Fermi surface of these systems by a cylinder of radius p_f, and a Fermi velocity of constant magnitude, v_f, along the layers. We also assume isotropic (s-wave) pairing and isotropic impurity scattering. Thus, the materials parameters of the model superconductor are: T_c, v_f, the 2D density of states $N_f = p_f/2\pi v_f$, and the mean free path $\ell = v_f\tau$. The model superconductor is type II with a large Ginzburg-Landau parameter $\kappa \gtrsim 100$, so the magnetic field is to good approximation constant in the region of the vortex core.

We first present results for the equilibrium vortex. The order parameter, $\Delta_0(\mathbf{R})$, is calculated by solving the transport equation (49), the gap equation (53), and the self-energy equation (55) self-consistently. These calculations are done at Matsubara energies ($\epsilon \to i\epsilon_n = i(2n+1)\pi T$). Details of the numerical schemes for solving the transport equation self-consistently are

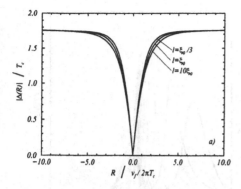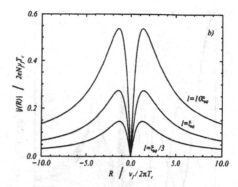

Figure 9. Results of self-consistent calculations of the modulus of order parameter (a) and the current density (b) at $T = 0.3T_c$ for superconductors with electron mean free path $\ell = 10\xi_0$ (clean) to $\ell = \frac{1}{3}\xi_0$ (dirty). R is the distance from the vortex center measured in units of the coherence length.

given elsewhere [22]. Charge conservation for the equilibrium vortex follows from the circular symmetry of the currents. Nevertheless, self-consistency of the equilibrium vortex is important; the equilibrium self-energies ($\check{\Delta}_0$, $\check{\sigma}_0$) and propagators ($\check{g}_0$) are input quantities in the transport equation (51) for the linear response. Charge conservation in linear response is non-trivial and requires self-consistency of both the equilibrium solution and the solution in first order in the perturbation.

Fig.1 shows the order parameter and the current density in the vortex core of an equilibrium vortex for different impurity scattering rates τ. As expected, scattering reduces the coherence length and thus the size of the core, and has a strong effect on the current density. Numerical results for the excitation spectrum of bound and continuum states at the vortex together with the corresponding spectral current densities are shown in Fig.2. The local density of states (per spin) and the spectral current density are defined by

$$N(\mathbf{R}, \epsilon) = N_f \frac{i}{4\pi} \mathrm{Tr} \langle \hat{\tau}_3 \left(\hat{g}^R(\mathbf{p}'_f, \mathbf{R}, \epsilon) - \hat{g}^A(\mathbf{p}'_f, \mathbf{R}, \epsilon) \right) \rangle \qquad (59)$$

$$\mathbf{j}(\mathbf{R}, \epsilon) = eN_f \frac{i}{2\pi} \mathrm{Tr} \langle \hat{\tau}_3 \mathbf{v}_f(\mathbf{p}'_f) \left(\hat{g}^R(\mathbf{p}'_f, \mathbf{R}, \epsilon) - \hat{g}^A(\mathbf{p}'_f, \mathbf{R}, \epsilon) \right) \rangle . \quad (60)$$

The zero-energy bound state is remarkably broadened by impurity scattering. Its width is well approximated by the scattering rate $1/(v_f \ell)$, which is $0.63T_c$ for $\ell = 10\xi_0$. The broadening of the bound states decreases with increasing energy. The results for the spectral current density (Fig.2b) show that nearly all of the current density of the equilibrium vortex resides in the energy range of the bound states. This reflects the observation that the supercurrents in the vortex core are predominantly carried by the bound

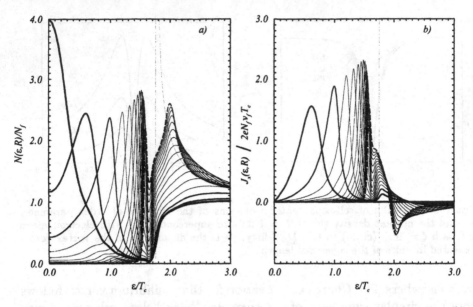

Figure 10. Local density of states $N(\mathbf{R}, \epsilon)$ (a) and local spectral current density $j_x(\mathbf{R}, \epsilon)$ (b) in the vortex core of a superconductor with $\ell = 10\xi_0$ at $T = 0.3T_c$. Results are shown for a series of spatial points on the y-axis, at distances $0, .25\pi\xi_0, .5\pi\xi_0, \ldots, 4\pi\xi_0$ from the vortex center. The thickest full line corresponds to the vortex center, and decreasing thickness indicates increasing distance from the center. Results for the outermost point $(4\pi\xi_0)$ are shown as dashed lines. The thin dotted lines show the density of states of the homogeneous superconductor (a) and the value of the bulk gap (b) respectively.

states [11, 48]. The physics of the core is dominated by the bound states, in particular also, as we will show, the response of the core to an electric field.

4.3. DISTRIBUTION FUNCTIONS

The set of formulas for calculating the quasiclassical linear response of a superconductor is given in a compact notation in (48)-(57). In this section we transform these formulas into a more suitable form for analytical and numerical calculations. The central differential equation of the quasiclassical response theory is the transport equation (51). It comprises 12 differential equations for the components of the three 2×2-Nambu matrices, $\delta\hat{g}^{R,A,K}$. The number of differential equations can be reduced significantly by using general symmetry relations and the normalization conditions of the quasiclassical theory. We focus on the simplifications which follow from the normalization equations (50) and (52). We use the projection operators introduced by Shelankov [51],

$$\hat{P}_+^{R,A} = \frac{1}{2}\left(\hat{1} + \frac{1}{-i\pi}\hat{g}_0^{R,A}\right) , \quad \hat{P}_-^{R,A} = \frac{1}{2}\left(\hat{1} - \frac{1}{-i\pi}\hat{g}_0^{R,A}\right) . \tag{61}$$

Obviously, $\hat{P}_+^{R,A} + \hat{P}_-^{R,A} = \hat{1}$. The algebra of the projection operators follows from the normalization conditions.

$$(\hat{P}_+^{R,A})^2 = \hat{P}_+^{R,A}, \; (\hat{P}_-^{R,A})^2 = \hat{P}_-^{R,A},$$
$$\hat{P}_+^{R,A}\hat{P}_-^{R,A} = \hat{P}_-^{R,A}\hat{P}_+^{R,A} = 0. \tag{62}$$

A key result is that the Nambu matrices $\delta\hat{g}^{R,A,K}$ can be expressed, with the help of Shelankov's projectors, in terms of 6 scalar *distribution functions*, $\delta\gamma^{R,A}$, $\delta\tilde{\gamma}^{R,A}$, δx^a and $\delta\tilde{x}^a$, each of which is a function of $\mathbf{p}_f$, $\mathbf{R}$, ϵ, t, and satisfies a scalar transport equation. The distribution functions are defined by

$$\delta\hat{g}^{R,A} \tag{63}$$
$$= \mp 2\pi i\left[\hat{P}_+^{R,A}\otimes\begin{pmatrix} 0 & \delta\gamma^{R,A} \\ 0 & 0 \end{pmatrix}\otimes\hat{P}_-^{R,A} - \hat{P}_-^{R,A}\otimes\begin{pmatrix} 0 & 0 \\ -\delta\tilde{\gamma}^{R,A} & 0 \end{pmatrix}\otimes\hat{P}_+^{R,A}\right],$$

and

$$\delta\hat{g}^a \tag{64}$$
$$= -2\pi i\left[\hat{P}_+^R\otimes\begin{pmatrix} \delta x^a & 0 \\ 0 & 0 \end{pmatrix}\otimes\hat{P}_-^A + \hat{P}_-^R\otimes\begin{pmatrix} 0 & 0 \\ 0 & \delta\tilde{x}^a \end{pmatrix}\otimes\hat{P}_+^A\right] .$$

where the *anomalous response*, $\delta\hat{g}^a$, is defined in terms of $\delta\hat{g}^K$, $\delta\hat{g}^R$, $\delta\hat{g}^A$ by

$$\delta\hat{g}^K = \delta\hat{g}^R\otimes\tanh(\beta\epsilon/2) - \tanh(\beta\epsilon/2)\otimes\delta\hat{g}^A + \delta\hat{g}^a , \tag{65}$$

The transport equations for the various distribution functions follow from (49) and (51) and one finds [22],

$$i\mathbf{v}_f\cdot\boldsymbol{\nabla}\delta\gamma^{R,A} + 2\epsilon\delta\gamma^{R,A}$$
$$+(\gamma_0^{R,A}\tilde{\Delta}^{R,A} - \Sigma^{R,A})\otimes\delta\gamma^{R,A} + \delta\gamma^{R,A}\otimes(\tilde{\Delta}^{R,A}\gamma_0^{R,A} + \Sigma^{R,A}) \tag{66}$$
$$= -\gamma_0^{R,A}\otimes\delta\tilde{\Delta}^{R,A}\otimes\gamma_0^{R,A} + \delta\Sigma^{R,A}\otimes\gamma_0^{R,A} - \gamma_0^{R,A}\otimes\delta\tilde{\Sigma}^{R,A} - \delta\Delta^{R,A},$$

$$i\mathbf{v}_f\cdot\boldsymbol{\nabla}\delta\tilde{\gamma}^{R,A} - 2\epsilon\delta\tilde{\gamma}^{R,A}$$
$$+(\tilde{\gamma}_0^{R,A}\Delta^{R,A} - \tilde{\Sigma}^{R,A})\otimes\delta\tilde{\gamma}^{R,A} + \delta\tilde{\gamma}^{R,A}\otimes(\Delta^{R,A}\tilde{\gamma}_0^{R,A} + \Sigma^{R,A}) \tag{67}$$
$$= -\tilde{\gamma}_0^{R,A}\otimes\delta\Delta^{R,A}\otimes\tilde{\gamma}_0^{R,A} + \delta\tilde{\Sigma}^{R,A}\otimes\tilde{\gamma}_0^{R,A} - \tilde{\gamma}_0^{R,A}\otimes\delta\Sigma^{R,A} - \delta\tilde{\Delta}^{R,A},$$

$$iv_f \cdot \nabla \delta x^a + i\partial_t \delta x^a + (\gamma_0^R \tilde{\Delta}^R - \Sigma^R) \otimes \delta x^a + \delta x^a \otimes (\Delta^A \tilde{\gamma}^A + \Sigma^A)$$
$$= \gamma_0^R \otimes \delta\tilde{\Sigma}^a \otimes \tilde{\gamma}_0^A - \delta\Delta^a \otimes \tilde{\gamma}_0^A - \gamma_0^R \otimes \delta\tilde{\Delta}^a - \delta\Sigma^a , \tag{68}$$

$$iv_f \cdot \nabla \delta\tilde{x}^a - i\partial_t \delta\tilde{x}^a + (\tilde{\gamma}_0^R \Delta^R - \tilde{\Sigma}^R) \otimes \delta\tilde{x}^a + \delta\tilde{x}^a \otimes (\tilde{\Delta}^A \gamma_0^A + \tilde{\Sigma}^A)$$
$$= \tilde{\gamma}_0^R \otimes \delta\Sigma^a \otimes \gamma_0^A - \delta\tilde{\Delta}^a \otimes \gamma_0^A - \tilde{\gamma}_0^R \otimes \delta\Delta^a - \delta\tilde{\Sigma}^a . \tag{69}$$

We have used the following short-hand notation for the driving terms in the transport equations, which includes external potentials, perturbations and self-energies:

$$-\frac{e}{c}\mathbf{v}_f \cdot \mathbf{A}\hat{\tau}_3 + \hat{\Delta}_{mf0} + \hat{\sigma}_{i0}^{R,A} = \begin{pmatrix} \Sigma^{R,A} & \Delta^{R,A} \\ -\tilde{\Delta}^{R,A} & \tilde{\Sigma}^{R,A} \end{pmatrix}, \tag{70}$$

$$\delta\hat{h} = \delta\hat{\Delta}_{mf} + \delta\hat{\sigma}_i + \delta\check{v} , \tag{71}$$

$$\delta\hat{h}^K = \delta\hat{h}^R \otimes \tanh(\beta\epsilon/2) - \tanh(\beta\epsilon/2) \otimes \delta\hat{h}^A + \delta\hat{h}^a , \tag{72}$$

$$\delta\hat{h}^{R,A} = \begin{pmatrix} \delta\Sigma^{R,A} & \delta\Delta^{R,A} \\ -\delta\tilde{\Delta}^{R,A} & \delta\tilde{\Sigma}^{R,A} \end{pmatrix}, \quad \delta\hat{h}^a = \begin{pmatrix} \delta\Sigma^a & \delta\Delta^a \\ \delta\tilde{\Delta}^a & -\delta\tilde{\Sigma}^a \end{pmatrix}. \tag{73}$$

The functions $\gamma_0^{R,A}$ and $\tilde{\gamma}_0^{R,A}$ in (66)-(69) are defined by the following convenient parameterization of the equilibriun. propagators.

$$\hat{g}_0^{R,A} = \mp i\pi \frac{1}{1 + \gamma_0^{R,A}\tilde{\gamma}_0^{R,A}} \begin{pmatrix} 1 - \gamma_0^{R,A}\tilde{\gamma}_0^{R,A} & 2\gamma_0^{R,A} \\ 2\tilde{\gamma}_0^{R,A} & -(1 - \gamma_0^{R,A}\tilde{\gamma}_0^{R,A}) \end{pmatrix} . \tag{74}$$

After elimination of the time-dependence in (66)-(69) by Fourier transform one is left with four sets of ordinary differential equations along straight trajectories in R-space. For given right-hand sides these equations are decoupled, and determine the distribution functions $\delta\gamma^{R,A}$, $\delta\tilde{\gamma}^{R,A}$, δx^a and $\delta\tilde{x}^a$. On the other hand, the self-consistency conditions relate the right-hand sides of (66)-(69) to the solutions of (66)-(69). Hence, equations (66)-(69) may be considered either as a large system of linear differential equations of size six times the chosen number of trajectories, or as a self-consistency problem. We solved the self-consistency problem numerically using special algorithms for updating the right hand sides. Details of our numerical schemes for a self-consistent determination of the response functions are given elsewhere [22].

4.4. RESPONSE TO AN A.C. ELECTRIC FIELD

We consider a pancake vortex in a layered s-wave superconductor and calculate the response of the electric current density, $\delta\mathbf{j}(\mathbf{R};t)$, to a small homogeneous a.c. electric field, $\delta\mathbf{E}^\omega(t) = \delta\mathbf{E}\exp(-i\omega t)$. The results presented in

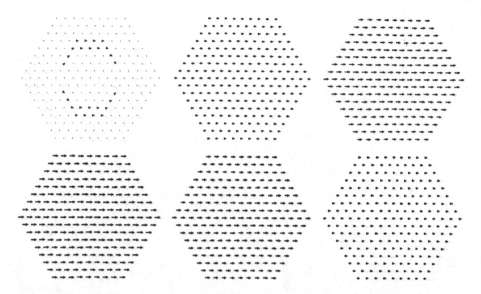

Figure 11. Snapshots of the time-dependent current pattern in the core of a pancake vortex at successive times, $t = 0, \frac{1}{12}\mathcal{T}, \frac{2}{12}\mathcal{T}, \frac{3}{12}\mathcal{T}, \frac{4}{12}\mathcal{T}, \frac{5}{12}\mathcal{T}$ (upper left pattern to lower right pattern). The arrows show the current density induced by an a.c. electric field in x-direction, $E_x(t) = \delta E \cos(\omega t)$, of frequency $\omega = 2\pi/\mathcal{T} = 1.5\Delta$. The data are calculated in linear response approximation for a clean superconductor ($\ell = 10\xi_0$) at $T = 0.3 T_c$. Current patterns in the second half-period, $t = \frac{1}{2}\mathcal{T} - \frac{11}{12}\mathcal{T}$, are obtained from the patterns in the first half-period by reversing the directions of the currents. The distance between two neighboring points on the hexagonal grid is $0.25\pi\xi_0$.

this section are obtained by solving numerically the quasiclassical transport equations (66)-(69) together with the self-consistency equations (54), (56), and the condition of local charge neutrality (57). The calculation gives the local conductivity tensor, $\sigma_{ij}(\mathbf{R}, \omega)$, defined by

$$\delta j_i(\mathbf{R}, t) = \sigma_{ij}(\mathbf{R}, \omega) \delta E_j^\omega(t) \ . \tag{75}$$

Figures 11 and 12 show the time development of the current pattern induced by an oscillating electric field in x-direction with time dependence $\delta E_x(t) = \delta E \cos(\omega t)$. Results are given for a medium range frequency ($\omega = 1.5\Delta$) and a low frequency ($\omega = 0.3\Delta$). Two features should be emphasized. At medium and higher frequencies the current flow induced by the electric field is to good approximation uniform in space and phase shifted by $\pi/2$ (non-dissipative currents). The phase shift of $\pi/2$ is the consequence of a predominantly imaginary conductivity at frequencies above Δ, as shown in Fig.5. The current pattern at low frequencies (Fig.4) is qualitatively different. At the vortex center the current is phase shifted by $\approx \pi/4$ in accordance with the conductivity at $\omega = 0.3\Delta$, which has about equal real (dissipative) and imaginary (non-dissipative) parts, whereas further away

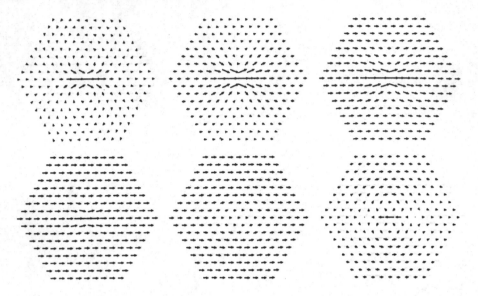

Figure 12. The same as in fig. 11, but for a smaller external frequency, $\omega = 0.3\Delta$. The length of the current vectors is scaled down by a factor 5 compared to those in fig. 11.

from the center the conductivity becomes more and more non-dissipative. Fig.4 shows that the current flow at low frequencies is non-uniform. The dissipative currents exhibit a dipolar structure with enhanced currents at the vortex center, and back-flow currents away from the center. On the other hand, non-dissipative currents are approximately uniform.

The frequency dependence of the local conductivity, $\sigma_{xx}(\mathbf{R}, \omega)$, for $\mathbf{R}$ along the x- and y-axis is shown in figures 13a,b. These figures include, for comparison, the Drude conductivity of the normal state and the conductivity of the homogeneous bulk superconductor. A significant feature of conductivity in the core is the strong increase of $\mathcal{R}e\,\sigma$ at low frequencies. The conductivity is in this frequency range much larger than the normal state Drude conductivity and the exponentially small conductivity of the bulk s-wave superconductor. The real part of conductivity scales at low frequencies like $1/\omega^2$. Its value at at the vortex center is $69.5e^2N_fv_f^2/\Delta$ (outside the range of the figure) for $\omega = 0.1\Delta$. The enhancement of the dissipative part at low frequencies is a consequence of the coupling of quasiparticles and collective order parameter modes, and cannot be obtained from non-selfconsistent calculations. Fig.13b shows that the real part of the conductivity becomes negative in the region of dipolar backflow on the y-axis. This leads locally to a negative time averaged power absorption, $< \mathbf{j} \cdot \mathbf{E} >_t < 0$, and a corresponding gain in energy, which is compensated by the strongly enhanced dissipation of energy in the center of the vortex. The dissipative part of the local conductivity at a distance R from the

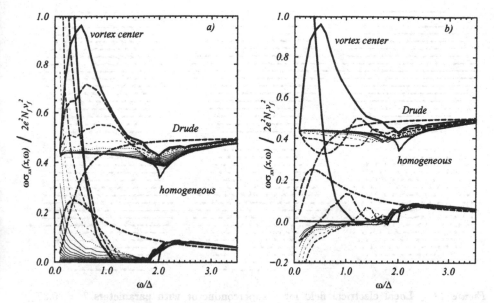

Figure 13. Frequency dependence of the real part ('lower' curves) and imaginary part ('upper' curves) of the local conductivity σ_{xx} for a superconductor with parameters $T = 0.3T_c$, $\ell = 10\xi_0$. For convenience, the conductivities are multiplied by ω. The full black curves give the conductivity at the vortex center, and the series of dashed lines with decreasing intensity the conductivity at increasing distance from the center. Fig.13a presents data at points along the x-axis (in steps of $(\pi/4)\xi_0$), and Fig.13b at points along the y-axis (in steps of $(\sqrt{3}\pi/4)\xi_0$). The dashed grey lines show the normal state Drude conductivity, and the full grey lines the conductivity of the homogeneous superconductor.

vortex center exhibits pronounced maxima whose frequencies increase with increasing R and are given by $2\times$ the energy at the maxima in the local density of states shown in Fig.2a. Hence, these features in the absorption spectrum must be identified as impurity assisted transitions between corresponding bound states at negative and positive energies. Impurities are required for breaking angular momentum conservation in these transitions. The applied electric field $\delta E(t)$ induces in the vortex core an internal field $-\nabla\delta\varphi(t)$, which is of the same order as the applied field. Fig.14 shows the total electric field, $\delta\mathbf{E}_{tot}(t) = \delta\mathbf{E}^\omega(t) - \nabla\delta\varphi(t)$, in the vortex core. The induced field is at low frequencies of dipolar form, and oscillates out of phase (phase shift $\pi/2$) with the applied field. This dipolar field originates from small charge fluctuation in the vortex core. At higher frequencies the dipolar field oscillates with a phase shift of $\approx \pi$, and screens part of the applied field.

We finally discuss the role of self-consistency in our calculation. Our results were obtained by iterating the self-consistency equations until the relative error stabilized below $\leq 10^{-10}$. Fig.15 compares the degree of violation of

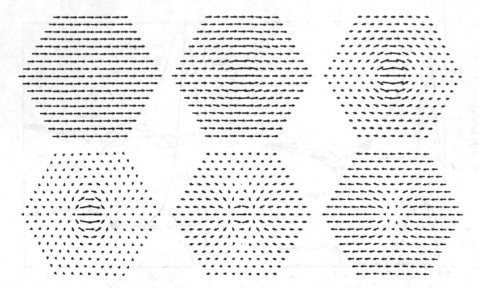

Figure 14. Local electrical field for a superconductor with parameters $T = 0.3T_c$, $\ell = 10\xi_0$, and an external frequency of $\omega = 0.3\Delta(T)$. Each field pattern is a snapshot at time t varying from 0 to half of time periode in time-steps of $\frac{1}{12}$ periode. The external electrical field $\delta\mathbf{E}^\omega(t) = \delta\mathbf{E}\cos(\omega t)$ is maximal and points in positive x-direction for the first picture ($t = 0$). In the first picture of the second line it is zero. The distance between two points in the grid corresponds to $0.25\pi\xi_0$.

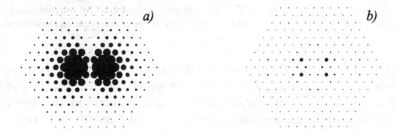

Figure 15. Degree of violation of charge conservation by the dissipative current flow for a non-selfconsistent calculation (Fig.15a), and a self-consistent calculation (Fig.15b). The largest deviation in the non-selfconsistent calculation amounts to $\delta\dot\rho + \boldsymbol{\nabla}\delta\mathbf{j} = 2.5e^2 N_f v_f \delta E^\omega$. The data are obtained for a superconductor with $\ell = 10\xi_0$ at $\omega = 0.4\Delta(T)$ and $T = 0.3T_c$.

charge conservation in a non-selfconsistent calculation (no iteration) with the self-consistent result. We measure the degree of violation at position $\mathbf{R}$ by $D(\mathbf{R}) = \max_t[\delta\dot\rho(\mathbf{R},t) + \boldsymbol{\nabla}\cdot\delta\mathbf{j}(\mathbf{R},t)]$. Charge conservation is obviously fulfilled if $D(\mathbf{R}) = 0$. The degree of violation at a point $\mathbf{R}$ is indicated in Fig.15a,b by the size of the filled circles around the grid points. The non-selfconsistent calculation (Fig.15a) results in a $D(\mathbf{R})$, which is much larger than the time derivative of the correct charge density, $\delta\dot\rho(\mathbf{R},t)$, obtained from a self-consistent calculation with $\delta\varphi(\mathbf{R}, t) = 0$ instead of charge neutrality condition. In Fig.15b we show the violation of charge conservation for our self-consistent calculation. The small remaining $D(\mathbf{R})$ is here a consequence of the finite grid size used in our calculations.

4.5. DISCUSSION

In this section we used the quasiclassical theory to calculate the electromagnetic response of a pancake vortex in a superconductor with finite but long mean free path. This complements previous calculations for perfectly clean systems, which were done self-consistently in the limit $\omega \to 0$ [27], and at finite frequencies without a self-consistent determination of the order parameter [28, 56]). The frequency range of interest in our calculations is of the order of the gap frequency, $\hbar\omega = \Delta$. We have shown that at low frequencies ($\hbar\omega < 0.5\Delta$) the electromagnetic dissipation is strongly enhanced in the vortex cores above its normal state value, and that this effect is a consequence of the coupled dynamics of low-energy quasiparticles excitations bound to the vortex core and collective order parameter modes. The induced current density has at low frequencies a dipole-like behaviour, which results from an oscillating motion of the vortex perpendicular to direction of the driving a.c. field. The response of the vortex in the intermediate frequency range, $.5\Delta \lesssim \hbar\omega \lesssim 2\Delta$, is dominated by bound states in the vortex core. We find peaks in the local dissipation at twice the bound state energies. At higher frequencies, $\hbar\omega > 2\Delta$, the conductivity approaches that of a very clean homogeneous superconductor, which is in good approximation given by the non-dissipative response of an ideal conductor.

References

1. Ambegaokar, V. and Tewordt, L. (1964), Phys. Rev. **134**, A805.
2. Ambegaokar, V. and Baratoff, A. (1963), Phys. Rev. Lett. **10**, 486; **11**, 104.
3. Ambegaokar, V., de Gennes, P.G. and Rainer, D. (1974), Phys. Rev. A **9**, 2676.
4. Andreev, A.F. (1964), Sov. Phys. JETP **19**, 1228.
5. Artemenko, S.N. and Volkov, A.F. (1979), Sov. Phys.-Usp. **22**, 295.
6. Ashauer, B., Kieselmann, G. and Rainer, D. (1986), J. Low Temp. Phys. **36**, 349.
7. Barash, Yu.S., Burkhardt, H. and Rainer D. (1996), Phys. Rev. Lett. **77**, 4070.
8. Bardeen, J., Cooper, L.N. and Schrieffer, J.R. (1957), Phys. Rev. **108**, 1175.
9. Bardeen, J., Rickayzen, G. and Tewordt, L. (1995), Phys. Rev. **113**, 982.
10. Bardeen, J. and Stephen, M.J. (1965), Phys. Rev. **140** A, 1197.
11. Bardeen, J., Kümmel, R., Jacobs, A.E. and Tewordt, L., (1969), Phys. Rev. **187**, 556.
12. Barone, A. and Paternò, G. (1982), *Physics and Applications of the Josephson Effect* (Wiley & Sons, N.Y.).
13. Betbeder-Matibet, O. and Nozieres, P. (1969), Ann. Phys. **51**, 329.
14. Buchholtz, L.J. and Zwicknagl, G. (1981), Phys.Rev. B **23**, 5788.
15. Buchholtz, L.J., Palumbo M., Rainer, D. and Sauls, J.A. (1995), J. Low Temp. Phys. **101**, 1079.

446

16. Caroli, C., de Gennes, P.G., Matricon, J. (1964), Phys. Lett. **9**, 307.
17. de Gennes, P.G. (1966) *Superconductivity in Metals and Alloys*,
    W.A. Benjamin, New York; reprinted by Addison–Wesley, Reading, MA (1989).
18. Eckern, U. (1981), Ann. Phys. **133**, 390.
19. Eilenberger, G. (1968), Z. Physik **214**, 195.
20. Eliashberg, G. (1962), Sov.Phys.-JETP **15**, 1151.
21. Eliashberg, G.M. (1972), Sov. Phys. JETP **34**, 668.
22. Eschrig, M. (1997), Ph.D. thesis, Bayreuth University.
23. Fogelström, M., Rainer, D. and Sauls, J.A. (1997), Phys. Rev. Lett. **79**, 281.
24. Geilikman, B.T. (1958), Sov. Phys. JETP **7**, 721
25. Gorkov, L.P. and Kopnin, N.B. (1975), Sov. Phys.-Usp. **18**, 496.
26. Hu, C.R. (1994), Phys. Rev. Lett. **72**, 1526.
27. Hsu, T.C. (1993), Physica C **213**, 305.
28. Jankó, Boldizsár and Shore, J.D. (1992), Phys. Rev. B **46**, 9270.
29. Keldysh, L.V. (1965), Sov. Phys. JETP **20**, 1018.
30. Kieselmann, G. (1987), Phys. Rev. B **35**, 6762.
31. Landau, L.D. (1957), Sov. Phys. JETP **3**, 920.
    Landau, L.D. (1957), Sov. Phys. JETP **5**, 101.
    Landau, L.D. (1959), Sov. Phys. JETP **8**, 70.
32. Larkin, A.I. and Ovchinnikov, Yu.N. (1968), Sov. Phys.-JETP **26**, 1200.
33. Larkin, A.I. and Ovchinnikov, Y.N. (1976), Sov. Phys. JETP **41**, 960.
34. Larkin, A.I. and Ovchinnikov, Yu.N. (1977), Sov. Phys. JETP **46**, 155.
35. Larkin, A.I. and Ovchinnikov, Yu.N. (1986), in *Nonequilibrium Superconductivity*,
    edited by D. N. Langenberg and A. I. Larkin (Elsevier Science Publishers), 493.
36. Larkin, A.I. and Ovchinnikov, Yu.N. (1976), Sov. Phys. JETP **41**, 960.
37. Leggett, A.J. (1965), Phys. Rev. Lett. **14**, 536.
    Leggett, A.J. (1965), Phys. Rev. **140**, 1869.
    Leggett, A.J. (1966), Phys. Rev. **147**, 119.
38. London, F. (1950), *Superfluids*, Vol.1; Wiley, New York.
39. Lüders, G. and Usadel, K.D. (1971), *The Method of the Correlation Function
    in Superconductivity Theory*, Springer Tracts in Modern Physics No.56
    (Springer, Berlin, 1971).
40. Mannhart, J., Chaudhari, P., Dimos, D., Tsuei, C.C. and McGuire, T.R. (1988),
    Phys. Rev. Lett. **61**, 2476.
41. Matsumoto, M. and Shiba, H. (1995), J. Phys. Soc. Jpn. **64**, 3384.
42. Mattis, D.C. and Bardeen, J. (1958), Phys. Rev. **111**, 412.
43. Migdal, A.B. (1958), Sov. Phys. JETP **7**, 996.
44. Ovchinnikov, Yu.N. (1969), Sov. Phys. JETP **29**, 853.
45. Parks, R.D. (ed.) (1969) *Superconductivity I+II*, Marcel Dekker, New York.
46. Rainer D., in *Recent Progress in Many-Body Theories*, Vol.1
    (Plenum Press New York 1988), 217.
47. Rainer, D. and Sauls, J. A. (1995), in *Superconductivity, From Basic Physics to
    the Latest Developments*, Lecture Notes of the Spring College in Condensed Matter
    Physics, I.C.T.P., Trieste, ed. by P.N. Butcher and L. Yu, World Sci., Singapore.
48. Rainer, D., Sauls, J.A. and Waxman, D. (1996), Phys. Rev. B **54**, 10094.
49. Schmid, A. and Schön, G. (1975), J. Low Temp. Phys. **20**, 207.
50. Serene, J.W. and Rainer, D. (1983), Phys. Reports **101**, 221.
51. Shelankov, A.L. (1984), Sov. Phys.-Solid State **26**, 981.
52. Shelankov, A.L. (1985), J. Low Temp. Phys **60**, 29.
53. Sigrist, M., Bailey, D.B. and Laughlin, R.B. (1995), Phys. Rev. Lett. **74**, 3249.
54. Tanaka, Y. and Kashiwaya, S. (1996), Phys. Rev. **53**, 9371.
55. Zaitsev, A.V. (1984), Sov. Phys. JETP **59**, 1015.
56. Zhu, Yu-Dong, Zhang, Fu-Chun and Drew, H.D. (1993), Phys. Rev. B **47**, 58.

MESOSCOPIC NORMAL-METAL CONDUCTIVITY IN THE MACROSCOPIC NS SYSTEM

YURI CHIANG AND OLGA SHEVCHENKO

*B. Verkin Institute for Low Temperature Physics and
Engineering, National Academy of Sciences of Ukraine,
47 Lenin ave., Kharkov 310164, Ukraine
e-mail:chiang@ilt.kharkov.ua*

In the range of helium temperatures, resistive properties of $3D$ systems "normal metal - superconductor" (Cu-Sn) have been investigated in clean limit. As temperature decreased below T_c the increase in resistance measured across the probes situated entirely in normal side, within macroscopic ballistic distances from NS boundary, was found. The effect is shown to be of mesoscopic nature and connected with non-equivalence of cross-section for scattering of electron and Andreev-reflected hole excitations along trajectories "impurity-boundary".

1. INTRODUCTION

In 1989, J.Herath and D.Rainer [1] have for the first time shown theoretically that in hybrid systems "normal metal - BCS superconductor" the NS boundary due to Andreev reflection should dramatically change the character of the elastic scattering of the electrons at the impurities in a normal metal that are located at ballistic distances from the boundary. The above authors' ideas have stimulated both performing a number of new experiments on metals a part of which I present here and development in theory.

The resistance of a metal is known to reflect the average of electron scattering over distributions of scatterers within the boundaries of length scale chosen, e.g., sample dimensions or separation between measuring probes.The scale within which coherence of the electron wave functions is retained is peculiar to manifestation of the quantum-interference phenomena (hereafter QIP). Phase-break length corresponds to inelastic mean free path, l_{inel}, which, in sufficiently perfect and pure enough metals at low temperatures, is mainly associated with electron-phonon collisions or fermi-liquid effects. In the above conditions, both inelastic and ballistic

447

S.-L. Drechsler and T. Mishonov (eds.), High-T$_c$ Superconductors and Related Materials, 447–453.
© 2001 *Kluwer Academic Publishers. Printed in the Netherlands.*

mean free path with respect to elastic collisions, l_{el}, may reach up to the macroscopic size. This makes it possible to locate the measuring probes for experimental studying the quantum-interference contribution to the metal conductivity within the range of l_{inel} and l_{el}. Up to the present, QIP have been studied in objects of micron size, which samples as well as the effects observed are commonly called mesoscopic. In accordance with $\Delta E/k_B T$-classification (ΔE, is level spacing), the system dimensions are considered macroscopic when that ratio is much more than unity, $\Delta E/k_B T \gg 1$, microscopic in the opposite case, $\Delta E/k_B T \ll 1$, and mesoscopic in the intermediate case, $\Delta E/k_B T$ is of order of unity or somewhat less. But, the total variety of possible quantum-interference effects, both in magnetic field and in the absence of it, should not be restricted to phenomena in micron-size objects. This especially concerns the ballistic scale which in such objects is too short ($l_{el} \sim 10^{-2}\mu$) due to specificity of preparing. Nevertheless, we will use the term "mesoscopic" when discussing QIP in samples of macroscopic size which are required for investigating such effects over a macroscopic phase-break length.

2. EXPERIMENT

We have investigated the samples with macroscopically large mean free paths ($l_{el} \approx 10 \div 100\mu$) and dimensions L characteristic of $3D$ systems ($L \gg l_{el}$). Therefore, using high precision of measurements, all the region where QIP could manifest themselves appeared to be accessible to examining since the separation between the probes could be made equal both inelastic and ballistic mean free path. Besides, the samples constituted bimetallic systems composed of a single-crystal normal metal (copper) contacted to a type I superconductor (tin) (for details see TABLE 1). That allowed us to investigate features peculiar to QIP due to impurity scattering of electrons and excitations Andreev-reflected from NS boundary. The initiation and scattering properties of the boundary could be controlled by means of changing temperature near superconducting transition of Sn. In the presence of NS boundary, there have been registered two new patterns in the temperature dependence of normal-metal resistance in QIP regime which we believe to have mesoscopic nature (the results were partially reported in [2]).

i) When the probes were placed entirely in the copper side at distances from the NS boundary comparable to ballistic mean free path, the resistance of copper sharply increased (up to 60 %) as temperature decreased below superconducting transition temperature of Sn whereas the reduction of regular type in the resistance of those copper layers was observed while Sn was normal (Fig. 1). The evidence for mesoscopic quantum-interference

TABLE 1. Schematic view of samples and position of the probes relative to NS boundary.

Specimen	Schematic view	NS boundary cross-section, μ^2	Distances from N–probe to NS boundary ($L_{N1}{:}L_{N2}$), μ	Distances from S–probe to NS boundary (Ls), μ
Sp1	Cu Sn L_{N2} L_{N1} L_S	200×200	$12,6:45$	31
Sp2	-----//-----	150×1500	$28:560$	375
Sp3	L_{N1} Cu Sn L_{N2} L_S	10×10	$20:2600$	<10
Sp4	Cu L_S L_{N2} L_{N1}	1000×2500	$70:100$	$<5;20$

nature of the effect is re-establishing classical temperature behavior as the probe is moved away from NS boundary inside the normal metal over the distance exceeded inelastic mean free path.

ii) When the probe in the normal-metal side was located at macroscopic distances from the boundary, but still less than inelastic mean free path and sufficiently exceeded ballistic length (approximately by a factor of 100), the resistance dramatically decreased as temperature reduced. Variation in the resistance over comparable temperature range exceeded by order of magnitude familiar change measured across the probes far beyond inelastic mean free path (Fig. 1: $Sp3$, lower curve; $Sp4$, upper curve).

Besides, a number of experiments has been performed across measuring probes disposed on each side of the NS boundary, and the excess boundary resistance (BR) has been studied. In such experimental geometry, which is widely applied in investigating NS-system conductivity, mesoscopic effects in normal region appear only slightly against the background of attendant effects. The main of those is BR, especially in the immediate vicinity of T_c where BR is maximal (Fig. 2). To reveal normal-side mesoscopic events, we offer the procedure for estimating contributions from the accompanying effects to the observations. Note that our special measurements across the probes located entirely in the supercoductor side over the distances within the limits of elastic mean free path from NS boundary do not reveal any resistive contribution from the superconductor into excess resistance of NS systems with large NS contact area (Fig. 3, triangles). This allows us to connect the nature of BR with NS boundary only.

450

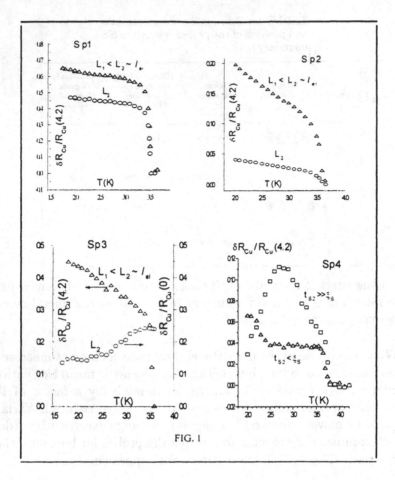

Figure 1. Temperature-dependent resistance of the near-boundary normal-metal layer of ballistic thickness for $Sp1$, $Sp2$, $Sp3$ (upper curve), diffusion thickness for $Sp3$ (lower curve), and intermediate thickness for $Sp4$ (upper curve).

3. DISCUSSION

Ballistic scale . As it can be seen from Fig. 1, temperature dependence of the resistance in the near-interface normal-metal layer, l_{el} in thickness, behaves in accordance with the law specific for that dependence of a superconductor gap. The existence of such an effect, as predicted in Ref. [1], results from non-equivalence of cross-section for impurity scattering of electronic and Andreev-reflected excitations in the normal metal. It was later on confirmed [5] that mesoscopic addition, δR_{andr}, with consideration for multiple Andreev reflections and average over angles, at $T = 0$ can double the resistance of that layer of thickness of thermal ballistic length

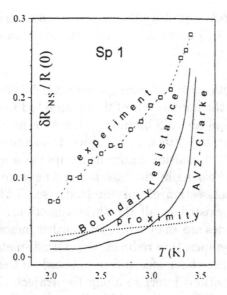

Figure 2. Temperature dependence of the resistance of Cu-Sn bisystem along with the *NS*-boundary resistance and proximity effect.

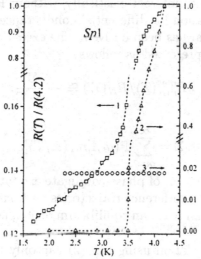

Figure 3. Total resistance of *NS* system (curve 1), superconductor resistance (curve 2), and normal-part resistance in the absense of *NS* boundary (curve 3) via temperature.

$\lambda_{ball} \sim \hbar v_F / k_B T$. For our samples, l_{el} is of order of λ_{ball}. δR_{andr} should obey the following relation at temperatures $T \sim \Delta$

$$\frac{\delta R_N^{andr}}{R_N} = \frac{l_{N,el}}{L_N} \{T_p\}, \qquad (1)$$

where $\{T_p\}$ is the effective probability for electron scattering by the layer of thickness of order of λ_{ball} considering Andreev reflection and the condition $\lambda_{ball} \leq l_{N,el} < L_N$.

Phase-coherence-length scale .A much different behavior of the resistance has been observed in the normal side of the sample $Sp3$ in which the spacing between the probe L_{N2} and NS boundary was of order of $l_{inel} \approx 10^2 l_{el}$ so that $L_{N2} \gg L_{N1}$. Similar data were obtained from the sample $Sp4$ below 3K. Its normal- region dimension along the sample axis was approximately $5l_{el}$, while those dimensions in the interface plane were of order of l_{inel}. So, due to small cross-section of current leads (see TABLE 1), non-zero current components existed in all the three directions. The results from those sample geometries are an evidence for another mesoscopic mechanism that diminishes, as temperature reduces, the normal-metal resistance.

The existence of the above mechanism is based on the assumption that phase- coherent excitations interfere along the trajectories resulting from multiple reflections at NS boundary and having relaxation (diffusive) length $\lambda_{diff} = (\hbar D/k_B T)^{1/2} \sim l_{inel}$ [6]. Along the trajectories, excitations and impurities are supposed to interact elastically, without losing phase memory. In Ref. [6], the expression for differential conductance of near-interface region of a normal metal has been derived. The excess resistance within this region can thus be represented as follows

$$\delta R_{inel}^{mes}(T)/R_N(T_c) \cong -\frac{\Sigma}{1+\Sigma}. \tag{2}$$

Here,

$$\Sigma = \sum_{m=1}^{\infty} P(m)I[m, (\Delta/T)], \tag{3}$$

m is a number of reflections of phase-conjugate excitations from NS boundary; $P(m)$ is a part of interference trajectories with m-reflections (hereafter m-trajectories) that can reach an equilibrium region, i.e. relax over the scale of order l_{inel}.

Although the calculation using Eq. (2) can only be performed numerically from the random-walk simulation, a qualitative outline is evident. When measuring across the probes $L_{N,ball}$ within a ballistic distance from NS boundary ($L_{N,ball} \sim l_{el} \ll l_{inel}$), the contribution to the change in conductivity from the mechanism involved should be negligible: It is reduced by a factor of $l_{inel}/L_{N,ball}$ as compared to the contribution from (1). This is the case for all the combinations from normal- side probe pairs in the samples $Sp1$ and $Sp2$, as well as for the probes $[L_{N1}, L_S]$ in the sample $Sp3$. (Even though one of the transverse dimensions in $Sp2$ is comparable to l_{inel}, it occurs in the direction normal to the current).

At $L_N \sim l_{inel}$, in which case, to the contrary, the contribution from (1) is negligible, the excess resistance of the metal slab, l_{inel} in thickness, should behave in accordance with (2) as the lower curve for $Sp3$ in Fig. 1 does. The most significant are the data from normal-part probes in the sample $Sp4$ with non-zero components of electric field in NS contact plane. Since $\bar{L}_N$ along the sample axis is of order of l_{el}, both effects, (1) and (2), should generally be added (see Fig. 1, $Sp4$, upper curve). Then the peak in the excess mesoscopic resistance at 2.8K should correspond to the temperature where $l_{el} \approx \hbar v_F / k_B T$.

4. CONCLUSION

In summary, it is reasonable to conclude that in NS bisystems investigated, the conditions are initiated for quantum phenomena, unknown before, of mesoscopic nature, to manifest themselves macroscopically. The first is connected with different efficiency of scattering from the nearest to NS boundary impurities (over a ballistic scale), for phase-coherent excitations, electrons and Andreev-reflected holes. The second results from the interference between those excitations along m-trajectories due to significant increase in the probability of Andreev reflection during multiple coherent reflections of electrons at the boundary provided the amount of such trajectories is sufficient. That is realized in a normal-metal layer with characteristic scale of order of inelastic mean free path. The fundamental condition for displaying and observing the above effects in a normal metal is large enough mean free paths (elastic and inelastic). In the case $l \ll \xi_N$, that is to be met in micron-size mesoscopic conductors, the above effects should be insignificant, exclusive of the boundary resistance as the calculation demonstrates.

References

1. J. Herath and D. Rainer, *Physica C* **161**, 209 (1989).
2. Yu.N. Chiang and O.G. Shevchenko, *JETP* **86**, 582 (1998).
3. S.N. Artemenko, A.F.Volkov, and A.V.Zaitsev, *J. Low Temp. Phys.* **30**, 487 (1978).
4. T.Y. Hsiang and J. Clarke, *Phys. Rev. B* **21**, 945 (1980).
5. A.M. Kadigrobov, *Low Temp. Phys.* **19**, 671 (1993).
6. B.J. van Wees, P. de Vries, P. Magnic, and T.M. Klapwijk, *Phys. Rev. Lett.* **69**, 510 (1992).

PINNING OF VORTICES AND LINEAR AND NONLINEAR AC SUSCEPTIBILITIES IN HIGH-T_C SUPERCONDUCTORS

E. H. BRANDT

Max-Planck-Institut für Metallforschung
D-70506 Stuttgart, Germany

1. Phenomenological Theories of Superconductivity
1.1. London Theory
1.2. Pippard Theory
1.3. Ginzburg-Landau Theory
1.4. Abrikosov's discovery Flux-Line Lattice
1.5. Observation of the FLL and its defects

2. Statics of the Flux-Line Lattice
2.1. Arbitrary Arrangements of Flux Lines
2.2. Nonlocal Elasticity of the FLL
2.3. Anisotropic and Layered Superconductors

3. Flux-Line Dynamics
3.1. Moving Flux Lines and Pinning
3.2. Thermally Activated Depinning and Flux Creep
3.3. Nonlinear Resistivity
3.4. Linear Complex Resistivity

4. Geometry Effects
4.1. Demagnetization Factors
4.2. Ideal Screening, Meissner State
4.3. Bean Model in Various Geometries

5. Nonlinear and Linear AC Susceptibilities
5.1. Nonlinear AC Susceptibility
5.2. A Scaling Law
5.3. Linear AC Susceptibility

6. Computations
6.1. Equation for the Current Density
6.2. One Dimensional Geometries
6.3. Thin Films of Arbitrary Shape
6.4. Strips and Disks of Finite Thickness
6.5. Computations with Arbitrary $H(B)$ and Finite H_{c1}

S.-L. Drechsler and T. Mishonov (eds.), High-T_c Superconductors and Related Materials, 455–486.
© 2001 *Kluwer Academic Publishers. Printed in the Netherlands.*

Abstract. The flux-line lattice in type-II superconductors can be calculated from various phenomenological theories. Static properties like the energy of arbitrary vortex arrangements and the nonlocal elasticity of the flux-line lattice are discussed and some consequences of the strong anisotropy and layered structure of high-temperature superconductors given. Vortex dynamics deals with moving flux lines, pinning, thermal depinning, flux creep, and nonlinear and linear complex resistivities. The concept of a demagnetization factor is discussed. Solutions of the Bean model in various geometries and some analytic expressions for nonlinear and linear ac susceptibilities are given. A numerical method is presented by which the magnetic response of superconductors can be computed in any geometry. This method is then applied to thin and thick superconductors in a perpendicular magnetic field and to the geometric barrier for flux penetration.

1. Phenomenological Theories of Superconductivity

Long before the microscopic origin of superconductivity was explained as a spontaneous pairing of electrons and the appearance of an energy gap, very powerful *phenomenological* theories of superconductivity were available. The still useful theory of Fritz and Heinz London of 1935 [1] proposes a linear second-order differential equation for the vector potential $\mathbf{A}(\mathbf{r})$ or its magnetic field $\mathbf{B}(\mathbf{r}) = \nabla \times \mathbf{A}(\mathbf{r})$ or the supercurrent density $\mu_0 \mathbf{J}(\mathbf{r}) = \nabla \times \mathbf{B}(\mathbf{r})$. The London theory was later generalized in various directions:

(i) Pippard [2] extended it to a *nonlocal* relation between $\mathbf{J}$ and $\mathbf{A}$.

(ii) Ginzburg and Landau (GL) [3] conceived a *nonlinear* theory which reduces to the London theory in the limit where the modulus of the complex GL function $\psi(\mathbf{r})$ is constant.

(iii) Numerous authors extended and applied London theory to *anisotropic* superconductors.

Later all these phenomenological theories were shown to follow from the microscopic BCS theory [4] in some approximation. They are particularly useful for describing the properties of Abrikosov flux-lines which occur in type-II superconductors.

1.1. LONDON THEORY

The London theory [1] supplements the Maxwell equations by the relation $\mu_0 \mathbf{J} = -\lambda_L^2 \mathbf{A}$ where λ_L is the London magnetic penetration depth (in the following denoted by λ) and the "London gauge" $\nabla \cdot \mathbf{A} = 0$ is used. Inserting here the Maxwell equation $\mu_0 \mathbf{J} = \nabla \times \mathbf{B} = -\nabla^2 \mathbf{A}$, one obtains

$$(1 - \lambda^2 \nabla^2) \, \mathbf{B}(\mathbf{r}) = 0 \, . \tag{1}$$

A simple one-dimensional solution of (1) describes the penetration of $\mathbf{B} = \hat{\mathbf{z}} B(x)$ and $\mathbf{J} = \hat{\mathbf{y}} J(x)$ into a superconducting half-space $x \geq 0$ when a magnetic field is applied along $\hat{\mathbf{z}}$, $B(x) = B_a \exp(-x/\lambda)$ and $J(x) = -(B_a/\lambda) \exp(-x/\lambda)$. In the presence of an Abrikosov flux line, oriented along $\hat{\mathbf{z}}$, positioned at $x = y = 0$, and carrying one quantum of magnetic flux Φ_0 (see below), the London equation (1) has to be modified by adding a singularity at the vortex core,

$$(1 - \lambda^2 \nabla^2) \, \mathbf{B}(\mathbf{r}) = \hat{\mathbf{z}} \, \Phi_0 \, \delta_2(\mathbf{r}) \, , \tag{2}$$

where $\delta_2(\mathbf{r}) = \delta(x)\delta(y)$ is the two-dimensional delta function. This inhomogenous equation is easily solved by Fourier transformation noting that $\mathbf{B}(\mathbf{r}) = \hat{\mathbf{z}} B(x, y)$,

$$B(\mathbf{r}) = \int \frac{\mathrm{d}^2 k}{4\pi^2} \, \tilde{B}(\mathbf{k}) \, e^{i\mathbf{k}\mathbf{r}}, \quad \delta_2(\mathbf{r}) = \int \frac{\mathrm{d}^2 k}{4\pi^2} \, e^{i\mathbf{k}\mathbf{r}} \, . \tag{3}$$

From this one obtains $\tilde{B}(\mathbf{k}) = \Phi_0/(1 + \lambda^2 k^2)$ and the magnetic field of an isolated flux line becomes

$$B(\mathbf{r}) = \Phi_0 \int \frac{\mathrm{d}^2 k}{4\pi^2} \frac{e^{i\mathbf{k}\mathbf{r}}}{1 + \lambda^2 k^2} = \frac{\Phi_0}{2\pi\lambda^2} K_0\!\left(\frac{r}{\lambda}\right) . \tag{4}$$

Here $K_0(x)$ is a modified Bessel function having the limits $\ln(1/x)$ for $x \ll 1$ and $(\pi/2x)^{1/2} \exp(-x)$ for $x \gg 1$. Since the Eq. (2) is linear, the magnetic field of any arrangement of flux lines is a linear superposition of solutions (4) centered at the vortex positions, see Sec. 2.1.

1.2. PIPPARD THEORY

Inspired by Chamber's nonlocal generalization of Ohm's law, Pippard [2] introduced a superconductor coherence length ξ by generalizing the London equation $\mu_0 \mathbf{J} = -\lambda_L^2 \mathbf{A}$ to a nonlocal relationship

$$\mu_0 \mathbf{J}(\mathbf{r}) = -\lambda_P^2 \frac{3}{4\pi\xi} \int \frac{\mathbf{r}'(\mathbf{r}'\mathbf{A}(\mathbf{r} - \mathbf{r}'))}{r'^3} \, e^{-r'/\xi} \mathrm{d}^3 r'. \tag{5}$$

In the presence of electron scattering with mean free path l, the Pippard penetration depth $\lambda_P = (\lambda_L^2 \xi_0 / \xi)^{1/2}$ exceeds the London penetration depth λ_L of a pure material with coherence length ξ_0, since the effective coherence length ξ is reduced by scattering, $\xi^{-1} \approx \xi_0^{-1} + l^{-1}$ [5]. In the limit of small

$\xi \ll \lambda_P$, Eq. (5) reduces to the local relation $\mu_0 \mathbf{J}(\mathbf{r}) = -\lambda_P^2 \mathbf{A}(\mathbf{r})$. In Fourier space Eq. (5) reads $\mu_0 \tilde{\mathbf{J}}(\mathbf{k}) = -Q_P(k)\tilde{\mathbf{A}}(\mathbf{k})$ with the Pippard kernel

$$Q_P(k) = \lambda_P^2 h(k\xi), \quad h(x) = \frac{3}{2x^3}\left[(1+x^2)\operatorname{atan} x - x\right], \quad h(0) = 1. \quad (6)$$

The microscopic BCS theory (in the Green function formulation of Gor'kov) for weak magnetic fields yields a very similar nonlocal relation $\mu_0 \tilde{\mathbf{J}} = -Q\tilde{\mathbf{A}}$ as predicted by Pippard, replacing the Pippard kernel $Q_P(k\xi)$ (6) by [6,7]

$$Q_{\mathrm{BCS}}(k) = \lambda^{-2}(T) \sum_{n=1}^{\infty} \frac{h[k\xi_K/(2n+1)]}{1.0518\,(2n+1)^3}. \quad (7)$$

Here $h(x)$ is given in Eq. (6), $\lambda(T) = Q_{\mathrm{BCS}}(0)^{-1/2} \approx \lambda(0)(1 - T^4/T_c^4)^{-1/2}$ is the temperature dependent magnetic penetration depth, and $\xi_K = \hbar v_F/(2\pi k_B T) \approx 0.844\lambda(T)T_c/(\kappa T)$ (v_F = Fermi velocity, κ = Ginzburg-Landau parameter). The range of the BCS-Gorkov kernel is of the order of the BCS-coherence length $\xi_0 = \hbar v_F/\pi\Delta_{BCS}$ (Δ_{BCS} is the superconducting energy gap at $T = 0$). See also recent papers on nonlocal electrodynamics in superconductors [8,9].

With the nonlocal relation $\mu_0\tilde{\mathbf{J}}(\mathbf{k}) = -Q(k)\tilde{\mathbf{A}}(\mathbf{k})$ the Eq. (2) for the vortex line now becomes $[1 + k^2/Q(k)]\tilde{B}(k) = \Phi_0$, thus

$$\tilde{B}(k) = \frac{\Phi_0 Q(k)}{Q(k) + k^2}, \quad B(r) = \frac{\Phi_0}{2\pi} \int_0^{\infty} \frac{Q(k)}{Q(k) + k^2} J_0(kr)\, k\, \mathrm{d}k. \quad (8)$$

The Pippard field $B(r)$ (8) of an isolated vortex is no longer monotonic as compared to the London field (4), but it exhibits a field reversal with a negative minimum at large distances $r \gg \lambda_P$ from the vortex core [6]. This effect should be observable if $\xi \approx \lambda$, i.e., for clean superconductors with small GL parameter κ at low temperatures.

The field reversal of the vortex magnetic field field is partly responsible for the attractive interaction between flux lines at large distances, which was observed in clean Niobium at temperatures not too close to T_c [10,11] and which follows from BCS theory at $T < T_c$ for pure superconductors with GL parameter κ close to $1/\sqrt{2}$ [12]. This attraction leads to abrupt jumps in the magnetization curve [10] and to an agglomeration of flux lines that can be observed in superconductors with demagnetization factor $N \neq 0$ (see Sec. 4.1) as FLL islands surrounded by Meissner state, or Meissner islands surrounded by FLL [11]. See also the reviews [13,14].

1.3. GINZBURG-LANDAU THEORY

The Ginzburg-Landau (GL) theory [3,5] introduced a complex order parameter $\psi(\mathbf{r})$. The GL function $\psi(\mathbf{r})$ later was shown to be proportional to the energy-gap function $\Delta(\mathbf{r})$, and its square $|\psi(\mathbf{r})|^2$ to the density of Cooper pairs. The GL theory exhibits two characteristic lengths: the magnetic penetration depth λ and the superconducting coherence length ξ over which $\psi(\mathbf{r})$ can vary. Both λ and ξ diverge at the superconducting transition temperature T_c according to $\lambda \propto \xi \propto (T_c - T)^{-1/2}$, but the GL parameter $\kappa = \lambda/\xi$ is nearly independent of the temperature T. The GL theory reduces to the London theory in the limit $\xi \to 0$ (or $\xi \ll \lambda$), which means constant magnitude $|\psi(\mathbf{r})| = $ const. As in London theory, in the Meissner state a weak applied magnetic field is expelled from the interior of a superconductor by screening currents flowing in a surface layer of thickness λ.

The GL equations of an isotropic superconductor are obtained by minimizing a free energy functional $F\{\psi, \mathbf{A}\}$ with respect to the GL function $\psi(\mathbf{r})$ and the vector potential $\mathbf{A}(\mathbf{r})$, yielding $\delta F/\delta\psi = 0$, $\delta F/\delta\psi^* = 0$, and $\delta F/\delta \mathbf{A} = 0$. With the length unit λ and magnetic field unit $\sqrt{2}B_c$ where $B_c = \Phi_0/(\sqrt{8}\pi\lambda\xi)$ is the thermodynamic critical field, the GL functional reads

$$F\{\psi, \mathbf{A}\} = \frac{B_c^2}{\mu_0} \int \left[-|\psi|^2 + \frac{1}{2}|\psi|^4 + |(-i\nabla/\kappa - \mathbf{A})\psi|^2 + (\nabla \times \mathbf{A})^2 \right] \mathrm{d}^3r . \quad (9)$$

The modified London equation (2) for an isolated vortex line may be obtained from the GL theory in the following way: From the second GL equation $\delta F/\delta \mathbf{A} = 0$ follows the supercurrent density in the form $\mathbf{J} = \nabla \times \mathbf{B} = -|\psi|^2 \mathbf{Q}$. Here $\mathbf{Q}(\mathbf{r})$ is the gauge-invariant supervelocity and $\varphi(\mathbf{r})$ is the phase of $\psi = |\psi|\exp(i\varphi)$. Writing this second GL equation explicitly with the term $[1 - \lambda^2\nabla^2]B$ put on the l.h.s. and all remaining terms on the r.h.s., and then inserting on the r.h.s. the approximate solutions for the isolated vortex, one finds that this r.h.s. has the shape of a 2D delta function with weight Φ_0 and with a finite width of the order of ξ, the vortex-core radius. Thus, if this r.h.s. is treated as an inhomogeneity (i.e. held constant during an iteration step) it corresponds to the delta function in Eq. (2). Here I have used the facts that near the center of a vortex positioned at $x = y = 0$, one has $\psi \propto x + iy$ and $\mathbf{Q} \approx (y, -x)\xi/(x^2 + y^2)$. A useful model for the isolated vortex line with $\kappa \gg 1$ is the approximate solution given by Clem [15] (see also its extensions [16,17,18]),

$$|\psi|^2 \approx \frac{r^2}{r^2 + 2\xi^2}, \quad B(r) \approx \frac{\Phi_0}{2\pi\lambda^2} K_0\left(\frac{\sqrt{r^2 + 2\xi^2}}{\lambda}\right). \quad (10)$$

1.4. ABRIKOSOV'S DISCOVERY OF THE FLUX-LINE LATTICE

The existence of a flux-line lattice was predicted in 1957 by A. A. Abrikosov, who found that a two-dimensional (2D) periodic solution of the GL equations exists for materials with $\kappa > 1/\sqrt{2}$, now called type-II superconductors. Niobium ($T_c = 9$ K) and its alloys, most other superconducting alloys or compounds, and the high-temperature superconductors (HTSs) like YBa$_2$Cu$_3$O$_{7-\delta}$ ($T_c = 92.5$ K), exhibit $\kappa > 1/\sqrt{2}$ or even $\kappa \gg 1$. Magnetic flux can penetrate into these type-II superconductors in form of a more or less regular triangular lattice of Abrikosov flux-lines, i.e., vortices of supercurrent, each exhibiting $\psi = 0$ at its center and each carrying a quantum of magnetic flux $\Phi_0 = h/2e = 2 \cdot 10^{-15}$ Tm^2. The average induction is then $B = 2\Phi_0/\sqrt{3}a^2$ where a is the flux-line spacing.

Type-II superconductors exhibit three critical field values B_{c1}, B_{c2}, and B_{c3}. In weak applied fields $B_a < B_{c1} \approx (\Phi_0 \ln \kappa)/(4\pi\lambda^2)$ (lower critical field) the superconductor is in the Meissner state (internal induction $B = 0$), for $B_{c1} < B_a < B_{c2} = \Phi_0/(2\pi\xi^2)$ (upper critical field) one has the flux-line lattice (FLL, $0 < B < B_{c2}$), and for $B_a \geq B_{c2}$ the bulk superconductivity has disappeared ($B = B_a$). A thin surface layer of thickness ξ remains superconducting up to a field $B_{c3} = 1.69B_{c2}$. All three critical fields go linearly to zero when $T \to T_c$ [5].

The periodic solutions minimizing the free energy (9) and describing the ideal flux-line lattice (FLL) were recently computed with high precision using a Fourier series ansatz for the real and periodic functions $|\psi(\mathbf{r})|^2$ and $\mathbf{Q}(\mathbf{r}) = \mathbf{A} - \nabla\varphi/\kappa$ [18] and computing the Fourier coefficients by an iteration procedure. The FLL solution only depends on the parameters B/B_{c2} and κ, with $0 \leq B/B_{c2} \leq 1$ and $\kappa \geq 1/\sqrt{2}$, and on the prescribed lattice symmetry. Within isotropic GL theory, the triangular FLL has the lowest energy for all values of B and κ.

1.5. OBSERVATION OF THE FLL AND ITS DEFECTS

The flux-line lattice was first observed directly at the surface of a Niobium disk by a Bitter decoration method using "magnetic smoke" which was generated by evaporating an iron wire in an atmosphere of 1 Torr Helium gas [19], see also Ref. [11], the review [13], and more recent decoration experiments [20]. The observed FLL exhibits structural defects: dislocations, stacking faults, vacancies, interstitials, and even disclinations [19]. A first theory of such defects was based on the London theory (valid at low $B \ll B_{c2}$) and the linearized GL theory (valid at high $B \approx B_{c2}$) [21].

Recently the statistical physics of point defects in the FLL was discussed [22] and their energy and interaction computed [22,23]. One can show analytically that, within continuum theory and linear elasticity theory, the interaction between point defects in the two-dimensional (2D) FLL is ideally screened, i.e., the elastic relaxation of the background lattice exactly compensates the long-range interaction between the defects [24]. In the real FLL, a small residual short-range interaction between point defects results from the discreteness of the lattice and from the non-linear elastic displacements near the core of the defects. The self-energy of point defects in the FLL is also strongly reduced by the lattice relaxation. All these compensation effects are due to the smooth repulsive interaction between the flux-lines, which ranges over many flux-line spacings a and which thus requires an external pressure (the applied magnetic field) to keep the FLL together. Other interesting consequences of this long-range interaction are a very small shear modulus (the FLL almost behaves as an incompressible solid) and a pronounced nonlocality of the elastic response, as discussed in Sec. 2.2.

2. Statics of the Flux-Line Lattice

2.1. ARBITRARY ARRANGEMENTS OF FLUX LINES

In the London limit $B/B_{c2} \ll 1$ and $\kappa \gg 1$, the magnetic field $\mathbf{B}(\mathbf{r})$ of arbitrary arrangements of straight or curved flux lines may be obtained from the London equation (1) modified by adding singularities, i.e. 2D or 3D δ-functions $\delta_2(x, y)$ or $\delta_3(x, y, z)$, at the prescribed vortex centers $\mathbf{r}_\nu(z) = [x_\nu(z), y_\nu(z), z]$. Generalizing Eq. (2) to curved flux lines one obtains, with $\mathbf{r} = (x, y, z)$,

$$\mathbf{B}(\mathbf{r}) - \lambda^2 \nabla^2 \mathbf{B}(\mathbf{r}) = \Phi_0 \sum_\nu \int d\mathbf{r}_\nu \, \delta_3(\mathbf{r} - \mathbf{r}_\nu) \,. \tag{11}$$

The line integral in (11) taken along the νth flux line may be written as an integral over a line parameter z',

$$\int d\mathbf{r}_\nu \, f(\mathbf{r}_\nu) = \int f[x_\nu(z'), y_\nu(z'), z'] \left(\frac{dx_\nu}{dz'}, \frac{dy_\nu}{dz'}, 1 \right) dz' \,. \tag{12}$$

For straight, parallel flux lines oriented along z the general solution of (12) is

$$\mathbf{B}(\mathbf{r}) = \hat{\mathbf{z}} \frac{\Phi_0}{2\pi \lambda^2} \sum_\nu K_0 \left(\frac{|\mathbf{r} - \mathbf{r}_\nu|}{\lambda} \right) \,, \tag{13}$$

where $\mathbf{r} - \mathbf{r}_\nu = (x - x_\nu, y - y_\nu, 0)$ and $K_0(x)$ is the modified Bessel function of Eq. (4). For arbitrarily curved flux lines one finds by Fourier transforming Eq. (11),

$$\mathbf{B}(\mathbf{r}) = \frac{\Phi_0}{4\pi \lambda^2} \sum_\nu \int d\mathbf{r}_\nu \frac{\exp(-|\mathbf{r} - \mathbf{r}_\nu|/\lambda)}{|\mathbf{r} - \mathbf{r}_\nu|} \,, \tag{14}$$

In Eqs. (13) and (14) one may cut off the infinities at $\mathbf{r} = \mathbf{r}_\nu$ by introducing a finite core radius $\sqrt{2}\xi$, e.g., by replacing $|\mathbf{r} - \mathbf{r}_\nu|$ with $[(\mathbf{r} - \mathbf{r}_\nu)^2 + 2\xi^2]^{1/2}$, cf. Eq. (10). With this (or any other) cutoff, the energy F_{2D} or F_{3D} of arbitrary straight or curved vortex arrangements is obtained as a sum over the finite magnetic field values in the vortex centers (L is the flux-line length),

$$F = L\frac{\Phi_0}{2\mu_0} \sum_\mu B(\mathbf{r}_\mu) \quad \text{or} \quad F = \frac{\Phi_0}{2\mu_0} \sum_\mu \int d\mathbf{r}_\mu B(\mathbf{r}_\mu). \tag{15}$$

This can be shown as follows. The energy of a superconductor with volume V within London theory is composed of the magnetic field energy and the kinetic energy of the currents,

$$F = \frac{1}{2\mu_0} \int_V [\mathbf{B}^2 + \lambda^2 (\nabla \times \mathbf{B})^2] \, d^3r. \tag{16}$$

Using the formula $\nabla \cdot (\mathbf{B} \times \mathbf{J}) = \mathbf{J} \cdot (\nabla \times \mathbf{B}) - \mathbf{B} \cdot (\nabla \times \mathbf{J})$ one has $(\nabla \times \mathbf{B})^2 = \mathbf{B} \cdot [\nabla \times (\nabla \times \mathbf{B})] + \nabla \cdot [\mathbf{B} \times (\nabla \times \mathbf{B})] = -\mathbf{B}\nabla^2\mathbf{B} + \mu_0 \nabla \cdot (\mathbf{B} \times \mathbf{J})$ since $\nabla \cdot \mathbf{B} = 0$ and $\nabla \times \mathbf{B} = \mu_0 \mathbf{J}$. After a partial integration this gives

$$F = \frac{1}{2\mu_0} \int_V \mathbf{B} \cdot (\mathbf{B} - \lambda^2 \nabla^2 \mathbf{B}) \, d^3r + \frac{1}{2} \int_S (\mathbf{B} \times \mathbf{J}) \cdot d\mathbf{s} = F_V + F_S. \tag{17}$$

The volume integral F_V in (17) may be evaluated by inserting the modified London equation (11) and integrating the delta functions; this yields the expressions (15). The surface integral F_S in (17) in general is not zero but describes surface effects, which often may be calculated by introducing image vortices.

Thus, the energy of vortices far from the surface equals the sums (15) over the peak fields. Since these field values themselves are given by lattice sums, Eqs. (13) and (14), the total energy F of arbitrary vortex arrangements is a double sum over all vortex positions $\mathbf{r}_\mu$ and $\mathbf{r}_\nu$. This means that all flux lines or line elements interact with each other by 2D or 3D pair potentials $V_2 \propto K_0(\tilde{r}_{\mu\nu}/\lambda)$ or $V_3 \propto d\mathbf{r}_\mu \cdot d\mathbf{r}_\nu \exp(-\tilde{r}_{\mu\nu}/\lambda)/r_{\mu\nu}$ where $\tilde{r}_{\mu\nu} = [(\mathbf{r}_\mu - \mathbf{r}_\nu)^2 + 2\xi^2]^{1/2}$, namely,

$$F_{2D} = L\frac{\Phi_0^2}{4\pi\mu_0\lambda^2} \sum_\mu \sum_\nu K_0(\tilde{r}_{\mu\nu}/\lambda), \tag{18}$$

$$F_{3D} = \frac{\Phi_0^2}{8\pi\mu_0\lambda^2} \sum_\mu \sum_\nu \int d\mathbf{r}_\mu \cdot \int d\mathbf{r}_\nu \frac{\exp(-r_{\mu\nu}/\lambda)}{r_{\mu\nu}}. \tag{19}$$

A more sophisticated interaction includes also a small attractive term originating from the gain in superconducting condensation energy of the overlapping vortex cores. This expression for F and other approximate expressions for $\mathbf{B}(\mathbf{r})$, $\psi(\mathbf{r})$, and F valid also at large inductions B are given in Refs. [14,25,26].

From (13) and (15) follows that for $B \ll B_{c2}$ and $\kappa \gg 1$ the peak magnetic field at the center of an isolated flux line is $B(0) = 2B_{c1}$ and the self energy of the flux line is $F_{\text{self}} = \Phi_0 B_{c1}/\mu_0$. The same line energy follows also from the definition of B_{c1} as the applied field $B_a = B_{c1}$ at which the Gibbs free energy $G = F - BB_a V/\mu_0$ vanishes and flux lines start to penetrate.

2.2. NONLOCAL ELASTICITY OF THE FLL

The flux-line displacements caused by pinning forces and by thermal fluctuations may be calculated using the elasticity theory of the FLL. The linear elastic energy F_{elast} of the FLL is obtained by expanding the energy F (19) with respect to small displacements $\mathbf{u}_\nu(z) = \mathbf{r}_\nu(z) - \mathbf{R}_\nu = (u_{\nu x}, u_{\nu y})$ of the flux lines from their ideal lattice positions $\mathbf{R}_\nu$ and keeping only the quadratic terms. In Fourier space one writes [25]

$$\mathbf{u}_\nu(z) = \int_{\text{BZ}} \frac{d^3k}{8\pi^3}\, \mathbf{u}(\mathbf{k})\, e^{i\mathbf{k}\mathbf{R}_\nu}, \quad \mathbf{u}(\mathbf{k}) = \frac{\Phi_0}{B} \sum_\nu \int dz\, \mathbf{u}_\nu(z)\, e^{-i\mathbf{k}\mathbf{R}_\nu}, \quad (20)$$

$$F_{\text{elast}} = \frac{1}{2} \int_{\text{BZ}} \frac{d^3k}{8\pi^3}\, u_\alpha(\mathbf{k})\, \Phi_{\alpha\beta}(\mathbf{k})\, u_\beta^*(\mathbf{k}) \quad (21)$$

with $(\alpha, \beta) = (x, y)$ and $\mathbf{k} = (k_x, k_y, k_z)$. Due to the discreteness and periodicity of the FLL, the k-integrals in (16,17) are over the first Brillouin zone (BZ) of the FLL and the "elastic matrix" $\Phi_{\alpha\beta}(\mathbf{k})$ is periodic in the k_x, k_y plane; the finite vortex core radius restricts the k_z integration in (21) to $|k_z| \leq \xi^{-1}$. But else, the elastic energy F_{elast} (21) looks like the corresponding expression for an elastic continuum. For a medium with uniaxial symmetry the elastic matrix reads

$$\Phi_{\alpha\beta}(\mathbf{k}) = (c_{11} - c_{66})k_\alpha k_\beta + \delta_{\alpha\beta}[\,(k_x^2 + k_y^2)c_{66} + k_z^2 c_{44}\,]. \quad (22)$$

The coefficients c_{11}, c_{66}, and c_{44} are the elastic moduli of uniaxial compression, shear, and tilt, respectively. For the FLL, $\Phi_{\alpha\beta}(\mathbf{k})$ was calculated from GL and London theories [25]. The result, a sum over reciprocal lattice vectors, should coincide with expression (22) in the continuum limit, i.e., for $|\mathbf{k}| \ll k_{BZ}$, where $k_{BZ} = (4\pi B/\Phi_0)^{1/2}$ is the radius of the circularized (actually hexagonal) Brillouin zone with area πk_{BZ}^2. In the London limit one finds

$$c_{11}(k) \approx \frac{B^2/\mu_0}{1+k^2\lambda^2}, \quad c_{66} \approx \frac{B\Phi_0/\mu_0}{16\pi\lambda^2}, \quad c_{44}(k) \approx c_{11}(k) + 2c_{66}\ln\frac{\kappa^2}{1+k_z^2\lambda^2}. \quad (23)$$

The **k** dependence (dispersion) of the compression and tilt moduli $c_{11}(k)$ and $c_{44}(\mathbf{k})$ means that the elasticity of the FLL is *nonlocal*, i.e., strains with short wavelengths $2\pi/k \ll 2\pi\lambda$ typically have a much lower elastic energy than a homogeneous compression or tilt (corresponding to $\mathbf{k} \to 0$). This elastic nonlocality comes from the fact that the magnetic interaction between the flux lines typically has a much longer range λ than the flux-line spacing a, therefore, each flux line interacts with many other flux lines.

The compressional modulus c_{11} and the typically much smaller shear modulus $c_{66} \ll c_{11} \approx c_{44}$ originate from the flux-line interaction, but the last term in c_{44} (10) originates from the line tension of isolated flux lines, defined by $P = \lim_{B\to0} c_{44}\Phi_0/B$. In isotropic superconductors like Nb and its alloys, the line tension coincides with the self energy of a flux line, $P = F_{\text{self}}$, $F_{\text{self}} = \Phi_0 B_{c1}/\mu_0 = (\Phi_0^2/4\pi\mu_0\lambda^2)\ln\kappa$. In anisotropic materials the line tension and line energy of flux lines in general are different, see Sec. 2.3. Note that in isotropic superconductors the uniaxial symmetry of the ideal FLL, i.e., the appearance of a preferred axis, is induced by the applied magnetic field B_a or by the line shape of the vortices. This induced anisotropy leads to the (small) difference between the compressional and tilt moduli c_{11} and c_{44}, but not to a difference between line energy and line tension.

As a consequence of nonlocal elasticity, the flux-line displacements $\mathbf{u}_\nu(z)$ caused by local pinning forces, and also the thermal fluctuations $\langle \mathbf{u}_\nu(z)^2 \rangle$, are much larger than they would be if $c_{44}(\mathbf{k})$ had no dispersion, i.e., if $c_{44}(0) \approx B^2/\mu_0$ were used. The time and space averaged thermal fluctuations $\langle u^2 \rangle \propto k_B T$ and the maximum displacement $u(0) \propto f$ caused at $\mathbf{r} = 0$ by a point force of density $f\delta_3(\mathbf{r})$ are given by similar expressions [14,27-30],

$$\frac{2u(0)}{f} \approx \frac{\langle u^2 \rangle}{k_B T} \approx \int_{BZ} \frac{d^3 k}{8\pi^3} \frac{1}{(k_x^2 + k_y^2)c_{66} + k_z^2 c_{44}(\mathbf{k})} \approx \frac{k_{BZ}^2 \lambda}{8\pi[c_{66}c_{44}(0)]^{1/2}}. \quad (24)$$

In (24) a large factor $[c_{44}(0)/c_{44}(k_{BZ})]^{1/2} \approx k_{BZ}\lambda \approx \pi\lambda/a \gg 1$ originates from the elastic nonlocality.

2.3. ANISOTROPIC AND LAYERED SUPERCONDUCTORS

The high-T_c superconductors (HTSs) are highly anisotropic, approximately uniaxial materials, characterized by two penetration depths $\lambda_a = \lambda_b = \lambda_{ab}$ and $\lambda_c \gg \lambda_{ab}$ for currents flowing in the ab planes or along the c axis. The anisotropy ratio $\Gamma = \lambda_c/\lambda_{ab}$ is $\Gamma \approx 5$ for $YBa_2Cu_3O_{7-\delta}$ but may be $\Gamma > 100$ for Bi, Tl, and Hg based HTSs. These HTSs are well described by anisotropic London theory [14,30]. In formula (24) λ is then replaced by the larger length λ_c, which means the anisotropy enhances the thermal fluctuations by a factor of $\Gamma = \lambda_c/\lambda_{ab} \gg 1$.

Due to the large thermal fluctuations of the vortex positions in HTSs, one expects that the FLL in HTSs "melts" at temperatures considerably below T_c, resulting in a "vortex liquid" [27,30]. This first order melting transition of the FLL recently has been observed as a small but distinct jump in the local flux density B when measured versus the applied field or temperature [31], and as a jump in the specific heat of the FLL [32], see also [33]. These three effects were recently related to each other and to the jump of the entropy at this melting transition, which then was calculated explicitly from London theory using Monte Carlo simulation [34].

The "softening" of the FLL in anisotropic superconductors formally originates from the expression for the tilt modulus when B is along the c axis [35],

$$c_{44}(\mathbf{k}) \approx \frac{B^2/\mu_0}{1+(k_x^2+k_y^2)\lambda_c^2+k_z^2\lambda_{ab}^2} + \frac{2c_{66}}{\Gamma^2}\ln\frac{\kappa^2\Gamma^2}{1+k_z^2\lambda_{ab}^2}, \quad c_{66} \approx \frac{B\Phi_0/\mu_0}{16\pi\mu_0\lambda_{ab}^2}. \quad (25)$$

Here $\kappa = \lambda_{ab}/\lambda_c$, $\Gamma = \lambda_c/\lambda_{ab} = \xi_{ab}/\xi_c$, and ξ_{ab} and ξ_c are two coherence lengths. One can see that for large anisotropy ($\Gamma \gg 1$, $\lambda_c \gg \lambda_{ab}$) the first term in $c_{44}(\mathbf{k})$ (the interaction term) becomes small as soon as the deformation is inhomogeneous ($\mathbf{k} \neq 0$), but the second term (line-tension term) is also small due to the factor $1/\Gamma^2$. The line tension P of a flux line along c is now much smaller than its line energy F_{self}. In general,

$$P = \lim_{B\to 0}\frac{c_{44}\Phi_0}{B} = F_{\text{self}} + \frac{\partial^2}{\partial^2\alpha^2}F_{\text{self}}, \quad (26)$$

where α is the tilt angle of the flux lines away from the c axis. For isotropic superconductors the term $\partial^2 F_{\text{self}}/\partial\alpha^2$ is zero, but in anisotropic materials this term may *nearly compensate* the term F_{self}. From anisotropic London theory one has

$$F_{\text{self}}(\alpha) = F_{\text{self}}(0)\,(\cos^2\alpha + \Gamma^{-2}\sin^2\alpha)^{1/2} \quad (27)$$

with $F_{\text{self}}(0) = \Phi_0^2\ln\kappa/(4\pi\mu_0\lambda_{ab}^2)$. Thus, for $B\|c$ ($\alpha = 0$) one obtains $P(0) = F_{\text{self}}/\Gamma^2$, i.e., the anisotropy *reduces* the tension of flux lines oriented along c by a factor $1/\Gamma^2 \ll 1$. For flux lines perpendicular to the c axis ($\alpha = \pi/2$) the effect is opposite: From (26) one has $F_{\text{self}}(\pi/2) = F_{\text{self}}(0)/\Gamma$, i.e., the flux lines want to lie in the ab plane and the line tension $P(\pi/2) = F_{\text{self}}(\pi/2)\Gamma^2 = P(0)\Gamma^3$ is *larger* than the self energy.

The FLL in HTSs is further complicated by the layered structure of these oxides: The supercurrents flow mainly in the ab planes built from Cu and O atoms. Each flux line is thus a stack of $2D$ vortex disks or "pancake vortices" in the planes. This stack may evaporate into a 3D gas of pancake vortices; the 2D pancake lattices in the CuO planes may melt (dislocation-mediated melting, equivalent to a Kosterlitz-Thouless transition); and the 2D pancake lattices in neighboring CuO planes may decouple. These (and other) phase transitions predicted for the "vortex matter" in layered superconductors [30], in the B-T plane define a "melting line", a "decoupling line", and several branches of an "irreversibility line" above which a HTS in a magnetic field looses its ideal conductivity due to vortex motion.

3. Flux-Line Dynamics

3.1. MOVING FLUX LINES AND PINNING

A transport current density $\mathbf{J} = \nabla \times \mathbf{B}/\mu_0$ (averaged over several vortex spacings) exerts a Lorentz force density $\mathbf{J} \times \mathbf{B}$ on the flux lines. If the flux lines are not pinned by material inhomogeneities, they will move with velocity $\mathbf{v} = \mathbf{B} \times \mathbf{J}/\eta$, where η describes the viscous friction caused by eddy currents and by relaxation of ψ near the moving vortex cores. This drift generates an electric field $\mathbf{E} = \mathbf{B} \times \mathbf{v}$ inside the superconductor or (when $\mathbf{J} \perp \mathbf{B}$) a flux-flow resistivity $\rho_{FF} = E/J = B^2/\eta$. To a good approximation one has $\rho_{FF}(B) = \rho_n B/B_{c2}$, where ρ_n is the normal resistivity observed just above $B_{c2}(T)$. Here and in the following a small Hall effect is disregarded, which originates from a component of the vortex velocity parallel to the current [36]. It should be mentioned that actually the force density $\mathbf{J} \times \mathbf{B}$ is not the Lorentz force (exerted on a magnetic flux by a current) but has opposite sign, see Ref. [37] for a recent discussion.

A further subtlety disregarded in the following, is that actually the current density which drives the flux lines is not the spatial average $\mathbf{J} = \nabla \times B/\mu_0$ but the average of the values of $\mathbf{J}(\mathbf{r})$ at the vortex centers, since the force acts on the singularities in Eq. (2), cf. also the F, Eq. (15). This results in an effective driving current density $\mathbf{J}_H = \nabla \times \mathbf{H}$, where $\mathbf{H} = \mathbf{H}(\mathbf{B}) = \partial(F/V)/\partial\mathbf{B}$ is the "reversible field" obtained from the free energy density F/V of the ideal FLL [38]. The detail that in general $\mathbf{B}$ differs from $\mu_0\mathbf{H}$, only is important at low inductions $B < B_{c1}$ and in the so-called transverse geometry (Sec. 6.2, 6.3), where it may lead to a "geometric barrier" which delays the penetration of the first magnetic flux [38,39,40] even in the absence of bulk or surface pinning.

The vortex motion usually is impeded by pinning of the flux lines at inhomogeneities of the material. For strong pinning a successful description is Bean's critical state model [41], which assumes that vortex motion starts when $J = |\mathbf{J}|$ exceeds a critical value $J_c(B)$ that may depend on the induction B. In the Bean model, J only may take the values 0 or J_c, since when J exceeds J_c the flux lines rearrange such that J is reduced. This picture strictly applies only at zero temperature.

3.2. THERMALLY ACTIVATED DEPINNING AND FLUX CREEP

At finite temperatures T the HTSs exhibit thermally activated depinning. This means that flux motion will occur even at $J < J_c$. Thermally activated flux flow is a complicated statistical problem [14,30]. The theory of collective creep [42] and the vortex-glass scaling idea [43] both predict that the activation energy U for flux jumps depends on the current density J and diverges for $J \to 0$. This is so since at lower J larger "bundles" of flux lines have to jump collectively in order to gain the thermal energy $k_B T$ from the Lorentz force density $\mathbf{J} \times \mathbf{B}$.

A general interpolation formula for the activation energy $U(J)$ and for the current–voltage law $E(J)$ reads

$$U(J) = U_0[(J/J_c)^\alpha - 1]/\alpha, \tag{28}$$
$$E(J) = E_c \exp[-U(J)/k_B T]. \tag{29}$$

For $\alpha > 0$ this formula approximates the results of collective creep and vortex glass theories. When $\alpha = -1$, the same expression (28) formally coincides with the prediction of the Kim-Anderson model [44,45], which was the first theory of thermal depinning and in which $U(J)$ is a linear function which does not diverge at $J = 0$. Note that for all choices of α the activation energy (28) vanishes at $J = J_c$. Thus, $E_c = E(J_c)$ has the meaning of a "voltage criterion" which defines the critical current density J_c. Typically, E_c is chosen as 1 μV/m.

In many experiments one observes a logarithmic dependence $U(J) = U_c \ln(J_c/J)$, which corresponds to the limit $\alpha \to 0$ in Eq. (28). Inserting this $U(J)$ into the Arrhenius law (29) for the electric field, one obtains a power law

$$E(J) = E_c(J/J_c)^n \tag{30}$$

with a "creep exponent" $n = U_c/k_B T$. In the limit $n \gg 1$ this power law reproduces the Bean model since then $E(J) = 0$ for $J < J_c$ and $E(J) = \infty$ for $J > J_c$. Finite $n < \infty$ leads to flux creep [46-49], i.e., to non-zero electric resistivity, and in the limit $n = 1$ this power law means normal Ohmic conductivity. In general, both $J_c(B)$ and $n(B)$ may depend on the local induction B.

Flux creep means the relaxation of the magnetic moment of a type-II superconductor after the applied magnetic field B_a was held constant or was switched off. The pinned flux lines then relax from a metastable state towards the equilibrium state by an approximately logarithmic time law. As shown by Gurevich [44], during flux creep (i.e. when the applied field does not change in time) the electric field E caused by the creeping flux lines takes a *universal* shape and decays approximately as $1/t$. From this known $E(\mathbf{r}, t) \approx f(\mathbf{r})/t$ and from a (highly nonlinear) model law $E(J)$ or $J(E)$ one then obtains the current distribution and thus the magnetic moment. This universality of creep not only applies to longitudinal geometry [46] but also to thin superconductors in perpendicular magnetic field [47] and to arbitrary geometry [48]. More precisely, with the power law (30) one obtains for the creep of electric field, current, and magnetic moment [48]

$$E \propto (t_1 + t)^{-n/(n-1)}, \quad J \propto m \propto (t_1 + t)^{-1/(n-1)}, \tag{31}$$

where t_1 is an integration constant. For large creep exponents $n \gg 1$ and times $t \gg t_1$ this reduces to $E \propto 1/t$ and $J \propto m \propto (\tau/t)^{1/(n-1)} \approx 1 - \frac{1}{n-1} \ln(t/\tau)$ with constant τ. Interestingly, during flux creep the magnetic response to a small ac field is linear, with the only time scale being the time which has elapsed since flux creep has started, i.e., since the ramping of $B_a(t)$ was stopped [49].

3.3. NONLINEAR RESISTIVITY

A useful *nonlinear* resistivity model which combines the power law (30) with the transition to the usual flux-flow state ($\rho \to \rho_{FF}$) at high current densities $J \gg J_c$, reads for isotropic superconductors:

$$\mathbf{E}(\mathbf{J}) = \rho(J)\mathbf{J}, \quad \rho(J) = \rho_{FF}(B) \frac{(J/J_c)^{n-1}}{1 + (J/J_c)^{n-1}}. \tag{32}$$

The B dependence of the flux-flow resistivity is approximately $\rho_{FF}(B) = \rho_n B/B_{c2}$, where $\rho_n(T)$ is the normal resistivity at the upper critical field $B_{c2}(T)$. This means $\rho_{FF}(B)$ may become very small at low inductions $B \ll B_{c2}$, where only a few flux lines are present which can dissipate energy by motion.

3.4. LINEAR COMPLEX RESISTIVITY

At sufficiently high temperatures T and inductions B, above the irreversibility line in the BT plane, the resistivity of type-II superconductors in a magnetic field becomes *linear* (independent of J) and in general will be complex and frequency dependent, $\rho = E/J = \rho_{ac}(\omega) = \rho'_{ac} + i\rho''_{ac}$. Here the usual complex time dependence $\exp(i\omega t)$ is assumed for the ac fields E and J.

A linear visco-elastic model which accounts for the viscous motion of flux lines [drag force density $-\eta\dot{\mathbf{u}}(\mathbf{r}, t)$, $\mathbf{u} =$ flux-line displacements], for elastic pinning [restoring force density $-\alpha_L \mathbf{u}(\mathbf{r}, t)$, $\alpha_L =$ Labusch parameter], thermal depinning [elastic pinning decays with time, $\alpha_L(t) = \alpha_L(0)\exp(-t/\tau)$], and for the London magnetic penetration depth λ, yields [14,50,51]

$$\rho_{ac}(\omega) = i\omega\mu_0\lambda^2 + \rho_{TAFF}\frac{1+i\omega\tau}{1+i\omega\tau_0}. \tag{33}$$

Here $\rho_{TAFF} = \rho_{FF}\exp(-U/k_BT) \ll \rho_{FF}$ is the linear resistivity for thermally activated flux flow [52] and $\tau_0 = \eta/\alpha_L = B^2/(\rho_{FF}\alpha_L)$ and $\tau = B^2/(\rho_{TAFF}\alpha_L) = \tau_0\exp(+U/k_BT) \gg \tau_0$ are relaxation times. In the Meissner state (no flux lines, $\rho_{TAFF} = 0$) one has a purely imaginary resistivity $\rho_{ac}(\omega) = i\omega\mu_0\lambda^2$, i.e., the electric response is inductive ("kinetic inductance" [53]). Real (Ohmic) resistivity ρ results from (33) in the limit $\lambda \to 0$ (or $\lambda^2 \ll \rho/\mu_0\omega$) and for either $\omega \gg \tau^{-1} > \tau_0^{-1}$ ($\rho \to \rho_{FF}$) or $\omega \ll \tau_0^{-1} < \tau^{-1}$ ($\rho \to \rho_{TAFF}$). In general, $\rho_{ac}(\omega)$ is complex and yields a complex ac penetration depth

$$\lambda_{ac}(\omega) = [\rho_{ac}(\omega)/i\omega\mu_0]^{1/2}. \tag{34}$$

For Ohmic ρ this complex penetration depth λ_{ac} is related to the skin depth $\delta = (2\rho/\mu_0\omega)^{1/2}$ by $\lambda_{ac} = \delta/(1+i) = (1-i)\delta/2$.

In some experiments other dependencies $\rho_{ac}(\omega)$ deviating from formula (33) were observed, e.g., in ceramic BSCCO a dependence [54] corresponding to an algebraic decay of the Labusch parameter $\alpha_L(t) = \alpha_L(0)/(1+t/\tau)^\beta$ with $\beta = 1/(1+U/k_BT) = 0.07 \cdots 1$ and nearly temperature-independent $\tau \approx 4 \cdot 10^{-12}$ sec (for $B = 1$ T). Vortex-glass scaling of $\rho_{ac}(\omega)$ was found in Refs. [55,56].

4. Geometry Effects

4.1. DEMAGNETIZATION FACTORS

When the shape of the superconductor is an ellipsoid and the magnetic field inside the superconductor is homogeneous, then the magnetic response in a homogeneous applied field B_a can be expressed in terms of a demagnetization factor $0 \le N \le 1$ ($N = 0$ for long specimens in parallel field, $N = 1/3$ for spheres, $N = 1$ for thin films in perpendicular field). In this case the ellipsoid sees an effective applied field $B_{\text{eff}} = B_a - NM$ where $M = B - B_{\text{eff}}$ is the homogeneous magnetization of the specimen [57] (here, all fields and M are in units Tesla). Thus, the induction (flux density) is $B = B_a + (1-N)M$ and $B_{\text{eff}} = (B_a - NB)/(1-N)$. When the magnetization curve is known for the case $N = 0$, $B_{N=0}(B_a) = f(B_a)$, then one obtains $B(B_a)$ for arbitrary N by solving the implicit equation

$$B = f(B_{\text{eff}}) = f(\frac{B_a - NB}{1 - N}) \tag{35}$$

for $B = B(B_a)$. In a pin-free type-II superconductor with $N < 1$, flux lines start to penetrate at $B'_{c1} = B_{c1}(1 - N)$, but the upper critical field is not changed, $B'_{c2} = B_{c2}$ since $M \to 0$ at $B = B_a = B_{c2}$. The complete magnetization curve is then obtained from (35) by inserting the ideal curve $B_{N=0}(B_a) = f(B_a)$, which may be calculated from the free energy $F(B)$ by inverting the relation

$$B_a(B) = \mu_0 \partial(F/V)/\partial B. \tag{36}$$

When the GL theory applies, $B_a(B)$ may be obtained without taking a derivative, namely, from a virial theorem which was discovered only recently [58],

$$B_a(B) = \frac{\langle |\psi(\mathbf{r})|^2 - |\psi(\mathbf{r})|^4 + 2B(\mathbf{r})^2 \rangle}{\langle 2B(\mathbf{r}) \rangle}. \tag{37}$$

Here $\langle \ldots \rangle$ means average over the superconductor. Relation (37) was checked and applied to the ideal periodic FLL in Ref. [18].

4.2. IDEAL SCREENING, MEISSNER STATE

A special case where the demagnetization factor is a useful concept is when the applied homogeneous field B_a is completely screened from the interior of the specimen by surface currents. Ideal screening may be realized (a) in the ideal Meissner state of superconductors, (b) in superconductors with strong pinning in a small increasing B_a, and (c) in normal metals in an ac magnetic field with sufficiently high frequency. In all three cases the penetration depth should be much smaller than the specimen size, namely, (a) the London depth λ, (b) the Bean depth $B_a/\mu_0 J_c$, and (c) the skin depth $\delta = (2\rho/\mu_0\omega)^{1/2}$. Since $\mathbf{B} = 0$ inside the specimen in this case, the homogeneous magnetization of an ellipsoid is then $M = (B - B_a)/(1 - N) = -B_a/(1 - N)$. The total magnetic moment is $m = MV/\mu_0$, where V is the specimen volume.

Formally, a demagnetization factor N may be defined also for non-ellipsoidal specimens, though here the internal induction in general is not homogeneous and the external field (B_a plus the stray field caused by the currents) is not a dipolar field. In the ideal screening case, however, one always has $\mathbf{B} = 0$, which is homogeneous. Therefore, a natural definition is $N = 1 - B_a/|M| = 1 - H_a V/|m|$ ($H_a = B_a/\mu_0$). The magnetic moment m for ideal screening was recently calculated for infinitely long strips with rectangular cross section $2a \times 2b$ in perpendicular B_a (oriented along the extension $2b$) [59] and for cylinders with diameter $2a$ and height $2b$ in axial B_a [60]. Inserting these m values into the above definition for N, one obtains the useful approximate formulae valid for all aspect ratios $0 < b/a < \infty$:

$$N_{\text{strip}} = 1 - \left\{1 + \frac{\pi a}{4b} + 0.64\frac{a}{b}\tanh\left[0.64\frac{b}{a}\ln\left(1.7 + 1.2\frac{a}{b}\right)\right]\right\}^{-1}, \quad (38)$$

$$N_{\text{disk}} = 1 - \left\{1 + \frac{4a}{3\pi b} + \frac{2a}{3\pi b}\tanh\left[1.27\frac{b}{a}\ln\left(1 + \frac{a}{b}\right)\right]\right\}^{-1}. \quad (39)$$

In particular, for $a = b$ one finds from this for infinite strips with square cross section $N_{\text{strip}} = 0.538$ (while $N = 1/2$ for a circular cylinder in perpendicular field) and for a short cylinder with quadratic cross section $N_{\text{cyl}} = 0.365$ (while $N = 1/3$ for a sphere).

4.3. BEAN MODEL IN VARIOUS GEOMETRIES

In the presence of flux pinning the concept of a demagnetizing factor does not work since the induction $\mathbf{B}$ in general will not be homogeneous inside the material. Therefore, for 30 years the Bean model (assuming that everywhere $J \leq J_c$ and $\mathbf{B} = \mu_0\mathbf{H}$) only had been solved for long cylinders or slabs in parallel B_a where the demagnetization factor is $N = 0$. In this longitudinal geometry $\mathbf{B}$ has only one component, say B_z, whose slope is piecewise constant, namely, $|\nabla B_z| = \mu_0 J_c$ everywhere where magnetic flux has penetrated. The magnetic moment $\mathbf{m} = m\hat{\mathbf{z}}$, in general defined as

$$\mathbf{m} = \frac{1}{2}\int \mathbf{r}\times\mathbf{J}\, d^3r, \quad (40)$$

here simply is $m = \langle B_a - B(\mathbf{r})\rangle V/\mu_0$ (average over specimen volume V). The virgin magnetization curve of slabs with width $2a$ and area A and of cylinders with radius a and length L is

$$m_{\text{slab}} = -J_c a^2 A\left(2h - h^2\right), \quad m_{\text{cyl}} = -\pi J_c a^3 L\left(h - h^2 + h^3/3\right) \quad (41)$$

for $0 \leq h \leq 1$ with $h = H_a/H_p$, where $H_a = B_a/\mu_0$ and $H_p = J_c a$ is the field of full penetration. For $H_a \geq H_p$, i.e. $h \geq 1$, m_{slab} and m_{cyl} stay constant since the current density has saturated to $J = J_c$ in the entire sample.

Recently the Bean model was solved analytically for thin disks [61] and strips [62] in perpendicular field B_a, see also the review [14]. For a thin strip of width $2a$, thickness $d = 2b \ll a$, and length $L \gg a$ and a thin disk with radius a the magnetization curve does no longer saturate at a finite H_a but approaches saturation exponentially (in the limit $d \ll a$). One has

$$m_{\text{strip}} = -J_c d a^2 L\tanh h, \quad m_{\text{disk}} = -J_c d a^3 \frac{2}{3}\left(\text{acos}\frac{1}{\cosh h} + \frac{\sinh|h|}{\cosh^2 h}\right) \quad (42)$$

with $h = H_a/H_c$ where $H_c = J_c d/\pi$ for strips and $H_c = J_c d/2$ for disks. The two curves $m(h)$ (42) differ by less than 0.012 if normalized to the same initial slope $m'(0) = 1$ and same saturation value $m(\infty) = 1$. For strips (bars) and disks (cylinders) with finite thickness $2b$, the magnetization curves saturate at a finite penetration field [59,63]

$$H_p = J_c \frac{b}{\pi} \left[\frac{2a}{b} \arctan \frac{b}{a} + \ln \left(1 + \frac{a^2}{b^2} \right) \right] \quad \text{(bar)}, \tag{43}$$

$$H_p = J_c b \ln \left[\frac{a}{b} + \left(1 + \frac{a^2}{b^2} \right)^{1/2} \right] \quad \text{(disk)}. \tag{44}$$

From the virgin curves $m(H_a)$ (41) to (42) and from the magnetic moment $m(H_a; a, b)$ computed for bars [59] and cylinders [60], the complete hysteresis loop of amplitude H_0 in cycled $H_a(t)$ is obtained by the construction

$$m_\downarrow = m(H_0) - 2m\left(\frac{H_0 - H_a}{2} \right), \quad m_\uparrow = -m(H_0) + 2m\left(\frac{H_0 + H_a}{2} \right). \tag{45}$$

The magnetic moment (45) of the ascending ($m_\uparrow$) and descending ($m_\downarrow$) branches of $m(H_a)$ is quasistatic, i.e., it does not depend on the sweep rate or time. This construction does not (or only approximately) apply when J_c depends on B or when flux creep is accounted for.

5. Nonlinear and Linear AC Susceptibilities

5.1. NONLINEAR AC SUSCEPTIBILITY

Fourier transforming the hysteresis loop $m(t)$ caused by a periodic applied ac field $H_a(t) = H_0 \sin \omega t$, one obtains the complex ac susceptibility of the superconductor in the form

$$\chi(H_0, \omega) = \chi - i \chi'' = \frac{i}{\pi H_0} \int_0^{2\pi} m(t) e^{-i\omega t} \, d(\omega t). \tag{46}$$

Nonlinear susceptibilities for higher harmonics, $\chi_n(\omega)$ with $n = 2, 3, \ldots$, may be defined by replacing in (46) $e^{-i\omega t}$ by $e^{-in\omega t}$.

To fix ideas, I give a simple example for $\chi(H_0)$ in the Bean model. Hollow cylinders and narrow rings in axial field $H_a(t)$, and in general any loop or short-cut coil of a superconducting wire, within Bean's model exhibit a hysteresis loop $m(H_a)$ which has the shape of a parallelogram with two horizontal sides. The virgin curve $m(H_a)$ has constant slope until the supercurrent in the loop $I \propto m \propto H_a$ reaches a critical value I_c at $H_a = H_p$. For $H_a \geq H_p$ the current saturates to $I = I_c$, and the magnetic moment thus stays constant, $m = m_{sat}$. This saturation may also be forced by introduction of a constriction (weak link) in the ring. In this case the $\chi(H_0) = \chi' - i\chi''$ (46), normalized to $\chi(0) = -1$ by dividing it by the slope $|m'(0)|$, reads [64,65] $\chi(H_0) = -1$ for amplitudes $H_0 \leq H_p$ (i.e., there is no dissipation, $\chi'' = 0$) and

$$\chi(H_0) = -\frac{1}{2} - \frac{1}{\pi}\arcsin s - \frac{s}{\pi}\sqrt{1 - s^2} - i\frac{1 - s^2}{\pi} \tag{47}$$

for $H_0 \geq H_p$ with $s = 2H_p/H_0 - 1$.

5.2. A SCALING LAW

For the Bean model the nonlinear $\chi(H_0)$ depends only on the amplitude H_0 (quasistatic theory), while any linear susceptibility $\chi(\omega)$ depends only on the frequency ω. In the general case, $\chi(H_0, \omega)$ depends on both H_0 and ω. If the superconductor satisfies $\mathbf{B} = \mu_0\mathbf{H}$ and $\mathbf{E} = \rho(J)\mathbf{J}$ with $\rho(J) = \rho_c(J/J_c)^\sigma$ [$\sigma = n - 1$, cf. Eq. (30)], then one can show that χ does not depend on H_0 and ω separately but only on the ratios $H_0/\omega^{1/\sigma}$ or ω/H_0^σ or any combination thereof. This scaling law [60,65] thus states that when $\chi(H_0, \omega)$ is measured at one frequency ω and many amplitudes H_0, then $\chi(H_0, \omega')$ at a different frequency ω' is obtained by shifting the amplitude scale according to

$$\chi(H_0, \omega') = \chi(H_0 \cdot (\omega/\omega')^{1/\sigma}, \omega). \tag{48}$$

5.3. LINEAR AC SUSCEPTIBILITY

Linear complex ac susceptibilities conveniently are defined slightly different from the definition (46). Choosing $H_0(t) = \mathrm{Re}\{H_0 e^{i\omega t}\}$ with in general complex amplitude H_0 (Re denotes the real part) one writes

$$\chi(\omega) = \chi - i\chi'' = \frac{1}{\pi H_0}\int_0^{2\pi} m(t)e^{-i\omega t}\, d(\omega t). \tag{49}$$

The linear susceptibilities $\chi(\omega) = \chi' - i\chi''$ or permeabilities $\mu(\omega) = \chi(\omega) + 1$ of conductors with arbitrary complex ac resistivity $\rho_{ac}(\omega)$, e.g. Eq. (33), in a homogeneous magnetic field $H_a(t) = H_0 e^{i\omega t}$ may be obtained by solving a linear diffusion equation for $J(\mathbf{r}, t)$ or $B(\mathbf{r}, t)$ with diffusivity $D = \rho_{ac}/\mu_0$ and appropriate boundary conditions. In this way one finds for infinite slabs (width $2a$) in parallel field [52] and spheres [2] and long cylinders (radius a) in arbitrarily oriented field [60]:

$$\mu_{\mathrm{slab}}(\omega) = \frac{\tanh u}{u}, \quad \mu_{\mathrm{cyl}}(\omega) = \frac{2I_1(u)}{uI_0(u)}, \quad \mu_{\mathrm{sphere}}(\omega) = \frac{3\coth u}{u} - \frac{3}{u^2}, \tag{50}$$

where $u = a/\lambda_{ac} = [i\omega\mu_0 a^2/\rho_{ac}(\omega)]^{1/2}$ is a complex variable and the definition $\chi(\omega) = m(\omega)/|m(\omega \to \infty)| = \mu(\omega) - 1$ was used. I_0 and I_1 are modified Bessel functions. The linear ac susceptibility for any geometry in principle may be written as an infinite sum, or approximated by a finite sum, of the form

$$\chi(\omega) = -w \sum_\nu \frac{\Lambda_\nu b_\nu^2}{w + \Lambda_\nu} \Big/ \sum_\nu \Lambda_\nu b_\nu^2 . \tag{51}$$

Here Λ_ν ($\nu = 1, 2, \ldots$) are the (real) eigenvalues of an eigenvalue problem, the b_ν ("dipole moments") are integrals over the (real) eigenfunctions $f_\nu(\mathbf{r})$, and the complex variable w is proportional to $i\omega/\rho_{ac}(\omega)$. The sum in the denominator of Eq. (51) normalizes $\chi(\omega)$ such that $\chi(\omega \to \infty) = -1$ (case of ideal diamagnetic screening by the skin effect). For thin disks and strips with diameter or width $2a$ and thickness $2b$ in perpendicular field, the positive numbers Λ_ν and b_ν are tabulated in Refs. [55,66], and the complex variable is $w = i\omega\mu_0 ab/[\pi\rho_{ac}(\omega)] = ab/[\pi\lambda_{ac}^2]$. By inverting the relationship (51) between the two complex functions $\chi(\omega)$ and $\rho_{ac}(\omega)$ numerically, the complex resistivity $\rho_{ac}(\omega)$ may be obtained from measured ac susceptibilities as done by Kötzler et al. [55].

6. Computations

6.1. EQUATION FOR THE CURRENT DENSITY

The nonlinear magnetic response of superconductors in realistic geometries, i.e., specimens of various shapes in perpendicular or inclined magnetic field $B_a(t)$, can be computed by time integration of an equation of motion for the current density $\mathbf{J}(\mathbf{r}, t)$. As opposed to the usual method of solving differential equations with boundary condition $\mathbf{B}(\mathbf{r}, t) \to \mathbf{B_a}(t)$ for $\mathbf{r} \to \infty$, this novel method has the advantage that one does not have to compute the magnetic induction $\mathbf{B}(\mathbf{r}, t)$ in the infinite space outside the specimen, but all required integrals only extend over the specimen volume V since $\mathbf{J}(\mathbf{r}, t)$ is zero outside. In the general 3D case with applied field $\mathbf{B_a}(\mathbf{r}, t) = \nabla \times \mathbf{A_a}(\mathbf{r}, t)$ (gauge $\nabla \cdot \mathbf{A} = 0$) and material laws $\mathbf{E} = \mathbf{E}(\mathbf{J}, \mathbf{B}, \mathbf{r})$ (electric field) and $\mathbf{B} = \mu_0 \mathbf{H}$, this equation of motion reads [40]

$$\dot{\mathbf{J}}(\mathbf{r}, t) = \int_V d^3r' \, K(\mathbf{r}, \mathbf{r}') \, [\, \mathbf{E_t}(\mathbf{J}, \mathbf{B}, \mathbf{r}') + \dot{\mathbf{A}}_a(\mathbf{r}', t)\,] . \tag{52}$$

Here the dot denotes $\partial/\partial t$ and $\mathbf{E_t}$ is the divergence-free ("transverse") part of $\mathbf{E}(\mathbf{J}, \mathbf{B}, \mathbf{r}) = \mathbf{E_t} + \mathbf{E_l}$ with $\nabla \cdot \mathbf{E_t} = 0$ and $\nabla \times \mathbf{E_l} = 0$. The integral kernel $K(\mathbf{r}', \mathbf{r})$ depends on the specimen shape. It is the kernel which inverts the relation between the vector potential $\mathbf{A}(\mathbf{r})$ and the current density

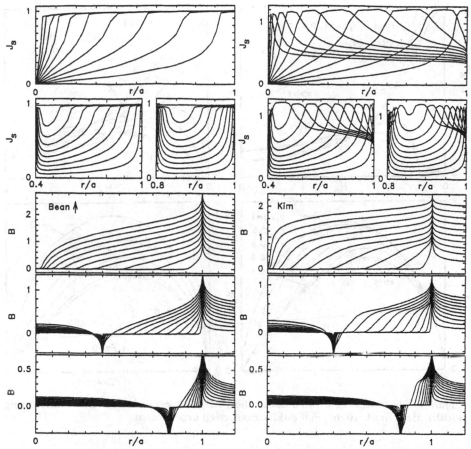

Figure 1. Profiles of sheet current (top) and magnetic field (bottom) in thin superconductor disks and rings in the Bean model (left) and Kim model (right).

$$J(r) = -\nabla^2(A - A_a),$$

$$A(r) = -\int_V d^3r' \, Q(r, r') \, J(r') + A_a(r), \tag{53}$$

$$J(r) = -\int_V d^3r' \, K(r, r') \, [A(r') - A_a(r')]. \tag{54}$$

Both Q and K are *scalar*, with $Q = -\mu_0/4\pi|r - r'|$ and $\int_V d^3r' K(r, r')$ $Q(r', r'') = \delta(r, r'')$. K may be obtained by inverting a matrix.

In problems with translational or rotational symmetry, the integration in Eqs. (52) to (54) is reduced to 2D or even 1D, with the new kernel Q obtained by integrating the 3D kernel Q over 1 or 2 coordinates. The inverse kernel may be calculated by tabulating $Q(r_i, r_j) = Q_{ij}$ on a grid r_i $(i = 1 \ldots N)$ and then inverting the $N \times N$ matrix Q_{ij}. By inverting this matrix one computes the surface current which screens the applied field B_a

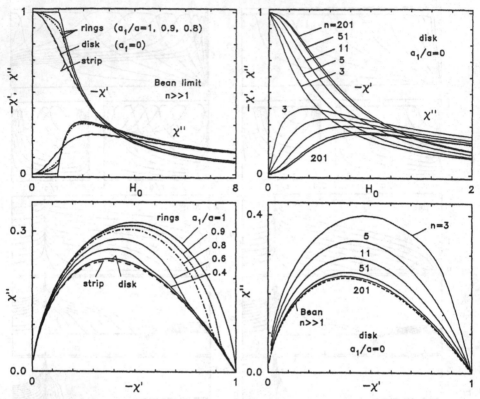

Figure 2. Nonlinear susceptibilities of Bean strips, disks, and rings. Left: Various ring widths, Bean limit. Right: Full disk, various creep exponents n.

from the interior of the specimen, cf. the second term in Eq. (54). One may say that the boundary condition for the magnetic field, e.g. $\mathbf{B}(\mathbf{r}) \to \mathbf{B}_a$ at $r \to \infty$, is now transformed into a surface condition for the current density. This method works also when the applied field is inhomogeneous.

For superconductors with given nonlinear relation $\mathbf{E}(\mathbf{J}, \mathbf{B}, \mathbf{r})$ in an applied field $\mathbf{B_a}(\mathbf{r_a}, t)$ increasing with time from $\mathbf{B_a}(\mathbf{r}, 0) = 0$, time integration of Eq. (52) is straightforward, starting with $\mathbf{J} = 0$ at $t = 0$. From the resulting current density $\mathbf{J}(\mathbf{r}, t)$ one then obtains the induction $\mathbf{B}(\mathbf{r}, t) = \nabla \times \mathbf{A}$ via Eq. (53) and the magnetic moment via Eq. (40). If after some time $B_a(t)$ is held constant, the term $\dot{\mathbf{A}}_a$ in Eq. (52) becomes zero and $\mathbf{J}$ relaxes, describing flux creep. For linear complex resistivity $\rho_{ac}(\omega)$, Eq. (52) leads to an eigenvalue equation from which the coefficients Λ_ν and b_ν of the linear ac susceptibility, Eq. (51), are calculated [55,60,66]. In the following some geometries are listed to which this method was applied.

Figure 3. Stream lines of sheet current (left) and contours of magnetic field (right) in a thin rectangular plate in increasing applied field (from top to bottom).

6.2. ONE DIMENSIONAL GEOMETRIES

Long thin strips (along y) in perpendicular field $B_a(t)$ (along z) and thin circular disks or rings in axial field (along z) are 1D problems. Flux penetration into thin strips and disks was calculated in Refs. [67,68,69], their linear ac susceptibility in Ref. [55,66], and a detailed theory of superconducting rings is presented in Ref. [65], see also the experiments on rings [70]. In the thin-film limit (thickness $d = 2b$, $-b \leq z \leq b$) only the sheet current $J_s = \int_{-b}^{b} J \, dz$ enters the equations. Using the appropriate symmetries for strips with width $2a$ ($-a \leq x \leq a$) and disks with radius a ($r \leq a$), Eq. (53) yields the perpendicular field B_z in the film plane in the form

$$B_z(x) = B_a + \frac{\mu_0}{2\pi} \int_0^a J_s(u) \left(\frac{1}{x-u} - \frac{1}{x+u} \right) du \quad \text{(strip)}, \quad (55)$$

$$B_z(r) = B_a + \frac{\mu_0}{2\pi} \int_0^a J_s(u) \left(\frac{E(k)}{r-u} - \frac{K(k)}{r+u} \right) du \quad \text{(disk)}, \quad (56)$$

where $E(k)$ and $K(k)$ are elliptic functions with argument $k = \sqrt{4ur}/(r + u)$. Combining this with Faraday's induction law one finds the equations of motion for J in strips and disks,

$$J_s(x,t) = \tau\left[2\pi x \dot{B}_a(t) + \int_0^1 \dot{J}_s(u,t) \ln\left|\frac{x-u}{x+u}\right| du\right] \quad \text{(strip)}, \quad (57)$$

$$J_s(r,t) = \tau\left[\pi r \dot{B}_a(t) + \int_0^1 \dot{J}_s(u,t) Q_{\text{disk}}(r,u) du\right] \quad \text{(disk)}, \quad (58)$$

where now the unit length is a and $\tau = \mu_0 a d/(2\pi\rho)$ is a time constant which in general depends on J via $\rho(J)$. The kernel $Q_{\text{disk}}(r,u)$ may be expressed by elliptic functions or written as an integral over the angle φ, cf. Eq. (31) of Ref. [65]. In Eqs. (57) and (58) $\dot{J}_s$ is still under the integral. Straightforward time integration becomes possible by introducing the inverse kernels $Q_{\text{strip}}^{-1}(x,u)$ or $Q_{\text{disk}}^{-1}(r,u)$ obtained by inverting a matrix. This yields for disks,

$$\dot{J}_s(r,t) = \mu_0^{-1} \int_0^a Q_{\text{disk}}^{-1}(r,u) \left[E\left(\frac{J_s(u,t)}{d}\right) - \pi u \dot{B}_a(t)\right] du. \quad (59)$$

The equation of motion (59) applies to any nonlinear or linear current–voltage law $E(J)$. But in contrast to the static Eqs. (57) and (58), in this dynamic equation the current density was assumed constant over the film thickness. If this assumption does not hold, one has to use the finite-thickness formulation of Sec. 6.4.

For flat rings or perforated disks with inner radius a_1 and outer radius a the integral in (59) ranges from a_1 to a. Figure 1 shows the profiles $J_s(r)$ and $B_z(r)$ during penetration of magnetic flux into a superconducting disk and two rings (with $a_1/a = 0.4$ and 0.8) computed for the Bean model (left, $J_c = $ const, creep exponent $n = 101$ in $E \propto J^n$) and Kim model [right, $J_c(B) = J_{c0}/(1 + |B|/B_1)$] with $B_1 = 0.8\mu_0 J_{c0}d$. For the disk these computed profiles of sheet current and magnetic induction practically coincide with the analytic expressions of Ref. [61].

Figure 2 shows the nonlinear complex susceptibility $\chi(H_0, \omega)$, Eq. (46), for disks, rings, and strips in the Bean model (left, $n \gg 1$) and for disks with finite creep exponent $n = 201, 51, 11, 5,$ and 3 in the power law $E(J) = E_c(J/J_c)^n$. Shown are the real and imaginary parts χ' and χ'' (top) and the polar plot χ'' versus $-\chi'$. The unit of H_0 is $J_c d/\pi$ for the strip and $\pi J_c d/8$ for the disk. For the rings H_0 is in units of a penetration field H_p' which makes the sharp rises of χ'' at $H_0 = H_p'$ coincide, cf. Ref. [65]. The frequency was $\omega = 2E_c/(\mu_0 J_c da)$.

6.3. THIN FILMS OF ARBITRARY SHAPE

When the thin strip is not infinitely long or the disk not circular symmetric, the problem becomes 2D. The theory of arbitrarily shaped thin flat

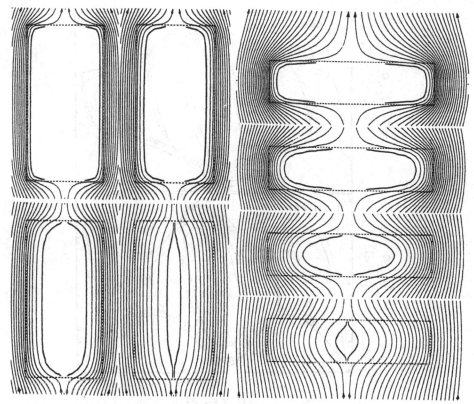

Figure 4. Magnetic field lines around superconducting cylinders. Bean model.

superconductors, e.g. rectangles or squares, is given in Ref. [71] and experiments in Ref. [72]. The sheet current now has two components which may be obtained from one scalar function $g(x, y)$ via $\mathbf{J_s}(x, y) = (J_{sx}, J_{sy}) = (\partial g/\partial y, -\partial g/\partial x)$ since $\nabla \cdot \mathbf{J} = 0$. Here $g(x, y)$ has the meaning of a local magnetic moment, or density of tiny current loops. At the edge of the film one has $g(x, y) = 0$. The perpendicular magnetic field is related to $g(x, y)$ by a nonlocal relation obtained from Eq. (53),

$$B_z(x, y) = B_a + \iint Q(x, y; x', y') \, g(x', y') \, \mathrm{d}x' \mathrm{d}y', \tag{60}$$

where the kernel $Q(x, y; x', y')$ describes the magnetic field of a small current loop positioned at (x', y'). The equation of motion for $g(x, y, t)$ reads

$$\dot{g}(\mathbf{r}, t) = \int Q^{-1}(\mathbf{r}, \mathbf{r}') \Big[\nabla[\rho_s \nabla g(\mathbf{r}', t)] - \dot{B}_a(t) \Big] \mathrm{d}^2 r', \tag{61}$$

where $\rho_s = \rho(J_s/d)/d$ is the sheet resistivity, $\mathbf{J_s}(x, y, t) = -\hat{z} \times \nabla g(x, y, t)$ the sheet current, and $\rho(J)$ is the isotropic resistivity defined by $\mathbf{E} = \rho(J)\mathbf{J}$.

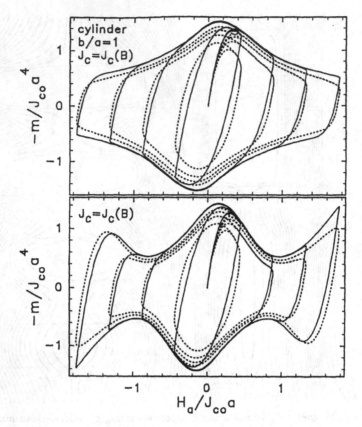

Figure 5. Magnetization loops of short cylinders for two $J(B)$ models.

Equation (61), like Eq. (59), is a nonlocal (and in general also nonlinear) diffusion equation which assumes constant current density **J** across the film thickness. When $\mathbf{J}(x, y, z)$ varies across the thickness, then Eq. (61) yields only the static results (e.g., the penetration of magnetic flux [71,72]), but the dynamics then requires full 3D computations.

Figure 3 shows the stream lines of the sheet current (left) and the contour lines of the perpendicular magnetic field $B_z(x, y)$ for a rectangular superconductor plate with side ratio $b/a = 1.4$ and thickness $d \ll a$ in the Bean model ($n = 19$) for applied fields $H_a = 0.25$, 0.55, and 1.55 in units of the critical sheet current $J_c d$.

6.4. STRIPS AND DISKS OF FINITE THICKNESS

A further class of problems where 2D computations are sufficient are infinitely long strips or bars of finite thickness in perpendicular field, and disks or cylinders of finite length in axial field $B_a \| y$. In general the cross section in the (x, y) or (r, y) plane may be arbitrary, e.g., the bar may be round (a long cylinder in transverse field) or the short cylinder may be replaced by any axially symmetric body, e.g. a cone, ellipsoid, or sphere. This theory is presented in Refs. [59,60]. Here I describe this method for the example of a cylinder with radius a and length $2b$.

From Eq. (53) one obtains the vector potential $\mathbf{A}(\mathbf{r}, t) = \hat{\varphi} A(r, y, t)$ caused by the current density $\mathbf{J}(\mathbf{r}, t) = \hat{\varphi} J(r, y, t)$ circulating in the cylinder and by the axial field $B_a(t)$ (applied along y),

$$A(\mathbf{r}) = -\int_0^a dr' \int_0^b dy' \, Q_{\text{cyl}}(\mathbf{r}, \mathbf{r}') \, J(\mathbf{r}') - \frac{r}{2} B_a. \tag{62}$$

Here $\mathbf{r} = (r, y)$, $r = (x^2 + z^2)^{1/2}$, and $\mathbf{r}' = (r', y')$. The integral kernel

$$\begin{aligned} Q_{\text{cyl}}(\mathbf{r}, \mathbf{r}') &= \mu_0 [\, f(r, r', y - y') + f(r, r', y + y') \,], \\ f(r, r', \eta) &= \frac{r'}{2\pi} \int_0^\pi \frac{\cos \varphi \, d\varphi}{\sqrt{\eta^2 + r^2 + r'^2 + 2rr' \cos \varphi}}, \end{aligned} \tag{63}$$

was obtained by integrating the 3D kernel of Eq. (59), $Q(\mathbf{r}, \mathbf{r}') = -\mu_0/(4\pi |\mathbf{r}_3 - \mathbf{r}'_3|)$ with $\mathbf{r}_3 = (x, y, z)$, over the angle $\varphi = \arctan(z/x)$. If desired, $f(r, r', \eta)$ may be expressed in terms of elliptic integrals, but it is more convenient to evaluate the φ integral numerically. With $\mathbf{B} = \nabla \times \mathbf{A}$ the induction law $\dot{\mathbf{B}} = -\nabla \times \mathbf{E}$ takes the form $\dot{A} = -E(J)$ since $\mathbf{E}(\mathbf{r}, t) = \hat{\varphi} E(r, y, t)$. Thus one has with $\mathbf{r} = (r, y)$

$$E[J(\mathbf{r}, t)] = -\dot{A}(\mathbf{r}, t) = \int d^2 r' \, Q_{\text{cyl}}(\mathbf{r}, \mathbf{r}') \dot{J}(\mathbf{r}', t) + \frac{r'}{2} \dot{B}_a(t). \tag{64}$$

For the particular choice $E(J) = E_c (J/J_c)^n$ the equation for $J(r, y, t)$ becomes

$$\dot{J}(\mathbf{r}, t) = \int_0^a dr' \int_0^b dy' \, Q_{\text{cyl}}^{-1}(\mathbf{r}, \mathbf{r}') \left[\frac{E_c}{J_c^n} J(r', y', t)^n - \frac{r'}{2} \dot{B}_a(t) \right]. \tag{65}$$

The inverse kernel $Q_{\text{cyl}}^{-1}(\mathbf{r}, \mathbf{r}')$ may be computed by tabulating $Q_{\text{cyl}}(\mathbf{r}, \mathbf{r}')$ (63) on a discrete grid $\mathbf{r}_i$, $\mathbf{r}_j$ and then inverting this matrix. Equation (65) is easily time-integrated on a Personal Computer, e.g., starting with $J = 0$ at time $t = 0$. It applies also to spheres and any superconductor with axial symmetry if in the y'-integral the upper boundary $b = b(r)$ depends on r.

Figure 4 shows the magnetic field lines during penetration of flux into cylinders with aspect ratio $b/a = 2$ (left) and $b/a = 0.25$ (right) within the Bean model ($n = 51$ in $E \propto J^n$) at applied fields $H_a/H_p = 0.1, 0.2, 0.4, 0.8$ (from top to bottom) where H_p is the field of full penetration, Eq. (44). The depicted lines are the contours of $rA(r, a)$ at nonequidistant levels, cf. Ref.

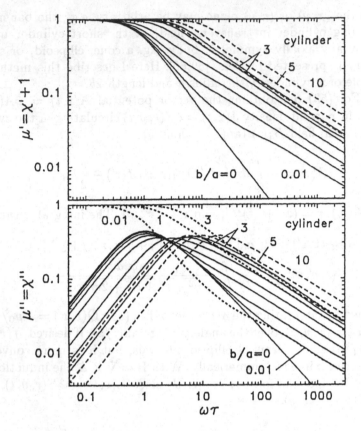

Figure 6. Linear ac susceptibility of Ohmic disks and cylinders of various aspect ratios $b/a = 0 \ldots 7$ in axial magnetic ac field.

[60]. The bold line (contour line $J = J_c/2$) marks the flux front separating the field- and current-free central core from the outer zone where $J = J_c$.

Figure 5 shows magnetization loops of superconductor cylinders with $b/a = 1$ and applied fields $H_a(t) = H_0 \sin \omega t$ with amplitudes $H_0/H_p = 2$, 1.5, 1, and 0.5 ($H_p = 0.8815 J_{c0} a$ is the Bean field of full penetration) for creep exponents $n = 51$ (solid lines) and $n = 5$ (dotted lines). The sweep frequency was $\omega = 2E_c/(\mu_0 J_{c0} a^2)$. Two models are chosen: $J_c(B) = J_{c0}/(1+3\beta)$ with $\beta(\mathbf{r}) = |B|/(\mu_0 H_p)$ (top, Kim model) and $J_c(B) = J_{c0}(1 - 3\beta + 3\beta^2)$ (bottom, a model for the "fishtail effect").

Figure 6 shows the linear ac susceptibility $\chi(\omega) = \chi' - i\chi'' = \mu - 1$ for Ohmic cylinders with aspect ratios $b/a = 0.01, 0.03, 0.1, 0.3, 1, 2$, and 3 as computed from Eq. (51) for axial ac field. The dashed lines for $b/a = 3$, 5, 7 are from the analytic expression (50) for infinite cylinders. The short dashes in the lower plot repeat some χ' curves from the upper plot for better

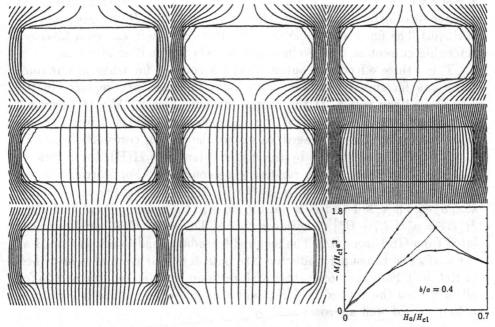

Figure 7. Penetration and exit of perpendicular magnetic flux in a pin-free superconductor with finite H_{c1} at applied fields (left to right, top to bottom) $H_a/H_{c1} = 0.25$, 0.35, 0.37, 0.41, 0.7, 0.45, 0.25. Due to the geometric barrier, flux penetration starts delayed at $H_a = H_{en} = 0.36H_{c1}$ for the depicted aspect ratio $b/a = 0.4$. The resulting magnetization loop $M(H_a)$ is also depicted.

comparison. Note that at sufficiently high frequencies even for rather thin disks ($b \ll a$) the curves $\mu' = \chi' + 1$ and $\mu'' = \chi''$ coincide and decrease $\propto 1/\sqrt{\omega}$ like in longitudinal geometry ($b \gg a$). This is so since at large ω the skin effect forces the magnetic field lines to flow almost parallel to the disk surface even when the disk is thin and the magnetic field is applied axially. Thus, with increasing frequency the magnetic response of thin disks changes from perpendicular to parallel.

6.5. COMPUTATIONS WITH ARBITRARY $H(B)$ AND FINITE H_{C1}

In all the above computations the existence of the lower critical field $B_{c1} = \mu_0 H_{c1}$ was disregarded by assuming $\mathbf{B} = \mu_0\mathbf{H}$. This approximation is good when B is sufficiently large. Consideration of a realistic $B(H) \neq \mu_0 H$ for type-II superconductors causes three novel effects:

(i) Finite H_{c1} leads to the appearance of a "geometric barrier" for the penetration of magnetic flux into thin superconductors of constant thickness [14,39,73]. This edge-shape barrier is absent when the specimen is elliptical or is tapered at the edges. (ii) Accounting for the fact that $B = 0$ in regions where $H < H_{c1}$, leads to the appearance of a "current string" at the penetrating flux front, where B jumps to zero while H changes smoothly [74]. (iii) The finite reversible magnetization $B - \mu_0 H$ causes a Meissner screening current at the specimen surface where B is discontinuous.

These three effects are automatically accounted for when in our computations for superconductors with finite thickness one replaces in the current–voltage law $\mathbf{E}(\mathbf{J})$ the current density $\mathbf{J} = \mu_0^{-1} \nabla \times \mathbf{B}$ (averaged over a few vortex spacings) by the current density which drives the vortices $\mathbf{J}_H = \nabla \times \mathbf{H}$ (averaged values at the vortex cores, cf. Sec. 3.1). The field $\mathbf{H}$ is determined by the material law $\mathbf{H} = \mathbf{H}(\mathbf{B})$, the reversible magnetization curve of the pin-free superconductor. For isotropic superconductors one has $\mathbf{H} = H(B)\mathbf{B}/B$ with $H(B)$ having the approximate shape, e.g., if $\kappa \gg 1$ and B is not too close to the upper critical field B_{c2}): $H(B) = \mu_0^{-1}[B_{c1}^\alpha + B^\alpha]^{1/\alpha}$ with $\alpha = 2$ or 3. The exact $H(B)$ may be calculated from GL theory [18]. The required boundary condition is $\mathbf{B} = \mu_0 \mathbf{H}$ at the surface of the superconductor (like outside). This method is described in Ref. [40]. For superconductors with rectangular cross section $2a \times 2b$ for all ratios b/a the applied field H_{en} where magnetic flux starts to penetrate is found to be well approximated by

$$H_{en} \approx H_{c1} \tanh \sqrt{cb/a}, \tag{66}$$

where $c = 0.36$ for long strips or bars and $c = 0.67$ for disks or cylinders.

Figure 7 shows the penetration and exit of perpendicular magnetic flux in a pin-free superconducting bar with aspect ratio $b/a = 0.4$. Depicted are the magnetic field lines at applied field values (from left to right, top to bottom) $H_a/H_{c1} = 0.25, 0.35, 0.37, 0.41, 0.45, 0.7, 0.45,$ and 0.25. These field values are marked by dots on the magnetization curve (bottom, left). Note that the flux first piles up at the corners. When H_a reaches the entry field $H_{en} = 0.36 H_{c1}$, the penetrating flux lines join at the equator and then contract and jump to the specimen center where the flux piles up and gradually fills the entire bar. When H_a is decreased again, the flux exits delayed. This *geometric barrier* thus causes a hysteretic magnetization curve even in the absence of pinning. An equidistant grid of 28×14 points was used. The amplitude of the applied field was $H_0/H_{c1} = 0.7$ (solid lines) and 0.45 (dashed lines). Inside the specimen, the depicted magnetic field lines may be interpreted as Abrikosov flux lines.

References

1. F. and H. London, *Proc. Roy. Soc.* **A149**, 71 (1935); F. London, *Superfluids II. Macroscopic Theory of Superconductivity* (Dover, New York, 1961).
2. A. Pippard, *Proc. R. Soc.* A **216**, 547 (1953).
3. V. L. Ginzburg and L. D. Landau, *Zh. Eksp. Teor. Fiz.* **20**, 1064 (1950).

485

4. J. Bardeen, L. N. Cooper, and J. R. Schrieffer, *Phys. Rev.* **108**, 1175 (1957).
5. P. G. DeGennes, *Superconductivity of Metals and Alloys* (Benjamin, New York, 1966); M. Tinkham, *Introduction to Superconductivity* (McGraw-Hill, New York, 1975).
6. E. H. Brandt, *phys. stat. sol.* (b) **57**, 277, 465 (1973); *Phys. Lett.* **43A**, 539 (1973).
7. A. A. Abrikosov, L. P. Gorkov, and I. E. Dzyalozhinskii, *Quantum Field Theoretical Methods in Statistical Physics* (Prentice Hall, Englewood Cliffs, NJ, 1963).
8. Ch.-Y. Mou, R. Wortis, A. T. Dorsey, and D. A. Huse, *Phys. Rev. B* **51**, 6575 (1995).
9. V. G. Kogan et al., *Phys. Rev. B* **54**, 12386 (1996); S. J. Phillipson, M. A. Moore, and T. Blum, *Phys. Rev. B* **57**, 5512 (1998).
10. U. Kumpf, *phys. stat. sol.* (b) **44**, 577 (1971).
11. U. Krägeloh, *phys. stat. sol.* **42**, 559 (1970).
12. E. H. Brandt, *phys. stat. sol.* (b) **77**, 105 (1976); *J. Low Temp. Phys.* **24**, 409, 427 (1976).
13. E. H. Brandt and U. Essmann, *phys. stat. sol.* (b) **144**, 13 (1987).
14. E. H. Brandt, *Rep. Prog. Phys.* **58**, 1465 (1995).
15. J. R. Clem, *J. Low Temp. Phys.* **18**, 427 (1975).
16. Z. Hao, J. R. Clem, M. W. Mc Elfresh, L. Civale, A. P. Malozemoff, and F. Holtzberg, *Phys. Rev. B* **43**, 2844 (1991).
17. A. Yaouanc, P. Dalmas de Réotier, and E. H. Brandt, *Phys. Rev. B* **55**, 11107 (1997).
18. E. H. Brandt, *Phys. Rev. Lett.* **78**, 2208 (1997).
19. U. Essmann and H. Träuble, *Phys. Lett.* **24A**, 526 (1967); *Sci. Am.* **224**, 75 (1971); H. Träuble and U. Essmann, *J. Appl. Phys.* **39**, 4052 (1968).
20. M. Marchevsky, L. A. Gurevich, and P. H. Kes, *Phys. Rev. Lett.* **75**, 2400 (1995); M. Marchevsky et al., *Phys. Rev. Lett.* **78**, 531 (1997); *Phys. Rev. B* **57**, 6061 (1998).
21. E. H. Brandt, *phys. stat. sol.* (b) **35**, 1027 (1969); **35**, 371, 381, 393 (1969).
22. E. Frey, D. R. Nelson, and D. S. Fisher, *Phys. Rev. B* **49**, 9723 (1994).
23. E. Olive and E. H. Brandt, *Phys. Rev. B* **57**, 13861 (1998).
24. E. H. Brandt, *Phys. Rev. B* **56**, 9071 (1997).
25. E. H. Brandt, *J. Low Temp. Phys.* **26**, 709 (1977); **28**, 735, 263, 291 (1977).
26. E. H. Brandt, *Phys. Rev. B* **34**, 6514 (1986).
27. D. R. Nelson, *Phys. Rev. Lett.* **60**, 1973 (1988).
28. A. Houghton, R. A. Pelcovits, and A. Sudbø, *Phys. Rev. B* **40**, 6763 (1989); E. H. Brandt, *Phys. Rev. Lett.* **63**, 1106 (1989).
29. M. A. Moore, *Phys. Rev. B* **39**, 136 (1989); M. V. Feigel'man and V. M. Vinokur, *Phys. Rev. B* **41**, 8986 (1990).
30. G. Blatter, M. V. Feigel'man, V. B. Geshkenbein, A. I. Larkin, and V. M. Vinokur, *Rev. Mod. Phys.* **66**, 1125 (1994).
31. E. Zeldov, D. Majer, M. Konczykowski, V. B. Geshkenbein, V. M. Vinokur, and H. Shtrikman, *Nature* **375**, 373 (1995); U. Welp, J. A. Fendrich, W. K. Kwok, G. W. Crabtree, and B. W. Veal, *Phys. Rev. Lett.* **76**, 4809 (1996).
32. A. Schilling, R. A. Fisher, N. E. Phillips, U. Welp, D. Dasgupta, W. K. Kwok, and G. W. Crabtree, *Nature* **382**, 791 (1996); M. Roulin, A. Junod, and E. Walker, *Phys. Rev. Lett.* **80**, 1722 (1998).
33. T. Sasagawa, K. Kishio, Y. Togawa, J. Shimoyama, and K. Kitazawa, *Phys. Rev. Lett.* **80**, 4297 (1998).
34. M. J. W. Dodgson, V. B. Geshkenbein, H. Nordborg, G. Blatter, *Phys. Rev. Lett.* **80**, 837 (1998); *Phys. Rev. B* **57**, 14498 (1998); H. Nordborg and G. Blatter, *Phys. Rev. Lett.* **79**, 1925 (1997).
35. A. Sudbø and E. H. Brandt, *Phys. Rev. B* **43**, 10482 (1991); E. H. Brandt and A. Sudbø, *Physica C* **180**, 426 (1991).
36. E. B. Sonin, *Phys. Rev. B* **55**, 485 (1997).
37. D.-X. Chen, J. J. Moreno, A. Hernando, A. Sanchez, and B.-Z. Li, *Phys. Rev. B* **57**, 5059 (1998).

38. R. Labusch and T. B. Doyle, *Physica C* **290**, 143 (1997).
39. E. Zeldov, A. I. Larkin, V. B. Geshkenbein, M. Konczykowski, D. Majer, B. Khaykovich, V. M. Vinokur, and H. Shtrikman, *Phys. Rev. Lett.* **73**, 1428 (1994).
40. E. H. Brandt, *Phys. Rev. B* **59** (Febr. 1999) in print.
41. C. P. Bean, *Rev. Mod. Phys.* **36**, 31 (1964).
42. M. V. Feigel'man et al., *Phys. Rev. Lett.* **63**, 2303 (1989).
43. D. S. Fisher, M. P. A. Fisher, and D. A. Huse, *Phys. Rev. B* **43**, 130 (1991).
44. P. W. Anderson, *Phys. Rev. Lett.* **9**, 309 (1962).
45. P. W. Anderson and Y. B. Kim, *Rev. Mod. Phys.* **36**, 39 (1964).
46. A. Gurevich and H. Küpfer, *Phys. Rev. B* **48**, 6477 (1993).
47. A. Gurevich and E. H. Brandt, *Phys. Rev. Lett.* **73**, 178 (1994).
48. E. H. Brandt, *Phys. Rev. Lett.* **76**, 4030 (1996).
49. A. Gurevich and E. H. Brandt, *Phys. Rev. B* **55**, 12706 (1997); E. H. Brandt and A. Gurevich, *Phys. Rev. Lett.* **76**, 1723 (1996).
50. M. Coffey and J. R. Clem, *Phys. Rev. Lett.* **67**, 386 (1991).
51. E. H. Brandt, *Phys. Rev. Lett.* **67**, 2219 (1991).
52. P. H. Kes, J. Aarts, J. van der Berg, C. J. van der Beek, and J. A. Mydosh, *Supercond. Sci. Technol.* **1**, 242 (1989).
53. J. Gilchrist and E. H. Brandt, *Phys. Rev. B* **54**, 3520 (1996).
54. R. Behr, J. Kötzler, A. Spirgatis, and M. Ziese, *Physica A* **191**, 464 (1992).
55. J. Kötzler, G. Nakielski, M. Baumann, R. Behr, F. Goerke, and E. H. Brandt, *Phys. Rev. B* **50**, 3384 (1994).
56. J. Kötzler and M. Kaufmann, *Phys. Rev. B* **56**, 13734 (1997).
57. E. H. Brandt and A. Seeger, *Adv. Physics* **35**, 189 (1986).
58. M. M. Doria, J. E. Gubernatis, and D. Rainer, *Phys. Rev. B* **39**, 9573 (1989); U. Klein and B. Pöttinger, *Phys. Rev. B* **44**, 7704 (1991).
59. E. H. Brandt, *Phys. Rev. B* **54**, 4246 (1996).
60. E. H. Brandt, *Phys. Rev. B* **58**, 6506, 6523 (1998).
61. P. N. Mikheenko and Yu. E. Kuzovlev, *Physica C* **204**, 229 (1993); J. R. Clem and A. Sanchez, *Phys. Rev. B* **50**, 9355 (1994).
62. E. H. Brandt, M. Indenbom, and A. Forkl, *Europhys. Lett.* **22**, 735 (1993); E. Zeldov et al., *Phys. Rev. B* **49**, 9802 (1994); E. H. Brandt and M. Indenbom, *Phys. Rev. B* **48**, 12893 (1993).
63. A. Forkl, *Physica Scripta* **T49**, 148 (1993).
64. T. Ishida and H. Mazaki, *J. Appl. Phys.* **52**, 6798 (1981).
65. E. H. Brandt, *Phys. Rev. B* **55**, 14513 (1997).
66. E. H. Brandt, *Phys. Rev. B* **50**, 13833 (1994).
67. E. H. Brandt, *Phys. Rev. B* **49**, 9024 (1994); **50**, 4034 (1994).
68. E. H. Brandt, *Physica C* **235-240**, 2939 (1994).
69. Th. Schuster et al., *Phys. Rev. Lett.* **73**, 1424 (1994); *Phys. Rev. B* **50**, 16684 (1994).
70. F. Mrovka, M. Wurlitzer, P. Esquinazi, E. H. Brandt, M. Lorentz, and K. Zimmer, *Appl. Phys. Lett.* **70**, 898 (1997); Th. Herzog, H. A. Radovan, P. Zieman, and E. H. Brandt, *Phys. Rev. B* **56**, 2871 (1997).
71. E. H. Brandt, *Phys. Rev. B* **52**, 15442 (1995); *Phys. Rev. Lett.* **74**, 3025 (1995).
72. Th. Schuster, H. Kuhn, E. H. Brandt, and S. Klaumünzer, *Phys. Rev. B* **56**, 3413 (1997); Th. Schuster et al., *Phys. Rev. B* **52**, 10375 (1995).
73. M. V. Indenbom, H. Kronmüller, T. W. Li, P. H. Kes, and A. A. Menovsky, *Physica C* **222**, 203 (1994); M. V. Indenbom and E. H. Brandt, *Phys. Rev. Lett.* **73**, 1731 (1994).
74. M. V. Indenbom, Th. Schuster, H. Kuhn, H. Kronmüller, T. W. Li, and A. A. Menovsky, *Phys. Rev. B* **51**, 15484 (1995).

WETTING PHASE TRANSITIONS AND SUPERCONDUCTIVITY

The Role of Surface Enhancement of the Order Parameter in the GL Theory

J.O.INDEKEU, F.CLARYSSE AND E.MONTEVECCHI

Laboratorium voor Vaste-Stoffysica en Magnetisme
Katholieke Universiteit Leuven, B-3001 Leuven, Belgium

1. Introduction and motivation

The principal subject of these lectures is the precise analogy between a wetting phase transition in an adsorbed fluid and the interface delocalization transition in a type-I superconductor. Although there is up to now very sparse experimental evidence of this phenomenon in superconductors, the Ginzburg-Landau (GL) theory gives a detailed prediction of the "wetting" phase diagram for type-I materials that can in principle be verified directly by magnetization or resistivity measurements. In addition, the *mechanism* responsible for the transition can occur equally well in type-II superconductors, for which it leads to interesting modifications of the phase diagram of critical field versus temperature, including an increase of T_c in zero field. A microscopic identification and derivation of the mechanism is still lacking, but at the phenomenological level of the GL theory the mechanism is the enhancement of the superconducting order parameter at the surface.

We start with the identification of the three phases and interfaces that play a prominent role in our problem. In Figure 1 we see a liquid drop on a substrate, under a gaseous atmosphere. The three interfaces, substrate/liquid, substrate/gas, and liquid/gas, are characterized by their interfacial tensions. These act as forces per unit length on the contact line where gas, liquid and substrate meet. Young's equation (1805) expresses the mechanical equilibrium of this line, relating the interfacial tensions to the dihedral contact angle.

S.-L. Drechsler and T. Mishonov (eds.), High-T$_C$ Superconductors and Related Materials, 487–504.

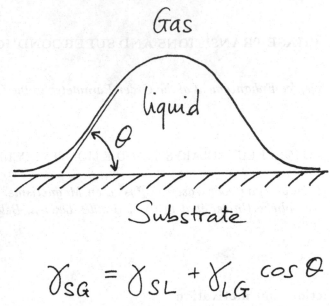

Figure 1. Liquid drop on a substrate and Young's law for the contact angle θ that the liquid/gas interface makes with the substrate.

Since 1977 statistical physicists have devoted major attention to systems in which, as a function of temperature, pressure, or chemical composition, the contact angle approaches zero at a special value of this "control parameter". The study of the singularity that occurs in the surface excess free energy at this "wetting" point, is known as the field of wetting phase transitions, and has up to now led to a wealth of experimental and theoretical findings (Moldover and Cahn, 1980; de Gennes, 1985; Dietrich, 1988).

Figure 2 shows the interior, near the surface, of a type-I superconductor at its critical field H_c. The magnetic field is parallel to the surface and perpendicular to the figure. In analogy with Figure 1, we recognize a drop, or rather semi-cylindrical "sausage" of superconducting (SC) Meissner phase extending along the direction of the field, and embedded in a normal phase region (N). On the outside (below the surface or "wall" in the figure) one usually assumes vacuum, but other choices of "boundary conditions" turn out to be more interesting for our purposes, as will become clear in Section 2. At this stage we only note that also for this system Young's law relates the three interfacial tensions (wall/superconducting, wall/normal, normal/superconducting (Boulter and Indekeu, 1996)) to the contact angle θ. The precise "S"-shape of the normal/superconducting interface near the wall can be calculated (Blossey and Indekeu, 1996), as is outlined in the Lectures of Ralf Blossey in these Proceedings, and like for the system of Figure 1 we can ask whether there exist "wetting" points at which θ becomes zero.

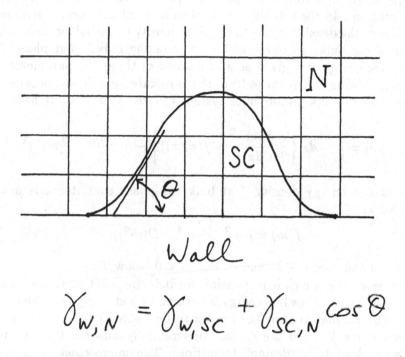

$$\gamma_{W,N} = \gamma_{W,SC} + \gamma_{SC,N} \cos\theta$$

Figure 2. Superconducting domain inside a type-I material at $H_c(T)$, and Young's law for the contact angle θ, which the normal/superconducting interface makes with the wall.

Although it was believed for a long time that, generally, the superconducting phase does not "wet" the surface (Livingston and DeSorbo, 1969; Dietrich, 1988), it was recently established that a proper choice of boundary condition in the GL theory does lead to a rich variety of interface delocalization transitions in type-I superconductors, including first-order wetting transitions (with the associated prewetting phenomenon), and critical wetting transitions (Indekeu and van Leeuwen, 1995; Indekeu and van Leeuwen, 1997). So, the analogy with adsorbed fluids holds precisely and in much detail.

In the following sections we will address the underlying mechanism of the new surface phase transitions, discuss how they manifest themselves theoretically, indicate how they could be detected experimentally, and outline implications for type-II materials.

2. Enhancement of superconductivity near the surface

In order to understand the mechanism for the interface delocalization transition in superconductors, it is instructive to recall the basic Landau theory for wetting in adsorbed fluids. Let m denote a suitable order parameter, for instance, the density minus the critical density of the adsorbate, so that $m < 0$ in the bulk gas phase and $m > 0$ in the bulk liquid phase. The surface free energy γ is given as a functional of the order parameter profile $m(x)$, x being the distance from the substrate into the adsorbate. The geometry is that of a semi-infinite system situated at $x \geq 0$. We have

$$\gamma[m] = \int_0^\infty dx \left\{ \frac{c}{2} \left(\frac{dm}{dx} \right)^2 + f(m(x)) \right\} - h_1 m(0) - \frac{g}{2} m(0)^2 \qquad (1)$$

The bulk free energy density f at bulk two-phase coexistence is usually expanded as

$$f(m) = \alpha m^2 + \frac{\beta}{2} m^4 + \mathcal{O}(m^6), \qquad (2)$$

with $\beta > 0$ and $\alpha \propto T - T_c$ and so that $\alpha < 0$ below T_c.

The parameter which is responsible for inducing wetting is the "surface field" h_1. For $h_1 > 0$ the free energy is lowered by a strongly positive surface value $m(0)$, so that the surface prefers to be wetted by the liquid phase. Conversely, for $h_1 < 0$ the surface preferentially adsorbs the gas phase (which can lead to a "drying" transition). The microscopic origin of h_1 is not hard to find. The intermolecular forces between adsorbate particles compete with those between substrate and adsorbate particles, and the net effect is a chemical potential difference near the surface, which either causes an accumulation of adsorbate particles or a depletion zone. Therefore, as a rule, $h_1 \neq 0$. In this system, wetting transitions can occur when h_1 reaches a certain threshold, or when the temperature is increased (towards the bulk critical point T_c). The parameter g plays a secondary role, in changing the character of the wetting transition from first-order to critical. For a detailed calculation of the wetting phase diagram in this theory, see, e.g., (Indekeu, 1995).

In superconductors with a surface, the GL surface free-energy functional reads

$$\gamma[\psi, \vec{A}] = \int_0^\infty dx \left\{ \alpha \mid \psi \mid^2 + \frac{\beta}{2} \mid \psi \mid^4 + \frac{1}{2m} \left| \left(\frac{\hbar}{i} \vec{\nabla} - q\vec{A} \right) \psi \right|^2 \right.$$

$$\left. + \left| \frac{\vec{\nabla} \times \vec{A} - \mu_0 \vec{H}}{2\mu_0} \right|^2 \right\} + \frac{\hbar^2}{2mb} \mid \psi(0) \mid^2 \qquad (3)$$

The order parameters in this case are the wave function ψ and the vector potential $\vec{A}$. The applied magnetic field $\vec{H} = H\vec{e}_z$ is parallel to the $x = 0$ surface. Also here $\alpha < 0$ for $T < T_c$, and $\beta > 0$. The term which merits special attention is the surface contribution, proportional to $|\psi(0)|^2$. This term is absent for the usual situation when the surface is against vacuum or an insulating material. However, in order for the wetting transitions to occur, this term must be included, and, moreover, the coefficient $1/b$ must be taken negative.

The significance of the surface contribution is apparent at the phenomenological level. Minimization of the surface free energy functional with respect to variations of the order parameters, including those of the wall value $\psi(0)$, leads to the wall boundary condition,

$$\left.\frac{d\psi}{dx}\right|_{x=0} = \frac{\psi(0)}{b} \tag{4}$$

This means that b corresponds to the *surface extrapolation length* of the wave function profile near the wall, as illustrated in Figure 3.

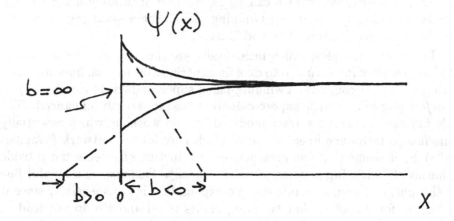

Figure 3. Superconducting wave function profiles near the surface at $x = 0$. The surface extrapolation length b is indicated for three different surfaces.

The case $b = \infty$ is well known and corresponds to surfaces against vacuum or insulators, as mentioned before. Also the case $b > 0$ occurs frequently, and represents suppression of superconductivity near the surface in the case of a contact with a normal metal or (even more extreme) a ferromagnet. Now, to induce wetting of the surface by the superconducting phase, $b < 0$ is needed. It remains to be seen precisely which surfaces are candidates for *enhancing* superconductivity in this manner.

At a microscopic level, the origin of b is not very clear. In contrast with the availability of a microscopic derivation of the GL free energy, starting from BCS theory, as reviewed, e.g., in (de Gennes, 1966), there is no precise microscopic identification yet of the surface term (Rainer, 1998). Therefore, we must at present treat b as a phenomenological parameter, and can only discuss *qualitatively* which surfaces give rise to $b < 0$ in the GL theory.

The first mention, to our knowledge, of a surface with $b < 0$, was in 1969 when Fink and Joiner reported that after *cold working* of the sample surface an increase in the critical field for surface superconductivity, $H_{c3}(T)$, was measured (Fink and Joiner, 1969). The experimental results were shown to be consistent with GL-theory calculations of H_{c3}, assuming $b < 0$, in which case the ratio H_{c3}/H_{c2} is *increased* with respect to its value for $b = \infty$, which is 1.69.

In the 80's the experiments on twinning-plane superconductivity (TPS) by Khlyustikov showed that in certain systems a twinning plane displays localized superconductivity at higher fields and higher temperatures than the bulk of the sample. In particular, even in zero field the transition temperature is increased with respect to the bulk T_c. Buzdin showed that the $H - T$ phase diagram of TPS can be reproduced accurately using the GL theory assuming $b < 0$ at the twinning plane. For a good review of this development, see (Khlyustikov and Buzdin, 1987).

These two examples, which incidentally are the only ones we are aware of that explicitly invoke $b < 0$ to explain experimental results, have an interesting feature in common. Twinning-plane superconductors clearly feature a *defect* plane, at which superconductivity is apparently enhanced. The other system features a surface modified by cold working, which essentially amounts to introduce linear *defects*, which may form a network (Wagner, 1998). So it seems that the presence or introduction of defects can provide a means of enhancing superconductivity locally. This can be a helpful line of thought, but even so a microscopic explanation is indispensible, since it is known, for instance, that twinning planes in Al and Pb do not lead to TPS, whereas in Sn, In, Nb, Re, and Tl they do (Khlyustikov and Buzdin, 1987).

Now that modern techniques of controlled surface modification have become available (such as Molecular Beam Epitaxy) it is almost evident that surfaces with $b < 0$ could be obtained in many more ways than the two examples above suggest. For instance, de Gennes argued (de Gennes, 1994) that the deposition of a suitably chosen thin layer (of thickness small compared with the coherence length ξ) on the surface of the material, should result in the required modification of the boundary condition at the GL-theory level. One possibility would be to choose for the deposited layer a material which has a higher transition temperature in bulk than the superconductor under study.

This brings us naturally to another (special) situation in which the wetting transition can be induced quite directly. This is the case in which the surface enhancement of superconductivity is caused by the *proximity effect*. Indeed, if one brings superconductor 2 with a higher T_c in contact with superconductor 1, a surface superconducting sheath is induced in the latter. Then, as the temperature T is increased while always remaining at the critical field $H_c(T)$, a transition point is predicted at which the thickness of the sheath becomes macroscopic (much larger than ξ) (Clarysse and Indekeu, 1998). This is precisely the wetting transition we are looking for. Note that in this setting there is no explicit surface term containing the parameter b, but there is another similar parameter which pertains to the transparency for electrons of the contact between the superconductors. It would also be useful to have at our disposal calculations of this parameter using the microscopic theory.

Finally, while these lectures were being given, Maki made the intriguing suggestion that another way of obtaining a surface with $b < 0$ could be to make a contact between the superconductor and a *semi-conductor*, with a suitable overlap of the band gap with the gap Δ of the superconductor (Maki, 1998). This suggestion is to be pursued further within the microscopic theory as well.

To summarize this section, there should be many more ways nowadays to achieve surface enhancement of superconductivity (corresponding to $b < 0$ in the GL theory), and therefore to open the way for observing the interface delocalization transition, than by applying cold working or studying twinning-plane systems. However, a microscopic derivation of the surface term in the GL functional would be most helpful to elucidate the precise role of defects, and is still lacking.

3. Theoretical manifestations of wetting in superconductors

At the level of the GL theory possible wetting behaviour is most clearly detected in (i) the variation of the effective order parameter, and (ii) the topology of the phase diagram. The appropriate effective order parameter in the theory of surface phenomena is the thickness l of the superconducting surface sheath, illustrated in Figure 4, which shows the wave function and magnetic induction profiles of this structure.

Below T_w, or if the wetting transition is absent (e.g., for $b = \infty$), l is at most of the order of the bulk coherence length ξ. However, if a wetting transition occurs, at some T_w and thus at $H_c(T_w)$, then for temperatures $T > T_w$, l *diverges* upon approaching the critical field $H_c(T)$ from above. This divergence signifies in practical terms that the sheath acquires a *macroscopic*, rather than mesoscopic, thickness. In other words, $l \gg \xi$.

494

This implies that the superconducting/normal interface *delocalizes* from the surface into the interior of the sample. We shall see in the next section that connected with l is the "experimental" order parameter, the total diamagnetic moment of the specimen.

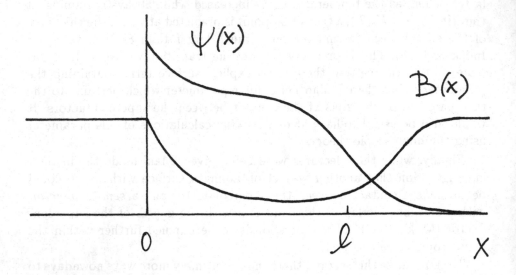

Figure 4. Superconducting wave function $\psi(x)$ and magnetic induction $B(x)$ profiles near the surface, assuming $b < 0$. The thickness l of the superconducting surface sheath is indicated.

Inspection of the topology of the phase diagram allows us most convincingly to distinguish systems with a wetting transition from other systems, and to clarify the difference between the surface critical nucleation field H_{c3} and the "prewetting" transition line. In the following figures we will sketch a number of cases, and also indicate how l varies along various isotherms in the phase diagram.

Figure 5 shows the $H-T$ phase diagram of a type-I superconductor with GL parameter $\kappa > 0.41$ and surface extrapolation length $b = \infty$ (surface against vacuum), as found by (Saint-James and de Gennes, 1963). The bulk critical field $H_c(T)$ and the surface nucleation field H_{c3} meet at the bulk critical temperature in zero field, T_c. Following the isotherm indicated by the arrow in part (a) of the figure, the sheath thickness l varies as shown in part (b). We emphasize that, approaching H_c from above, the sheath thickness remains *finite* for all $T < T_c$. There is no wetting.

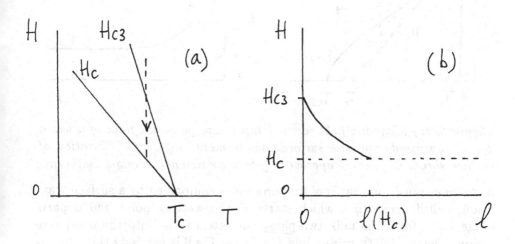

Figure 5. (a) Magnetic field versus temperature phase diagram of a type-I superconductor with a surface transition H_{c3}, for the case $b = \infty$, and for GL parameter $\kappa > 0.41$. (b) The behaviour of the surface sheath thickness l as the field is lowered towards H_c.

In Figure 6 the $H - T$ phase diagram is presented for a surface-enhanced superconductor ($b < 0$), with $\kappa \approx 0.3$. In contrast with the case shown in the previous figure, the surface transition line now extends between a *wetting point*, at T_w and $H_w \equiv H_c(T_w)$, and a surface nucleation point in zero field, at T_{cs}. The fact that $T_{cs} > T_c$ is due to the assumption $b < 0$. Indeed, the location of T_{cs} is determined by the condition

$$\frac{\xi(T_{cs})}{b} = -1 \tag{5}$$

This condition was first found in the context of twinning-plane superconductivity (Khlyustikov and Buzdin, 1987), but is valid more generally (Indekeu and van Leeuwen, 1995). Incidentally, the phase diagram of Fig.6 is very similar, at least for sufficiently low κ, to that of TPS.

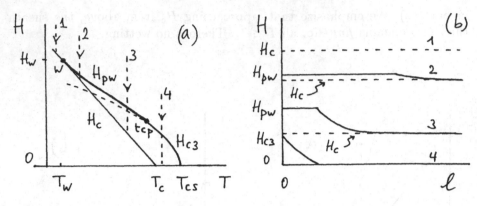

Figure 6. (a) Magnetic field versus temperature phase diagram of a low-κ type-I superconductor with surface enhancement (b < 0). (b) Variation of surface sheath thickness l approaching bulk coexistence, along 4 isotherms.

A wetting transition that is of first order is accompanied by a surface transition, called *prewetting*, which starts at the wetting point and departs tangentially from the bulk two-phase coexistence line into the one-phase region. The surface transition line $H_{pw}(T)$ in Fig.6 is precisely this prewetting phenomenon, departing from the line $H_c(T)$ into the bulk normal phase region. The prewetting transition is of first order, and is marked by a *jump* of l from a small (or, in our case, zero) value to a mesoscopic value. The magnitude of the jump in l diverges upon approach of the wetting point, and vanishes upon approach of the *tricritical point* tcp, where the prewetting transition changes to a second-order surface nucleation transition, $H_{c3}(T)$. The latter continues until $H = 0$, where it terminates at T_{cs}. It is important to note that the nucleation line $H_{c3}(T)$ has a continuation (dashed line) for $T < T_{tcp}$, which marks the limit of stability of undercooling of the surface normal state. Clearly, the first-order $H_{pw}(T)$ line preceeds the $H_{c3}(T)$ line and the distinction between prewetting and surface nucleation is easy to see.

Fig.6(b) gives the variation of l along 4 different isotherms. Along the first path there is no surface sheath ($l = 0$). The second path crosses the prewetting line close to the wetting point. Therefore, the jump in l at H_{pw}, from zero to a mesoscopic value, is almost macroscopic. Upon further lowering the field towards H_c, l diverges continuously in the manner (Indekeu and van Leeuwen, 1995)

$$l \propto \ln\left(\frac{1}{H - H_c}\right) \tag{6}$$

This divergence, which applies to the whole interval $T_w < T < T_c$, is called *approach to complete wetting*, while the term *partial wetting* is used to describe the behaviour for $T < T_w$ approaching $H_c(T)$. Likewise, along path 3 the prewetting line is crossed, and l jumps. The jump is smaller here, because one is significantly away from T_w already. Lowering the field further, l diverges in the complete wetting manner, given by Eq.(6). Finally, along path 4 a surface nucleation transition is encountered at H_{c3}, at which point l gradually increases from zero to a (small) value, as H is further decreased towards zero.

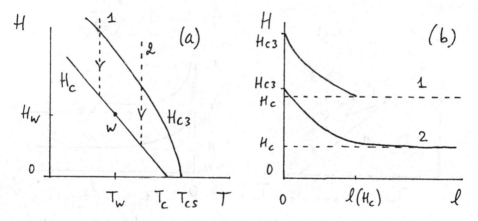

Figure 7. (a) Critical-wetting phase diagram for an "intermediate"-κ type-I superconductor with surface enhancement (b < 0). (b) Two variations of l along isotherms below and above T_w.

The interface delocalization transition in type-I superconductors with $b < 0$ at the surface changes from first-order to critical at $\kappa = 0.374$ (Indekeu and van Leeuwen, 1995). In Figure 7 we present a critical-wetting phase diagram, for $\kappa \approx 0.5$. The wetting point is now an isolated point on the line $H_c(T)$, fully separated from the surface nucleation line $H_{c3}(T)$. There is no prewetting transition connected with it. The characteristic signature of the critical wetting transition is the divergence of the sheath thickness l, upon approach of the wetting point, coming from lower temperatures, and moving along $H_c(T)$. This divergence is logarithmic, in the manner $l \propto \ln\{1/(T_w - T)\}$. In addition there is the aforementioned divergence of l upon approach of H_c at temperatures above T_w (complete wetting). Thus, Fig.7(b) shows two isotherms, one for the approach to partial wetting ($T < T_w$) and another for the approach to complete wetting ($T > T_w$). Note that for partial wetting $l(H_c)$ remains finite.

For systems in which the surface superconducting sheath is induced by the proximity to another "stronger" superconductor, the nucleation transition $H_{c3}(T)$ is redundant. This is in fact fully analogous to wetting in adsorbed fluids, where there is no nucleation transition either. Consequently, assuming that the wetting transition is of first order, the prewetting transition has the significance of two-phase coexistence between sheaths of different mesoscopic thicknesses, l_1 and l_2. The difference $l_1 - l_2$ vanishes at the *prewetting critical point*, which has the properties of an ordinary critical point, but in reduced dimensionality ($d \to d - 1$). On the other hand, $l_1 - l_2$ diverges at the first-order wetting point. The $H - T$ phase diagram for this situation is sketched in Figure 8, which, as we already anticipated, is strongly reminiscent of the standard wetting phase diagram for fluids.

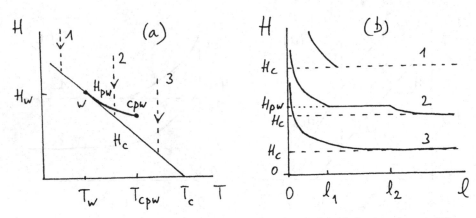

Figure 8. (a) Phase diagram for a proximity-induced first-order wetting transition. The prewetting line does not merge into a surface nucleation line, but terminates at a surface critical point cpw, as for adsorbed fluids. (b) Isothermal variations of l reminiscent of "adsorption isotherms" for fluids at substrates.

In Fig.8(b) three different isothermal variations of the sheath thickness l are shown. The first exemplifies the approach to partial wetting. The second crosses the prewetting line, where l makes a finite jump from l_1 to l_2, and subsequently displays the approach to complete wetting. The third isotherm features the approach to complete wetting at a temperature above the prewetting critical point.

4. Expected experimental signatures of wetting

Considering the amount of effort and time that has been spent in the experimental search for the elusive prewetting transition in adsorbed fluids before the first successful attempt (Kellay et al., 1993), it is all the more remarkable that the analogue of the prewetting transition in type-I superconductors has been found experimentally already about a decade before (Khlyustikov and Buzdin, 1987). Indeed, in view of our present knowledge concerning the close analogy of the interface delocalization transition in superconductors to the wetting transition in fluids, it is certain that the phase diagram of twinning-plane superconductivity measured in Sn by Khlyustikov is a prewetting phase diagram. Of course, it was at that time not recognized as one, and, moreover, the most interesting wetting point in the diagram, where the prewetting line meets the $H_c(T)$ line, was not given proper attention. In spite of its incompleteness, the TPS diagram is to our knowledge the only experimental evidence to date of wetting phenomena in superconductors.

Recently, an attempt was made in our Laboratory to study the approach to complete wetting for a Sn foil coated with a thin layer of Pb (Metlushko and Strunk, 1996). Very recently, new efforts have been made to fabricate suitable "sandwiches" for measuring interface delocalization phenomena (Arutyunov et al., 1998). Up to now these studies are not enough detailed to be conclusive.

In principle, two basic techniques are adequate for detecting the transition. Resistivity measurements across the system (perpendicular to the surface) would allow to observe gradual or sudden changes in the thickness of the normal phase region, following changes in the superconducting fraction. Alternatively, measurements of the total magnetic moment should allow to see variations in the diamagnetic response as the thickness of the superconducting layer changes.

In the following we will adopt the viewpoint of magnetization measurements (akin to those employed by Khlyustikov to study TPS) and present the expected experimental signals corresponding to the theoretical variations of l discussed in the previous section. The magnetic moment per unit area M corresponds in the GL theory to the expression

$$-M \propto \int_0^\infty dx (\dot{A}_y(x) - H), \qquad (7)$$

where $\dot{A}_y(x)$ equals the magnetic induction $B(x)$. For thick surface sheaths this quantity is proportional to the theoretically defined thickness l, so that M can be regarded as the experimental counterpart of the effective order parameter suitable for studying wetting transitions.

Figure 9 shows a hypothetical plot of the magnetic moment versus field, at constant temperature, for (a) the approach to partial wetting, and (b) the approach to complete wetting.

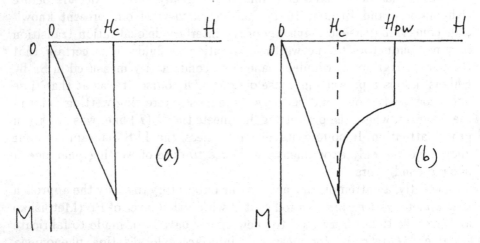

Figure 9. (a) Magnetic moment M versus field H for a type-I superconductor with surface enhancement (b < 0), corresponding to Fig.6: (a) $T < T_w$ (partial wetting), (b) For $T > T_w$ there is a prewetting jump followed by the approach to complete wetting.

We have assumed a low-κ superconductor with surface enhancement, as in Fig.6. Since in case (a) no surface sheath is nucleated ($l = 0$), the magnetic moment remains zero down till the bulk transition to the Meissner state at H_c. For $H < H_c$ the absolute value of the magnetization decreases linearly to zero. In case (b) the magnetic moment makes a jump from zero to a finite value at the prewetting field H_{pw}, after which it diverges logarithmically upon approach of H_c. This divergence is of course cut off by the finite system size.

Assuming that a surface superconducting sheath is present a priori, e.g., by the intervention of the proximity effect, we expect magnetic moment versus field plots similar to those in Figure 10, which is the experimental counterpart of Fig.8(b). The situation drawn in Fig.10(a) pertains to temperatures below the (first-order) wetting transition, while Fig.10(b) applies to $T > T_w$, while $T < T_{cpw}$, so that the system traverses the prewetting line. Fig.10(c) describes the approach to complete wetting for $T > T_{cpw}$.

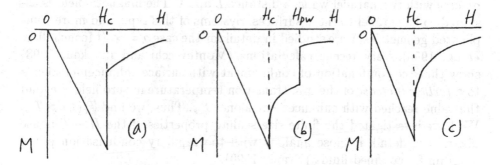

Figure 10. (a) Magnetic moment M versus field H for a type-I superconductor in which a surface sheath is induced by the proximity effect, corresponding to Fig.8: (a) $T < T_w$ (partial wetting), (b) $T > T_w$ (complete wetting) and $T < T_{cpw}$, (c) $T > T_{cpw}$ (complete wetting).

As soon as experimental magnetization curves are available, they can be compared directly with calculations based on the GL theory, and using Eq.(7). The only parameters in the theory are the value of κ and, for thin film geometry, the system size. The value of the surface extrapolation length can in principle be deduced from the experimental determination of T_{cs}, the transition temperature in zero field. Indeed, using Eq.(5), and assuming that the zero-field coherence length $\xi(T)$ for the bulk material can be obtained from other calculations or measurements, the value of b is given by $-\xi(T_{cs})$.

It is worth emphasizing that the status of the GL theory for superconductivity is very different from that of the Landau theory for fluids. Both being mean-field theories, the importance of thermal fluctuations must be assessed. Classical superconductivity is one of the very few areas (long polymers being another) for which thermal fluctuations can be safely neglected. Consequently, in this field one should expect quantitative agreement between experiment and the GL theory.

5. Effect of geometrical confinement and implications for type-II superconductors

Up to now we have devoted our attention to semi-infinite geometry, with a single surface. Recently, (Fomin *et al.*, 1998) calculated how the ratio H_{c3}/H_{c2} is enhanced in a *wedge*-geometry, as a function of the opening angle. They restricted their attention to surfaces against vacuum or insulators ($b = \infty$), so that the wedge transition in zero field remains at the bulk T_c. It would be interesting to study how the nucleation field is further increased by assuming $b < 0$ in this geometry.

One further step towards a realistic description is to consider the *film geometry* with two parallel walls, a distance L apart. The magnetic field is assumed to be parallel to these surfaces. Systems of this type, and more complicated geometries, were studied in detail for the case $b = \infty$ (Moshchalkov *et al.*, 1995). Very recent calculations (Montevecchi and Indekeu, 1998) show that the combination of confinement with surface enhancement leads to a *further increase* of the film transition temperature in zero field, beyond the value reached with enhancement alone, T_{cs}. Thus, we find $T_c(L) > T_{cs}$. We have investigated the finite-size scaling properties of the $H - T$ phase diagram in detail, in close analogy with the capillary condensation phase diagram for confined fluids (Evans, 1990).

Although the wetting transition itself is restricted to semi-infinite geometry, the prewetting phenomenon can persist for *thick* films of type-I materials. For *thin* films with surface enhancement on both sides, the most conspicuous effect is the substantial increase of T_c. This effect is unrelated to wetting and occurs for type-I and type-II superconductors alike. Within the limits of validity of the GL theory, the maximum increase $T_c(L) - T_c(\infty)$ that can be predicted is of the order of $T_c(\infty)$ itself.

As far as high-T_c superconductors are concerned, for which twinning-plane superconductivity of closely spaced twins was once thought to be a possible mechanism (Brandt, 1998), the notion of surface enhancement ($b < 0$) may well be easier to achieve than for classical superconductors. Indeed, the profile of charge carrier density $n(x)$ shows a characteristic decrease near the surface of the material. The bulk T_c, on the other hand, displays a maximum as a function of n. If the material is suitably "overdoped" in the bulk, $n(x)$ will attain an "optimal" value near the surface, so that surface enhancement of superconductivity may result (Hilgenkamp, 1998). It remains to be seen, preferably based on a microscopic calculation, if this gives rises to a negative extrapolation length for the wave function.

Acknowledgements

The first author would like to thank Todor Mishonov for his kind invitation to lecture at the School, and Valya Mishonova for her exquisite organizational efforts and continuous kind attention. We are grateful to Konstantin Arutyunov, Vitali Metlushko, Victor Moshchalkov, Christoph Strunk, Kristiaan Temst, Margriet Van Bael, and Lieve Van Look, for their interest and efforts towards an experimental verification of wetting in superconductors. Ralf Blossey made useful remarks on the manuscript and was a source of constant support in Albena.

J.O.I. is a Research Director of the Fund for Scientific Research of Flanders, and has been supported by the Belgian Concerted Action Programme (GOA) and Interuniversity Poles of Attraction Programme (IUAP).

References

K.Arutyunov, L. Van Look, M. Van Bael, and K.Temst (1998), unpublished

R.Blossey and J.O.Indekeu (1996) Interface potential approach to surface states in type-I superconductors, *Physical Review B* **53**, p. 8599

For a recent study of this quantity, see C.J. Boulter and J.O. Indekeu (1996) Accurate analytic expression for the surface tension of a type-I superconductor, *Physical Review B* **54**, p. 12407

We learned this historical note from E.H.Brandt (1998) private communication (Albena)

F.Clarysse and J.O.Indekeu (1998), unpublished

See, e.g., P.-G. de Gennes (1966) *Superconductivity of metals and alloys*, Benjamin, New York

For a review, see P.-G. de Gennes (1985) Wetting: statics and dynamics, *Reviews of Modern Physics* **57**, p. 827

P.-G. de Gennes (1994) private communication (Amsterdam)

For a review, see S. Dietrich (1988) Wetting phenomena, in C.Domb and J.Lebowitz (eds.),*Phase Transitions and Critical Phenomena*, Academic, London, Vol.12, p. 1

R.Evans (1990) Fluids adsorbed in narrow pores: phase equilibria and structure, *Journal of Physics: Condensed Matter* **2**, p. 8989

H.J.Fink and W.C.H.Joiner (1969) Surface nucleation and boundary conditions in superconductors, *Physical Review Letters* **23**, p. 120

V.M. Fomin, J.T. Devreese and V.V.Moshchalkov (1998) Surface superconductivity in a wedge, *Europhysics Letters* **42**, p. 553

H. Hilgenkamp (1998) private communication (Albena)

J.O.Indekeu (1995) Introduction to wetting phenomena, *Acta Physica Polonica B* **26:6**, p. 1065

J.O. Indekeu and J.M.J.van Leeuwen (1995) Interface delocalization transition in type-I

504

superconductors, *Physical Review Letters* **75**, p. 1618; Wetting, prewetting and surface transitions in type-I superconductors, *Physica C* **251**, p. 290

For a tutorial article, see J.O. Indekeu and J.M.J. van Leeuwen (1997) "Wetting" phase transitions in type-I superconductors, *Physica A* **236**, p. 114

H. Kellay, D. Bonn and J. Meunier (1993) Prewetting in a binary liquid mixture, *Physical Review Letters* **71**, p. 2607

I.N.Khlyustikov and A.I.Buzdin (1987) Twinning-plane superconductivity, *Advances in Physics* **36**, p. 271

J.D.Livingston and W.DeSorbo (1969) The intermediate state in type I superconductors, in R.D.Parks (ed.), *Superconductivity*, Marcel Dekker, New York, Vol.2, p. 1235

K.Maki (1998) private communication (Albena)

V. Metlushko and C. Strunk (1996), unpublished

M.R.Moldover and J.W.Cahn (1980) An interface phase transition: complete to partial wetting, *Science* **207**, p. 1073

E.Montevecchi and J.O.Indekeu (1998), unpublished

V.V. Moshchalkov, L. Gielen, C. Strunk, R. Jonckheere, X. Qiu, C. Van Haesendonck, and Y. Bruynseraede (1995) Effect of sample topology on the critical fields of mesoscopic superconductors, *Nature* **373**, p. 319

D. Rainer (1998) private communication (Albena)

D.Saint-James and P.-G. de Gennes (1963) Onset of superconductivity in decreasing fields, *Physics Letters* **7**, p. 306

H. Wagner (1998) private communication (Leuven)

BERNOULLI EFFECT IN THE QUANTUM HALL REGIME AND TYPE-II SUPERCONDUCTORS IN MAGNETIC FIELD

A new superfluid effect in two dimensional electron gas and a method for determination of the effective mass and vortex charge in superconductors

T. MISHONOV[1,2]
[1]*Laboratorium voor Vaste-Stoffysica en Magnetisme, Katholieke Universiteit Leuven, Celestijnenlaan 200 D, B-3001 Leuven, Belgium*

AND

D. NIKOLOVA, K. KALOYANOV, SL. KLENOV
[2]*Faculty of Physics, University of Sofia, 5 J. Bourchier Blvd., Bg-1164 Sofia, Bulgaria*

Abstract. The current dependence of the chemical potential of a two dimensional electron gas in the quantum Hall effect regime is considered theoretically. It is shown that a new effect related with dissipationless flowing of the electric current can be interpreted as a Bernoulli effect of the incompressible superfluid. An experiment with a field effect transistor (FET) is proposed for the observation of this effect. The systematic investigation of the Bernoulli effect in FET can give a new low-frequency electronic method for determination of the effective mass of charge carriers. The analogy with the Bernoulli effect in superconductors is shortly analysed and an experiment how to measure the electric charge related with superconducting vortices is suggested.

1. Introduction

Due to the very small low-frequency energy dissipation, the quantum Hall effect (QHE) [1, 2, 3] in a two dimensional electron gas (2DEG) is a phenomenon analogous to superfluidity and superconductivity. If energy conservation is a good approximation for the mechanics of the fluid the Bernoulli theorem is applicable, actually proposed a hundred years ear-

S.-L. Drechsler and T. Mishonov (eds.), High-T$_c$ Superconductors and Related Materials, 505–518.

lier by Toricheli [4]. The aim of the present paper is to analyse how the Bernoulli-Torichelli effect can be observed in a 2DEG in the QHE regime, and what kind of applications could be expected for this effect. In the next section we shall examine the simplest possible case of the integer QHE (IQHE) with completely filled Landau levels. Later in Sec. 3 we shall use this simple model for a short consideration of the chemical potential in IQHE regime. After that we shall apply this result to the case of small field gradients, and in the spirit of electrodynamics of continuous media we will derive the Bernoulli potential in Sec. 4 by averaging of the electric potential with respect of the electron motion. In order to analyse the Bernoulli effect of the 2DEG in the QHE regime we briefly discuss in Sec. 5 the Bernoulli effect in superconductors and the problem for determination of the vortex charge. Finally in Sec. 6 we consider several setups and numerical examples for suggested experiments and come to the conclusion that we could expect appearance of an effective mass spectroscopy for semiconductors based on the Bernoulli effect in the quantum Hall regime.

2. Model and notations

In order to introduce standard notations we shall reproduce some well-known results from the quantum mechanics of an electron in magnetic field. Our starting point is the Hamiltonian of a 2D electron in external electric and magnetic field

$$\hat{H} = \frac{1}{2m}(\mathbf{p} - e\mathbf{A})^2 + e\varphi, \quad \varphi(x) = -Fx, \quad A_y = Bx, \quad A_z = A_x = 0, \quad (1)$$

see, for example, the textbook on quantum mechanics by Landau and Lifshitz [5] (in the present paper we will use SI units). The electrons move in the plane $z = 0$. The electric field $\mathbf{E} = -\nabla\varphi = F(1,0,0)$ is parallel to 2DEG, we choose the x-axis along it, while the magnetic field $\mathbf{B} = \nabla \times \mathbf{A} = B(0,0,1)$ is perpendicular to 2DEG. The y variable is cyclic and we suppose periodic boundary conditions for electron wave functions

$$\psi(x,y) = \psi(x, y + L_y) = \frac{1}{\sqrt{L_y}}\psi_{n_x}(x)\exp\left(\frac{i}{\hbar}p_y y\right), \quad p_y L_y = 2\pi\hbar n_y, \quad (2)$$

where n_y is integer. The substitution in the Schrödinger equation $\hat{H}\psi(x,y) = E\psi(x,y)$ reduces the problem to a one dimensional (1D) problem

$$\hat{H} = \frac{1}{2m}\hat{p}_x^2 + \frac{1}{2}m\omega_c^2(x - x_0)^2 + e(-Fx), \quad \hat{H}\psi_{n_x}(x) = E_{n_x, n_y}\psi_{n_x}(x), \quad (3)$$

where $\omega_c = -eB/m$ is the cyclotron frequency and $x_0(p_y) = p_y/eB$ is the centre of the cyclotron orbit at zero electric field, and n_x is the eigen function index of this 1D motion. Let $v_y = F/B$ be the velocity at which the electric field is zero after a Lorentz transformation. Later on we shall verify that this is charge carriers drift velocity $\mathbf{v}_{dr} = (0, v_y, 0)$ in QHE regime. Let

$$u_x(F) = eF/m\omega_c^2 = mF/eB^2 \tag{4}$$

be the displacement of the centre of cyclotron orbit under the action of an external electric field and $\tilde{x} \equiv x - x_0 - u_x$ be the x-coordinate taken into account from that shifted centre of cyclotron motion

$$x_c(p_y, F) = x_0(p_y) + u_x(F) = p_y/eB - mF/eB^2 = -\frac{1}{\omega_c}\left(\frac{p_y}{m} - v_y\right). \tag{5}$$

In these new variables the Schrödinger equation takes the form

$$\left(\frac{1}{2m}\hat{p}_x^2 + \frac{1}{2}m\omega_c^2\tilde{x}^2 - eFx_0 - \frac{1}{2}mv_y^2\right)\psi_{n_x}(\tilde{x}) = E_{n_x,n_y}\psi_{n_x}(\tilde{x}). \tag{6}$$

The substitution of oscillator wave functions $\psi_{n_x}^{(osc)}$ cf. Ref. [5]

$$\left(\frac{1}{2m}\hat{p}_x^2 + \frac{1}{2}m\omega_c^2\tilde{x}^2\right)\psi_{n_x}^{(osc)}(\tilde{x}) = \hbar\omega_c\left(n_x + \frac{1}{2}\right)\psi_{n_x}^{(osc)}(\tilde{x}), \tag{7}$$

using $mv_y^2 = m\omega_c^2 u_x^2 = eFu_x$ cf. Ref. [6] gives the spectrum

$$E_{n_x,p_y} = \hbar\omega_c\left(n_x + \frac{1}{2}\right) - eFx_c + \frac{1}{2}mv_y^2, \quad n_x = 0, 1, 2, \ldots \tag{8}$$

and wave functions

$$\psi_{n_x,p_y}(x, y) = \psi_{n_x}^{(osc)}(\tilde{x})\frac{1}{\sqrt{L}}\exp\left(\frac{i}{\hbar}p_y y\right). \tag{9}$$

We shall use this spectrum in the next section in the derivation of the chemical potential.

For zero electric field the degeneracy per unit area of each Landau level

$$g = \frac{eB}{2\pi\hbar} = 0.241 \times 10^{15}\, B(\text{T})\,\text{m}^{-2} \tag{10}$$

can be obtained by the condition that the electrons are moving in a restricted area $x_0(p_y) \in (0, L_x)$, which can be rewritten as $n_y \in (0, gL_xL_y)$. In the IQHE regime all levels from the ground $n_x = 0$ to $n_x = n_{\text{HOMO}}$ are completely filled with electrons. For indexes here we shall use the chemical notations, Highest Occupied Molecular Orbital (HOMO) for highest occupied Landau level and analogously Lowest Unoccupied Molecular Orbital

(LUMO) for the lowest unoccupied Landau orbital. For the 2D electron density in IQHE regime we have

$$n^{(2D)} = (2s+1)gn_{\text{LUMO}} = \frac{eB}{h}i_{\text{IQHE}}, \qquad (11)$$

where the multiplier $2s + 1$ takes into account the two spin projections $h = 2\pi\hbar$, and $i_{\text{IQHE}} = 2n_{\text{LUMO}}$. Due to the relatively small effective masses $m_{\text{eff}} = m/m_0 \ll 1$ of technological semiconductors, for example $m_{\text{GaAs}} \approx 0.07m_0$, the spin splitting is almost negligible with respect of the cyclotron energy $\hbar\omega_c$ and critical currents are significantly higher for even number i_{IQHE} of filled Landau levels. In order to obtain the contribution to 2D current from each electron orbital we shall use the well-known quantum mechanical formula for the current density [5]

$$j_y(x, y) = \frac{e}{m}\left(-i\hbar\nabla\theta - e\mathbf{A}\right)_y |\psi|^2 = ev_y|\psi|^2, \qquad \psi(x, y) = |\psi|e^{i\theta}. \quad (12)$$

It is easy to check that for zero electric field the Hall current is also zero. The action of the electric field is reduced to a displacement of centres of cyclotron orbits to a distance u_x. This shift causes a change of the vector-potential $\Delta A_y = -Bu_x$, and according to the upper formula for the current creates the drift velocity $v_{dr} = -e\Delta A_y/m$. After elementary substitutions we derive that this drift velocity is equal to the mentioned above velocity parametrizing the Lorentz transformation $v_{dr} = v_y \equiv F/B$. So for the Hall current we have $I_y = en^{(2D)}v_yL_x$, and the taking into account the Hall voltage $U_x = E_xL_x$ we arrive to the quantized Hall resistance $R_{xy}(i_{\text{IQHE}}) = U_x/I_y = R_H/i_{\text{IQHE}}$, $R_H = h/e^2 = 25812\Omega$.

3. Chemical potential in IQHE regime

Let us now determine the chemical potential of the 2DEG at low temperatures, $T \ll \hbar\omega_c$, for the middle of Hall plateaux. For thermally activated electrons in LUMO we have the Fermi filling factor

$$N_{\text{LUMO}} = \left[\exp\left(\frac{E_{\text{LUMO}} - \mu}{T}\right) + 1\right]^{-1} \approx \exp\left(-\frac{E_{\text{LUMO}} - \mu}{T}\right) \ll 1. \quad (13)$$

We again ignore the small energy of spin splitting

$$E_{\text{spin}} = m_{\text{eff}}\hbar\omega_c\left(n_{\text{spin}} - \frac{1}{2}\right), \qquad n_{\text{spin}} = 0, 1 \quad \text{for} \quad \uparrow, \downarrow. \quad (14)$$

Similarly for the holes in HOMO we have

$$1 - N_{\text{HOMO}} = 1 - \left[\exp\left(\frac{E_{\text{HOMO}} - \mu}{T}\right) + 1\right]^{-1} \approx \exp\left(-\frac{\mu - E_{\text{HOMO}}}{T}\right) \ll 1. \quad (15)$$

The thermal activation for the rest orbitals is negligible and the condition for the charge neutrality $N_{\text{LUMO}} = 1 - N_{\text{HOMO}}$ determines the chemical potential at zero current density

$$\mu_0 = \frac{1}{2}(E_{\text{HOMO}} + E_{\text{LUMO}}) = \hbar\omega_c N_{\text{LUMO}}. \tag{16}$$

This formula is analogous to the well-known result from the physics of semiconductors, for a pure semiconductor the chemical potential (the Fermi level) is in the middle of the forbidden band at low temperatures.

In case of nonzero electric field the chemical potential $\mu(x)$ is space dependent. In order to derive the Bernoulli potential, we shall derive in the next the chemical potential averaged with respect of electron motion.

4. Bernoulli effect in IQHE regime

For small electric fields a space dependent electrochemical potential could be introduced considering, in the first approximation, the electric potential as a constant. We have to substitute in the formula for chemical potential Eq. (16) the electric field dependent spectrum Eq. (9), so we have

$$\zeta = \frac{1}{2}(E_{\text{HOMO}}(F) + E_{\text{LUMO}}(F)) = \hbar\omega_c N_{\text{LUMO}} - eFx_c + \frac{1}{2}mv_y^2, \tag{17}$$

where the change of the notation $\mu \to \zeta$ denotes that for charged particles this is the electrochemical potential. Now we have to take into account that in Eq. (9) the term $e\varphi(x_c) = -eFx_c$ is the potential energy of an electron localised at x_c and $\frac{1}{2}mv_y^2$ is the kinetic one for the homogeneous drift of all electrons with velocity v_y. In the spirit of electrodynamics of continuous media (cf. Ref. [7]) we have to substitute the centre of the shifted cyclotron orbit $x_c = \langle n_x p_y | x | n_x p_y \rangle$ where the electron is localised with the local coordinate $x_c \to x$, so we finally get

$$\zeta(x) = \hbar\omega_c N_{\text{LUMO}} + e\varphi(x) + \frac{1}{2}mv_y^2. \tag{18}$$

This is our central result, the first term in this expression is the chemical potential at zero electric field, the second is the potential energy of an electron at x, and the last one is formally equal to the kinetic energy of a drifting electron with velocity $\mathbf{v}_{dr} = (0, v_y, 0)$. In the case of negligible entropy production we have a local thermodynamic equilibrium along the stream lines, i.e. along the drift velocity for $x = $ const the electrochemical potential is also constant: $\zeta = \mu_0 + e\varphi(x) + \frac{1}{2}mv_y^2 = $ const. Let us multiply this condition for the zero ohmic resistivity with the density of the 2DEG in the IQHE regime $n^{(2D)} = 2gn_{\text{LUMO}} = $ const. We obtain the equation

$$\frac{1}{2}mv_{dr}^2 n^{(2D)} + e\varphi n^{(2D)} = \text{const}, \tag{19}$$

which is just the Bernoulli theorem in IQHE regime; for each volume of the liquid along a stream line (cf. Ref. [8]) $x = $ const, the sum of the kinetic and the potential energy is a constant. This result does not depend on how correlated the electron motion is and it could be applicable for the fractional QHE (FQHE) too (we have only to substitute i_{IQHE} with the fractional filling ν_{FQHE}). The Bernoulli theorem is only a consequence of the energy conservation applied to the magneto-hydrodynamics of a 2DEG in the QHE regime. Here we have made a straightforward microscopic derivation which is simple for the electron gas in the IQHE regime. We have reproduced standard results because actually the Bernoulli effect in IQHE regime is hidden in the solution of elementary quantum-mechanical problems [5, 6]. Returning back to the Bernoulli theorem we have to mention, however, that the pressure term p is absent. The same situation arises with the microscopic BCS derivation of the Bernoulli effect in superconductors; the microscopic BCS theory is a self consistent approximation for noninteracting quasiparticles moving in a self-consistent field. Roughly speaking the pressure term disappears in the self-consistent approximation of noninteracting particles. In order to understand the Bernoulli effect in QHE regime we will review the basic statements of the Bernoulli effect in superconductors in the next section.

5. Bernoulli effect in superconductors

For superconductors the Bernoulli theorem stays almost the same as in Eq. (19). In framework of the BCS self-consistent theory we have free particles after the Bogoliubov (u, v)-transformation and again $p = 0$

$$\frac{1}{2}m\mathbf{v}_{dr}^2 n(T) + \varphi\rho_{tot} = \text{const}, \qquad \rho_{tot} = e^* n(T = 0). \tag{20}$$

Let us point out only the differences. 1) For a superconductor the voltmeter displays a zero voltage because the electrochemical is constant for all the sample, while for the QHE it is a constant only along the stream lines and Hall voltage can be measured by a voltmeter. 2) For metals the incompressibility of the electron fluid comes from the conditions for the charge neutrality while for IQHE this is a consequence of the Pauli principle applied to the Landau subbands. 3) Only the temperature dependent superfluid density $n(T)$ participates in the kinetic energy but both components (the superfluid and the normal) take part in the potential energy (cf. for example the references in Ref. [9]). According to the BCS theory $\rho_{tot} = e^* n(T = 0)$ is the volume charge density of all electrons from the conduction band, $|e^*| = 2|e|$, and the proportion of the superfluid liquid $n(T)/n(0)$ is given by the penetration depth $\lambda(T)$ or by the superconducting gap $\Delta(T)$. Let us also remind ourselves of the well-known formulas [10] for pure s-type superconductors

$$\frac{n(T)}{n(0)} = \frac{\lambda^2(0)}{\lambda^2(T)}, \quad \frac{n(0)}{n(0) - n(T)} = 1 - \frac{d \ln T}{d \ln \Delta(T)}, \quad \frac{1}{\lambda^2(T)} = \frac{4\pi n(T) e^{*2}}{m^* c^2},$$
(21)

and $m^* \approx 2m_{eff} m_e$. The same two-fluid model for the Bernoulli theorem is applicable for the superfluid He, as well cf. Ref. [8], but with pressure p included in Eq. (20) and gravitation potential energy in the second term of Eq. (20). Thus, the Bernoulli theorem has common form for all superfluids including the QHE incompressible one. For charged particles, however, the current induced Bernoulli potential $\Delta\varphi$ gives a correction to the work function $\Delta W = -e\Delta\varphi$, which is actually a contact potential difference [7]. The contact potential difference could be measured by electrical methods and in the next section we shall propose a possible experimental setup for IQHE. Here we briefly consider 2 possible experiments with superconductors in geometry similar for those proposed for QHE measurements.

5.1. BERNOULLI EFFECT IN MAGNETISED TYPE II SUPERCONDUCTORS - HOW TO MEASURE THE VORTEX CHARGE

Let us analyse the charging effect related to the Bernoulli potential in a plane capacitor with plates perpendicular to the magnetic field. One of the plates is the investigated superconductor with superconducting surface, the other is a normal metal, and the insulator layer with thickness d_{ins} should be extremely thin. The quality of superconductor-insulator interface should be good enough for the preparation of Josephson junctions, but the insulator layer should be slightly thicker to exclude the tunnelling currents. Nevertheless, the thickness of the insulator layer d_{ins} should be smaller than the penetration depth at zero temperature $\lambda(0)$. The normal metal layer should be just thick enough to be homogeneous. We have to calculate the distribution of the electric potential $\varphi(r)$ around a vortex line. For distances r larger than the coherent length $\xi(T)$ but smaller than penetration dept we have the Bohr's condition $m^* v r = \hbar$ for the velocity of the superfluid $v(r)$. The substitution in the Bernoulli theorem Eq. (20) gives

$$\varphi(r) = -\frac{1}{2e^*} \frac{\hbar^2}{m^* r^2} \frac{n(T)}{n(0)}, \quad r \in (\xi(T), \lambda(T)).$$
(22)

Far from the vortex core $r > d_{ins}$ we can consider the electric field E_z as homogeneous and to express the induced charge density $\sigma(r)$ at the $xy-$surface of normal metal by Bernoulli potential

$$E_z = \frac{\varphi}{d_{ins}} = \frac{\sigma}{\epsilon_0 \epsilon},$$
(23)

where ϵ is the relative dielectric constant of the material, and for the numerical value in SI we have $\epsilon_0 = 1/(4\pi \times 10^{-7}c^2)$. The integration of this density gives for the total vortex charge q_v induced in the normal plate

$$\frac{q_v}{e} = \int_{r_{min}}^{r_{max}} \sigma(r)\mathrm{d}\,\pi r^2 = A_v \frac{n(T)}{n(0)} \ln \kappa_{eff}, \qquad A_v = \frac{\pi \epsilon_0 \epsilon \hbar^2}{d_{ins} ee^* m^*}, \qquad (24)$$

where $\kappa_{eff} = r_{max}/r_{min} > 1$,

$$r_{min} = \max\left(\xi(T), d_{ins}\right), \qquad r_{max} = \min\left(\lambda(T), \sqrt{\Phi_0/B}\right), \qquad (25)$$

and $\Phi_0 = 2\pi\hbar/|e^*|$ is the flux quantum. The density of vortex lines per unit area is $\nu_v = B/\Phi_0$, so a drift velocity $v_v = F/B$ of the vortices in the superconducting layer will create a 2D current density in the normal one

$$j^{(2D)} = \nu_v q_v v_v, \quad I_y/U_x = q_v e^*/2\pi\hbar, \quad I_y = L_x j_y^{(2D)}, \quad U_x = -FL_x. \quad (26)$$

This Hall current in the normal strip could be measured directly in case of a short circuit, or a corresponding Hall voltage could be detected. The simplest way to measure this vortex charge is to use a cross shape structure of a metal strip with oxidised surface and another strip evaporated in perpendicular direction; the setup is shown in Fig. 1. This can be an Al-AlO$_x$Au structure. We have to apply a voltage in the superconducting strip in order to create drift of the vortices and to measure the small current induced in the normal strip. Due to very the small charge of vortices this can be a very difficult experiment. According to Eq. (24) we can use for order evaluation $q_v \simeq 10^{-6}$ and probably a SQUID voltmeter will be indispensable for the observation of the vortex charge Hall effect in MIS structures. For cuprates we can use scotch tape method to obtain in high vacuum an atomically clean surface of $Bi_2Sr_2CaCu_2O_8$, but the technological problem is how to affix the normal metal with oxidised surface. In principle it is possible to use layer-by-layer grown structures but technologists are never interested by fundamental science and new effects. A thin $SrTiO_3$ insulator layer on the atomically clean $Bi_2Sr_2CaCu_2O_8$ surface will be an ideal structure for the suggested experiment. Here we want to add that the Bernoulli effect gives a simple explanation of the change in the sign of the Hall coefficient in cuprates. Every pancake vortex has a charge $2q_v$ induced in the neighbouring CuO_2 slabs. One can easily check that the sign of Hall effect for the of pancake liquid is just opposite to the sign of the Hall effect in the normal phase. It is necessary to take into account the direction of the pancake drift, the pancake charge induced in the neighbouring layers has the sign of the normal carriers.

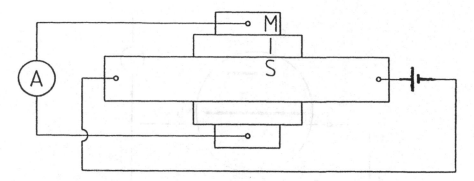

Figure 1. MIS-type structure for measurement of vortex charge. In strong magnetic field the voltage applied trough superconducting (S) layer creates a flux flow. The electrostatic interaction across the insulator (I) layer creates the polarisation charges and the ammeter detects this flow of electrostatic images in the normal metal (M) layer. According to Eq. (26) the Hall conductance is $I_y/U_x = \sigma_{yx} = q_v/\Phi_0$.

The considered 4 terminal structure is nevertheless more complicated than a 2 terminal plane capacitor. That is why we will discuss what will be happen in a typical situation when the space between plates d_{ins} is larger than the intervortex distance $\sqrt{\Phi_0/B}$. In this case we have to average the Bernoulli potential Eq. (20) at the superconductor surface. The integration between $\xi(T)$ and r_{max} gives

$$\langle \varphi \rangle = -\frac{\pi \hbar^2}{e^* m^*} \frac{n(T)}{n(0)} \nu_v \ln \frac{r_{max}}{\xi(T)}. \tag{27}$$

This potential can be observed as charge $Q = C\langle \varphi \rangle$ induced in a capacitor C in an impulse magnetic field, or directly as magnetic field induced change of the electron chemical-potential. Actually some experiments are already done, see for example, Ref. [11] but the interpretation is still unclear. For $YBa_2Cu_3O_{7-\delta}$ the surface is not so clean as for $Bi_2Sr_2CaCu_2O_8$ and the picture is complicated by pinning effects. If the energy dissipated by vortex motion during the magnetisation is negligible with respect of the energy of creation of vortices, and $\kappa \gg 1$, we can evaluate the energy of the magnetisation as kinetic energy of the superfluid

$$\Delta F = -\int_0^{B^{(ext)}} M dH \approx \int_V n(T) \frac{1}{2} m^* v^2 dV = \frac{1}{2}(c^2 \epsilon_0) \int_V \lambda^2(T)(\nabla \times \mathbf{B})^2 dV, \tag{28}$$

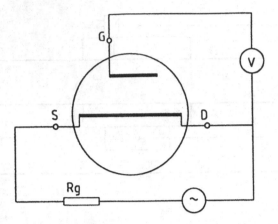

Figure 2. Principle scheme for observation of Bernoulli effect in FET: S is the source, G is the gate and D is the drain electrode. The SD current with frequency f and amplitude I_{SD} is created by an AC current generator with internal resistance R_g. The voltmeter with internal resistance $R_{\text{Voltmeter}} \gg 1/fC_{GD}$ detects the Bernoulli GD voltage at the doubled frequency $V_{GD}^{(2f)} \propto (I_{SD}^{(f)})^2$, and the coefficient of this proportionality law gives the effective mass of charge carriers.

where $B^{(\text{ext})}$ is the external magnetic field. In this case the averaged Bernoulli potential due to a distribution of vortices reads as

$$-\frac{1}{V} \int_0^{B^{(\text{ext})}} M \, dB \simeq -\rho_{tot}\langle\varphi\rangle. \tag{29}$$

For experimental data processing we expect a high correlation coefficient in a magnetisation versus chemical potential plot, even if the magnetic hysteresis loops are significant.

6. Experimental setup and numerical example

Probably the simplest method for observation of the Bernoulli effect in 2DEG in semiconductors is to measure the contact potential difference between the source and the gate of a metal-oxide-semiconductor field-effect transistor (MOSFET) when an alternating current (AC) is applied between the equipotential source and drain in IQHE regime. The proposed experimental setup is presented in Fig. 2 (cf. Fig. 1.7 in Ref. [2]). The gate electrode and the 2DEG of the MOSFET form a capacitor with capacitance C. Applying an AC current through the 2DEG will cause charging and discharging of the gate electrode if it is connected with the source electrode. This gate-drain (GD) voltage U_{GD} created by the Bernoulli effect can be detected by a voltmeter. Let us make an order evaluation of the effect using acceptable parameters. In a real case if we want to extract quantitative information from the experiment we have to solve a simple electrodynamic problem.

6.1. SIMPLE ELECTRODYNAMIC PROBLEM

In order to calculate the static current distribution of a superfluid

$$j_y(x) = \left(i_{\text{IQHE}} \frac{e^2}{2\pi\hbar} \right) E_x(x) \tag{30}$$

we have to minimise the sum of the kinetic energy of the superfluid and the energy of electromagnetic field. For superconductors [12] the energy of the magnetic field with density $\frac{1}{2}c^2\epsilon_0\mathbf{B}^2$ dominates while for the QHE we have to take into account the energy of the electric field [13] with density $\frac{1}{2}\epsilon_0\mathbf{E}^2$. So an electrostatic problem of self-consistent calculation of the electric potential φ arises. For zero current across the strip with width $L_x = 2w$ we have a constant density of the charge carriers

$$n^{(2D)}(x) = n^{(2D)}\theta(w - |x|), \quad w \equiv L_x/2. \tag{31}$$

The current is related to a shift u_x of these charges, and simultaneously a variation of the charge density $\delta\sigma(x)$ is also created by the deformation

$$\begin{aligned}
\delta\sigma &= \frac{\partial}{\partial x}\left(en^{(2D)}(x)u_x(x) \right) \\
&= en^{(2D)}\left[-\frac{\partial}{\partial x}u_x(x) + u_x(w)\delta(x - w) - u_x(-w)\delta(x + w) \right],
\end{aligned} \tag{32}$$

where δ-functions are created by derivatives of the density at edges of the strip. The charge conservation requirement $\int_{-\infty}^{\infty}\sigma(x)dx = 0$ is automatically fulfilled for arbitrary space redistribution $u_x(x)$. The density variation creates the electric potential and the Hall voltage

$$\varphi(x) = -\frac{1}{2\pi\epsilon_0\epsilon}\int \delta\sigma(x\prime) \left[\ln|x' - x| - \ln l_{\text{images}} \right] dx', \tag{33}$$

$$l_{\text{images}} \equiv \sqrt{(x' - x)^2 + (2d_{ins})^2},$$

where we added the potential of images induced on the gate electrode spaced at a distance d_{ins} from the 2DEG. In the small gradient approximation the electric field $E_x(x) = -\partial\varphi(x)/\partial x$ creates the polarisation shift $u_x(x) \approx eE_x(x)/m\omega_c^2$ according Eq. (4), and finally the self-consistent equation for the electric potential with a boundary condition: $-\varphi(w) + \varphi(-w) =$

U_x reads as

$$\varphi(x) = -\Lambda_{QHE}^{(2D)} \{ \ \int_{-w}^{w} [\ln|x'-x| - \ln l_{\text{images}}] \frac{\partial^2}{\partial x'^2} \varphi(x') dx' \tag{34}$$
$$-\frac{\partial}{\partial x}\varphi(+w) + \frac{\partial}{\partial x}\varphi(-w) \ \},$$

where

$$\Lambda_{QHE}^{(2D)} \equiv \frac{e^2 n^{(2D)}}{2\pi\epsilon\epsilon_0 m\omega_c^2} = \frac{i_{\text{IQHE}}}{\pi}\frac{l_B^2}{a_{\text{Bohr}}}, \qquad l_B = \sqrt{\frac{\hbar}{eB}}, \qquad a_{\text{Bohr}} = \frac{4\pi\epsilon_0\epsilon\hbar^2}{e^2 m}$$

are respectively the 2D screening dept, magnetic length and effective Bohr radius. An analogous Bohr radius participates in the formula for the vortex charge Eq. (24). The equation for the electric potential Eq. (34) has a lot of common details with the equations presented in Refs. [13, 14]. For a thin superconducting strip (with thickness $d_{\text{film}} < \lambda(T)$) spaced on the insulated surface of a bulk superconductor we have similar singular integral equation

$$A_y(x) = \frac{1}{\Lambda_{SC}^{(2D)}} \int_{-w}^{w} [\ln|x'-x| - \ln l_{\text{images}}] A_y(x'), \tag{35}$$

where

$$\Lambda_{SC}^{(2D)}(T) = \frac{2\pi\lambda(T)^2}{d_{\text{film}}} = \frac{2\pi\epsilon_0 c^2 m^*}{e^{*2} n(T) d_{\text{film}}}$$

is the 2D screening length which is related with the kinetic inductance. In both cases for a fixed current the skin effect in the strip increases the Bernoulli signal so for qualitative consideration we can neglect the redistribution of the current density. One can check that the naive assumption for the homogeneous current gives the minimal value of the Bernoulli signal which we are going to evaluate.

6.2. NUMERICAL EXAMPLE

Let us take a gate region with $L_x \times L_y = 1$ mm^2 spaced 1000 Å from 2DEG. For the dielectric constant we take $\epsilon = 13.6$ and for the GD-capacity we get $C = 1.20$ nF. We consider for illustration a square domain 1mm×1mm. Let the density of electrons be $n^{(2D)} = 2 \times 10^{11}$ cm^{-2}, and we want to use $i_{\text{IQHE}} = 4$ Hall plateau. At these conditions: $R_{xy}(4) = 6453\Omega$, $B = 2.07$ T, $l_B = (\hbar/m\omega_c)^{1/2} = (\hbar/eB)^{1/2} = 178$ Å. For GaAs $m_{eff} = 0.068$, $\epsilon = 13.6$ and $a_{\text{Bohr}} = 106$ Å. Substituting $m = m_{eff}m_0$, where m_0 is the mass of free electron, we have for the cyclotron energy $\hbar\omega_c = h854$ GHz $= k_B 41$ K $= 3.5$ meV. For 2D skin depth we have $\Lambda_{QHE}^{(2D)} = 384$ Å. Let us apply a source-drain current density $j(t) = j_0\cos(\omega t)$ with frequency $f = \omega/2\pi = 50$ kHz and amplitude $j_0 = 40$ mA/m which is almost an order of magnitude smaller than the typical cited break-down current densities 600-900 mA/m (see, for example Ref. [15]). Significantly higher critical currents could be reached in Corbino geometry using ring electrodes capacitively joined with 2DEG. For the amplitude v_0 of the drift velocity $v(t) = v_0 \cos(\omega t)$ we have $v_0 = j_0/en^{(2D)} = F/B = 125$ m/s, $F = 258$ mV/m, $u_x = 0.25$ Å. For the considered square domain we have Hall voltage $U_x = 258$ mV, and $I_{SD} = 40$ μA. At helium temperatures $T \simeq 4$ K we have the law of mass action $N_{LUMO}(1 - N_{HOMO}) \approx \exp(-\hbar\omega_c/k_BT) \ll 1$ and we see that the number of exited particles is small. The Bernoulli potential $\Delta\varphi_B = -\frac{1}{2}mv_{dr}^2/e$ can also be expressed in terms of current density

$$-\Delta\varphi_B(t) = \frac{1}{2}\frac{m}{e^3\left(n^{(2D)}\right)^2}j^2 = \varphi_0\cos(2\omega t) + \varphi_0, \qquad (36)$$

$$\varphi_0 = \frac{1}{4}\frac{m_0}{e}m_{eff}\left(\frac{I}{en^{(2D)}L_x}\right)^2$$

where the amplitude of this GD Bernoulli voltage is $\varphi_0 = 1.6$ nV. This Bernoulli voltage $\varphi_B(t) = U_{GD}$ could create a polarisation GD-current $I_{GD} = I_{GD}^{(0)}\cos(2\omega t)$, where for the GD-current amplitude we have $I_{GD}^{(0)} = (2\omega)\,C\varphi_0 = 1.21$ pA; for a few Hz this polarisation current will be unobservable small. This Bernoulli GD-current could be detected by phase sensitive method at doubled frequency $2f$. Another possibility is a direct detection of the Bernoulli potential (between source and gate) by a low noise preamplifier and lock-in voltmeter working at $2f$. For the internal resistance of the ammeter and the voltmeter we must have $R_A \ll 1/\omega C = 2.8$ k$\Omega \ll R_V$.

6.3. DISCUSSION AND CONCLUSIONS

Probably there will be substantial difficulties related with cleaning of the parasite signals having the basic frequency f of source-drain current. Now the Bernoulli effect can realise a drift-velocity meter for FET in QHE regime, analogous to many technical applications in the fluid dynamics [4]. Static measurements are also possible, in principle, but we have to measure the polarisation charge of several e. One can still wonder whether the band mass or the many-body renormalised mass will be measured. In the latter case the mass of composite fermions [19] could be determined.

Acknowledgements. One of the authors (TM) is thankful to Prof. J. Indekeu for the stimulating discussions and hospitality in KU Leuven, where the present work began in 1997. He also would like to thank to Atanas Groshev for the warning that this simple picture can be significantly complicated by the edge currents [16, 17] and states [18]. The writing of this paper was provoked by Dr. Manus Hayne during the 60th birthday party of

518

Prof. F. Herlach. The same author is very much indebted to Dr. Hayne for suggestions, interest and critical reading of the manuscript. He appreciates the stimulating discussions with Dr. M. J. Van Bael, Dr. K. Temst, and Prof. Y. Bruynseraede on the perspectives for performance of experiments with superconducting films.

This paper was partially supported by the Bulgarian NSF No. 627/1996. The research has been supported by the DWTC, the Programme VIS/97/01, the IUAP and the GOA.

References

1. Klitzing v. K., Dorda G., and Pepper M. (1980) *Phys. Rev. Lett.* **45**, 494.
2. *The Quantum Hall Effect*, edited by Prange R.E. and Girvin S.M. (1987, Springer-Verlag, New York).
3. Klitzing v. K. (1981) *Festkörperprobleme XXI*, (Friedr. Vieweg&Sohn, Muenchen), p.1.
4. Giancoli D.C. (1984) *General Physics*, (Prentice-Hall, New York), Chap. 13.3, Figs. 13.6 and 13.12, Venturi and Pitot tubes.
5. Landau L.D., Lifshitz E.M. and Pitaevskii L.P. (1990) *Course on Theoretical Physics*, Vol. 3, *Quantum Mechanics*, (Pergamon, New York), Sec. 19, and Chap. XV, Sec. 112.
6. R.B. Laughlin R.B. (1981) *Phys. Rev. B* **23**, 5632; Galitzkii V.M., Karnakov B.M. and Kogan V.I. (1984) *Problems on Quantum Mechanics*, (Nauka, Moskow), problem 6.11. (in Russian).
7. Landau L.D., Lifshitz E.M. and Pitaevskii L.P. (1984) *Course on Theoretical Physics*, Vol. 8, *Electrodynamics of Continuous Media*, 6th ed. (Pergamon, New York), Sec. 1, Chap. III, Sec. 23.
8. Landau L.D., Lifshitz E.M. and Pitaevskii L.P. (1988), *Course on Theoretical Physics*, Vol. 6, *Hydrodynamics*, (Pergamon, New York).
9. Mishonov T.M. (1994) *Phys. Rev. B* **50**, 4009.
10. Landau L.D., Lifshitz E.M. and Pitaevskii L.P. (1979) *Course on Theoretical Physics*, Vol. 9, *Statistical Physics*, Part 2, (Pergamon, New York), Chap. V, Eq. (40.14).
11. Nizhankovskii V.I. and Zybsev S.G. (1994) *Phys. Rev. B* **50**, 1111.
12. London F. and London H. (1935) *Proc. R. Soc. London Ser. A* **149**, 71; (1935) *Physica* **2**, 341.
13. MacDonald A.H., Rice T.M. and Brinkman W.F. (1983) *Phys. Rev. B* **28**, 3648.
14. Balaban N.Q., Meirav U., Shtrickman H., and Levinson Y. (1993), *Phys. Rev. Lett.* **71**, 1443.
15. Cage E., in Ref. 2, Sec. 2.12, and references therein.
16. Kouwenhoven L.P., van Wees B.J., van der Vaart N.C., Harmans C.J.P.M., Timmering C.E., and Foxon C.T. (1990) *Phys. Rev. Lett.* **64**, 685.
17. Watts J.P., Usher A., Matthews A.J., Zhu M., Elliot M., Herrenden-Harker W.G., Morris P.R., Simmons M.Y., and Ritchie D.A. (1998) *Phys. Rev. Lett.* **81**, 4220.
18. Takaoka S., Oto K., Kurimoto H., Murase K., Gamo K., and Nishi S. (1994) *Phys. Rev. Lett.* **72**, 3080.
19. Jain J.K. (1989) *Phys. Rev. Lett.* **63**, 199.

GRAIN BOUNDARIES AND OTHER INTERFACES IN CUPRATE HIGH-T_c SUPERCONDUCTORS

HANS HILGENKAMP and JOCHEN MANNHART
Center for Electronic Correlations and Magnetism
Institute of Physics, University of Augsburg
Universitätsstrasse 1, 86135 Augsburg
Germany.

1. Introduction

A better understanding and control of the electronic transport properties of interfaces in high-T_c superconductors is of great importance for a further successful development of applications based on these materials. A rapid development of large-current applications, for example, has been hampered by the critical current-limiting properties of high-angle grain boundaries, demanding special efforts to circumvent their presence in cables and tapes. In addition, for such applications interfaces between superconductors and normal metal connections are important, as large contact-resistances can give rise to excessive heating. Also in micro-electronic circuits, interfaces are of central importance, e.g. for the performance of Josephson junctions, the basic building blocks for many superconducting sensors or circuits. Besides their relevance for applications of high-T_c superconductors, interfaces are also crucial in experimental studies aimed to gain understanding of the physics of high-T_c superconductors, concerning e.g. the symmetry of the superconducting order parameter or the electronic band structure of these materials, as such experiments are often based on surface-sensitive techniques, such as tunneling spectroscopy.

A useful model-system to study the physics of interfaces is given by thin film grain boundaries. By depositing epitaxial films on bicrystalline substrates, well defined grain boundaries of designed misorientation are prepared [1]. Transmission Electron Microscopy (TEM) has revealed that for various materials, including $YBa_2Cu_3O_{7-\delta}$, such thin film grain boundaries can be prepared clean and free of second phases [2]. In this contribution, the electronic properties of thin film grain boundaries are discussed and fundamental mechanisms controlling these properties are considered. A number of these mechanisms are not only of interest for grain boundaries, but control more generally the characteristics of various kinds of interfaces involving high-T_c cuprates, including surfaces and superconductor-normal metal contacts.

519

S.-L. Drechsler and T. Mishonov (eds.), High-T$_c$ Superconductors and Related Materials, 519–528.
© 2001 *Kluwer Academic Publishers. Printed in the Netherlands.*

520

2. Grain Boundary Properties

Grain boundaries are characterized by their misorientation configuration, for which different possibilities exist (see e.g. Fig. 1). In a series of pioneering studies, Dimos *et al.* [1] have investigated the transport properties of grain boundaries of various misorientations on SrTiO$_3$ bicrystals. They found that the critical current density J_c is strongly dependent on the grain boundary misorientation angle θ and decreases for all grain boundary configurations if θ is increased from 0 to 45°. These results were later confirmed by other groups, such as by Ivanov *et al.* [3] who showed that for [001]-tilt boundaries the θ-dependence follows closely an exponential behavior..

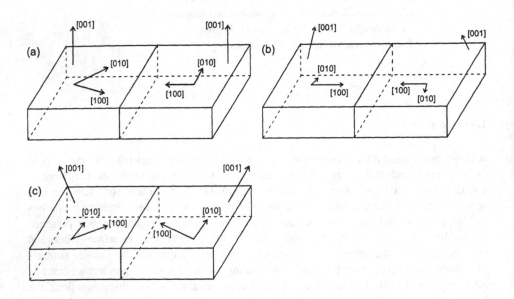

Figure 1. Various grain boundary configurations, (a) [001]-tilt boundary, (b) [100]-twist boundary, (c) [010]-tilt boundary

The angular dependence of J_c, and that of the normal-state resistivity R_nA and the I_cR_n-product for YBa$_2$Cu$_3$O$_{7-\delta}$ [001]-tilt boundaries are illustrated by our recent data in Fig. 2. As is seen in this figure, for an increase of θ from 0° to 45°, J_c decreases by about a factor of 3×10^3 at 4.2 K, as $J_c \approx 2\times10^7\exp(-\theta/\theta_0)$ A/cm^2, with $\theta_0 \approx 6.3°$ [4]. Remarkably, [001]-tilt boundaries with θ larger than $\approx$ 10-15° behave as overdamped Josephson junctions [5], with I_cR_n-values smaller than expected from the presumed energy-gap values of the high-T_c cuprates. For smaller θ, the grain boundaries act as weak links with rounded current-voltage $V(I)$-characteristics, indicating flow of Abrikosov vortices. It is to be noted that grain boundaries in conventional metallic superconductors do not behave as weak links or Josephson junctions, illustrating the extraordinary nature of such interfaces in high-T_c superconductors.

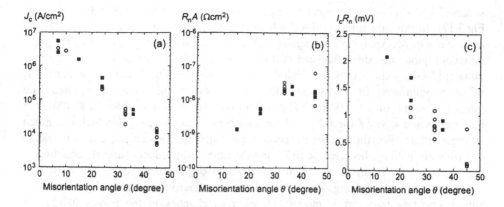

Figure 2. Dependencies of (a) the critical current density J_c, (b) the normal state interface resistivity R_nA, and (c) the I_cR_n-product for symmetric (filled squares) and asymmetric (open circles) [001]-tilt grain boundaries in YBa$_2$Cu$_3$O$_{7-\delta}$ thin films on the misorientation-angle θ (T = 4.2 K). For the asymmetric boundaries, one grain was oriented with [110] normal to the boundary-plane. From Ref. [4].

In addition to the peculiar superconducting properties of the grain boundaries, an interesting issue is provided by the rather high values of the normal-state resistivity R_nA, being typically in the range 10^{-9}-10^{-7} Ωcm^2 (Fig. 2b). These values are several orders of magnitude larger than the Sharvin-resistance, which yields a lower limit to the resistance in metallic contacts. With a noticeable spread in values, R_nA rises by about a factor of 20, as θ is increased from 15° to 45°. The flux-flow behavior for smaller angles disallows a useful measurement of R_n. The different θ-dependencies of J_c and R_nA result in a decrease of I_cR_n with increasing θ, as is shown in Fig. 2c.

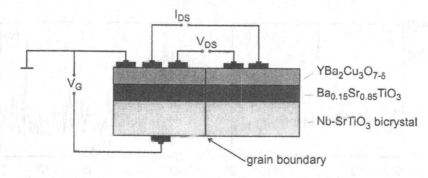

Figure 3. Sketch of an inverted Metal-Insulator-Superconductor Field Effect Transistor (MISFET) -structure. After [7].

The influences of applied electric fields on the transport properties of the grain boundaries has been investigated by using three-terminal configurations [6], mostly with the purpose to develop high-T_c superconducting transistor-like devices. A successful configuration for this aim is presented by the inverted MISFET-structure of Fig.3 [7]. In this, use is made of a conducting Nb-doped $SrTiO_3$ bicrystalline substrate as gate electrode. Applying a voltage V_G to the gate, the transport properties of the grain boundary junctions are influenced, which reflects itself as changes in the current–voltage (I_{DS}-V_{DS}) characteristics. With the configuration of Fig. 3, the critical current I_c of grain boundaries in $YBa_2Cu_3O_{7-\delta}$ films could be modulated by an applied gate voltage to values of ≈ 1 %/V for 24° [001]-tilt boundaries, ≈ 4 %/V for 36.8° [001]-tilt boundaries and ≈ 8%/V for 45° [001]-tilt boundaries respectively [7], which is a much stronger effect than that on single crystalline films. In contrast, R_n did not change significantly with applied electric field, and thus the I_cR_n-product changed according to the change in I_c. From the different electric-field dependencies of I_c and R_n it was suggested that by applying an electric field the effective width of the barrier for superconducting transport is modulated, whereas changes in the barrier height are insignificant [7].

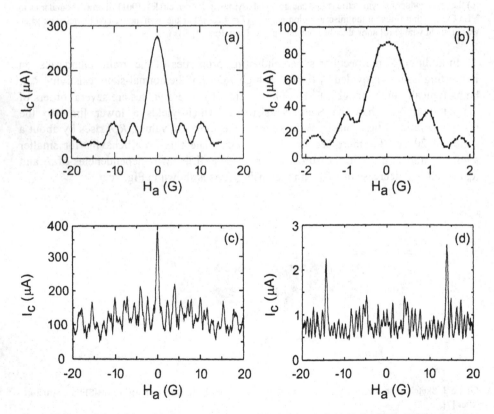

Figure 4. Dependence of the critical current I_c on the applied magnetic field H_a for [001]-tilt grain boundary junctions of different misorientation-angle θ: (a) $\theta = 24°$ (asymmetric: 45°-21°), $t = 180$ nm, $w = 8$ μm, (b) $\theta = 36.8°$ (symmetric), $t = 15$ nm, $w = 16$μm, from [9], (c) $\theta = 45°$ (symmetric), $t = 180$ nm, $w = 20$ μm, (d), $\theta = 45°$ (asymmetric: 45°-0°), $t = 22$ nm, $w = 16$ μm, from [10]. In this, w is the junction width and t the film thickness. Measurement (a) is taken at $T = 80$ K, all other at $T = 4.2$ K

The magnetic field (H_a) dependence of I_c of grain boundaries also provides very useful and exciting information about the physics of high-T_c superconductors. As is shown in Fig. 4, for increasing θ the H_a-dependence deviates more and more from the expected Fraunhofer-like characteristic, which is known very well from Josephson-contacts in conventional superconductors [8]. For 45° grain boundaries the magnetic field dependencies are particularly anomalous, with the maxima of I_c not necessarily occurring at zero applied magnetic field, but at some finite field value [9]. Below, it will be shown that this results from the combination of grain boundary faceting and the predominant $d_{x^2-y^2}$ order parameter symmetry in most high-T_c cuprates [10].

3. Mechanisms controlling electronic transport properties of grain boundaries

Of central importance for the superconducting transport properties of interfaces in the high-T_c cuprates are the small values of the superconducting coherence lengths ξ, due to which the boundary-conditions imposed on the pair-potential by e.g. interfaces, structural defects or impurities, may cause a drastic reduction of the order parameter [11]. Combined with this, various mechanisms have been suggested to influence the behavior of grain boundaries in high-T_c superconductors, of which several may act simultaneously. In the following such mechanism are listed, for clarity grouped into five families.

1. Mechanisms based on the structural properties of the grain boundaries.
Grain boundaries are structural defects which, by definition, interrupt the lattice structure of the adjacent crystals. Transmission Electron Microscopy (TEM)-studies have revealed that the structure of most grain boundaries can be described by standard dislocation theory [2]. At small angle grain boundaries an array of separate dislocations is created, which merge with increasing θ. This was recognized already in the early works of Dimos *et al.* [1] as an important aspect in the θ-dependence of the grain boundary transport properties, including the transition from weak link to Josephson behavior for θ larger than $\approx 10°$-$15°$.

An important parameter characterizing the structural properties of the grain-boundaries is the effective structural width d of the dislocation-layer. Recent TEM-measurements have shown that d increases linearly from ≈ 0.2 nm to ≈ 0.9 nm for an increase of θ from 11° to 45° [12]. The structural distortions, and their associated stress-fields, give rise to a weakened pairing interaction, modify the electronic structure of the superconductor and cause enhanced quasiparticle scattering, all together leading to a reduction of the superconducting order parameter at the grain boundary interface.

2. Mechanisms based on deviations from the ideal stoichiometry.
Second, deviations of the stoichiometry from the ideal values have been suggested as a prime cause of the weak-link behavior of grain boundary interfaces. Deviations from the ideal cation-stoichiometry can be ruled out as the controlling factor for the grain boundary properties, as it was shown that even boundaries with excellent cation composition are weak-links [2]. On the other hand, if present, defects in the oxygen sub-

lattice will depress the superconducting order parameter, and, if the oxygen concentration is low enough, will drive the material insulating.

3. Order parameter symmetry based mechanism.

Third, the predominant d_{x2-y2}-symmetry of the order parameter of the high-T_c cuprates is expected to have an important influence on the grain boundary transport properties, for several reasons. Owing to the spatial coherence of the wave function describing the superconducting state, a boundary region with a depressed order parameter is expected at the interface between two d_{x2-y2}-wave superconductors, which form a contact at a misorientation angle θ [13]. Typically, such a boundary region will stretch over distances of the order of ξ from the boundary. The depression of the order parameter in the boundary region is due to the frustration caused by the different crystallographic orientations of the superconductors on either side of the boundary and by the finite value of ξ. It is expected that the depression of the order parameter gives rise to quasiparticle states, bound by Andreev reflections, at the boundary plane, comparable to the quasiparticle states occurring in vortex cores [14]. The magnitude of this order parameter depression depends among other things on θ, the grain boundary symmetry and the misorientation-configuration, on temperature and on the materials involved. For [001]-tilt boundaries it is strongest for $\theta = 45°$, where the maxima of the order parameter of one superconductor coincide with the nodes in the gap-function of the other.

In addition, depending on their trajectories, quasiparticles may experience a sign change in the pair-potential when they are reflected at interfaces or crossing a grain boundary in a d_{x2-y2} superconductor [15]. This effect can give rise to bound quasiparticle states at the midgap-energy (zero-energy-states), which are seen as zero-bias-anomalies in tunneling spectroscopy [16,17]. These bound states are expected to lead to a reduction of the order parameter [18,19], which influences supercurrent transport across the boundary.

Further, independent of the details of the current transfer across the junction (such as e.g. the degree of directionality of the charge-transfer process) the order parameter orientations can, in specific configurations, cause an additional π-phase shift over the junction ('π–junction'). This is, for example, the principle underlying the spontaneous generation of half-flux quanta in specially designed tricrystal rings [20]. Intrinsic π-phase shifts also play an important role in the properties of single grain boundaries, because thin-film grain boundaries are not straight, but consist in general of facets of various orientations and with typical dimensions < 100 nm (see e.g. [21,22]). This faceting in combination with a d_{x2-y2} symmetry component of the order parameter leads to an inhomogeneous distribution of the Josephson current in the grain boundary junctions [10], including regions for which the additional π-shift arising from the order parameter orientation leads to a counterflow of the Josephson current ('negative' J_c). The inhomogeneity, which is most prominent for asymmetric 45° [001] tilt grain boundaries, is the cause for the anomalous magnetic field dependencies of the critical current depicted in Fig.4, and has been shown to lead e.g. to spontaneously generated magnetic flux within grain boundary junctions [23] and to deviations of the current-phase relationship of the Josephson junctions from the standard sinusoidal dependence [24].

4. Bending of the electronic band structure.

Recently, we have pointed out that at interfaces in high-T_c cuprates, bending of the electronic band structure [25-27] can occur, causing depletion- or enhancement layers next to the interface [28,29] (Fig. 5), analogous to grain boundaries in dielectric or ferroelectric oxides [30] and in semiconductors [31]. In these layers with modified carrier density the order parameter will be reduced and, for strong enough depletion, the cuprates will be driven into the antiferromagnetic insulating state.

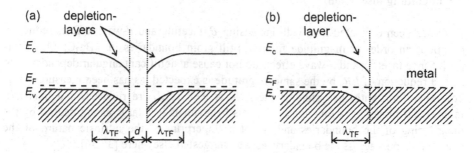

Figure 5: Sketch of a possible scenario based on a semi-conductor band-picture, for band-bending at (a) a grain boundary and (b) a superconductor-normal metal contact, leading to the formation of depletion-layers at the interface (From [27]).

In contrast to most conventional superconductors, bending of the band structure, for example due to local variations in the work function, to electrostatic charging or to surface states, is an effect strong enough to control superconducting properties of the high-T_c cuprates. As a result of the large dielectric constant ε_r and the small carrier density n of the cuprates, the Thomas-Fermi screening lengths λ_{TF}, the length scales over which band bending occurs, are in the range of ≈ 0.5-1 nm, and thus are comparable to ξ. In addition, the transport properties of the high-T_c cuprates depend much more sensitively on the carrier concentration than the transport properties of conventional metallic superconductors, for which it is virtually impossible to induce a phase transition into an insulating state.

An exact quantitative description of the effects of band bending on the electronic properties of interfaces is hard to make. As the typical length-scales involved, such as λ_{TF}, are comparable to the lattice-spacing and to ξ, such a description requires detailed knowledge of the microscopic electronic properties including the pairing mechanism, which is not available. Nevertheless, it has been shown by Gurevich *et al.* that band bending can explain the reduction of J_c for low angle grain boundaries [32]. A further indication for its importance can be obtained from its contribution to R_nA. Treating the grain boundary as a double Schottky barrier it was estimated that band bending leads to R_nA values of the order of 10^{-8} Ωcm^2, increasing with an increasing width of the dislocation layer formed at the grain boundary interface [4,26,27]. For the grain boundary capacitance $C \approx 1 \times 10^{-7}$ F/cm^2 was estimated. Both values agree well with the measured ones [33].

5. Other mechanisms, based on a direct suppression of the pairing mechanism

Depending on the pairing mechanism, the misorientation and the interruption of the periodic lattice structure depress the pairing interaction [34]. In an analogous way, a preferred direction for the spins of the charge carriers forming a Cooper-pair, caused for example by finite spin-orbit coupling, leads to spin-flip processes which lower the order parameter of grain boundaries of various misorientations. [25,26].

4. Concluding discussion

It has been estimated that, with increasing θ, faceting and d-wave effects reduce J_c by about an order of magnitude for [001]-tilt grain boundaries in YBa$_2$Cu$_3$O$_{7-\delta}$ films [10]. Since faceting and d-wave effects do not cause a significant angular dependence of R_nA, a reduction of I_cR_n by the same magnitude is expected, as has been measured (Fig. 2c). Therefore, with increasing θ, the I_cR_n as well as J_c are reduced, and thus I_cR_n increases if plotted as a function of J_c. This behavior is to a large extent caused by the meandering of the boundaries and is not a fingerprint of the intrinsic nature of the charge transport across the boundary, as was suggested elsewhere [35,36].

Many of the mechanisms presented above do not rely on properties specific to certain high-T_c compounds, but are generic to most high-T_c cuprates and provide, without claiming completeness, a consistent explanation for the universal behavior in these materials. More generally, some of the issues discussed here for grain boundaries, for example the possibility of significant bending of the electronic band structure, are expected to be relevant to other interfaces involving the high-T_c cuprates, e.g. contacts with normal metals.

In summary, it was shown that high-T_c copper-oxide superconductors exhibit unconventional interface properties, which are not found for conventional superconductors. The interface properties are controlled by the short coherence length, in combination with various other aspects, such as the unconventional order parameter symmetry, and the possibility for significant bending of the electronic band structure, leading to changes in the density of mobile charge carriers in the crystals abutting the interface. With the improved understanding of the interface properties gained in the last couple of years it is anticipated that techniques will advance which will enhance the control over the interface properties and allow a improved tailoring of their properties for specific applications

The authors thank A. Kleinsasser for valuable discussions. This work was supported by the BMBF (13N6918/1). H.H. further thanks the Royal Dutch Academy of Sciences and the University of Twente for their support.

5. References:

[1]: Dimos, D., Chaudhari, P., Mannhart, J., and Legoues, F.K. (1988), *Phys. Rev. Lett.* **61**, 1653; Chaudhari, P., Dimos, D., Mannhart, J. (1989), *IBM J. Res. Develop.* **33**, 299; Dimos, D., Chaudhari, P., and Mannhart, J. (1990), *Phys. Rev. B* **41**, 4038.

[2]: Chisholm, M.F., and Smith, D.A. (1989), *Phil. Mag.* **59**, 181.

[3]: Ivanov, Z.G., Nilsson, P.Å., Winkler, D., Alarco, J.A., Claeson, T., Stepantsov, E.A., and Tzalenchuk, A.Ya. (1991), *Appl. Phys. Lett.* **59**, 3030.

[4]: Hilgenkamp, H., and Mannhart, J. (1998), *Appl. Phys. Lett.* **73**, 265.

[5]: Mannhart, J., Chaudhari, P., Dimos, D., Tsuei, C.C., and McGuire, T.R. (1988), *Phys. Rev. Lett.* **61**, 2476.

[6]: For a recent review, see: Mannhart, J. (1996), *Supercond. Sci. Technol.* **9**, 49.

[7]: Mayer, B., Mannhart, J., Hilgenkamp, H. (1996), *Appl. Phys. Lett.* **68**, 3031.

[8]: Barone, A., Paternò, G., *Physics and Applications of the Josephson Effect*, (Wiley & Sons, 1982 New York).

[9]: Mannhart, J., Mayer, B., and Hilgenkamp, H. (1996), *Z. Phys. B* **101**, 175.

[10]: Hilgenkamp H., Mannhart. J., and Mayer, B. (1996), *Phys. Rev. B* **53**, 14586.

[11]: Deutscher, G., and Müller, K.A. (1987), *Phys. Rev. Lett.* **59**, 1745.

[12]: Browning, N.D., Buban, J.P., Nellist, P.D., Norton, D.P., Chisholm, M.F., and Pennycook, S.J. (1998), *Physica C* **294**, 183.

[13]: Hilgenkamp, H., and Mannhart, J. (1997), *Appl. Phys. A* **64**, 553.

[14]: Hess, H.F., Robinson, R.B., and Waszczak, J.V. (1991), *Physica B* **169**, 422.

[15]: Hu, C.R. (1994), *Phys. Rev. Lett.* **72**, 1526.

[16]: Alff, L., Takashima, H., Kashiwaya, S., Terada, N., Ihara, H., Tanaka, Y., Koyanagi, M., and Kajimura, K. (1997), *Phys. Rev. B* **55**, R14757.

[17]: Lesueur, J., Greene, L.H., Feldmann W.L., and Inam, A. (1992), *Physica C* **191**, 325.

[18]: Tanaka, Y., and Kashiwaya, S. (1995), *Phys. Rev. Lett.* **74**, 3451 and (1996), *Phys. Rev. B* **53**, R11957.

[19]: Barash, Yu. S., Burkhardt, H., and Rainer, D. (1996), *Phys. Rev. Lett.* **77**, 4070.

[20]: Tsuei, C.C. *et al.* (1994), *Phys. Rev. Lett.* **73**, 593; Kirtley, J.R., *et al.* (1995) *Nature* **373**, 225; Tsuei C.C., *et al.* (1996), *Science* **271**, 329.

[21]: Alarco, J.A., Olsson, E., Ivanov, Z.G., Nilsson, P.Å., Winkler, D., Stepantsov, E.A., and Tzalenchuk, A.Ya. (1993), *Ultramicroscopy* **51**, 239.

[22]: Træholt, C., Wen, J.G., Zandbergen, H.W., Shen, Y., and Hilgenkamp, J.W.M. (1994), *Physica C* **230**, 425.

[23]: Mannhart, J., Hilgenkamp, H., Mayer, B., Gerber, Ch., Kirtley, J.R., Moler, K.A., and Sigrist, M. (1996), *Phys. Rev. Lett.* **77**, 2782.

[24]: Il'ichev, E., Zakosarenko, V., IJsselsteijn, R.P.J., Schultze, V., Meyer, H.-G., Hoenig, H.E., Hilgenkamp, H., and Mannhart J. (1998), *Phys. Rev. Lett.* **81**, 894.

[25]: Mannhart, J., and Hilgenkamp, H. (1997), *Supercond. Science and Technol.* **10**, 880.

[26]: Mannhart J., and Hilgenkamp H. (1998), *Mater. Sci. Engr. B* **56**, 77-86 (1998).

[27]: Hilgenkamp, H., and Mannhart, J. (1998), *Proc. 4th Twente HTS Workshop*, in press.

[28]: Browning, N.D., Chisholm, M.F., Pennycook, S.J., Norton, D.P., and Lowndes, D.H. (1993), *Physica C* **212**, 185.

[29]: Babcock, S.E., Cai, X.Y., Larbalestier, D.C., Shin, D.H., Zhang, N., Zhang, H., Kaiser, D.L., and Gao, Y. (1994), *Physica C* **227**, 183.

528

[30]: Vollmann, M., Hagenbeck, R., and Waser, R. (1997), *J. Am. Ceram. Soc.* **80**, 2301.

[31]: Werner, J. (1985), in *'Polycrystalline Semiconductors'*, *Springer Series in Solid-State Sciences* **57**, Harbeke, G., ed., Springer, p. 76.

[32]: Gurevich, A., and Pashitkii, E.A. (1998), *Phys. Rev. B* **57**, 13878.

[33]: Tarte, E.J., Wagner, G.A., Somekh, R.E., Baudenbacher, F.J., Berghuis, P., and Evetts, J.E. (1997), *IEEE Trans. Appl. Supercond.* **7**, 3662.

[34]: Chaudhari, P., Dimos, D., and Mannhart, J., in *'Earlier and Recent Aspects of Superconductivity'*, Bednorz, J.G., and Müller, K.A., eds. (Springer, Berlin 1990), p. 201.

[35]: Gross, R., and Mayer, B. (1991), *Physica C* **180**, 235; Gross, R., in *'Interfaces in High-T$_c$ Superconducting Systems'*, Shindé, S.L. and Rudman, D.A., eds. (Springer, New York 1994), p.176.

[36]: Halbritter, J. (1992), *Phys. Rev. B* **46**, 14861 and (1993), *Phys. Rev. B* **48**, 9735.

ENHANCEMENT OF THE CRITICAL CURRENT DENSITY OF YBA₂CU₃O₇₋δ–FILMS BY SUBSTRATE IRRADIATION

S. LEONHARDT,   J. ALBRECHT,   R. WARTHMANN AND
H. KRONMÜLLER
Max-Planck-Institut für Metallforschung
Heisenbergstr. 1, D-70569 Stuttgart, Germany

Abstract.
    One of the most striking properties of $YBa_2Cu_3O_{7-\delta}$ (YBCO) thin films
is their high critical current density. This is due to strong pinning centers
which are generated during film growth. As YBCO grows epitactically, the
film growth mechanism depends strongly on the roughness of the substrate
surface. With a Focused Gallium Ion Beam (FIB) the structure of the substrate surface is modified. Magneto-optical measurements show the local
flux densities in the superconductor and the local current density distribution can be calculated by a numerical inversion of Biot-Savart's law [1]. It
is shown that the critical current density in the superconductor is enhanced
by specific substrate modification.

    The critical current density of a high-temperature superconductor is
a decisive factor for application. In this context epitactical thin films of
YBCO exhibit the desired high current densities of several 10^{11} A/m². In
this paper we show, that the current density can be further increased by
modifying the substrate on which the superconducting film is grown.
    For this purpose a Focused Gallium Ion Beam (FIB) is used to irradiate
different sections of $SrTiO_3$-substrates. With the FIB, microstructures with
lateral dimensions of approximately 1 μm can be generated, which directly
influence the growth mechanism of superconducting thin films. Under appropriate conditions enhancement of the current carrying capability of the
film up to 30 % are achieved.
    To examine the superconducting properties of the modified samples, the
flux penetration in an external magnetic field after zero-field-cooling is investigated by the magneto-optical Faraday effect. This effect is an excellent
tool to visualize the local flux density distribution in superconductors with

S.-L. Drechsler and T. Mishonov (eds.), High-Tc Superconductors and Related Materials, 529–534.

a high spatial resolution of approximately 3 μm. A sketch of the measuring technique is shown in Figure 1.

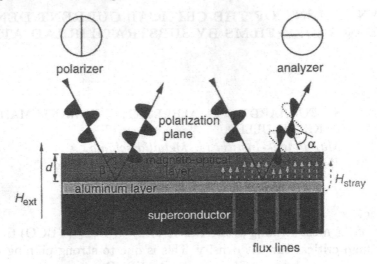

Figure 1.   Detection of the spatially resolved flux density distribution using the magneto-optical Faraday-Effect.

Both, a magneto-optically active and a reflective layer, are put on top of the superconducting film. By passing through the the magneto-optical layer, the polarization plane of polarized light is rotated by an angle which depends on the local magnetic field. A gray scale image of the magnetic field distribution is obtained by a crossed polarizer–analyzer pair. In addition to this, the local current density distribution can be calculated from the observed magnetic flux density by an inversion of Biot-Savart's law [1]. This numerical method yields the current density directly, without assuming any model; it works therefore for any geometric boundary conditions. Finally, the spatially resolved distribution of the critical current in the superconducting thin film is available. This local information allows us to compare modified and unmodified regions of the same film and to separate the real effect of the substrate modification from ordinary quality variations of the superconducting films.

To demonstrate the achieved improvements, an unmodified superconducting film in an external magnetic field after zero field cooling is depicted in the left part of Figure 2. It shows a gray scale plot of the flux density distribution of a YBCO film with a thickness of 130 nm in an external flux density of $B_{ext} = 70$ mT. In addition to this the calculated current stream lines are depicted as black lines.

Obviously the magnetic flux has penetrated the sample in a four-fold symmetry because of the quadratic shape of the film. The quadratic shape

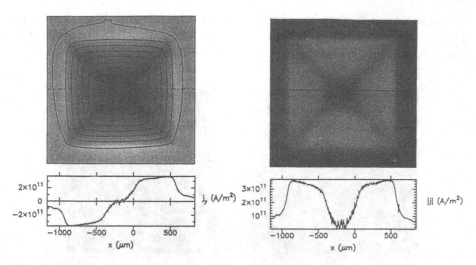

Figure 2. Left side: Flux density distribution and current stream lines of an unmodified YBCO thin film in an external magnetic field after zero field cooling. Right side: The calculated current density distribution in the same sample.

of the current stream lines exists for the same reason and is directly correlated to the boundary condition and to div $\mathbf{j} = 0$ and $|\mathbf{j}|$ = const. inside the sample.[2] The single closed current loop outside the film is related to artefacts which are generated by the numerical analysis of the measured data. Considering the current densities inside the sample the error which is made is smaller than 5 % [1]. In the right part of Figure 2 a grey scale plot of the absolute value of the current density is depicted. Bright parts refer to high current densities of over 3×10^{11} A/m^2. This image shall serve as a basis for a comparison with the following modified superconducting films.

To compare the effects that different geometries of substrate surface modifications have on the superconducting quality of the films, the FIB was used to make arrays of dots, lines and scans over rectangular parts of the substrate. These structures have been characterized by Atomic Force Microscopy (AFM) in order to correlate the modifications of the substrate surface with their effects on the superconducting films.

On the left hand side of Figure 3, part of a line is shown which was written on the substrate surface. This line is about 1.5 μm wide and has a height of approximately 50 nm. The right side of Figure 3 shows a sector of an array of 100 $\times$ 100 dots. The four brighter patches that can be seen in the four corners are the irradiated parts of the substrate. Their diameters are between 1.5 and 3 μm and their heights are about 10 to 15 nm. The profiles below show the exact geometric properties. Meanwhile the contents of the substrate surface have been investigated by Energy Dispersive X-ray Diffraction and it has been found that a part of approximately 20 % of Ga was implanted in the surface layers of the $SrTiO_3$.

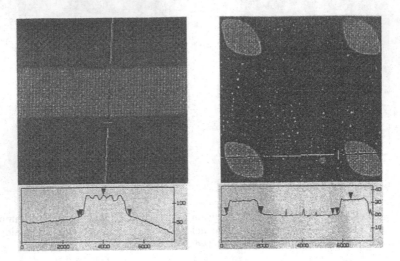

Figure 3. Atomic force micrographs of a line and of four dots on the $SrTiO_3$-surface written with the FIB. The length scales are given in nm. The height of the irradiated patches is between 10 and 50 nm.

As the substrate has to be heated before and during filmgrowth, examinations of the substrate have been made after an annealing process. By this treatment the height of the modified patches is reduced to less than 1 nm but their shapes are still visible very clearly and the surface in these areas is rougher than in the unmodified part.

Structuring of the substrate influences the growth of the superconducting material drastically. In Figure 4 the surface of a superconductor at the border between a scanned area and an unmodified part of the substrate is depicted. The upper half of the image shows the modified part. The image, taken by a light microscope, shows stronger contrast and larger irregularities in the unmodified part of the superconductor. This indicates a more ordered growth of the superconductor above the substrate surface, which was scanned with the FIB.

The effects on the critical current density due to these modifications in YBCO thin films are shown in Figure 5. The greyscale plot represents the absolute value of the local current density which is calculated from the measured flux density distribution. Brighter parts of the image are related to higher values of the current density.

The bright rectangular part indicates the part of the superconductor that was grown on the scanned area of the substrate surface. The profile

Figure 4. Micrograph of the surface of the superconducting film showing the difference of the surface roughness between modified (top) and unmodified (bottom) substrate area.

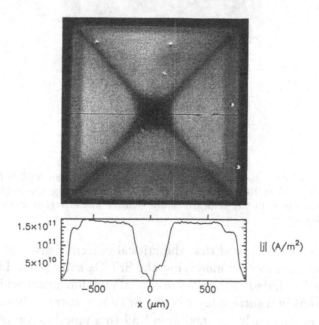

Figure 5. Calculated current density distribution of a YBCO thin film on a substrate partly scanned by the FIB. The bright rectangular part refers to the increased current density of the film in the modified region.

below shows an increase of the current density by 2.5×10^{10} A/m^2 in the modified area. This corresponds to an improvement of about 20 %.

Similar results can be achieved by modifying the substrate surface with an array of dots. Figure 6 shows the distribution of the absolute value of the current density in a YBCO sample that has been grown on such a substrate.

534

The bright square on the right side of the sample indicates an enhancement of the critical current density in the modified part of the superconductor. The profile below shows an increase by more than 3.5×10^{10} A/m^2. This again is a rise of more than 20 %.

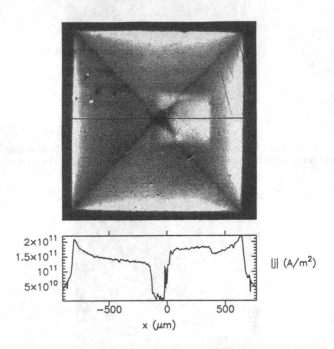

Figure 6. Absolute value of the critical current density of an YBCO film on a substrate with an array of 100 × 100 dots. In the grayscale image the brighter area indicates an increasement of the critical current density in the region above the modified SrTiO$_3$-substrate.

In this work we showed that the critical current density of YBCO thin films can be enhanced by modifying the SrTiO$_3$ substrates. The substrate surface was irradiated with a Focused Gallium Ion Beam which implants Gallium atoms in a surface layer of the SrTiO$_3$ substrate. Both an array of dots and a continuously scanned area lead to a modification of the growth of the YBCO thin film. The modified films show an enhancement by the critical current density of up to 30 %.

References

1. Ch. Jooß, R. Warthmann, A. Forkl and H. Kronmüller, Physica C **299**, 215 (1998).
2. A. Forkl and H. Kronmüller, Phys. Rev. B **52**, 16130 (1995).

DESIGN OF HTS DC SQUID GRADIOMETER FOR APPLICATION IN NONDESTRUCTIVE EVALUATION AND BIOMAGNETISM

F. SCHMIDL, S. WUNDERLICH, P. SEIDEL

Institut für Festkörperphysik, Friedrich-Schiller-Universität Jena,
Helmholtzweg 5, D-07743 Jena, Germany

Abstract.
We investigated dc SQUID structures for operations in planar high T_C-gradiometer structures for different applications in unshielded evironments. The developement of dc SQUIDs for practical applications is determined by the type of sensor in which the SQUID will be used. Generally we can distinguish between different instance between magnetically and galvanically coupled sensor layouts.
The layout of galvanically coupled dc superconducting quantum interference devices (SQUIDs) is investigated with respect to their application in planar thin film gradiometers. The devices were made of $YBa_2Cu_3O_{7-X}$ thin films prepared by laser deposition on $SrTiO_3$ bicrystal substrates. Different superconducting properties, like the critical, the normal resistance and the noise properties of the Josephson junctions were realized by defined variation of the grain boundary angle (24°, 30° and 36.8°), of the film thickness and of the junction width. At first the influence of the dc SQUID geometry on the SQUID inductance and on the effective area will be discussed. We measured the change of the mutual inductance as well as the change of the parasitic inductance of our dc SQUIDs due to variation of different geometric parameters e.g. the length and the width of the SQUID loops and the width of the incoupling lines or of the Josephson junctions. The experimental results at 77 K were compared with numerical simulations. We will show the possibilities and limitations of dc SQUID layouts with respect to the noise behaviour.

1. Introduction

High temperature superconducting (HTS) superconducting quantum interference device (SQUID) gradiometers operating at 77 K are applied in systems [1] for biomagnetic measurements [2, 3] as well as for nondestructive evaluation (NDE)

S.-L. Drechsler and T. Mishonov (eds.), High-Tc Superconductors and Related Materials, 535–552.

[4, 5, 6]. For a planned employ of superconducting magnetic field sensors in unshielded environment, different concepts are possibly to suppress the electromagnetic disturbances. Depending on the way of effort, one must consider both strength and frequency range of the disturbances which are to be expected in order to adapt the sensor and the measuring system. An example is the relatively uniform earth magnetic field of about 50 μT which is around 6 to 9 orders of magnitude larger than the biomagnetic signals to be measured. Therefore a large dynamic range of the used superconducting sensor and a high slew rate of the control electronics are necessary. One possible solution is the compensation of the homogeneous magnetic field with a gradiometer arrangement. J. Vrba could show the influence of disturbances on magnetometer and gradiometer sensors vividly [7]. The realisation of such gradiometer structures on the basis of HTS materials is possible in different variants. Here we concentrate on the design of thin film planar dc SQUID gradiometers.

2. Principles of Gradiometer Sensors

The advantage of superconductive antennas consists in the possibility to convert an external magnetic field or better the magnetic induction B_{ext} fitting normally to the antenna area with a small amount of additional noise into a screening current I_A flowing in the antenna itself. This allows measurements of static magnetic fields in comparison to normal conducting coils. The ratio between the magnetic flux Φ_A in antenna structures and the magnetic induction B_{ext} is designated as the active area of antenna A_A. This value characterizes the effectiveness of a specific antenna geometry in the case of the transformation of the external magnetic field into a corresponding magnetic flux in the antenna. The conversion of the magnetic flux Φ_A into a screening current I_A in the antenna is determined by the antenna inductance $L_A = \Phi_A/I_A$. The so called efficiency E of the antenna is established as a quotient from the active area of the antenna A_A and the antenna inductance L_A.

A main parameter the field gradient resolution $\sqrt{S_G}$ is defined by the flux noise spectral density $\sqrt{S_\Phi}$, by the base length b of the gradiometer and by the effective area A_{eff} of the gradiometer in the equation

$$\sqrt{S_G} = \sqrt{S_\Phi}/bA_{eff}. \tag{1}$$

Considering the white noise of the gradiometer sensor this noise is dominated by the white noise of the dc SQUID and strongly dependent on the dc SQUID inductance L_S. The product of base length and effective area corresponds to the quotient from the magnetic flux in the SQUID sensor Φ_{SQUID} and the external magnetic field gradient $B_{ext,i}/\Delta i$ (2). In this case, the index i indicates the considered field direction.

$$bA_{eff} = \Phi_{SQUID}/(\Delta B_{ext,i}/\Delta i) \tag{2}$$

The effective area of the gradiometer can be determined experimentally with the base lengths together. From this point of view different possibilities to optimize the sensor design of gradiometer structures exist. On the one hand, it is necessary to minimize the spectral noise density depending on Josephson junction and SQUID type, on the other hand, the product of base length and effective area should be chosen as large as possible. The dimension of the substrates represent the natural limit for the enlargement of the base length in this case. We can distinguish between two different types of gradiometer type antennas in respect to the antenna layout itself [8, 9]. In the first case the two antennas were connected in series (Figure 1b), in the second type in parallel together (Figure 1a). The induced screening current I_A is either proportional to the measured magnetic field strength (for instance B_Z) or to the gradient of this field ($\partial B_Z/\partial x$). Both types have been successfully used for low temperature superconductor dc SQUID gradiometers. In serial connected gradiometer structures does not flow a supercurrent in the case of homogeneous magnetic field in both antennas. This is the great advantage of this antenna type for operation in unshielded environment especially in larger magnetic fields. But we need for the practically realisation of this layout in thin film technology multilayer structures with insulating crossovers and superconducting vias. In opinion to this behaviour we find in parallel connected antennas a screening current in the hole antenna corresponding to the external magnetic field. the application of a uniform external field gives no current through the incoupling coil or line connected with the dc SQUID in any way. In higher fields this principle can leads to problems if the screening current reaches the critical current of the superconducting lines. But it is possible to realize such structures only with one superconducting layer.

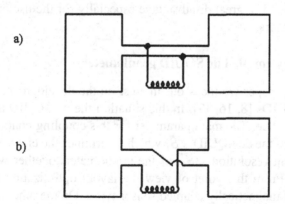

Figure 1. a) Scheme parallel loop gradiometer b) Scheme series loop gradiometer.

There are three different ways to realize the magnetic coupling between antenna structures and the dc SQUID itself shown in Figure 2, the so called direct coupling

(Figure 2a or Figure 2b), the transformer coupling (Figure 2c) or the galvanic coupling (Figure 2d).

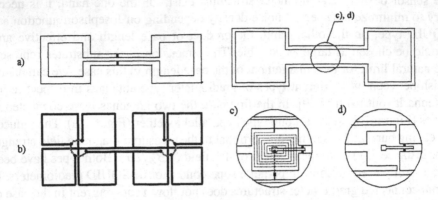

Figure 2. Different coupling schemes of the flux antenna to the SQUID converter. a) direct coupling, b) direct coupling with multiloop pickup coil (Drung-type), c) transformer coupling (Ketchen-type), d) galvanic coupling.

With transformer coupling [10, 11] or multiloop pick up coupling [12, 13, 14] can be realised the best matching between antenna and SQUID inductance resulting in a high field gradient sensitivity of the described geometries. But only the coupling scheme in Figure 2a or Figure 2d can be realized with one superconducting layer much more easier to produce by HTS thin film technology. That's why we will focus our interest on this special type of planar HTS gradiometer structures. In the direct coupled gradiometer the antenna structure is a part of the SQUID itself. The inductance of the antenna limits in such a way the practically realized size of the sensor [4, 15] a great disadvantage especially for the use of dc SQUIDs in gradiometer sensors.

3. Galvanically coupled dc SQUID gradiometer

In the following paper we focus our interest on the development of galvanically coupled dc SQUIDs [8, 16, 17]. In this situation the dc SQUID acts as a current measurement device. The main parameter for this coupling conditions is the current resolution of the dc SQUID $\sqrt{S_I}$ which determines the effective area A_{eff} and the field gradient resolution $\sqrt{S_G}$ of the gradiometer together with the antenna layout mainly. From this point of view the layout optimization differs strongly from the one of magnetically coupled sensor types. On the other side the application in gradiometer structures brings different demands which must be taken into consideration. In single layer arrangements we have the problem with HTS sensors that the Josephson junctions, the dc SQUID structure and the antenna itself are sensitive to magnetic flux or magnetic fields in the same direction [17, 18]. For this reason we get different conditions for sensor optimization:

small width of the Josephson junctions shall lower the influence of the applied magnetic field to the junctions itself,

i) The relation between the dc SQUID and the antenna structures differs for magnetometers and gradiometers.

ii) For a magnetometer the dc SQUID is an additional field sensitive element in the same field direction with a field resolution several orders lower, but it does not influence the magnetometer operation drastically. In the case of gradiometers this situation changes. The gradiometer is used to suppress the homogeneous magnetic field or the disturbances coming from long-distance sources. The dc SQUIDs itself measure the absolute magnetic field and create therefore an additional imbalance in the gradiometer structure. A large effective area of the dc SQUIDs itself can result in stronger disturbances during the operation in unshielded environment.

3.1. NUMERICAL SIMULATIONS

The structure of the planar gradiometer and of the inner part with the galvanically coupled dc SQUID is shown in figure 3 a) and b) respectively. In our experiments we change the free parameters loop length D and the part of the loop length C (in our description only the length beginning on the end of the incoupling line), the loop width B, the width of the incoupling lines W, the junction width K and the length of the superconducting lines around the Josephson junctions $P1$ (Figure 3b). Our first step was the numerical simulation to determine the SQUID inductance L as the sum of the coupling or mutual inductance M and the parasitic inductance L_P. Only the mutual inductance is responsible for the incoupling of the signal in galvanically coupled dc SQUIDs.

The parasitic inductance around the Josephson junctions gives an additional part of the full inductance of the dc SQUID and influences the transfer function $dV/d\Phi$ and the flux noise properties $\sqrt{S_\Phi}$ of the dc SQUID in this way.

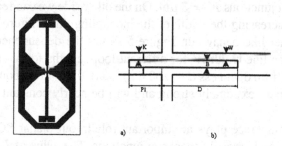

Figure 3. a) Scheme of the dc SQUID gradiometer on 10×10 mm² substrates. b) Scheme of the dc SQUID with abbreviations for the geometrical parameters.

For this investigations we employ a simulation programme using a volume integration method bases on the Maxwell equations and on the London theory [19].

This method was tested for different layouts and it showed a good agreement with experiments [20]. The temperature dependent penetration depth λ is one free parameter. The experimental verification of this parameter will be discussed later.

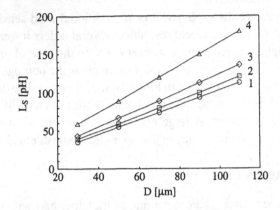

Figure 4. Numerical calculations of the dependence of the SQUID inductance L_S on the loop length D with SCIM 96 [19] for different film thicknesses (1: $d = 200$ nm, 2: $d = 150$ nm, 3: $d = 100$ nm, 4: $d = 50$ nm), parameter set: $\lambda_L = 0.26$ μm, $W = 5$ μm, $B = 4$ μm, $K = 3$ μm.

The results of these numerical simulations are plotted in the next figures. The strong influence of the loop length and of the width of the incoupling lines on the SQUID inductance L can be seen clearly. We can approximate the dependence of the mutual inductance from the loop length (figure 4) with a linear behaviour and achieve a rise in L of about 1.1 pH/μm for a film thickness of about 150 nm. In this case the precise data depend on the used film thickness with respect to the assumed penetration depth. The rise of the inductance is stronger for thinner films. For the calculations we used a London penetration depth λ of 0.26 μm at 77 K, the loop width $B = 4$ μm, the width of the incoupling lines $W = 5$ μm and the width of the Josephson junctions $K = 3$ μm. On the other side we can reduce the mutual inductance by increasing the width of the incoupling lines (figure 5) from narrow strips to a washer-like behaviour. Figure 5 shows the dependence of the mutual inductance on the line width W for different loop lengths for the same parameter set B, K, and λ (film thickness $d = 150$ nm). Started at narrow coupling structures the inductance rises extremely strong and will be nearly constant for washer like structures.

The parasitic inductance plays an important role in our planar SQUID structures which are based on bicrystal Josephson junctions. The value of L_P is determined by the length $P1$ and the junction width K (figure 3b). The first parameter depends on the precision of the resist mask preparation for the junctions (see next part). For bicrystal junctions we have an additional technology step, namely the placement of markers for the grain boundary. The optical resolution of the used microscope limits the minimization of the length $P1$. The second parameter, the

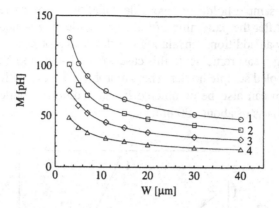

Figure 5. Numerical calculation of the mutual inductance M as function of the width of the incoupling line W for different loop length (1: $D = 110$ nm, 2: $D = 90$nm, 3: $D = 70$nm, 4: $D = 50$nm), parameter set like in figure 4.

junction width K depends on the aimed at critical current I_C of the dc SQUIDs. For bicrystal Josephson junctions we have a fixed critical current density on the same substrate. The exact value of the critical current density is generally determined by the value of the grain boundary angle [21]. But we can get also a deviation from this behaviour by the variation of the film thickness or by the inhomogeneities along the grain boundary. With current densities j_C between $2 \cdot 10^3$ A/cm^2 and $9 \cdot 10^4$ A/cm^2 the junction width K must be patterned in dimensions from 10 μm to 1 μm, respectively. Especially for small widths of the superconducting structures the loop inductance L_P increases very strong and reaches an important part of the total SQUID inductance L_S without any increase of the current resolution of dc SQUID.

3.2. EXPERIMENTAL SETUP

For our experiments we prepared YBa$_2$Cu$_3$O$_{7-x}$ (YBCO) thin films by laser deposition on 10×10 mm^2 SrTiO$_3$ bicrystal substrates with $24°$, $30°$ and $36.8°$ grain boundary angles. The laser energy of the KrF excimer laser (248 nm wavelength) between 400 mJ and 500 mJ, the substrate temperature of 680°C to 710°C and the oxygen partial pressure between 30 Pa and 40 Pa are used as standard preparation conditions. The resulting growth rate is located in the range between 0.5 nm/s and 0.8 nm/s. The film thickness is found to range from 80 nm to 200 nm. Sputtered gold or silver layers on the future contact areas provide low ohmic contacts, for instance for the noise measurements of the dc SQUIDs. The grain boundaries of the bicrystal substrates were marked by laser patterning. The quality of this preparation step determines the possible resolution of the following lithographic steps. We used electron beam lithography as well as standard photolithography to prepare photoresist masks for ion beam etching. All samples were etched on a liquid

542

nitrogen cooled sample holder to avoid degradation of the patterned Josephson junctions [22]. After the patterning of the superconducting structures all samples were covered by an additional insulating overlayer to protect all dc SQUIDs during the following measurements. In this case we used sputtered YBCO which was prepared on a cooled sample holder. This amorphous YBCO with a film thickness of about 200 nm can also be completed by an additional sputtered CeO_2 layer with a film thickness of about 100 nm.

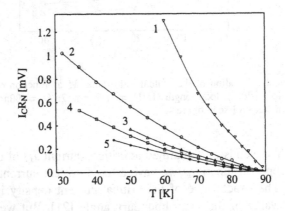

Figure 6. $I_C R_N$ product as function of temperature for different investigated Josephson junctions. Grain boundary angle: 1: 24°, 2: 30° and 3 to 5: 36.8°, film thickness: 1, 2 and 3: $d = 150$ nm, 4 : $d = 100$ nm and 5: $d = 50$ nm.

The product of the critical current I_c and the normal resistance R_N of the investigated Josephson junctions is presented in figure 6. The result is, that there are clearly different values for $I_C R_N$ products according to the used grain boundary angle. This relative high values for 24° bicrystal junctions are due to the high critical current densities. On the other hand Josephson junctions which are based on 30° bicrystals have a 5 to 10 times smaller critical current densities, but a clearly higher normal resistance R_N. An increase of the film thickness from 50 nm to nearly 150 nm results in an increase of the $I_C R_N$ product for all investigated grain boundary Josephson junctions too. For a higher film thickness $d > 200$ nm excess currents were observed often. Only dc SQUIDs with nearly identical superconducting parameters are included in the investigations of the dependence of geometry. To lower the influence of possible inhomogeneities of the used substrates or the influence of thickness variations of the YBCO films all dependencies were tested on different substrates and with different grain boundary angles. In all these cases the location where the dc SQUID structure is situated on the substrate was varied. Additional to that, all samples were investigated concerning to the structure transfer in a scanning electron microscope and with regard to keep the geometrical parameters.

3.3. EXPERIMENTAL DETERMINATION OF THE MUTUAL INDUCTANCE

In the beginning the important role of the mutual inductance M for operation of galvanically coupled dc SQUIDs was mentioned. This inductance can be obtained from direct experimental measurements [23, 24]. In this case the mutual inductance M is to be determined by the measurement of a current I_M flowing through the incoupling line

$$M = \Phi_0/\Delta I_M = M_1 + M_2. \tag{3}$$

According to the contacts for the current leads it is possible to measure the whole inductance M as well as both parts of the mutual inductance M_1 and M_2 corresponding to the relevant superconducting lines (Figure 3b). In the last case we can experimental determine the asymmetry of the dc SQUID structure itself after the patterning process. These asymmetries can be caused by differences in structure after the ion beam etching process as well as by local inhomogeneities of the superconducting properties in the strip line geometry. The best value for such an asymmetry is smaller than 0.3 percent, even if we used extreme long and narrow incoupling lines ($C > 120~\mu$ m, $W = 4~\mu$m, $B = 4~\mu$m). The effective area $A_{eff,SQUID}$ for external magnetic fields of the dc SQUID itself was determined too. Therefore a special measurement equipment is used in shielded environment, where the calibration sensor, the single dc SQUID, the magnetometer or the gradiometer are moved in different, but well defined, heights above a current driven wire. This scan of the magnetic field is fitted according to the Biot-Savart law. The result is a method by which we can determine the effect of different geometrical parameters to the effective area of the dc SQUIDs [5].

The investigations to determine the dependence of the mutual inductance M on the width of the incoupling line W are realized with constant loop lengths. The other parameter set $B = 5~\mu$m, $K = 3~\mu$m, $d = 150$ nm should be constant. The experimental determined mutual inductance M decrease with the increase of the line width W in good agreement with the results of our numerical calculations (Figure 7). The smaller mutual inductance of sample 3 results from the smaller loop length $D = 65~\mu$m, while $D = 95~\mu$m for set 1 and 2. The differences in the inductance of these two samples can be attributed to the different film thickness of the prepared YBCO layers which follow

The decrease of the inductance with the increase of the width of the incoupling lines is also connected with a very strong increase of the investigated dc SQUIDs effective area shown in figure 8 too [25] (sample 2 figure 7). This results in a significant lower current resolution $\sqrt{S_I}$ of the dc SQUIDs but in a higher imbalance if we use this layout in gradiometer structures. The investigations of Dantsker et al. have shown, that the noise properties of such washer like dc SQUID structures deteriorate strongly with an increasing loop width depending on the applied external magnetic field [26]. In external magnetic fields, that means under conditions typical for operation in unshielded environment, the $1/f$ noise level increases

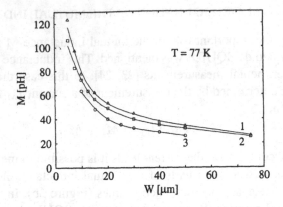

Figure 7. Measured mutual inductance M as function of the width of the incoupling line W. Loop length D: 1, 2: $D = 95$ μm, 3: $D = 65$ μm, parameter set: $B = 5$ μm, $K = 3$ μm, $d = 150$ nm.

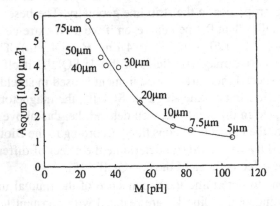

Figure 8. Measured effective area as function of mutual inductance with variation of the width of the incoupling line (inset). $T = 77$ K (sample 2, figure 7).

and the corner frequency shifts to higher frequencies if the line width extends to $W > 10 \ldots 20$ μm. The washer structure can be successfully modified by additional slits or holes, which suppresses the influence of the magnetic field [27], but the effective area $A_{eff,SQUID}$ remains constant in comparison to a non-modified washer. We prepared and measured such samples with washer structures with and without additional slits. We used 4 μm wide slits with a spacing of 5 μm for constant outer dimensions. Besides, the increase of the experimentally determined mutual inductance is not connected to a reduction of the effective area $A_{eff,SQUID}$ was found. For the washer structure with 4 slits we determined the mutual inductance $M = 35.6$ pH and the corresponding effective area $A_{eff} = 2800$ μm^2 (Figure 9).

In opposite to the increase of the line width an increase in loop length D results in an increase of the mutual inductance M as expected from numerical calculation.

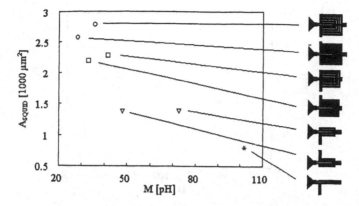

Figure 9. Effective area as a function of mutual inductance for washer structures with and without slits in the superconducting areas ($T = 77$ K). Schemes of the layouts connected by thin lines with measured values.

For two samples (curve 1 and 2 in figure 10) the loop length was varied for a line width $W = 5$ μm and a loop width $B = 3$ μm. In this case too, the deviation to the expected linear curve of the mutual inductance is determined by variations in the film thickness and by local inhomogeneities in the incoupling lines. The experimental determined value for the variation of mutual inductance M with the loop length as a part of the SQUID inductance L_S of about 1.05 pH/μm is in good agreement with the numerical simulations (Figure 4). A slight decrease of the line width W to 4 μm results in a very strong increase of the inductance M (curve 3 in figure 10) with about 1.8 pH/μm, while the loop width B is increased to 5 μm at the same time. These considerable differences show the important role of an exact structure transfer for this kind of SQUID structures plays. Therefore the masks of the investigated dc SQUIDs were produced with the help of electron beam lithography.

By means of curve 4 the strong influence of the film thickness gradient is demonstrated, as it can occur by laser deposition. The change of the inductance M and therefore the change of the SQUID inductance L_S, is balanced particularly by the great film thickness gradient in this case, when the investigated SQUIDs are situated at the edge of the sample.

In case of extended loop length the increase of the mutual inductance M results in an increase of the effective area of the investigated dc SQUIDs. The effective areas were determined for the dc SQUIDs shown in curve 3, figure 10. Even if we use high values for $M = 235$ pH, the measured amounts with a maximum of $A_{eff} = 1700$ μm^2 (Figure 11) remain smaller than absolute values of the effective areas for washer structures by a factor of 3 to 4. The variation of the loop width B changes the mutual inductance too, especially if small loop widths ($B < 6$ μm) are used. But the relative increase of M lowers clearly, if the structure width is

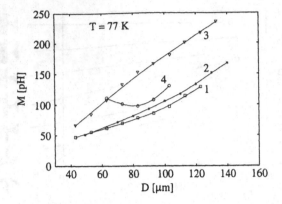

Figure 10. Mutual inductance measured as function of the loop length D. curve 1,2 and 3: $d = 150$ nm, curve 4: variation of d, parameter set: curve 1,2,4: $B = 3\,\mu m$, $W = 5\,\mu m$, curve 3: $B = 5\,\mu m$, $W = 4\,\mu m$.

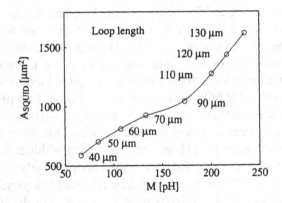

Figure 11. Effective area as function of mutual inductance M with variation of the loop length (inset). $D = 150$ nm, $B = 5\,\mu m$, $W = 4\,\mu m$ (sample 3, figure 10).

oversized. Additional to that, a larger loop width leads to a higher low-frequency noise of the investigated dc SQUIDs [28].

Finally the parasitic inductance was investigated experimentally. We produce structures with determined inductances for different width of the Josephson junctions. The grain boundary was placed in the range of the incoupling lines, so that the measured current is transmitted through the area $P1 = 10\,\mu m$. Figure 12 shows the curve which resulted from a measurement with a film thickness of $d = 150$ nm. DC SQUIDs which are used in gradiometer structures have a parasitic inductance $L_P = 20\ldots30$ pH adequate to the structure width of the Josephson junctions.

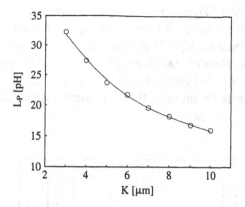

Figure 12. Measured parasitic inductance L_P as function of the width of the Josephson junctions K. Parameter set: $T = 77$ K, $d = 150$ nm, $B = 4$ μm.

3.4. NOISE MEASUREMENTS

The flux noise value $\sqrt{S_\Phi}$ and hence the correlated current resolution $\sqrt{S_I}$ are influenced by the investigated mutual inductance M as a part of the whole SQUID inductance L as well as by the superconducting properties of the used Josephson junctions like I_C [1],[29] and $I_C R_N$. For the investigations we prepared three different types of bicrystal junctions with different grain boundary angles (24°, 30° and 36.8°) To achieve informations about optimal parameters we changed the critical current I_C of the dc SQUIDs for a constant mutual inductance M by changing the width of the Josephson junctions on the different grain boundary angles. The film thickness was kept at $d = 150$ nm constant. Generally it can be seen, that for 36.8° grain boundaries the relative small $I_C R_N$ products at 77 K with values from 35 μV to 70 μV deteriorate the current resolution. Contrary to this 24° grain boundary Josephson junctions show higher $I_C R_N$ products in a range between 150 μV and 300 μV. Although the realized critical currents in that special case are with values above 180 μA considerably higher, the current resolution (white noise) could be improved compared to the 36.8° grain boundaries. This effect is caused by the higher $I_C R_N$ product which compensates the increase of the SQUID parameter L to values above 5. Values around 1 are realized for 36.8° grain boundary. Considering the low frequency noise in a flux locked loop mode and in an ac-bias operation this noise is mainly influenced by the flux motion. Comparing 24° and 36.8° grain boundary angles we observed for 24° an increase of the flux noise spectral density at 1 Hz with an increasing critical current (the inductance was kept constant). This effect can be referred to the increasing flux motion in the grain boundary region for high critical current density grain boundaries like 24°. High I_C values generate an additional flux motion by interacting with these vortices. This effect was not observed for 36.8°. The possibility of critical current reduction be reducing the film thickness in the area of the Josephson junctions only fits the SQUID parameter L better to optimal value. This is shown in Wunderlich et al. [30]. Figure 13 displays the spectral density of the field gradient resolution of a planar dc SQUID gradiometer on 5×10 mm^2 substrate as an example. We achieved values of 300 fT/cm$\sqrt{Hz}$ for the white noise level and 820 fT/cm$\sqrt{Hz}$ at 1 Hz in shielded environment. These field gradient resolutions enable these kind of sensors for investigations in biomagnetics and non-destructive testing.

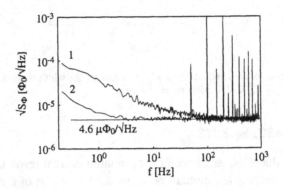

Figure 13. Flux noise spectral density of galvanically coupled dc SQUID at 77 K. curve 1: dc-bias mode, curve 2: ac-bias mode.

3.5. INVESTIGATION OF THE TEMPERATURE DEPENDENCE

We investigated the temperature dependence of the mutual inductance and the influence of this dependence on the noise properties for dc SQUIDs with different loop length on 30° bicrystal substrates. This bicrystal angle of 30° is very suitable for experiments with higher SQUID inductances. The operation temperature $T > 87$ K is high enough to test the SQUID behaviour at 77 K too. The critical current density is small enough to realize critical currents at 77 K in the order of several 10 μA with junction widths between 2 μm and 4 μm, also for a film thickness of about 150 nm. On the other hand the relative large normal resistance results in $I_C R_N$ products at 77 K between 110 μV and 130 μV and allows SQUID operation also with very large mutual inductances $M > 300$ pH. The variation of the operation temperature influences the mutual inductance M and consequently the SQUID inductance L_S too by a change of the kinetic part of the inductance corresponding to the decrease of the London penetration depth with lowered temperature (figure 14). The absolute value of this change in the mutual inductance depends on the loop length. If we normalize this value to the patterned loop length of the dc SQUID we found a good correlation between the temperature dependencies of all investigated dc SQUIDs.

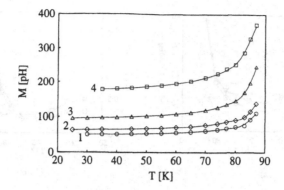

Figure 14. Temperature dependence of the mutual inductance M for variation of the loop length D curve 1: $D = 30$ μm, curve 2: $D = 40$ μm, curve 3: $D = 60$ μm, curve 4: $D = 120$ μm. Parameter set: loop width $B = 5$ μm, line width $W = 3$ μm, $d = 150$ nm.

In contrast the critical currents of the dc SQUID increase with decreasing temperature. We achieve with temperature variation a change in the SQUID parameter $\beta_L = I_C \cdot L/\Phi_0$. We investigated the the temperature dependent flux noise of two dc SQUIDs with extreme loop length $D = 40$ μm (SQUID 1) and $D = 120$ μm (SQUID 2, shown in Figure 15). Figure 15 presents the temperature dependent change of the mutual inductance and the change of the critical current of SQUID 2.

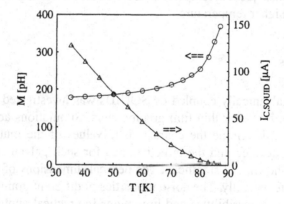

Figure 15. Temperature dependence of the mutual inductance and the critical current for a dc SQUID with a loop length of 120 μm. Parameter set: $W = 5$ μm, $d = 150$ nm.

All noise measurements were done in flux locked loop mode and ac-bias reversal technique.

The lowest measured values of the flux noise density were 6.8 $\mu\Phi_0/\sqrt{\text{Hz}}$ at 65 K (SQUID 1, Figure 16 a) and 28 $\mu\Phi_0/\sqrt{\text{Hz}}$ at 70 K (SQUID 2, Figure 16 b) in the white noise region. If we take the measured mutual inductance of both structures

550

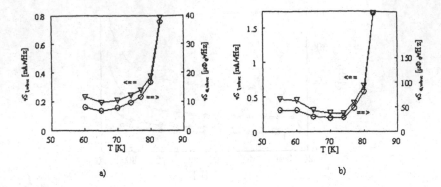

Figure 16. Temperature dependence of the flux noise density S and the correlated current resolution $\sqrt{S_I}$ for a dc SQUID with different loop length D: a) $D = 40\ \mu$m, b) $D = 120\ \mu$m.

into consideration we can calculate the temperature dependence of the current resolution $\sqrt{S_I}$. The minimum in the current resolution $\sqrt{S_I} < 0.2$ nA/$\sqrt{\text{Hz}}$ for the dc SQUID with the smaller loop length at 65 K correlates with a SQUID parameter $\beta_L \approx 1.2$. The equivalent value for the same sensor at 77 K was $\sqrt{S_I} = 0.28$ nA/$\sqrt{\text{Hz}}$ due to the higher SQUID inductance at 77 K (Figure 16). For the dc SQUID with the larger mutual inductance M we find the minimum of $\sqrt{S_I} = 0.27$ nA/$\sqrt{\text{Hz}}$ consequent at temperature $T = 70$ K, correlating with lower critical currents at this higher temperature.

4. Conclusions

The layout of galvanically coupled dc SQUIDs was investigated with respect to their application in planar thin film gradiometers. Simulations and experimental measurements to determine the characteristic values of the mutual inductance, the parasitic inductance and the effective area for such galvanically coupled dc SQUIDs were shown. All significant geometrical dimensions of the dc SQUIDs were varied systematically. The noise properties of different grain boundary junctions as well as the possibilities and limitations in practical applications of these gradiometer structures were discussed.

Acknowledgements

The authors wish to thank F. Schmidt and C. Steigmeier for thin film preparation. This work was partially supported by the German BMBF (contract No. 13 N 6864) and Studienstiftung des Deutschen Volkes (S. W.).

References

1. Koelle, D., Kleiner, R., Ludwig, F., Dantsker, E., Clarke, J. (1999) High-transition-temperature superconducting quantum interference devices, *Rev. Mod. Phys.* **71**, 631–686.
2. Weidl, R., Brabetz, S., Schmidl, F., Klemm, F., Wunderlich, S., Seidel, P. (1997) Heart monitoring with high-T_C d.c. SQUID gradiometers in an unshielded environment, *Supercond. Sci. Technol.* **10**, 95–99.
3. Tavrin, Y., Zhang, Y., Mück, M., Braginski, A.I., Heiden, C. (1993) $YBa_2Cu_3O_7$ thin film SQUID gradiometer for biomagnetic measurements, *Appl. Phys. Lett.* **62**, 1824–1826.
4. Zhang, Y., Soltner, H., Krause, H.-J., Sodtke, E., Zander, W., Schubert, J., Gruneklee, M., Lomparski, D., Banzet, M., Bousack, H., Braginski, A.I. (1997) Planar HTS Gradiometers with Large Baseline, *IEEE Trans. Appl. Supercond.* **7**, 2866–2869.
5. Wunderlich, S., Schmidl, F., Specht, H., Dörrer, L., Schneidewind, H., Hübner, U., Seidel, P. (1998) Planar gradiometers with high-T_C DC SQUIDs for non-destructive testing, *Supercond. Sci. Technol.* **11**, 315–321.
6. Carr, C., Cochran, A., Kuznik, J., McKirdy, D.A., Donaldson, G.B. (1996) Electronic gradiometry for NDE in an unshielded environment with stationary and moving HTS SQUIDs, *Cryogenics* **36**, 691–695.
7. Vrba, J. (1996) SQUID-Gradiometers in real environments, in: "SQUID Sensors: Fundamentals, Fabrication and Applications" Ed. H. Weinstock, Kluwer Academic Publisher, Dordrecht, 117–178.
8. Ketchen, M.B. (1985) Design of improved integrated thin-film planar dc SQUID gradiometers, *J. Appl. Phys.* **58**, 4322–4325.
9. Daalmans, G.M. (1995) HTS DC SQUIDs for Practical Applications, *Appl. Supercond.* **3**, 399–423.
10. Eidelloth, W., Oh, B., Robertazzi, R.P., Gallagher, W.J., Koch, R.H. (1991) $YBa_2Cu_3O_{7-\delta}$ thin-film gradiometers: Fabrication and performance, *Appl. Phys. Lett.* **59**, 3473–3475.
11. Keene, M.N., Satchell, J.S., Goodyear, S.W., Humphreys, R.G., Edwards, J.A., Chew, N.G., Lander, K. (1995) Low Noise HTS Gradiometers and Magnetometers Constructed From $YBa_2Cu_3O_{7-x}/PrBa_2Cu_3O_{7-y}$ Thin Films, *IEEE Trans. Appl. Supercond.* **5**, 2923–2926.
12. Drung, D. (1991) DC SQUID systems overview, *Supercond. Sci. Technol.* **4**, 377–385.
13. Ludwig, F., Dantsker, E., Kleiner, R., Koelle, D., Clarke, J., Knappe, S., Drung, D., Koch, H., Alford, N.McN., Button, T.W. (1995) Integrated High-T_C multiloop magnetometer, *Appl. Phys. Lett.* **66**, 1418–1420.
14. Reimer, D., Schilling, M., Knappe, S., Drung, D. (1995) Integrated $YBa_2Cu_3O_7$ multi-loop magnetometers at 77 K, *IEEE Trans. Appl. Supercond.* **5**, 2342–2345.
15. Zakosarenko, V., Berthel, K.-H., Blüthner, K., Seidel, P., Weber, P. (1993) Nb-based integrated SQUID gradiometer without flux-transformer as a prototype for high-T_C devices, *Appl. Supercond.*, Ed. H.C. Freyhardt, DGM Informationsgesellschaft Oberursel, 1339–1342.
16. Knappe, S., Drung, D., Schurig, T., Koch, H., Klinger, M., Hinken, J. (1992) A planar $YBa_2Cu_3O_7$ gradiometer at 77 K, *Cryogenics* **32**, 881–884.
17. Zakosarenko, V.M., Schmidl, F., Schneidewind, H., Dörrer, L., Seidel, P. (1994) Thin-film dc SQUID gradiometer using a single $YBa_2Cu_3O_{7-x}$ layer, *Appl. Phys. Lett.* **65**, 779–801.
18. Cantor, R., Lee, L.P., Teepe, M., Vinetskiy, V., Longo, J. (1995) Low Noise, Single-Layer $YBa_2Cu_3O_{7-x}$ DC SQUID Magnetometers at 77 K, *IEEE Trans. Appl. Supercond.* **5**, 2927–2930.
19. Hildebrand, G., SCIM 96 computer programm for numerical calculation of inductances.
20. Schmidl, F., Dörrer, L., Seidel, P., Schneidewind, H., Matthes, A., Heinz, E., Köhler,T., Töpfer, H. (1993) Modification of DC-SQUIDs with step-edge Josephson junctions, *Appl. Supercond.*, Ed. H.C.Freyhardt, DGM Informationsgesellschaft Oberursel, 1163–1166.
21. Gross, R. and Mayer, B. (1991) Transport processes and noise in $YBa_2Cu_3O_{7-\delta}$ grain boundary junctions, *Physica C* **180**, 235–242.
22. Schneidewind, H., Schmidl, F., Linzen, S., and Seidel, P. (1995) The possibilities and limitations of ion beam etching of $YBa_2Cu_3O_{7-x}$ thin films and microbridges, *Physica C* **250**, 191–201.

552

23. Forrester, M.G., Davidson, A., Talvacchio, J., Gavaler, J.R., and Przybysz, J.X. (1994) Inductance measurements in multilevel high T_c step-edge grain boundary SQUIDs, *Appl. Phys. Lett.* **65**, 1835–1837.

24. Il'ichev, E., Dörrer, L., Schmidl, F., Zakosarenko, V., Seidel, P., Hildebrandt, G. (1996) Current resolution, noise and inductance measurements on high-T_c dc SQUID galvanometers, *Appl. Phys. Lett.* **68**, 708–710.

25. ter Brake, H.J.M., Aarnink, W.A.M., van den Bosch, P.J., Hilgenkamp, J.W.M., Flokstra, J., and Rogalla, H. (1997) Temperature dependence of the effective sensing area of high-T_c dc SQUIDs, *Supercond. Sci. Technol.* **10**, 512–515.

26. Dantsker, E., Tanaka, S., Nilsson, P.-Å., Kleiner, R., Clarke, J. (1996) Reduction of $1/f$ noise in high-T_c dc superconducting quantum interference devices cooled in an ambient magnetic field, *Appl. Phys. Lett.* **69**, 4099–4101.

27. Dantsker, E., Tanaka, S., and Clarke, J. (1997) High-T_c superconducting quantum interference devices with slots or holes: Low $1/f$ noise in ambient magnetic fields, *Appl. Phys. Lett.* **70**, 2037–2039.

28. Schmidl, F., Wunderlich, S., Dörrer, L., Specht, H., Heinrich, J., Barholz, K.-U., Schneidewind, H., Seidel, P. (1997) HTSC-dc-SQUID gradiometer for a nondestructive testing system, Proc. 3rd EUCAS, Eds. H. Rogalla, D.H.A. Blank, *IOP Conf. Series* **158**, 651–654.

29. Marx, A. and Gross, R., (1997) Scaling behavior of $1/f$ noise in high-temperature superconductor Josephson junctions, *Appl. Phys. Lett.* **70**, 120–122.

30. Wunderlich, S., Schmidl, F., Dörrer, L., Schneidewind, H., Seidel, P. (1999) Improvement of Sensor Performance of High-T_C Thin Film Planar SQUID Gradiometers by Ion Beam Etching, *IEEE Trans. Appl. Supercond.* **9**, 71–76.

SUBJECT INDEX